PARALLEL DISTRIBUTED PROCESSING

Computational Models of Cognition and Perception

Editors

Jerome A. Feldman
Patrick J. Hayes
David E. Rumelhart

Parallel Distributed Processing: Explorations in the Microstructure of Cognition. Volume 1: Foundations, by David E. Rumelhart, James L. McClelland, and the PDP Research Group

Parallel Distributed Processing: Explorations in the Microstructure of Cognition. Volume 2: Psychological and Biological Models, by James L. McClelland, David E. Rumelhart, and the PDP Research Group

Neurophilosophy: Toward a Unified Science of the Mind-Brain, by Patricia S. Churchland

Qualitative Reasoning About Physical Systems, edited by Daniel G. Bobrow

Visual Cognition, edited by Steven Pinker

PARALLEL DISTRIBUTED PROCESSING

Explorations in the Microstructure
of Cognition

Volume 2: Psychological and Biological Models

James L. McClelland   David E. Rumelhart
and the PDP Research Group

Chisato Asanuma	Alan H. Kawamoto	Paul Smolensky
Francis H. C. Crick	Paul W. Munro	Gregory O. Stone
Jeffrey L. Elman	Donald A. Norman	Ronald J. Williams
Geoffrey E. Hinton	Daniel E. Rabin	David Zipser
Michael I. Jordan	Terrence J. Sejnowski	

Institute for Cognitive Science
University of California, San Diego

A Bradford Book

The MIT Press
Cambridge, Massachusetts
London, England

Third printing, 1987

© 1986 by The Massachusetts Institute of Technology

Printed and bound in the United States of America

Library of Congress Cataloging-in-Publication Data

Rumelhart, David E.
   Parallel distributed processing.

   (Computational models of cognition and perception)
   Vol. 2 by James L. McClelland, David E. Rumelhart, and the PDP Research Group.
   "A Bradford book."
   Includes bibliographies and indexes.
   Contents: v. 1. Foundations — v. 2. Psychological and biological models.
   1. Human information processing. 2. Cognition.
I. McClelland, James L. II. University of California, San Diego. PDP Research Group. III. Title. IV. Series.
BF455.R853 1986   153   85-24073
ISBN 0-262-18120-7 (v. 1)
      0-262-13218-4 (v. 2)
      0-262-18123-1 (set)

Rok Sosič

Contents

VOLUME 1
FOUNDATIONS

VOLUME 2
PSYCHOLOGICAL AND BIOLOGICAL MODELS

Preface

The chapters in this volume continue the explorations of parallel distributed processing that we began in Volume I. The accompanying table lays out the organization of both volumes. Part IV, which begins this volume, presents parallel distributed processing (PDP) models of psychological processes, and Part V explores the biological mechanisms of parallel distributed processing in the brain. Part VI provides a brief retrospective discussion of the whole book and our thoughts about future directions.

The various parts of this book are interrelated, but they can generally be read in any order, and indeed most of the chapters can be understood without reference to the rest of the book. However, readers may find it useful to read some of the chapters in Part I before delving into the specific topics covered in the chapters of Parts IV and V. A discussion of the contents of each chapter and the background that is relevant to it may be found in the introductory section at the beginning of each part of the book.

Primary support for both volumes of the book came from the System Development Foundation and the Office of Naval Research. The System Development Foundation has provided direct support for the PDP research group through a grant to Norman and Rumelhart, and has also supported several of the individual members of the group (Crick, Hinton, Sejnowski, and Zipser). ONR contracts that have contributed support include N00014-79-C-0323, NR 667-437; N00014-85-K-0450, NR 667-548; and N00014-82-C-0374, NR 667-483.

Several other sources have contributed to the support of individual members of the group. These include the National Institute of Mental

A CONDENSED TABLE OF CONTENTS

VOLUME I

VOLUME II

Health, through a Career Development Award—PHS-MH-00385—to McClelland and postdoctoral fellowships to Paul Smolensky and Paul Munro under Grant PHS-MH-14268 to the Center for Human Information Processing at UCSD. Smolensky received support in the form of a fellowship from the Alfred P. Sloan Foundation, and some of McClelland's work was supported by a grant from the National Science Foundation (BNS-79-24062). These and other sources of support for specific individuals or projects are acknowledged in the appropriate chapters.

Once again we would like to acknowledge the support of the Institute for Cognitive Science at UCSD and of all of the people at UCSD and elsewhere who have helped us bring this work to completion. These people are mentioned individually in the acknowledgments section of Volume I, which applies to both volumes.

December 1985

David E. Rumelhart
LA JOLLA, CALIFORNIA

James L. McClelland
PITTSBURGH, PENNSYLVANIA

Addresses of the PDP Research Group

Chisato Asanuma

Salk Institute
P.O. Box 85800
San Diego, CA 92138

Francis H. C. Crick

Salk Institute
P.O. Box 85800
San Diego, CA 92138

Jeffrey L. Elman

Department of Linguistics
University of California, San Diego
La Jolla, CA 92093

Geoffrey E. Hinton

Department of Computer Science
Carnegie-Mellon University
Pittsburgh, PA 15213

Michael I. Jordan

Department of Computer and
    Information Science
University of Massachusetts
Amherst, MA 01003

Alan H. Kawamoto

Department of Psychology
Carnegie-Mellon University
Pittsburgh, PA 15213

James L. McClelland

Department of Psychology
Carnegie-Mellon University
Pittsburgh, PA 15213

Paul W. Munro

Department of Information Science
University of Pittsburgh
Pittsburgh, PA 15260

Donald A. Norman

Institute for Cognitive Science
University of California, San Diego
La Jolla, CA 92093

Daniel E. Rabin

Intellicorp
1975 El Camino Real West
Mountain View, CA 94040

David E. Rumelhart

Institute for Cognitive Science
University of California, San Diego
La Jolla, CA 92093

Terrence J. Sejnowski

Department of Biophysics
Johns Hopkins University
Baltimore, MD 21218

Paul Smolensky

Department of Computer Science
University of Colorado
Boulder, CO 80309

Gregory O. Stone

Center for Adaptive Systems
Department of Mathematics
Boston University
Boston, MA 02215

Ronald J. Williams

Institute for Cognitive Science
University of California, San Diego
La Jolla, CA 92093

David Zipser

Insitute for Cognitive Science
University of California, San Diego
La Jolla, CA 92093

PARALLEL DISTRIBUTED PROCESSING

PART **IV**

PSYCHOLOGICAL PROCESSES

The chapters in this section are designed to show how parallel distributed processing has changed the way we think about the mechanisms of cognition. Each chapter describes a model of some mental process. Each shows how the model captures some basic aspects of the process. And each offers an alternative to other, earlier ways of thinking about aspects of the process.

The models described in these chapters were motivated by a number of different considerations. Quite often, they were motivated by the feeling that parallel distributed processing mechanisms provide a very natural and appropriate set of mechanisms for dealing with some aspect of cognition that has not been dealt with successfully by other approaches. Quite often, they were motivated by an attempt to meet computational challenges posed by the processes they address. And quite often, they were motivated by attempts to extend the domain of PDP models, to encompass processes and phenomena we did not know how to model at the outset.

The first chapter of the section, Chapter 14, explains how parallel distributed processing can take us beneath the surface of schemata, to a level of description that allows us to see how we can preserve the desirable characteristics of schemata and at the same time make them more flexible, more sensitive to context, more adaptable. This chapter also begins to show how PDP mechanisms can be used as the building blocks from which we may construct sequential thought processes, such as problem solving.

The next two chapters consider issues in perception. Chapter 15 presents a model of speech perception and shows how it provides a unified framework for capturing a number of aspects of speech. Here, the goal was to develop a model that accounts in detail for psychological data on the process of speech perception and, at the same time, to begin to deal with several computational problems that make the extension of PDP models to speech a challenging and stimulating task. The model introduces a processing structure called *the Trace*, a dynamic working memory, in which units that stand for hypotheses about the contents of an utterance at different points in time can interact. The parallel distributed processing that occurs in the Trace allows the model to account for contextual influences on phoneme identification, and the simple competitive interactions among hypotheses representing competing interpretations of the same portion of an utterance allow the model to segment and identify the words in an utterance in a simple and integrated fashion. The model accounts for a wide range of data in a direct and coherent way, and shows how PDP mechanisms offer new ways of interpreting several phenomena, such as categorical perception and the perception of phonologically regular nonwords.

The model described in Chapter 15 purchases its parallel processing capabilities by duplicating the same hardware to cover each time-slice of the Trace. This idea seems incorrect; in general, it seems more plausible to view the Trace not as a fixed processing structure, but as one that is dynamically configured in the course of processing, using knowledge of contingencies between hypotheses to construct the Trace on the fly. Chapter 16 develops a model called the Programmable Blackboard Model of Reading (PABLO for short) that does just this, though for printed, as opposed to spoken input. This model is based on the idea of *connection information distribution*—roughly, the idea is to use information stored in one part of a processing system to *set* or *program* connections in another part of the same system. Two related simulation models based on this idea are applied to a number of aspects of reading that could not be addressed by the interactive activation model of word recognition, which was described in Chapter 1.

Neither TRACE nor PABLO really extend above the word level to deal with the larger syntactic and semantic structures that organize words into sentences; these levels are considered in Chapter 19, which we will describe a bit more below.

Chapters 17 and 18 describe distributed models of different aspects of learning and memory. Chapter 3 (on Distributed Representations) provides useful background for these chapters, as well as for Chapter 19. Chapter 17 considers an existing dilemma for models of memory—whether to store summary representations in memory or whether to store an enumeration of specific experiences. The chapter points out that with distributed representations, you can have it both ways. In the model, the (long-term) memory trace of an event is the change or *increment* to the connections that results from the event. Functional equivalents of summary representations (e.g., prototypes) emerge naturally from the superimposition of memory traces of specific events. Traces of recent or often-repeated events can coexist with the summary representation. The chapter also illustrates that the same composite memory trace can learn several different prototypes from exemplars, without ever being informed—and indeed, without having to "figure out"—which exemplars belong to each category. The model uses the delta rule (examined in Chapters 2, 8, and 11) for adjusting connection strengths and has some advantages over some distributed memory models, but it contains no hidden units, and so is not capable of overcoming what we call the "linear predictability constraint" on the set of patterns it can learn. At the end of the chapter, we illustrate with a simple example simulation how the model can be extended with hidden units that are trained with the aid of the generalized delta rule discussed in Chapter 8.

Chapter 18 draws out some other implications of distributed models of learning and memory. It considers how knowledge underlying the lawful use of language might be represented in a PDP model and how that knowledge might be acquired. More generally, it shows how distributed representations provide an alternative to the conventional view that linguistic knowledge is represented in the form of explicit (though inaccessible) rules. The chapter considers a paradigm case of rule learning from the language acquisition literature—the acquisition of the past tense by children acquiring English as their first language. This case is often cited as an instance of rule acquisition par excellence because of the fact that children "regularize" irregular verbs at one stage, often saying "goed," for example. The model we describe in this chapter exhibits this and many other aspects of the acquisition process, and it does this by using very simple learning mechanisms. Lawful behavior emerges from the superposition of changes to connection strengths. The representation of the rules of past-tense formation is implicit in the resulting connection strengths and is acquired without the aid of any device that relies on the formulation and testing of explicit but inaccessible rules.

Many of the chapters we have been describing deal with aspects of language, but none of them get much beyond the processing of individual words. Does this indicate that PDP models are inappropriate for capturing higher levels of language structure and processing? We think not. Indeed, one of our goals has been to work toward the development of PDP models of sentence processing. As Chapter 19 indicates, much of the groundwork has now been laid. That chapter describes a distributed model that brings the benefits of parallel distributed processing to the processing of simple sentences, and shows how a PDP network can be configured to produce a representation that can capture the underlying case structure of simple sentences. The model exhibits a number of very nice properties. It can assign arguments of sentences to the appropriate case roles based on word-order information, based on the mutual selectional constraints imposed by the different words in the sentence, and based on word-order and mutual constraints working together; it can choose the appropriate "case frame" for a verb on the basis of the content and configuration of the arguments in the sentence; it can choose the appropriate reading of a semantically ambiguous word based on constraints imposed by the other arguments in the sentence; it can fill in default values for missing arguments; and it can generalize its case-role assignments to novel verbs if it is given a representation of some of the semantic features of the verb. The model has not yet reached the stage where it can process sentences with embedded clauses, but we suggest three ways in which it might be extended to do so, including one that relies on the use of the connection information distribution mechanism described in Chapter 16 and one that involves "true" recursion. These suggestions indicate that PDP mechanisms are capable of processing recursively defined structures, contrary to prevalent belief.

Similarities and Differences Between Models

Each of the models described in these chapters differs in detail from all of the others. One model uses continuous-valued, asynchronous units; several others use continuous, sychronous units; still others use stochastic units. In some models, activation values can range from 1 to -1; in others, they can only range from 1 to 0, or slightly below 0. All the models, however, are examples of the class of PDP models, as described in Chapter 2, and their minor differences are in most cases incidental to their behavior. Where these detailed assumptions seem to make a difference, we point it out, but in general, we do not think that much hinges on these differences.

One characteristic that differentiates some of the models from the others requires some comment: The models in Chapters 15 and 16 use local representations, while the models in the other chapters use distributed representations. Even this does not reflect a fundamental difference in the philosophy of the models; rather, they reflect differences in the points we wanted to make and the issues we wanted to raise in the different chapters. In fact, we believe that both the TRACE model and the models described in Chapter 16 could be made more efficient by the use of distributed rather than local representations, for reasons discussed in Chapters 3 and 12. It is also true that nearly all of the models can be seen as either local or distributed, depending on one's point of view. They are local, in the sense that each unit generally stands for a particular conceptual object or feature, or perhaps a (coarse coded) conjunction of features. They are distributed, in the sense that any given stimulus object—a phoneme, a word, a sentence, a scene—produces a pattern of activation over a very large number of units. Whether we see the models as distributed or local is primarily a matter of whether we are looking at the forest or the trees.

Stepping Stones

Most of the models we present represent the result of a long series of explorations. In some cases, we learned a great deal from our earlier attempts, and the problems we had with them lead natually to better formulations. In other cases, our early attempts were failures we did not know at first how to overcome. For example, an early version of the model of past tense learning could be made to produce overgeneralization errors, but it could only process verbs consisting of a consonant, a vowel, and a consonant. It took us two years to formulate a better representation.

We think the models we describe in these chapters demonstrate progress in the development of PDP models of psychological processes. But none of the models is *the final word*. Each model has its weaknesses, as well as its strengths. We have not solved all the problems; rather, we have suggested an approach that shows promise of leading us closer to their solution. We offer these models in that spirit: as stepping stones along the way to a deeper understanding of the microstructure of cognition.

CHAPTER 14

Schemata and
Sequential Thought Processes in PDP Models

D. E. RUMELHART, P. SMOLENSKY, J. L. McCLELLAND
and G. E. HINTON

One of our goals for this book is to offer an alternative framework for viewing cognitive phenomena. We have argued that talk at the level of units and activations of units is the preferable way to describe human thought. There is, however, already an established language for discussing cognitive phenomena. In this chapter we wish to address the relationship between some of the key established concepts and our parallel distributed processing models. There are many important concepts from modern cognitive science which must be explicated in our framework. Perhaps the most important, however, is the concept of the *schema* or related concepts such as scripts, frames, and so on. These large scale data structures have been posited as playing critical roles in the interpretation of input data, the guiding of action, and the storage of knowledge in memory. Indeed, as we have argued elsewhere (cf. Rumelhart, 1980), the schema has, for many theorists, become the basic building block of our understanding of cognition. Yet, the PDP language we are proposing is devoid of terms such as schemata, scripts, frames, and so forth. Instead, we have proposed building blocks at a much more microlevel—at the level of units, activations, and similar "low-level" concepts. Interestingly, it was struggling with the concept of the schema and some of its difficulties that led one of us (DER) to an exploration of PDP models to begin with. It was therefore with some priority that we began to develop an interpretation of the schema in the language of parallel distributed processing.[1]

Perhaps the first thought that comes to mind is to map the notion of the schema onto the notion of the unit. This does, indeed, capture some of the important aspects of the schema. In particular, the unit is an element, like the schema, which monitors its inputs searching for a good fit and takes on a value which represents how well its inputs fits its own internal criteria. However, such an identification misses much of what makes the schema a powerful conceptual tool. In particular, there is no analog to the *variable* or *default values*. There is no notion of the internal structure of the schema nor many of the other important aspects of schemata. Moreover, the scale is wrong. Schema theorists talk of schemata for rooms, stories, restaurants, birthday parties, and many other high-level concepts. In our parallel distributed processing models, units do not tend to represent such complex concepts. Instead, units correspond to relatively simple features or as Hinton (1981a) calls them *microfeatures*. If we are to do justice to the concept of the schema, we are going to have to look beyond the individual unit. We are going to have to look for schemata as properties of entire networks rather than single units or small circuits. In the following sections we show how features of networks can capture the important features of schemata. Since our interpretation is clearest in the subset of PDP models that can be characterized as *constraint satisfaction* networks, it will be useful to first describe that class of models and provide a language for talking about their properties.

PARALLEL DISTRIBUTED PROCESSING MODELS AS CONSTRAINT SATISFACTION NETWORKS

It is often useful to conceptualize a parallel distributed processing network as a *constraint network* in which each unit represents a hypothesis of some sort (e.g., that a certain semantic feature, visual feature, or acoustic feature is present in the input) and in which each connection represents constraints among the hypotheses. Thus, for example, if feature B is expected to be present whenever feature A is,

---

[1] All of the authors have contributed to the ideas expressed in this chapter. Smolensky's slightly different framework is sketched in Chapter 6. Hinton's view of the microstructure of symbols is sketched in J. A. Anderson and Hinton (1981, pp. 29-32), and McClelland (1981) shows how PDP networks can be employed to fill default values (see the discussion in Chapter 1). While we all agree with the flavor of the current discussion not all of us endorse the exact details.

there should be a positive connection from the unit corresponding to the hypothesis that A is present to the unit representing the hypothesis that B is present. Similarly, if there is a constraint that whenever A is present B is expected *not* to be present, there should be a negative connection from A to B. If the constraints are weak, the weights should be small. If the constraints are strong, then the weights should be large. Similarly, the inputs to such a network can also be thought of as constraints. A positive input to a particular unit means that there is evidence from the outside that the relevant feature is present. A negative input means that there is evidence from the outside that the feature is not present. The stronger the input, the greater the evidence. If such a network is allowed to run it will eventually *settle* into a locally optimal state in which as many as possible of the constraints are satisfied, with priority given to the strongest constraints. [2] The procedure whereby such a system *settles* into such a state is called *relaxation*. We speak of the system *relaxing* to a solution. Thus, a large class of PDP models, including the interactive activation model of word perception, are constraint satisfaction models which settle on locally optimal solutions through the process of relaxation.

Figure 1 shows an example of a simple 16-unit constraint network. Each unit in the network represents a hypothesis concerning a vertex in a line drawing of a Necker cube.[3] The network consists of two interconnected subnetworks—one corresponding to each of the two global interpretations of the Necker cube. Each unit in each network is assumed to receive input from the region of the input figure—the cube—corresponding to its location in the network. Each unit in the Figure is labeled with a three letter sequence indicating whether its vertex is hypothesized to be front or back (F or B), upper or lower (U or L), and right or left (R or L). Thus, for example, the lower left-hand unit of each subnetwork is assumed to receive input from the lower left-hand vertex of the input figure. The unit in the left-hand network represents the hypothesis that it is receiving input from a lower left-hand vertex in the front surface of the cube (and is thus labeled FLL), whereas the one in the right subnetwork represents the hypothesis that it is receiving input from a lower left vertex in the back surface (BLL).

---

2 Actually, these systems will in general find a locally best solution to this constraint satisfaction problem. It is possible under some conditions to insure that the "globally" best solution is found through the use of stochastic elements and a process of annealing (cf. Chapters 6 and 7 for a further discussion).

3 J. A. Feldman (1981) has proposed an analysis of the Necker cube problem with a somewhat different network. Although the networks are rather different, the principles are the same. Our intention here is not to provide a serious account of the Necker cube phenomena, but rather to illustrate constraint networks with a simple example.

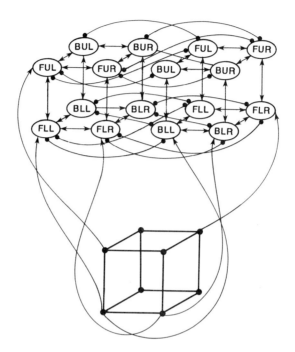

FIGURE 1. A simple network representing some of the constraints involved in perceiving the Necker cube.

Since there is a constraint that each vertex has a single interpretation, these two units are connected by a strong negative connection. Since the interpretation of any given vertex is constrained by the interpretations of its neighbors, each unit in a subnetwork is connected positively with each of its neighbors within the network. Finally, there is the constraint that there can only be one vertex of a single kind (e.g., there can only be one lower left vertex in the front plane FLL). There is a strong negative connection between units representing the same label in each subnetwork. Thus, each unit has three neighbors connected positively, two competitors connected negatively, and one positive input from the stimulus. For purposes of this example, the strengths of connections have been arranged so that two negative inputs exactly balance three positive inputs. Further, it is assumed that each unit receives an excitatory input from the ambiguous stimulus pattern and that each of these excitatory influences is relatively small. Thus, if all three of a unit's neighbors are on and both of its competitors are on, these effects would entirely cancel out one another; and if there was a small input from the outside, the unit would have a tendency to come on. On the other hand, if fewer than three of its neighbors were on and both of its competitors were on, the unit would have a tendency to turn off, even with an excitatory input from the stimulus pattern.

In the last paragraph we focused on the individual units of the networks. However, it is often useful to focus not on the units, but on entire *states* of the network. In the case of binary (on-off or 0-1) units, there is a total of 2^{16} possible states in which this system could reside. That is, in principle, each of the 16 units could have the value either 0 or 1. In the case of continuous units, in which each unit can take on any value between 0 and 1, the system can, in principle, take on any of an infinite number of states. Yet, because of the constraints built into the network, there are only a few of those states in which the system will settle. To see this, consider the case in which the units are updated asynchronously, one at a time. During each time slice, one of the units is chosen to update. If its net input exceeds 0 its value will be pushed toward 1, otherwise its value will be pushed toward 0, using the activation rule from the word perception model:

$$a_j(t+1) = a_j(t) + \begin{cases} net_j(1 - a_j(t)) & net_j > 0 \\ net_j a_j(t) & \text{otherwise.} \end{cases}$$

Here, $a_j(t)$ stands for the activation of unit j at time t, and $net_j(t)$ stands for the net input to unit j at t. $net_j(t)$ is simply the sum of the excitatory and inhibitory influences on unit j:

$$e_j(t) + \sum_{i \neq j} w_{ji} a_i(t)$$

where $e_j(t)$ is the external input to unit j at t and w_{ji} is the weight on the connection to unit j from unit i.

Imagine that the system starts with all units off. A unit is then chosen at random to be updated. Since it is receiving a slight positive input from the stimulus and no other inputs, it will be given a positive activation value. Then another unit is chosen to update. Unless it is in direct competition with the first unit, it too will be turned on. Eventually, a coalition of neighboring units will be turned on. These units will tend to turn on more of their neighbors in the same subnetwork and turn off their competitors in the other subnetwork. The system will (almost always) end up in a situation in which all of the units in one subnetwork are fully activated and none of the units in the other subnetwork are activated. That is, the system will end up interpreting the Necker cube as either facing left or facing right. Whenever the system gets into a state and stays there, the state is called a *stable state* or a *fixed point* of the network.

Figure 2 shows the output of three runs of a simulation based on this network. The size of the square indicates the activation values of each

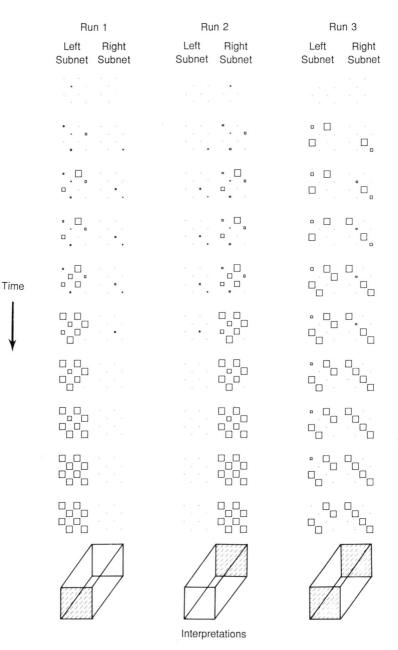

FIGURE 2. Three runs of a simulation based on this network. The size of the square indicates the activation value of each unit. The units are arranged in the shape of the sub-network with each square shown in its position corresponding to the vertex of the cube from which it is receiving input. The states are shown after every second update.

unit. The units are arranged in the shape of the subnetwork with each square shown in its position corresponding to the vertex of the cube from which it is receiving input. The system begins with a zero activation value on all units—represented by single dots. Then, once each time slice, at most one unit is changed. On each run the system winds up in a state in which each unit has a value of either 0 or 1 (designated by a large square). The first two runs are most typical of the system. In this case, the inputs are low relative to the strength of the constraints among units. When low inputs are involved, the system virtually always winds up either in the state in which all of the units in the left-hand network are turned on and all of the units in the right-hand are off or vice versa. These final stable states correspond to the interpretations of a left-facing and right-facing cube as illustrated in the figure for the first and second run respectively. The third example of simulation results is much more aberrant and was generated with a high input value. With a high input value, the system can occasionally get a third interpretation of the Necker cube. This is the "impossible" cube with two front faces illustrated in the figure. Thus, of the 2^{16} possible states of the system, only two are ever reached with low input values and only three are ever reached at all. The constraints implicit in the pattern of connections among the units determines the set of possible stable states of the system and therefore the set of possible interpretations of the inputs.

Hopfield (1982) has shown that it is possible to give a general account of the behavior of systems such as this one (with symmetric weights and asynchronous updates). In particular, Hopfield has shown that such systems can be conceptualized as minimizing a global measure which he calls the *energy* of the system through a method of *gradient descent* or, equivalently, maximizing the constraints satisfied through a method of *hill climbing*. In particular, Hopfield has shown that the system operates in such a way as to always move from a state that satisfies fewer constraints to a state that satisfies more constraints, where the measure of constraint satisfaction is given by[4]

$$ G(t) = \sum_i \sum_j w_{ij} a_i(t) a_j(t) + \sum_i input_i(t) a_i(t). $$

---

[4] Note, the question of what to call this constraint satisfaction function is difficult. Hopfield uses the negation of this function and, by analogy to thermodynamics, calls it *energy*. This system can thus be said to settle into states of minimum energy. Similarly, Hinton and Sejnowski (Chapter 7) use the same terminology. Smolensky (Chapter 6) has a similar function which he calls *harmony* to emphasize that increasing values correspond to more harmonious accounts of the inputs. In this chapter we have chosen to use the language of constraint satisfaction and call the function G for measure of the goodness-of-fit of the state to its constraints.

Essentially, the equation says that the overall goodness-of-fit is given by the sum of the degrees to which each pair of units contribute to the goodness plus the degree to which the units satisfy the input constraints. The contribution of a pair of units is given by the product of their activation values times the weights connecting them. Thus, if the weight is positive, each unit wants to be as active as possible—that is, the activation values for these two units should be pushed toward 1. If the weight is negative, then as least one of the units should be 0 to maximize the pairwise goodness. Similarly, if the input constraint for a given unit is positive, then its contribution to the total goodness-of-fit is maximized by being the activation of that unit toward its maximal value. If it is negative, the activation value should be decreased toward 0. Of course, the constraints will generally not be totally consistent. Sometimes a given unit may have to be turned on to increase the function in some ways while decreasing it in other ways. The point is that it is the sum of all of these individual contributions that the system seeks to maximize. Thus, for every state of the system—every possible pattern of activation over the units—the pattern of inputs and the connectivity matrix W determines a value of the goodness-of-fit function. The system processes its input by moving upward from state to adjacent state until it reaches a state of maximum goodness. When it reaches such a *stable state* or *fixed point* it will stay in that state and it can be said to have "settled" on a solution to the constraint satisfaction problem or alternatively, in our present case, "settled into an interpretation" of the input.

It is important to see, then, that entirely *local* computational operations, in which each unit adjusts its activation up or down on the basis of its net input, serve to allow the network to converge towards states that maximize a *global* measure of goodness or degree of constraint satisfaction. Hopfield's main contribution to our present analysis was to point out this basic fact about the behavior of networks with symmetrical connections and asynchronous update of activations.

In general, since there are so many states, it is difficult to visualize the goodness-of-fit function over which the system is moving. In the present case, however, we can get a reasonably good image of this landscape. To begin, we can limit our consideration to those states in which a particular unit is either on or off since the system always ends up in such states. We can consider the states arrayed along two dimensions. One dimension corresponds to the number of units turned on in the left subnetwork and the other dimension corresponds to the number of units turned on in the right subnetwork. Thus, at (0,0) we locate the state in which no units are turned on. Clearly, by the above equation such a state will have zero goodness of fit.[5] At (8,8) we have the state in which all of the units are turned on. At location (8,0) we have the state in which the units on the left network are all turned on and those on the right network are all off. At position (0,8) we have the state in which those in the left network are all off and those in the right network are all on. Each of those locations contain unique states. Now, consider the location (1,0) in which one unit from the left subnetwork and zero units in the right subnetwork are turned on. There are eight different states, corresponding to the eight different units in the left subnetwork that might have been turned on. In order to plot the goodness-of-fit landscape for this state space, we have plotted only the states at each location of the two-dimensional space with highest goodness-of-fit—i.e., the best state at each location. Figure 3 shows the landscape. In the figure, we are viewing the goodness landscape from about the (0,0) corner, the start state. Thus, the peak to the right corresponds to the goodness of the state in which all of the units in the left subnetwork are turned on and all in the right subnetwork are turned off. The peak at the upper left portion of the figure

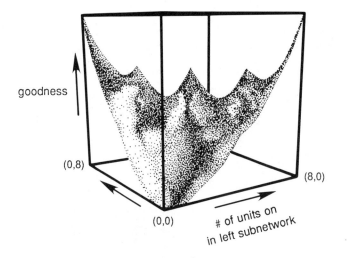

FIGURE 3. The goodness-of-fit surface for the Necker-cube network. The low point at the (0,0) corner corresponds to the start state. The peaks on the right and left correspond to the standard interpretations of the Necker cube, and the peak in the center corresponds to the impossible Necker cube illustrated in the previous figure.

---

[5] Note, zero goodness-of-fit is *not* the minimum goodness-of-fit attainable. In general, goodness-of-fit can be negative as well as positive. When there is negative goodness-of-fit, the system can always be made better by turning off all of the units.

corresponds to the state (0,8). The two peaks in the graph at (8,0) and (0,8) correspond to the two primary interpretations of the Necker cube. It should be clear that if we start a system at (0,0) and allow it to "hill climb" it will almost always end up at one of these two peaks. It might be noted, that there are three smaller peaks right in the middle of the surface. These local peaks are very hard to get to because the system is almost always swept from the start state uphill to one of the two major peaks. It is possible, by having large input values, to reach location (4,4). This peak corresponds to the impossible Necker cube illustrated in the previous figure.

The input to the system can be conceptualized as systematically modifying or *sculpting* the goodness landscape. This effect is illustrated in Figure 4. In this case, the same landscape has been plotted, except the units corresponding to the interpretation of the Necker cube as facing to the left receive more input than the corresponding units on the other subnetwork. (This could perhaps be done by slightly shading that face of the Necker cube.) What we see is a "sloping" goodness surface with the peak associated with the interpretation of the Necker cube as left facing.

To summarize, then, there is a large subset of parallel distributed processing models which can be considered constraint satisfaction models. These networks can be described as carrying out their information processing by climbing into states of maximal satisfaction of the

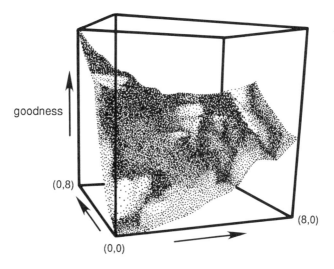

FIGURE 4. The distortions of the goodness landscape when a large input is given to the units corresponding to the front face of a left-facing cube. The figure shows only one major peak corresponding to the view of the left-facing cube.

constraints implicit in the network. A very useful concept that arises from this way of viewing these networks is that we can describe the behavior of these networks, not only in terms of the behavior of individual units, but in terms of properties of the network itself. A primary concept for understanding these network properties is the *goodness-of-fit landscape* over which the system moves. Once we have correctly described this landscape we have described the operational properties of the system—it will process information by moving uphill toward goodness maxima. The particular maximum that the system will find is determined by where the system starts and by the distortions of the space induced by the input. One of the very important descriptors of a goodness landscape is the set of maxima which the system can find, the size of the region that feeds into each maximum, and the height of the maximum itself. The states themselves correspond to possible interpretations, the peaks in the space correspond to the best interpretations, the extent of the foothills or skirts surrounding a particular peak determines the likelihood of finding the peak, and the height of the peak corresponds to the degree that the constraints of the network are actually met or, alternatively, to the goodness of the interpretation associated with the corresponding state.

CONSTRAINT SATISFACTION AND SCHEMATA

In the previous section we recounted a perspective on parallel distributed processing systems. In this section we address, again, the nature of the schema and relate it to constraint satisfaction systems and PDP models. We will proceed by first recounting some of the history of the concept of schemata, then by offering an interpretation of the schema in terms of PDP models, by giving a simple example, and finally showing how the various properties attributed to schemata are, in fact, properties of the PDP networks of the kind we have been discussing.

The schema, throughout its history, has been a concept shrouded in mystery. Kant's (1787/1963) use of the term has been provocative but difficult to understand. Bartlett's (1932) usage has long been decried for its vagueness. Piaget (1952) used the term schema, but it was difficult to come up with a consistent interpretation of Piaget's own views on the matter. Throughout most of its history, the notion of the schema has been rejected by mainstream experimental psychologists as being too vague. As a result, the concept of the schema was largely shunned until the mid-1970s. The concept was then revived by an attempt to offer a more clearly specified interpretation of the schema in terms of explicitly specified computer implementations or, similarly, formally specified implementations of the concept. Thus, Minsky (1975) postulated the concept of the frame, Schank and Abelson (1977) focused on the concept of the script, and Bobrow and Norman (1975) and Rumelhart (1975) developed an explicit notion of the schema. Although the details differed in each case, the idea was essentially the same. Perhaps Minsky (1975) was clearest in the motivation:

> It seems to me that the ingredients of most theories both in artificial intelligence and in psychology have been on the whole too minute, local, and unstructured to account—either practically or phenomenologically—for the effectiveness of common sense thought. The "chunks" of reasoning, language, memory, and "perception" ought to be larger and more structured, and their factual and procedural contents must be more intimately connected in order to explain the apparent power and speed of mental activities. (p. 211)

Minsky and the others argued that some higher-level "suprasentential" or, more simply, conceptual structure is needed to represent the complex relations implicit in our knowledge base. The basic idea is that schemata are data structures for representing the generic concepts stored in memory. There are schemata for generalized concepts underlying objects, situations, events, sequences of events, actions, and sequences of actions. Roughly, schemata are like models of the outside world. To process information with the use of a schema is to determine which model best fits the incoming information. Ultimately, consistent configurations of schemata are discovered which, in concert, offer the best account for the input. This configuration of schemata together constitutes the *interpretation* of the input.

Different theorists have proposed more or less concrete specifications of the exact nature of these higher-level structures, but somehow none of them has ever really been adequate. None of them ever captured all of the qualitative characteristics that schemata were supposed to have. For example, a schema is supposed to be a kind of generative thing, which is flexible but which can produced highly structured interpretations of events and situations. Many representational formats have been proposed in an attempt to meet these criteria. For example, Rumelhart (1975) chose as a representation for the schema, a notation rich in generative capacity, namely, the rewrite rules from generative linguistics. Although the generativity of the rewrite rules and the idea that the structure is "constructed" in the process of interpretation is well captured by the rewrite rules, the nonprocedural character of such a system seems wrong. Some more active representation seems necessary. Moreover, the important notions of "default values," variables, and so forth are poorly represented by the rewrite notation. Minsky (1975) and Schank and Abelson (1977) employed passive data structures with slots and explicit default values. These representations are better but are not active and seem to lack the flexibility and generativity that the schema requires. Rumelhart (1977) proposed a representation in which schemata are special kinds of procedures. This view was most completely explicated in Rumelhart (1980). Attempts to build explicit models employing this view, however, have proven unsuccessful. The representation is simply too unwieldy.

It should be clear from the foregoing that there are two distinct ways in which the term schema can be used. On the one hand, it is used to refer to an idea which is common to the work of Kant, Bartlett, Piaget, Minsky, Schank and Abelson, Norman and Bobrow, Rumelhart and Ortony, and many others. This is an idea that has evolved over the years and through the eyes of many different theorists. Many people have sought to clarify and further develop the idea. On the other hand, the term schema is used to refer to one of a large number of instantiations of the general idea of the schema. These explicit schema models are always only pale representations of the underlying intuitions. Whenever a new instantiation of the schema idea is developed, a new perspective is offered on the underlying idea. What we hope to do in this chapter is to propose an alternative to the conventional representation of the schema and at the same time, through the development of a new perspective on schemata, sharpen the idea and develop a system which better captures our intuitions of the nature of the human information-processing system.

One important feature of schemata proposed by Rumelhart and Ortony (1977) has never actually been included in any implementation of the idea. This involves the nature of variable constraints and the filling of default values. The variable constraints associated with each variable serve two functions. On the one hand, they are important for determining whether a particular candidate is an allowable assignment for a variable and, if the variable remains unfilled, are used in the assignment of a default value. These constraints should not be considered absolute. Rather it was proposed that variable constraints should be considered as distributions of possible values. The nearer to the mode of the distribution, the better the variable filler. Moreover, the mode could itself be considered the default value. Importantly, however, there are interdependencies among the possible slot fillers. If one variable is filled with a particular value then it changes the default for the other variables. It was therefore proposed that the variable constraints (and the fillers of the default values) should be considered *multivariate distributions* in which the default value for a particular variable is determined by the values filling the other slots. This idea was difficult to integrate with any of the conventional semantic networks or similar representational formats for schemata. As we shall see, this is a central feature of the PDP analog to schemata.

If schemata are to work as a basis for models of cognitive processing, they must be very flexible objects—much more flexible than they really ever have been in any actual implementations. This is a sort of dilemma. On the one hand, schemata are the structure of the mind. On the other hand, schemata must be sufficiently malleable to fit around most everything. None of the versions of schemata proposed to date have really had these properties. How can we get a highly structured schema which is sufficiently rich to capture the regularities of a situation and to support the kinds of inferences that schemata are supposed to support and at the same time is sufficiently pliable to adapt to new situations and new configurations of events?

On our current view, the answer is simple. Schemata are not "things." There is no representational object which is a schema. Rather, schemata emerge at the moment they are needed from the interaction of large numbers of much simpler elements all working in concert with one another. Schemata are not explicit entities, but rather are implicit in our knowledge and are created by the very environment that they are trying to interpret—as it is interpreting them. [6] Roughly, the idea is this: Input comes into the system, activating a set of units. These units are interconnected with one another, forming a sort of constraint satisfaction network. The inputs determine the starting state of the system and the exact shape of the goodness-of-fit landscape. The system then moves toward one of the goodness maxima. When the system reaches one of these relatively stable states, there is little tendency for the system to migrate toward another state.

The states themselves are the product of the interaction among many groups of units. Certain groups, or subpatterns of units tend to act in concert. They tend to activate one another and, when activated, tend to inhibit the same units. It is these coalitions of tightly interconnected units that correspond most closely to what have been called schemata. The stable pattern as a whole can be considered as a particular configuration of a number of such overlapping patterns and is determined by

---

[6] Hofstadter (1979) expresses essentially the same view in his book *Gödel, Escher, Bach* when the Anteater says:

My "symbols" are ACTIVE SUBSYSTEMS of a complex system, and they are composed of lower-level active subsystems . . . They are therefore quite different from PASSIVE symbols, external to the system, such as letters of the alphabet of musical notes, which sit there immobile, waiting for an active system to process them. (p. 324)

the dynamic equilibrium of all of these subpatterns interacting with one another and with the inputs. Thus, the maxima in the goodness-of-fit space correspond to interpretations of the inputs or, in the language of schemata, configurations of instantiated schemata. In short, they are those states that maximize the particular set of constraints acting at the moment. Depending on the context and the inputs, the system will be closer to one or another of the peaks in the goodness-of-fit function at the outset and will usually find the closest one. This interpretation, we believe, captures almost all of the important aspects of the schema with a view that is at once more flexible than the previous interpretations and yet highly structured. The degree of structure depends on the tightness of the coupling among the coalitions of units which correspond to the schemata in question. Thus, the language of schemata and schema theories should be considered an approximation to the language of PDP. In those cases in which there are coalitions of units that tend to work together, we have a rather close correspondence to the more conventional notion of a schema. In those cases in which the units are more loosely interconnected, the structures are more fluid and less schema-like. Often, knowledge is structured so that there are relatively tight connections among rather large subsets of units. In these cases, the schema provides a very useful description.

One important difference between our interpretation of schemata and the more conventional ones is that in the conventional story, schemata are stored in memory. Indeed, they are the major *content of memory*. In our case, *nothing stored corresponds very closely to* a schema. What is stored is a set of connection strengths which, when activated, have implicitly in them the ability to generate states that correspond to instantiated schemata. This difference is important—especially with regard to learning. There is no point at which it must be decided to create this or that schema. Learning simply proceeds by connection strength adjustment, according to some simple scheme such as those we discuss in various places in this book. As the network is reorganized as a function of the structure of its inputs, it may come to respond in a more or less schema-like way.

We now turn to an example to illustrate the various aspects of these PDP networks and show that many of those features that prompted the invention of schemata in the first place are present in these networks. At the same time, we show that certain features that are problematic with conventional representations of schemata are better dealt with in the PDP language.

An Example

Consider our knowledge of different kinds of rooms. We all have a clear idea of what a typical kitchen or bathroom or living room or bedroom or office looks like. We know that living rooms have sofas and easy chairs, but they don't usually have ovens or bathtubs and that offices have desks and typewriters, but they don't usually have beds. On the other hand, kitchens, living rooms, and offices might all very well have telephones, carpets, etc. Our default bathroom is very small, our default kitchen is somewhat larger but still probably small relative to our default living room. We chose our knowledge of rooms and types of rooms as the primary example to illustrate the PDP representation of schemata. To begin, we need a constraint network that embodies the constraints implicit in our knowledge of rooms. We built our constraint network in the following way. We chose a set of 40 descriptors of rooms. These descriptors are listed in Table 1. We asked two subjects to imagine an office and then, for each of the 40 descriptors asked if the descriptor was accurate of that office. We then asked subjects to imagine a living room and asked about the 40 descriptors again. We then asked about a kitchen, a bathroom, and a bedroom. After finishing these five types of rooms we asked subjects to imagine another office, etc. We collected a total of sixteen judgments of the 40 descriptors on each of the five room types. This data served as the basis for creating our network.[7] In principle, we could imagine presenting each of these 80 room descriptions to the system and have it

TABLE 1

THE FORTY ROOM DESCRIPTORS

ceiling	walls	door	windows	very-large
large	medium	small	very-small	desk
telephone	bed	typewriter	bookshelf	carpet
books	desk-chair	clock	picture	floor-lamp
sofa	easy-chair	coffee-cup	ashtray	fireplace
drapes	stove	coffeepot	refrigerator	toaster
cupboard	sink	dresser	television	bathtub
toilet	scale	oven	computer	clothes-hanger

[7] This was not designed to be a formal experiment of any kind. Rather it was conceptualized as a method of quickly getting a reasonable data base for building an example. Some slight modifications in the data base were made in order to emphasize certain points in our example.

learn according to one or another learning rule we have discussed. Rather than doing that, however, we simply set the weights according to the following equation:

$$w_{ij} = -\ln \frac{p(x_i = 0 \ \& \ x_j = 1)p(x_i = 1 \ \& \ x_j = 0)}{p(x_i = 1 \ \& \ x_j = 1)p(x_i = 0 \ \& \ x_j = 0)}.$$

This equation is derived from a Bayesian analysis of the probability that unit x_i should be on given unit x_j is on and vice versa (see Hinton & Sejnowski, 1983). Four aspects of the weight equation should be noted:

- If the two units tend to be on and off together (i.e., the probability that $x_i = x_j$ is much greater than the probability that $x_i \neq x_j$), then the weight will be a large positive value.

- If, on the other hand, the probability that the two units take on different values (i.e., $x_i \neq x_j$) is much greater than the probability that they take on the same values (i.e., $x_i = x_j$), then the weight takes on a large negative value.

- If the two units come on and off independently (i.e., if $p(x_i = v_1 \ \& \ x_j = v_2) = p(x_i = v_1)p(x_j - v_2)$), then the weight between the two units is zero.

- The weights are symmetric (i.e., $w_{ij} = w_{ji}$).

In addition, each unit has a bias (constant input) which is given by

$$bias_i = -\ln \frac{p(x_i = 0)}{p(x_i = 1)}.$$

Note that if the unit is usually off, it has a negative bias; if it is usually on, it has a positive bias; and if it is equally often on or off, it has a zero bias. [8] The weight matrix estimated by this means is shown in Figure 5. The figure uses the method of Hinton and Sejnowski (Chapter 7) to display the weights. Each unit is represented by a square. The name below the square names the descriptor represented by each square. Within each unit, the small black and white squares represent the weights from that unit to each of the other units in the

---

[8] With a finite data base some of the probabilities mentioned in these two equations might be 0. In this case the values of weights are either undefined or infinite. In estimating these probabilities we began by assuming that everything occurs with some very small probability (.00001). In this way the equation led to finite values for all weights.

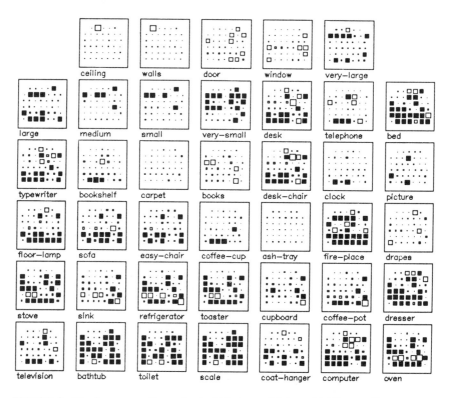

FIGURE 5. The figure uses the method of Hinton and Sejnowski (Chapter 7) to display the weights. Each unit is represented by a square. The name below the square names the descriptor represented by each square. Within each unit, the small black and white squares represent the weights from that unit to each of the other units in the system. The relative position of the small squares within each unit indicates the unit with which that unit is connected.

system. The relative position of the small squares within each unit indicates the unit with which that unit is connected. For example, the white square on the lower right-hand portion of the *refrigerator* units represents the strength of the connection between *refrigerator* and *oven*. White squares represent positive connections and black squares represent negative connections. The size of the square represents the strength of the connection. Thus, the fact that the square representing the connection from the *refrigerator* unit to the *stove* unit is large and white represents the fact that there is a strong positive weight between the two.

It should be noted that each of the units in this example is a visible unit in the sense that each can directly receive inputs from outside the network. There are no hidden units receiving inputs only from other units in the network. We consider this example to be a simplified case. It is possible to imagine hidden units which respond to patterns among the input units. In the general case, we, of course, recognize that hidden units would be required to give different coalitions enough coherence. As we have pointed out elsewhere in the book (cf. Chapter 2 and Chapter 8), multilayer systems containing hidden units are sometimes required to carry out certain computations. In the present instance, however, the existence of hidden units would not change the basic properties of the network which we wish to illustrate. Such higher-level units are not required for the basic schema-like behavior of these networks and, in no case should such a unit be confused with a schema.

It should also be noted that we have chosen a rather high level of abstraction for this example. We have taken such features as *has television* as a *microfeature*. In a more realistic example, we would expect *television* to itself be a particular pattern over a set of units that are used to represent many different varieties of television. There might be many variations on the *television* pattern corresponding to variations among televisions. Moreover, since televisions in bedrooms may be systematically different (perhaps smaller) than televisions in living rooms, we would expect that these correlations would be picked up and there would be a context dependency between the particular version of *television* and the remaining objects in the room. In such a case the units that participate in the representation of television would play the role of a slot in a schema, and the particular pattern of activation on these units would represent the characteristics of the slot filler.

Figure 6 shows several examples of the processing of this network. These runs started by "clamping" one of the descriptors on (that is, by setting the value to 1 and not letting it change) and then letting the system find a goodness-of-fit maximum. In the first example, the descriptor *oven* was clamped on. In such a case, we expect that the system will bring those units most tightly bound to the *oven* unit on and turn off those units negatively correlated to *oven* or other units that it turns on. On the assumption that *oven* is a central feature of the *kitchen* schema, the pattern the system eventually turns on is just that which might be said to correspond to the default *kitchen* schema. The strengths of each of the 40 units is shown along with the "goodness-of-fit" of the state after every 20 updates. The system begins with *oven* and *ceiling* on and then adds *coffee-cup* (weakly), then *sink* and *refrigerator*, concludes that the room is *small*, adds *toaster* and *coffeepot* and finally ends up at a maximum with *ceiling, walls, window, small, telephone, clock, coffee-cup, drapes, stove, sink, refrigerator, toaster, cupboard, coffeepot,* and *oven*. In other words, it finds the default or prototype kitchen. Similarly, runs of the system starting with *desk, bathtub, sofa,* or *bed* clamped lead to goodness maxima corresponding to the prototype or default office, bathroom, living room, or bedroom, respectively. It is, as previously noted, these maxima that we believe correspond roughly to instantiations of schemata for kitchens, offices, bathrooms, living rooms, and bedrooms. The system receives input in the form of having some of the descriptors clamped from the outside. It then finds the best interpretation of the input through this process of hill climbing. As it climbs, the system "fills in" the relevant descriptors of the scene in question.

In the case of the network we created for this example, there are essentially five maxima—one corresponding to each of the different room types. There are 2^{40} possible binary states in which the system could potentially settle, but, in fact, when it is started by clamping exactly one descriptor, it will only settle into one of five states. This roughly corresponds to the view that this data base contains five schemata defined over the set of 40 descriptors. There are, as we shall see, numerous subschemata which involve subpatterns within the whole pattern.

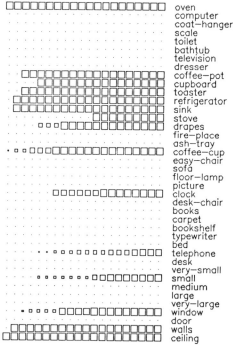

FIGURE 6. Five runs of the network from five different starting places. In each case, the unit *ceiling* and one other is clamped on. The clamping of *ceiling* represents information indicating that *room* is the domain of discussion. The other clamped units are *oven, desk, bathtub, sofa,* and *bed* in the five runs. In each case, the system settles on a prototype for the type of room most closely related to the clamped units.

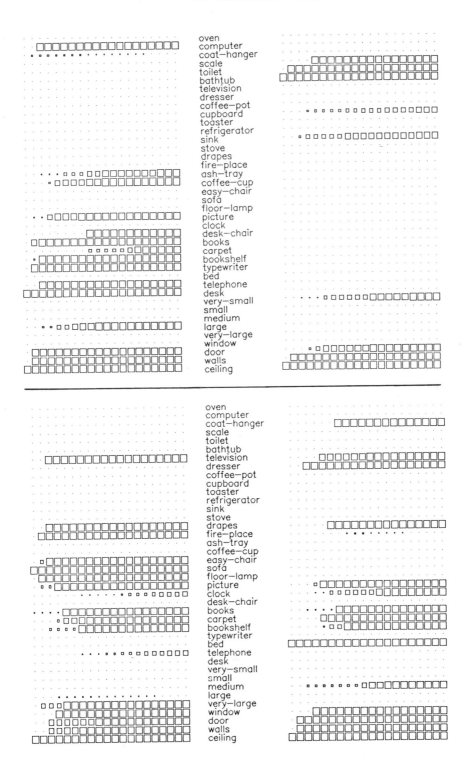

If it was difficult to visualize the landscape for the case of the Necker Cube model described above with 2^{16} states, it is even more difficult with the 2^{40} states of this example. It is, however, possible to get some idea of the landscape in the region of the maxima by plotting the goodness function over a small subset of the goodness landscape. It should be recalled that the states of a system with 40 units can be considered to be a vector in 40-dimensional space. If, as in the present case, the units have a minimum and maximum value, then each point in the 40-dimensional hypercube is a possible state of the system. The states of the system in which all units are at their minimum or maximum values (the binary states of the system) correspond to the corners of the hypercube. Now, each maximum of our network falls at one corner of this 40-dimensional cube. The intersection of a plane and the hypercube will pick out a two-dimensional subset of all of the possible states of the system. Finally, since three points determine a plane, we chose subsets of three of the maxima and plotted the goodness-of-fit landscape for those states falling on the plane passing through those three points. Figure 7 shows the landscape for the plane passing through the maxima for *bedroom, office,* and *kitchen.* Note that there are three peaks on the graph, one corresponding to each of the maxima on the plane. Note also that there are "ridges" connecting the two maxima. These correspond to simple mixtures of pairs of concepts. The fact that the ridge connecting *bedroom* and *office* is higher than those connecting *kitchen* to either of the others indicates that *kitchen* is more

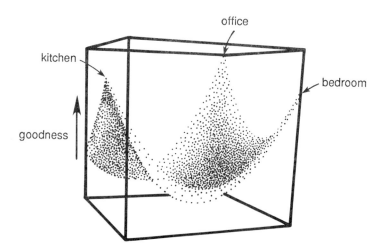

FIGURE 7. The value of the goodness function for the states on the plane passing through the three goodness maxima corresponding to the prototypes for *kitchen, bedroom,* and *office.*

distinctive from the other two than they are from each other. Figure 8 shows the plane containing *office, bedroom*, and *bathroom*. In this case we see that the goodness function sinks much lower between the *office-bedroom* axis and *bathroom*. This occurs because *bathroom* is much more different from the other two than is *kitchen*. In any of these cases, it should be recognized that given the starting configuration of the system it will simply find one of these maxima and thereby find one of these interpretations of the input. By contrast, Figure 9 shows the goodness-of-fit landscape on the plane passing through *bedroom, office*, and *living-room*. In order to get a clearer perspective on the surface, the angle of viewing was changed so we are looking at the figure from between *bedroom* and *office*. Clearly, the whole graph is greatly elevated. All points on the plane are relatively high. It is a sort of goodness plateau. These three points are essentially three peaks on a much larger mountain containing all three maxima. Finally, Figure 10 shows the goodness landscape on the plane containing the three most distinct prototypes—*bedroom, kitchen*, and *bathroom*. The goodness function dips well below zero in this plane. Mixtures of *kitchens, living-rooms*, and *bathrooms* are poor rooms indeed.

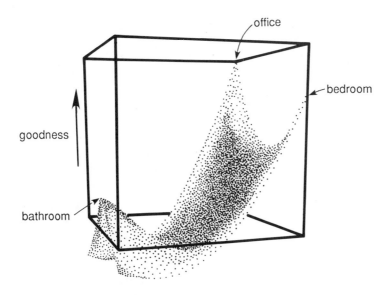

FIGURE 8. The value of the goodness function for the states on the plane passing through the three goodness maxima corresponding to the prototypes for *bathroom, bedroom*, and *office*.

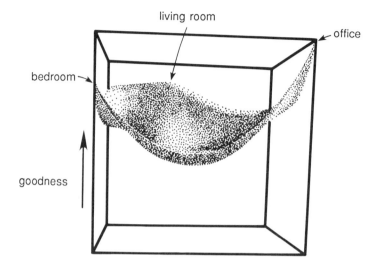

FIGURE 9. The value of the goodness function for the states on the plane passing through the three goodness maxima corresponding to the prototypes for *living-room*, *bedroom*, and *office*.

It should be mentioned that there are essentially two assumptions that can be made about the input. Under one assumption, inputs are clamped to either their minimum or maximum value and aren't allowed to move. That was the way inputs were treated in the present examples. Other times, it is convenient to imagine that inputs are merely biases feeding input into certain of the units. This is the way inputs were treated in the Necker cube example. These two ways of viewing inputs can be combined by assuming that the case of clamping corresponds to very, very strong biasing. So strong that internal constraints can *never* overcome the external evidence. The case of clamping is simpler, however. In this case there is no distortion of the goodness-of-fit landscape; certain states are simply not available. The system is forced to move through a different part of the state space. In addition to its effects on the region of the state space that is accessible, the input (along with the context) determines where the system begins its processing and therefore, often, which of the maxima it will find. Figure 11 illustrates this point. The figure shows the goodness function on the set of states on the plane passing through the start state, the *bedroom* maximum, and the *kitchen* maximum for two different inputs. In Figure 11A we have clamped the *bed* unit to be on. In Figure 11B we have clamped the *oven* unit on. It should be noted that to move from the start state to the *kitchen* peak in the first case involves climbing through a dip in goodness-of-fit. Since the system strictly goes

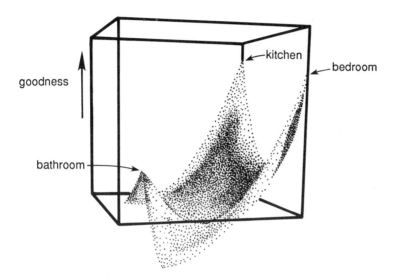

FIGURE 10. The value of the goodness function for the states on the plane passing through the three goodness maxima corresponding to the prototypes for *kitchen*, *bedroom*, and *bathroom*.

"uphill" it will be unable to reach the *kitchen* maximum and will move instead monotonically uphill toward the *bedroom* peak. Similarly, in Figure 11B with *oven* clamped there is a dip separating the start state from the *bedroom* maximum but a monotonically increasing slope flowing into the *kitchen* peak. Figure 11C shows, for comparison, the landscape from the start state when no units are clamped on. In this case, there is no dip separating the start state from either peaks, so the system can move to either maximum.

To summarize, we have argued that the maxima in the goodness-of-fit landscapes of our networks correspond to configurations of instantiated schemata. We have shown how these maxima are determined by coalitions among units and how the inputs determine which of the maxima the system will find. It should be clear that the multivariate distributions proposed by Rumelhart and Ortony are readily captured in the PDP framework. The values of each variable determine what values will be filled in for the other variables. We have yet to show that the kind of PDP system we have been describing really has all of the important properties of schemata. In the following section, we use the present example to illustrate these properties and discuss some of the advantages of our formulation over previous formulations of the schema idea.

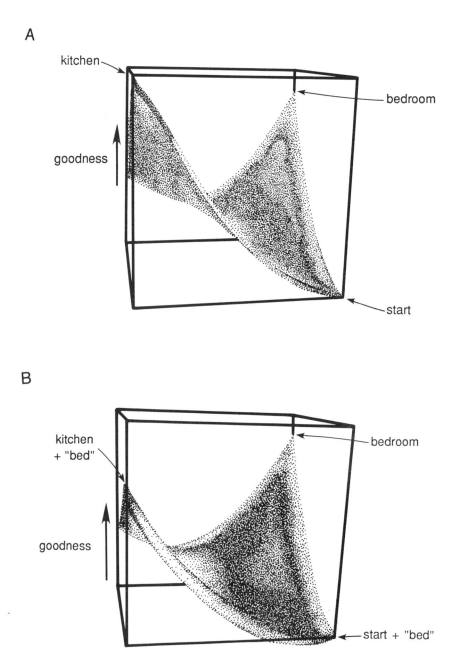

FIGURE 11.

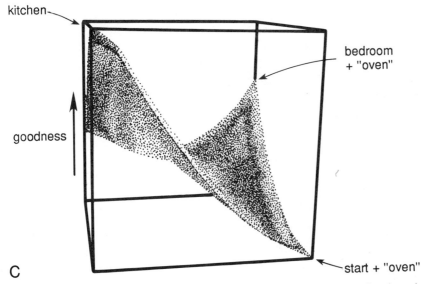

kitchen

goodness

bedroom
+ "oven"

C

start + "oven"

FIGURE 11. The goodness function over the set of states on the plane passing through the start state, the *bedroom* maximum, and the *kitchen* maximum for two different inputs. In *A* we have clamped the *bed* unit to be on. In *B* we have clamped the *oven* unit on. *C* shows the landscape from the start state when no units are clamped on.

Properties of Schemata and Maxima in Constraint Satisfaction Networks

Rumelhart and Ortony (1977; Rumelhart, 1980) have outlined a set of properties which characterize schemata. In this section we consider each of these properties and show how they map onto features of the PDP networks we are now outlining.

Schemata have variables. Essentially, the variables of a schema correspond to those parts of the pattern that are not completely determined by the remainder of the structure of the pattern itself. It is these parts that vary from one situation to another in which, nevertheless, the bulk of the units corresponding to the schema is active. On this account, the binding of a variable amounts to filling in a variable subpattern. Default values represent variable subpatterns that tend to get filled-in in the absence of any specific input. Since patterns tend to complete themselves, default values tend to be automatically filled in by the process of settling into an interpretation. In some cases, there are sets of units that are mutually inhibitory so that only one can be active at a time but any of which could be combined with most other units. Such a set of units can be considered to constitute a *slot* which is filled in in the processing of the input. Perhaps the best example from our current data base is what might be called the *size slot*. In this case, the *very-large*, *large*, *medium*, *small*, and *very-small* units are all mutually inhibitory. (See the weight matrix in Figure 6). The different maxima have different default values for these slots. The *bathroom* has a default value of *very-small*, the *kitchen* has a default value of *small*, the *bedroom* has a default value of *medium*, the *office* is *large*, and the default *living-room* is *very-large*. Interestingly, when the input contains information that descriptors other than the default descriptors apply, the default size changes as well. For example, Figure 12 shows a case in which *bed* and *sofa* were both clamped. What we get in such a case is a room which might best be described as a large, fancy bedroom. The size variable is filled in to be *large*, it also includes an *easy-chair*, a

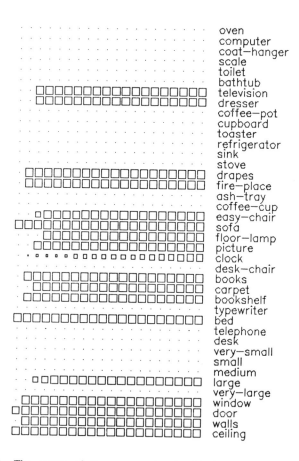

FIGURE 12. The output of the room network with *bed*, *sofa*, and *ceiling* initially clamped. The result may be described as a large, fancy bedroom.

floor-lamp, and a *fireplace*. Similarly, the setting of the size variable modifies the default values for the other descriptors. In this case, if we set the size variable to *large* we get a bedroom with a *fireplace*.

Schemata can embed. In structural interpretations of schemata, it is useful to think of a schema as a kind of tree structure in which subschemata correspond to subtrees that can fill variable slots. Under our interpretation, subschemata correspond to small configurations of units which cohere and which may be a part of many different stable patterns (and therefore constitute a schema on their own right). Each stable subset of cohering units can be considered a schema. Large schemata will often consist of patterns of coherence among these coherent subsets. There are several instances of this in our example. For example, the *easy-chair* and *floor-lamp* constitute a subschema, the *desk* and *desk-chair*, the *window* and *drapes*, and other similar combinations constitute small schemata that can be either present or absent in several different configurations. Consider, for example, the case of *window* and *drapes*. These two elements almost always appear together and either both appear or neither appear. We will refer to this pattern as the *window* schema. Figure 13 shows the effect of adding *drapes* and/or *window* to the *office* schema. The default value for this schema involves no windows. The highest peak, at the origin of the graph, corresponds to the *office* maximum. One axis corresponds to the

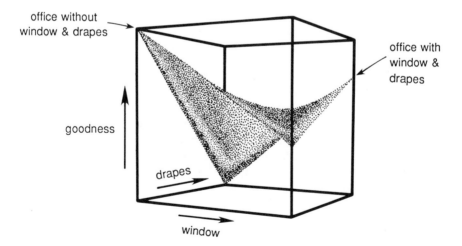

FIGURE 13. The goodness landscape for *office* as a function of the activation value of the *drapes* unit and the *window* unit. The function shows that it is maximum when either both are 0 or both are 1. This pattern of interaction is consistent with the view that the combination *window-drapes* form a subschema.

amount of *drapes* added to the schema (i.e., to the activation value for the *drapes* unit). The second axis corresponds to the amount of *window* added. The third axis, of course, corresponds to the goodness-of-fit for each of the states. It should be noted that the low points on the graph correspond to those cases in which one of the two units of the window subschema is on and the other is off. The high points on the graph, corresponding to goodness maxima, occur when either neither (at the origin) or both of the units are on. The case where neither is on corresponds to a slightly higher peak than when both are on. Thus, the default *office* probably doesn't have a window, but if the input indicates that either one of the units (*window* or *drapes*) is on, turning the other one on is best. To conclude, large schemata such as the *office* schema can be conceptualized as consisting, in part, of a configuration of subschemata which may or may not be present as wholes. Having parts of these subschemata is worse than having either the entire subschema or none of it.

Schemata represent knowledge at all levels. They should represent encyclopedic knowledge rather than definitional information. This amounts to the claim that all coherent subpatterns should be considered schemata as well as the whole stable pattern. It also suggests that knowledge of all sorts should be represented in the interconnections among the constituent units.

Schemata are active processes. This is obviously true of the the PDP system we are describing. They are a kind of organic element which grows and fulfills itself within its environment.

Schemata are recognition devices whose processing is aimed at the evaluation of their goodness-of-fit to the data being processed. This feature is obviously also a part of the idea outlined here. The goodness-of-fit is roughly determined by the height of the peak in goodness space. The processing of the system is aimed toward climbing uphill along the goodness-of-fit gradient. The stable points correspond to local maxima in this space. The height of the peak corresponds to the goodness-of-fit.

Some additional features of our interpretation. There are three major difficulties with the conventional representation of schemata that are naturally overcome in the PDP approach. In the conventional approaches, decisions must be made about which aspects of a given schema are constant and which are variable. The PDP solution is essentially that all aspects are variable; some aspects are simply more tightly constrained than others. Secondly, in a conventional representation, one has to decide exactly which aspects of the situation are part of the schema and which are not. In our PDP approach, units may cohere more or less strongly to their mates and in this sense be more or less a part of the schema. Finally, on the conventional view a decision must be made about whether a certain set of relationships should be put together to form a schema at all. Again, in the PDP formulation no such decision needs to be made. One can have schemata of varying degrees of existence. The rigidity of the schema is determined by the tightness of bonding among the units that constitute the schema. The tighter the bond, the more strongly the constituent elements activate one another, and the more rigid the structure. The weaker the bonds, the more fluid the structure, and the more easily a system can flow among states. This degree of fluidity depends on the shape of the goodness-of-fit landscape. Tightly rigid schemata have sharp peaks in goodness space; fluid schemata with many variable parts correspond to rounded hilltops. The goodness landscape, in turn, depends on the knowledge base that lies beneath it. If the knowledge is tightly inter-constrained so that one part strongly predicts other parts, then we have a rigid schema. We can't easily get just part of it active. If part of it becomes active, the part will pull in the whole and suppress the activity of aspects that are not part of it. On the other hand, if the knowledge is only loosely interrelated, the schema will be a relatively weak organizer of the information and will be a pressure for structuring the input, but it will flow easily from one pattern to another. Moreover, within a schema itself, some aspects will be tightly bound together while other aspects will be only more loosely tied to the body of the schema. Input situations that demand an interpretation that breaks up the tightly bound clusters are going to be more difficult for the system to attain than those that require breaking up much more loosely interconnected elements.

Finally, we point out one way in which these ideas about schemata might be elaborated to overcome one apparent deficiency of the network we have thus far been considering. The network uses a fixed set of units to represent each type of object that might be present in a particular instantiation of a schema. This is clearly an oversimplification since it is often necessary to be able to think about two different instantiations of the same subschema within a larger overall schema—for example, there is often more than one chair in a living room. To capture such situations, it is necessary to imagine that the network may contain several subsets of units, each capable of representing a different possible chair. The subsets would correspond to different roles the different chairs might play in the overall room schema. This would also allow the representation to capture the assignment of a particular object to a particular role.

PARALLEL DISTRIBUTED PROCESSING MODELS AND THINKING

In the previous section we offered an interpretation of the schema in terms of the emergent properties of simple PDP networks. From there, we believe that we can make contact with much of the cognitive science literature. There are central issues, however, which remain difficult to describe within the PDP framework. We have particularly in mind here, the process of thinking, the contents of consciousness, the role of serial processes, the nature of mental models, the reason for mental simulations, and the important synergistic role of language in thinking and in shaping our thought. These issues are especially important because these are issues that PDP approaches do not, on first blush, seem to have much to say about. In this section we attack some of those problem areas. We don't claim to have solutions to them, rather we can outline a story that represents our current understanding of these processes. The story is overly simplistic but it does give an idea of where we are in our thinking on these critical issues.

It should be noted here that the account of mental processing we have been developing offers an interesting perspective on the relations between parallel and serial processing. The "parallel" in "parallel distributed processing" is intended to indicate that, as a basic architectural design principle, processing is carried out, in so far as possible, in parallel. Parallel algorithms are employed rather than serial ones. At the same time, however, the "distributed" in "parallel distributed processing" brings a serial component to PDP systems. Since it is patterns of activations over a set of units that are the relevant representational format and since a set of units can only contain one pattern at a time, there is an enforced seriality in what can be represented. A given set of units can, however, be seen as representing a sequence of events. Since we assume that the system is moving toward a maximum goodness solution every time a new input comes into the system, the system operates in the following way. An input enters the system, the system relaxes to accommodate the new input. The system approaches a relatively stable state which represents the interpretation of the input by the system. The system then occupies this state until the stimulus conditions change. When a new input arrives, the system relaxes to a new state. Looking at the system over a short time frame, it is dominated by the relaxation process in which all units work cooperatively to "discover" an interpretation of a new input. Looking over a somewhat longer time frame, we see the system as sequentially occupying a series of relatively stable states—one for each change in input. Roughly, if we imagine that the relaxation process takes on the order of a few tenths of a second and that the time spent in essentially the same stable state is on the order of a half of a second or so, we could see events requiring less than about a half second to be essentially a parallel process, and those requiring several seconds to involve a series of such processes and therefore to have a serial component.

The Contents of Consciousness

It isn't necessary for the arguments that follow, but for the sake of concreteness, we suppose that there is a relatively large subset of the total units in the system whose states of activity determine the contents of consciousness. We imagine that the time average of the activities of these units over time periods on the order of a few hundred milliseconds correspond to the contents of consciousness. Since we imagine that our systems must be such that they reach equilibrium in about this amount of time, the contents of consciousness are dominated by the relatively stable states of the system. Thus, since consciousness is on the time scale of sequences of stable states, consciousness consists of a sequence of interpretations—each represented by a stable state of the system. Typically, consciousness contains a single interpretation (i.e., a single pattern representing its inputs) and consists of a sequence of such interpretations. On occasions in which the relaxation process is especially slow, consciousness will be the time average over a dynamically changing set of patterns and thus would be expected to lead to "fuzzy" or unclear phenomenal impressions.

The Problem of Control

One common critique of the kind of model we have sketched so far is that it can't really change without external prodding. Suppose that we are in a fixed stimulus environment. In this case, the system will relax into an interpretation for that environment and stay there. Our conscious experience will be of a fixed interpretation of a fixed stimulus. Until the world changes there is no change to the system nor to the contents of our consciousness. Obviously this is an incorrect conclusion. How can such a system do something? Perhaps the first thing that comes to mind is that the environment never really is fixed. It is always changing and therefore the contents of our consciousness must always be changing to interpret the current state of affairs. A good example of this might be the movies. We sit in the movies and watch. Our system reaches a sequence of interpretations of the events on the screen. But, since the movie is always continuing, we are driven to continue to interpret it. Surely, what is true of a movie is also true of life—to some extent. This may be part of the story, but it would appear to be rather more passive than we might want. We don't just sit passively by and let the world change and then passively monitor it. Rather we act on the world.

A second answer to the problem of a system fixated on a particular interpretation becomes apparent in realizing that our interpretation of an event often dictates an action which, in turn, changes the environment. The environmental change can then feed back into the system and lead to another interpretation and another action. Figure 14 illustrates how this feedback loop can continuously drive the system from state to state. A paradigm case for this is playing a game. We can imagine that we are playing a game with someone; our input consists of a board configuration, and we settle into a state which includes a specification of a response. It would be quite easy to build a relaxation network that would take as input a description of a current board situation and produce, as part of the state to which it relaxed, a specification of the response. It would simply require that, for each game situation, the system relaxes to a particular state. Certain units of the state represent the action (or class of actions) that should be taken. Upon taking these actions, the opponent makes a play which in turn leads to a new set of constraints to which the system relaxes. In this way, the system can make a sequence of moves. Indeed, as we describe below, we have built such a network that can play tic-tac-toe. Other more complex games, such as checkers or chess require rather more effort, of course, but can, in principle, be dealt with in the same way. Although this is a much more activist view, it is still a "data-driven" view. The system is entirely reactive—given I am in this state, what should I do?

Mental Models

Suppose, for arguments sake, that the system is broken into two pieces—two sets of units. One piece is the one that we have been discussing, in that it receives inputs and relaxes to an appropriate state that includes a specification of an appropriate action which will, in turn,

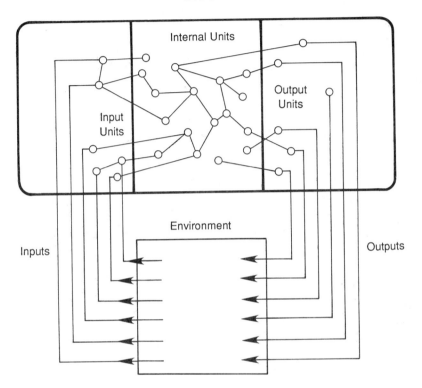

FIGURE 14. The inputs to the PDP system should be considered as partially due to the effects that the output of the system has had on the environment. In this way, the output of the system can drive it from state to state.

change the inputs to the system. The other piece of the system is similar in nature, except it is a "model" of the world on which we are acting. This consists of a relaxation network which takes as input some specification of the actions we intend to carry out and produces an interpretation of "what would happen if we did that." Part of this specification would be expected to be a specification of what the new stimulus conditions would be like. Thus, one network takes inputs from the world and produces actions; the other takes actions and predicts how the input would change in response. This second piece of the system could be considered a mental model of the world events. This second portion, the mental model of the world, would be expected to be operating in any case, in as much as it is generating expectations about the state of the world and thereby "predicting" the outcomes of actions.

Now, suppose that the world events did not happen. It would be possible to take the output of the mental model and replace the stimulus inputs from the world with inputs from our model of the world. In this case, we could expect that we could "run a mental simulation" and imagine the events that would take place in the world when we performed a particular action. This mental model would allow us to perform actions entirely internally and to judge the consequences of our actions, interpret them, and draw conclusions based on them. In other words, we can, it would seem, build an internal control system based on the interaction between these two modules of the system. Indeed, as we shall show, we have built a simple two-module model of tic-tac-toe which carries out exactly the process and can thereby "imagine" playing tic-tac-toe. Figure 15 shows the relationships between the interpretation networks, the inputs, the outputs, and the network representing a model of the world and the process of "mental simulations."

Mental Simulations and Mental Practice

One nice feature of this model is that it ties into the idea of mental simulations and learning through mental practice. Performance in the task involves two parts—a system that determines what to do in any given situation and a system that predicts what will happen if any given action is carried out. If we have a reasonably good model of the "world" we could learn from our model the various consequences of our actions—just as if we were carrying them out in the real world. It may very well be that such a feature accounts for the improvement that occurs in mentally practicing complex motor tasks.

Conversations: Actual and Imagined

Imagine a situation in which we had a relaxation network which would take as input a sentence and produce an interpretation of that sentence as well as the specifications for a response to that input. It is possible to imagine how two individuals each with such a network could carry out a conversation. Perhaps, under appropriate circumstances they could even carry out a logical argument. Now, suppose that we don't actually have another participant, but instead have a mental model of the other individual. In that case, we could imagine carrying

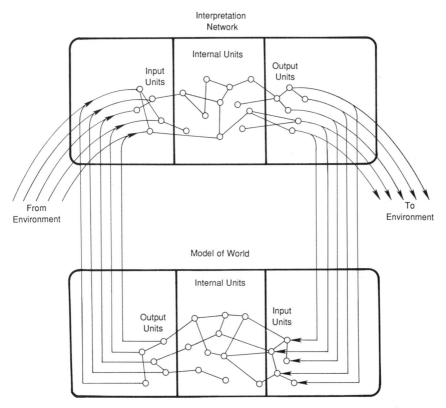

FIGURE 15. The relationships among the model of the world, the interpretation network, the inputs, and the outputs for the purpose of mental simulations.

out a conversation with someone else. We could hold an imaginary argument with someone else and perhaps even be convinced by it! Indeed, this brings up the last move that we wish to suggest. Suppose that we don't have a model of another person at all. Suppose that instead we simply use our one single model to both produce and react to imagined linguistic inputs. This description is, it would seem, consistent with Vygotsky's (1934/1962) view of thinking and is consistent with the introspections about a certain kind of thinking and "internal speech." Note, this does *not* suggest that thinking is simply internal speech. More generally, thinking, as we have argued, involves a sequence of states of consciousness. There are a number of ways of controlling that sequence. One way involves running "mental simulations." Another way involves recycling linguistic inputs. Note, this gives language an interesting, almost Whorfian, role however. Suppose that the interpretation that led to the production of the internal speech was much richer than the linguistic forms could possibly suggest. Thus, the linguistic forms pick out aspects of the entire interpretation to emphasize. Once this emphasis has taken place and the new input has been processed, the next state will be strongly affected by the new input and our new interpretation will be shaped, to some extent, by the words we chose to express our first idea. Thus, our thinking about a topic will be, sometimes strongly, affected by the language tools we have for expressing our ideas.

External Representations and Formal Reasoning

If the human information-processing system carries out its computations by "settling" into a solution rather than applying logical operations, why are humans so intelligent? How can we do science, mathematics, logic, etc.? How can we do logic if our basic operations are not logical at all? We suspect the answer comes from our ability to create artifacts—that is, our ability to create physical representations that we can manipulate in simple ways to get answers to very difficult and abstract problems.

The basic idea is that we succeed in solving logical problems not so much through the use of logic, but by making the problems we wish to solve conform to problems we are good at solving. People seem to have three essential abilities which together allow them to come to logical conclusions without being logical. It is these three abilities that have allowed us to accomplish those uniquely human achievements of formal reasoning. These abilities are:

- We are especially good at pattern matching. We seem to be able to quickly "settle" on an interpretation of an input pattern. This is an ability that is central to perceiving, remembering, and comprehending. Our ability to pattern match is probably not something which sets humans apart from other animals, but is probably *the* essential component to most cognitive behavior.

- We are good at modeling our world. That is, we are good at anticipating the new state of affairs resulting from our actions or from an event we might observe. This ability to build up expectations by "internalizing" our experiences is probably crucial to the survival of all organisms in which learning plays a key role.

- We are good at manipulating our environment. This is another version of man-the-tool-user, and we believe that this is perhaps the crucial skill which allows us to think logically, do mathematics and science, and in general build a culture.

Especially important here is our ability to manipulate the environment so that it comes to represent something. This is what sets human intellectual accomplishments apart from other animals.

Roughly speaking, the view is this: We are good at "perceiving" answers to problems. Unfortunately, this is not a universal mechanism for solving problems and thinking, but as we become more expert, we become better at reducing problem domains to pattern-matching tasks (of the kind best accomplished by PDP models).[9] Thus, chess experts can look at a chess board and "see" the correct move. This, we assume, is a problem strictly analogous to the problem of perceiving anything. It is not an easy problem, but it is one that humans are especially good at. It has proven to be extraordinarily difficult to duplicate this ability with a conventional symbol-processing machine. However, not all problems can be solved by immediately "seeing" the answer. Thus, few (if any) of us can look at a three-digit multiplication problem (such as 343 times 822) and see the answer. Solving such problems cannot be done by our pattern-matching apparatus, parallel processing alone will not do the trick; we need a kind of serial processing mechanism to solve such a problem. Here is where our ability to manipulate our environment becomes critical. We can, quite readily, learn to write down the two numbers in a certain format when given such a problem.

343
822
―――

Moreover, we can learn to see the first step of such a multiplication problem. (Namely, we can see that we should enter a 6 below the 3 and 2.)

343
822
―――
6

We can then use our ability to pattern match again to see what to do next. Each cycle of this operation involves first creating a representation through manipulation of the environment, then a processing of this (actual physical) representation by means of our well-tuned perceptual apparatus leading to a further modification of this representation.

―――――――――

[9] As we have argued before, it is because experts have such a powerful pattern-matching capability that expert systems that rely only on pattern matching (albeit not nearly as powerful as the human system) are as successful as they are.

By doing this we reduce a very abstract conceptual problem to a series of operations that are very concrete and at which we can become very good. Now this applies not only to solving multiplication problems. It applies as well to solving problems in logic (e.g., syllogisms), problems in science, engineering, etc. These dual skills of manipulating the environment and processing the environment we have created allow us to reduce very complex problems to a series of very simple ones. This ability allows us to deal with problems that are otherwise impossible. This is *real* symbol processing and, we are beginning to think, the primary symbol processing that we are able to do. Indeed, on this view, the external environment becomes a key extension to our mind.

There is one more piece to the story. This is the tricky part and, we think, the part that fools us. Not only can we manipulate the physical environment and then process it, we can also learn to internalize the representations we create, "imagine" them, and then process these imagined representations—just as if they were external. As we said before, we believe that we are good at building models of our environment so that we can anticipate what the world would be like after some action or event takes place. As we gain experience with the world created by our (and others') actions we develop internal models of these external representations. We can thus imagine writing down a multiplication problem and imagine multiplying them together. If the problem is simple enough, we can actually solve the problem in our imagination and similarly for syllogisms. Consider, for example, a simple syllogism: All A are B and no C are B. We could solve this by drawing a circle for A, a larger circle including all of the A's around the first circle to represent the B's, and a third disjoint circle standing for the C's. We could then "see" that no A's are C. Alternatively, we need not actually draw the circles, we can merely imagine them. We believe that this ability to do the problem in our imagination is derivative from our ability to do it physically, just as our ability to do mental multiplication is derivative from our ability to do multiplication with pencil and paper. The argument that external representations play a crucial role in thought (or, say, in solving syllogisms) is sometimes challenged on the ground that we don't really *have* to draw Venn diagrams (or whatever) to solve them since we *can* solve them in our head. We argue that the major way we can do that is to imagine doing it externally. Since this imagination is, we argue, dependent on our experience with such representations externally, the argument that we *can* solve them mentally loses its force against the view that external symbols are important for thought processes. Indeed, we think that the idea that we reason with mental models is a powerful one precisely because it is about this process of imagining an external representation and operating on that.

It is interesting that it is apparently difficult to invent new external representations for problems we might wish to solve. The invention of a new representation would seem to involve some basic insight into the nature of the problem to be solved. It may be that the process of inventing such representations is the highest human intellectual ability. Perhaps simply creating an external representation sufficient to support problem solving of a particular kind is evidence of a kind of abstract thinking outside of the simple-minded view sketched here. That may be, but it seems to us that such representational systems are not very easy to develop. Usually they are provided by our culture. Usually they have evolved out of other simpler such systems and over long periods of time. Newer ones, when they are developed, usually involve taking an older system and modifying it to suit new needs. One of the critical aspects of our school system would seem to be teaching such representational schemes. The insights into the nature of the problem become embedded in the representations we learn to use to solve the problems.

Language plays an especially tricky and interesting role in all of this. Perhaps the internal/external issue is not too important with language. The notion we have here is one of "self-instruction." This follows Vygotsky's (1934/1962) view, we believe. We can be instructed to behave in a particular way. Responding to instructions in this way can be viewed simply as responding to some environmental event. We can also remember such an instruction and "tell ourselves" what to do. We have, in this way, internalized the instruction. We believe that the process of following instructions is essentially the same whether we have told ourselves or have been told what to do. Thus, even here, we have a kind of internalization of an external representational format (i.e., language). We don't want to make too much of this point since we recognize that the distinction between external and internal when we ourselves produce the external representation is subtle at best, but we don't really think it differs too much from the case in which we write something down and therefore create a real, physical, viewable representation. Saying something out loud creates a hearable representation. There are interesting cases in which people talk to themselves (for example, solving difficult problems in noisy environments leads people to literally talk to themselves and instruct themselves on the problems they are solving).

Before leaving this topic, one more important aspect of external representations (as opposed to internal representations) should be noted. External representations allow us to employ our considerable perceptual/motor abilities in solving abstract problems. This allows us to break problems into a sequence of relatively simple problems. Importantly, once an external representation is created, it can be reinterpreted without regard to its initial interpretation. This freedom allows us to discover solutions to problems without "seeing" our way to the end. We can inspect intermediate steps and find alternative solutions which might be better in some ways. In this way, we can discover new features of our representations and slowly extend them and make them more powerful.

Goal Direction in Thinking

Our discussion thus far has left one central issue undiscussed, namely, the role of goals in thought and problem solving. Clearly it is not the case that the same perceptual stimulus always drives the system to react in a consistent way. Rather, our goals or intentions interact with the stimuli (internal and external) that provide the inputs to the thinking process. Further, goals organize whole sequences of thoughts into a coherent problem-solving activity, and the notion that there is a hierarchy of goals is certainly important for understanding these coherent sequences.

While we have not stressed the importance of goals, it is not difficult to see how they could be incorporated into our framework. Goals can be explicitly represented as patterns of activation and can thus provide one source of input to the thinking process. Nor is it difficult to imagine how a PDP network could learn to establish specific subgoal patterns in response to particular superordinate goals and inputs.

Summary

These ideas are highly speculative and detached from any particular PDP model. They are useful, we believe, because they suggest how PDP models can be made to come into contact with the class of phenomena for which they are, on the face of it, least well suited—that is, essentially sequential and conscious phenomena. Even in these cases, however, they lead us to view phenomena in new ways.

An Example

In these last few sections, we have been talking at a very general level. We often find it useful to be concrete about our ideas.

Therefore, to illustrate the notion of thought as mental simulation more concretely, we created two relaxation networks that can be connected together to mentally simulate playing a game of tic-tac-toe. The two networks are very similar. The first is a system which, given a pattern representing the board of a tic-tac-toe game, will relax to a solution state that fills in an appropriate response. The second module is nearly identical to the first; it takes as input a board position and a move and settles to a prediction of the opponent's responding move. In short, it is a "mental model" of the opponent. When the output of the first is fed, as input, to the second and the output to the second is fed, as input, to the first, the two networks can simulate a game of tic-tac-toe.

Figure 16 illustrates the basic structure of the tic-tac-toe playing network. The network consists of a total of 67 units. There are nine units representing the nine possible responses. These are indicated by the nine dots at the top of the figure. There is one unit for each of the nine possible moves in tic-tac-toe. Since only one response is to be

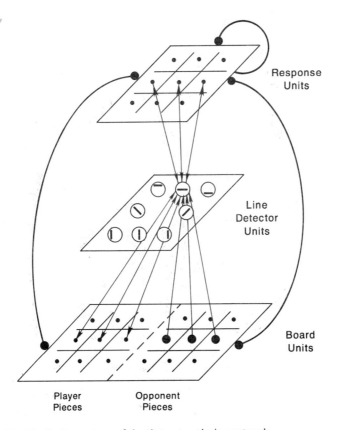

FIGURE 16. The basic structure of the tic-tac-toe playing network.

made at a time, these units are mutually inhibitory. This is indicated by the heavy black line feeding back from the top plane in the figure to itself. There are a total of 18 units representing the board configuration. These are divided into two groups of nine: one group for the positions of the friendly or player pieces on the board and one group representing the positions of the opponent's pieces. Since if any square is occupied, it is not a possible move, each board unit strongly inhibits its corresponding output unit. This strong inhibition is indicated by the heavy black lines connecting the board units to the response units. In addition to these 9 output units and 18 input units, there is a total of 40 hidden units which detect patterns in the board units and activate the various response alternatives. These 40 units can be divided into eight classes corresponding to the eight possible ways of getting three ×'s or ○'s in a row. In the figure, one of each such category of units is illustrated. The receptive field of each unit is indicated by the line inside the circle representing that unit. Thus, there is one unit for each of the three horizontal lines, one for each of the three vertical lines, and one for each of the two possible diagonal lines. For each of the eight classes of hidden units, there are five different pattern types that different units are responsive to. For example, some units are responsive to empty regions. That is, they respond just in case none of the board units from which it receives inputs is turned on. This is implemented by making them be inhibited by any activity in their receptive field and by giving them a negative threshold. We call this an *empty line detector*. All things being equal, it is better to move into an empty row, column, or diagonal; therefore these units weakly activate their respective output units. At the start of the game these are the only units which are active and therefore the sole criterion for the first move is the number of possible strings of three the square is a member of. Since the center square intersects the largest number, it will usually be chosen by the system for its first move. On later moves, there are other units feeding into the decision, but these units also contribute. Another unit type will respond whenever two or more units of the same kind occur in its regions. This is the kind of unit illustrated in the figure. It receives strong inhibitory input from one set of board units (in this case the opponent's pieces) and excitatory inputs from the other set. It has a rather high positive threshold so that it will not come on until at least two units are on in its row. We call this a *friendly doublet detector*. Whenever this unit comes on it means that the system can make a winning move by playing in that row. Therefore, it is strongly positively connected to its respective output units. If such a move is possible, the system will make it. There are similar units which respond to two or more units from the representation of the opponent's pieces are active in its receptive field. We call this an *opponent doublet detector*. If such a unit comes on, it means that the opponent could win by a move in this region, so it excites its response units very strongly as well. Unless there is a winning move somewhere else, the blocking response will become the most active. Finally, there are units which respond to one or more friendly pieces or one or more opponent pieces in an otherwise open line. We call these *friendly singleton* and *opponent singleton detectors.* It is generally good to extend your own singleton or block an opponent's singleton if there is nothing more pressing, so these units also activate their respective output units, only rather more weakly than the units detecting doublets. Thus, the net input arriving at any given unit is the weighted sum of all of these urgencies detected by the hidden units. Because of the direct inhibition from the board units, only those response units corresponding to open squares receive positive input. The mutual inhibition insures that the strongest unit will usually win. In order that this system truly climb in overall goodness-of-fit, all weights are symmetric and the update is done asynchronously at random. This means that when there isn't much difference between the possible response alternatives, a weaker one will sometimes get the upper hand and win the competition. This never happens, however, when a unit is receiving very strong input, as with the case of an open double for a win or to be blocked.

The simplest case is when the system simply is given a board position, settles on a move, is given the next board position incorporating the opponent's response, settles on a move for that position, etc., until a game is finished. This involves only one network and presumably would be the basic modality for the system to run in. Figure 17 shows a sample run of such a situation. The activation levels of each of the 67 units is shown in each column. Successive activation states are shown from left to right. At the start, it should be noted that there is a friendly piece in the center square and two enemy pieces, one in each corner of the upper row. It is the system's move. The system starts with the units corresponding to the board position clamped on and all other units off. The figure shows the strengths for each of the units after every 50 asynchronous updates. By the second time slice, the system is beginning to extract the relevant features of the board position. The five groups of eight units shown in each column following the response units and board position display the line-detector units. Each group of eight is laid out with three units across the top corresponding to the top, middle, and bottom lines from left to right. The second row of the eight corresponds to the left and right diagonals, and the bottom row of each eight corresponds to the left, middle, and right columns. Thus, we see that the system has, by the second time slice, begun to discover that the bottom row is empty, that it has a singleton in the middle row, that the opponent has a singleton in the left column, and

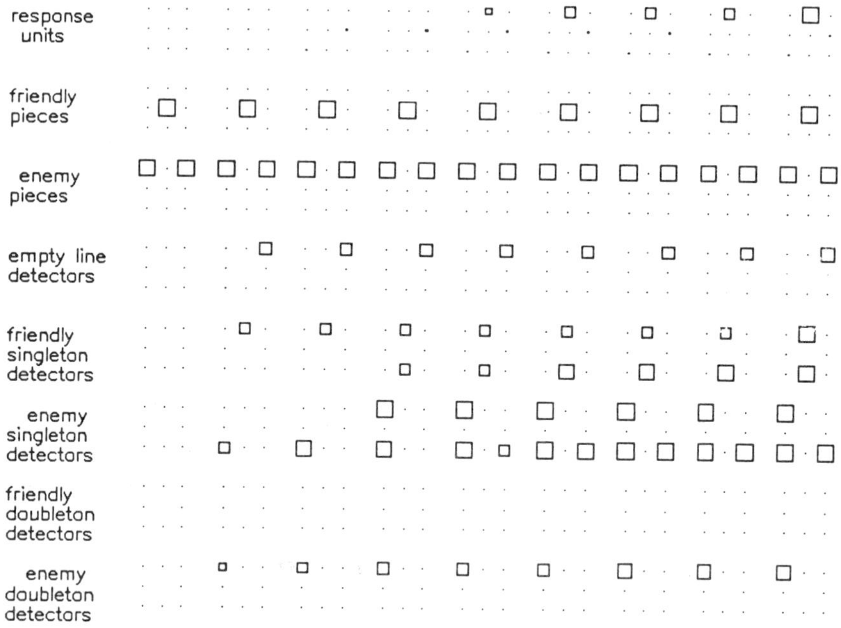

FIGURE 17. A sample run of the tic-tac-toe playing network.

that the opponent has a doubleton in the top row. Then the system discovers a friendly singleton in the middle column, an enemy singleton in the top row, and an enemy singleton in the right column. Based on this, the blocking response in the top-middle position begins to gain strength. As the observation that the opponent has an open double increases in strength, the activation level of the blocking response gains strength even faster. Finally, the strength of a response unit exceeds criterion and the system detects that it has settled on a move. At this point, we assume a motor system (or the like) would be invoked and the move actually made. Upon having made the move, the world has changed and (we can imagine) the opponent makes a response. In this case, a new board position could be presented to the system and the process could begin again.

Here we have a system that can take a board position and decide on a move. When we make our move and present the system with the new board position, it can make its next move, etc., until the game is finished. Here we see a kind of sequential behavior of our relaxation system driven by changes in the outside which are, in turn, the result of the activity of the system itself. Nothing is different; we see that sequential processing results simply from the addition of the systems own responding to the environmental process of generating a sequence of stimuli. This is a very common form of control of sequential behavior and involves no special mechanism. The case of "mental simulations" involves a bit more machinery. We need a "model of the world" to predict the environment's response to any action that might be taken. In the case of tic-tac-toe, this involves a network which models the opponent. In general, the model of the opponent may be arbitrarily different from that used by the system to make its own response. In our case, however, it is sufficient to build an identical network for the opponent. The only difference is that in the opponent's network, the board is interpreted differently: The opponent's board position of the original network setup drives the friendly board position in the model of the opponent, and the friendly board position in the original network drives the opponent board position in the model of the opponent. Figure 18 shows a run of the system "mentally simulating" the play of a game. The state of the units are shown after every 100 updates. First, the state of the original network is shown as it settles on its move, then the state of the model of the opponent is shown as it settles on its move. This continues until the game reaches a conclusion. In this instance, the "simulated player" makes a "mistake" in its response to the system's opening. Successive moves show the system taking advantage of its own mistake and winning the game.

CONCLUSION

We have argued in this book that the analysis of psychological phenomena in terms of units, activations of units, connections among units, and the action of large coalitions of such units leads us to many insights that have not been possible in terms of the language that has recently been more popular in cognitive science. In that sense, we may be perceived as throwing out the insights gained from the more conventional language and concepts. We are not throwing out such insights. In this chapter, we attempt to show the relationship between two such insights and our models. At start, we argue that the concept of the schema has a correspondence in the PDP framework. In particular, we argue that a schema is best viewed as a coalition of units which cohere and that configurations of such coalitions determine the interpretations that the system can attain. These stable states correspond to instantiated configurations of schemata that can be characterized in terms of goodness-of-fit maxima that the system can move into. Such processing systems, we argue, have all of the features of schemata and more. Among the advantages of this view is that the

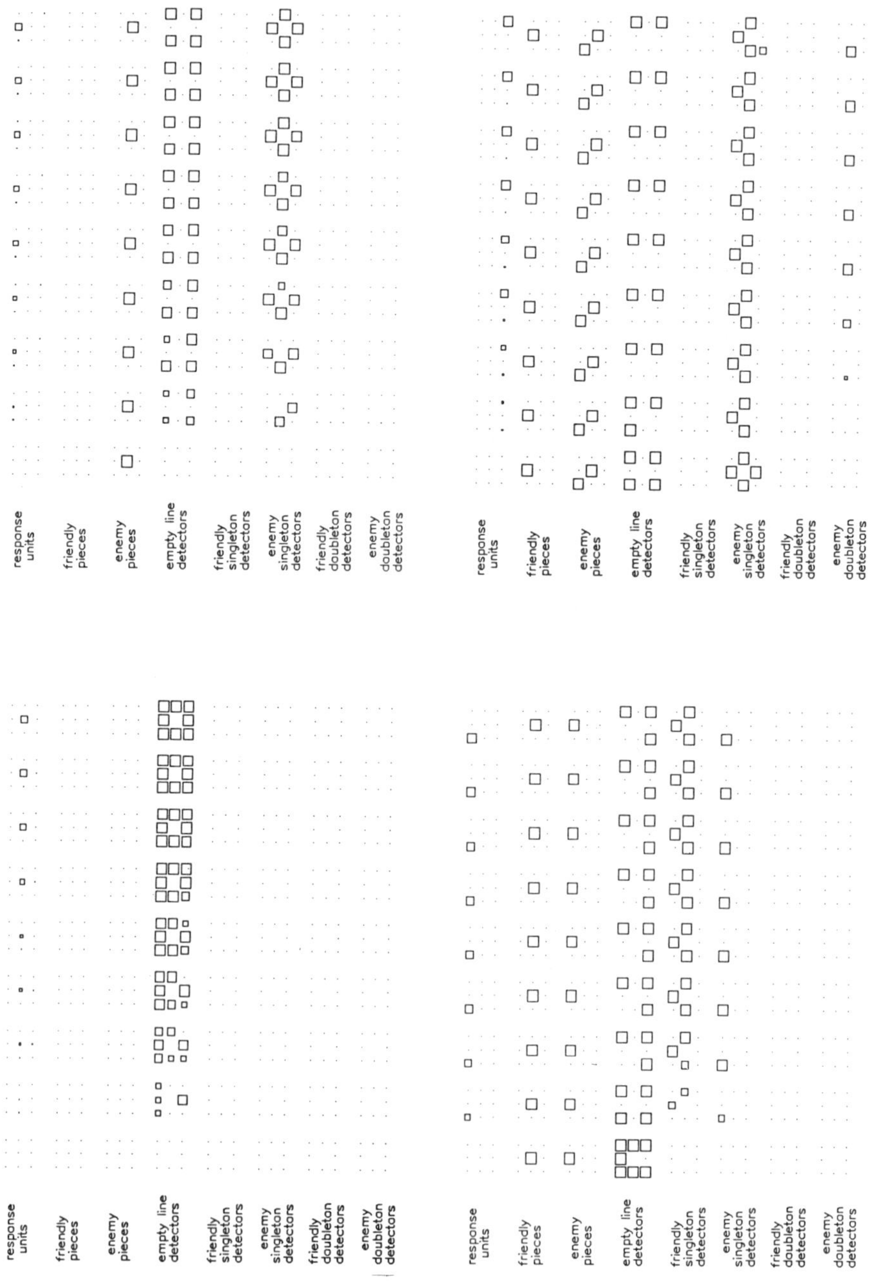

response
units

friendly
pieces

enemy
pieces

empty line
detectors

friendly
singleton
detectors

enemy
singleton
detectors

friendly
doubleton
detectors

enemy
doubleton
detectors

response
units

friendly
pieces

enemy
pieces

empty line
detectors

friendly
singleton
detectors

enemy
singleton
detectors

friendly
doubleton
detectors

enemy
doubleton
detectors

response
units

friendly
pieces

enemy
pieces

empty line
detectors

friendly
singleton
detectors

enemy
singleton
detectors

friendly
doubleton
detectors

enemy
doubleton
detectors

response
units

friendly
pieces

enemy
pieces

empty line
detectors

friendly
singleton
detectors

enemy
singleton
detectors

friendly
doubleton
detectors

enemy
doubleton
detectors

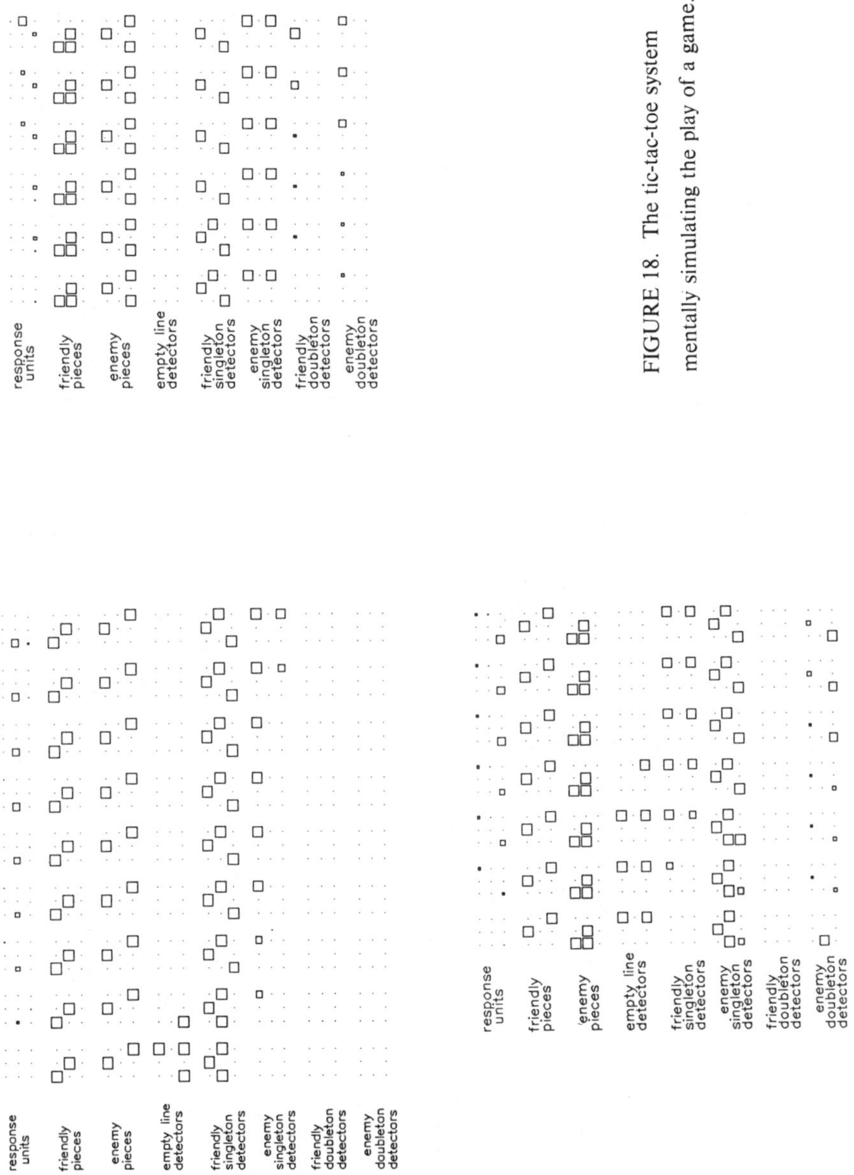

FIGURE 18. The tic-tac-toe system mentally simulating the play of a game.

schema becomes much more fluid and flexible and able to accommodate itself to inputs. In more conventional representations there is a strong distinction between variables and slots and the bulk of the schema. Under our interpretation an aspect of the schema is more or less a variable. Even central aspects of the schema can be missing and the system can still find a reasonably low energy stable state. If very rigid schemata are implicit in the knowledge base, this will show up as narrow peaks in the goodness landscape. If more fluid and variable schemata are required, the landscape will contain broad plateaus which allow for a good deal of movement in the region of the maximum.

We see the relationship between our models and schema theory as discussed by other researchers as largely a matter of levels of analysis. This is roughly analogous to the relationship between the level of discussion of fluid dynamics and an underlying level of description involving statistical mechanics. It is often useful to theorize at the level of turbulence and different kinds of turbulence, and such a description will do for many purposes. However, we can often run up against phenomena in which our high-level descriptions will not do, we must describe the system in terms of the underlying processes in order to understand its behavior. Another feature of this example is our understanding of the phenomena of emergent properties. Turbulence is not predicted by the knowledge of the elements of the system; it is inherent in the interactions among these elements. Similarly, we do not believe that single-unit activity is the appropriate level of analysis. Properties of networks "emerge" from the interactions of the elements. Indeed, such properties as goodness maxima, etc., are emergent in just this way. In general, we see cognitive phenomena as emergent from the interactions of many units. Thus, we take the symbolic level of analysis to provide us with an approximation to the underlying system. In many cases these approximations will prove useful; in some cases they will be wrong and we will be forced to view the system from the level of units to understand them in detail.

We also discussed the relationship between PDP models and the more conventional sequential processing systems. We believe that processes that happen very quickly—say less than .25 to .5 seconds—occur essentially in parallel and should be described in terms of parallel models. Processes that take longer, we believe, have a serial component and can more readily be described in terms of sequential information-processing models. For these processes, a process description such as a production would, we imagine, provide a useful approximate description. We would caution, however, that when one chooses a formalism such as production systems and attempts to use it not only to describe the conscious sequential processes that occur at this slow time scale, it is important not to fall into the trap of assuming that the microstructure of these sequential processes should also be described in the same terms. Production systems have the power of Turing machines and people often attempt to describe phenomena at this faster time scale in terms of the same sequential formalism that seems appropriate for the slower time scale. We believe that it will turn out that this approach is wrong, that the power of the formalism has led us astray. In these cases we suspect that the unit level of analysis will be required.

Finally, we showed how the important notion of mental models and the related notion of mental simulations play important roles in the sequential behavior of a PDP system. We illustrated this point with a system which could use a "model of its opponent" to "mentally simulate" a tic-tac-toe game. We suspect that this process will turn out to be important when we begin to apply our models to temporally extended reasoning and problem-solving tasks.

ACKNOWLEDGMENTS

This research was supported by Contracts N00014-79-C-0323, NR 667-437 and N00014-85-K-0450, NR 667-548 with the Personnel and Training Research Programs of the Office of Naval Research, by grants from the System Development Foundation, and by a NIMH Career Development Award (MH00385) to the third author.

CHAPTER 15

Interactive Processes in
Speech Perception: The TRACE Model

J. L. McCLELLAND and J. L. ELMAN

Consider the perception of the phoneme /g/ in the sentence *She received a valuable gift.* There are a large number of cues in this sentence to the identity of this phoneme. First, there are the acoustic cues to the identity of the /g/ itself. Second, the other phonemes in the same word provide another source of cues, for if we know the rest of the phonemes in this word, there are only a few phonemes that can form a word with them. Third, the semantic and syntactic context further constrain the possible words that might occur, and thus limit still further the possible interpretation of the first phoneme in *gift*.

There is ample evidence that all of these different sources of information are used in recognizing words and the phonemes they contain. Indeed, as R. A. Cole and Rudnicky (1983) have recently noted, these basic facts were described in early experiments by Bagley (1900) over 80 years ago. Cole and Rudnicky point out that recent work (which we consider in detail below) has added clarity and detail to these basic findings but has not lead to a theoretical synthesis that provides a satisfactory account of these and many other basic aspects of speech perception.

In this chapter, we describe a model that grew out of the view that the interactive activation processes that can be implemented in PDP

This chapter is a condensed version of the article "The TRACE Model of Speech Perception" by J. L. McClelland and J. L. Elman, which appeared in *Cognitive Psychology*, 1986, *18*, 1-86. Copyright 1986 by Academic Press, Inc. Adapted with permission.

models provide a natural way to capture the integration of multiple sources of information in speech perception. This view was based on the earlier success of the interactive activation model of word perception (McClelland & Rumelhart, 1981; Rumelhart & McClelland, 1982) in accounting for integration of multiple sources of information in recognizing letters in words.

In attempting to apply the ideas embodied in the interactive activation model of word perception to speech, it soon became apparent that speech provided many challenges. The model we have come up with, the TRACE model, is a response to many of these challenges and demonstrates how they can be met within the PDP framework. After we developed the model, we discovered many aspects of its behavior that are consistent with facts about speech. Thus, we were gratified to discover that the search for a mechanism that was sufficient to meet many of the challenges also lead to a model that provided quite close accounts of a number of basic aspects of the literature on speech perception.

In what follows, we begin by reviewing several facts about speech that played a role in shaping the specific assumptions embodied in TRACE. We then describe the structure of the TRACE model and the salient features of the two versions we have developed to handle different aspects of the simulations. Following this, we describe how the model accounts for a considerable body of psychological data and meets some of the computational challenges facing mechanisms of speech perception. The discussion section considers some reasons for the success of the model, explains its limitations, and indicates how we plan to overcome these in future work.

SOME IMPORTANT FACTS ABOUT SPEECH

Our intention here is not to provide an extensive survey of the nature of speech, but rather to point to several fundamental aspects of speech that have played important roles in the development of the TRACE model. A very useful discussion of several of these points is available in Klatt (1980).

Temporal Nature of the Speech Stimulus

It does not, of course, take a scientist to observe one fundamental characteristic of speech: It is a signal that is extended in time. This differentiates speech perception from most other perceptual applications of PDP models, which have generally been concerned with visual stimuli.

The sequential nature of speech poses problems for the modeling of contextual influences, in that to account for context effects, it is necessary to keep a record of the context. It would be a simple matter to process speech if each successive portion of the speech input were processed independently of all of the others, but, in fact, this is clearly not the case. The presence of context effects in speech perception requires a mechanism that keeps some record of that context, in a form that allows it to influence the interpretation of subsequent input.

Left and Right Context Effects

A further point, and one that has been much neglected in certain models, is that it is not only prior context, but also subsequent context, that influences perception. (This and related points have recently been made by Grosjean & Gee, 1984; Salasoo & Pisoni, 1985; and Thompson, 1984). For example, Ganong (1980) reported that the identification of a syllable-initial speech sound that was constructed to be between /g/ and /k/ was influenced by whether the rest of the syllable was /ɪs/ (as in *kiss*) or /ɪft/ (as in *gift*). Such *right context effects* (Thompson, 1984) indicate that the perception of what comes in now both influences and is influenced by the perception of what comes in later. This fact suggests that the record of what has already been presented cannot be a static representation but should remain in a malleable form, subject to alteration as a result of influences arising from subsequent input.

Lack of Boundaries and Temporal Overlap

A third fundamental point about speech is that the cues to successive units of speech frequently overlap in time. The problem is particularly severe at the phoneme level. A glance at a schematic speech spectrogram (Figure 1) clearly illustrates this problem. There are no separable packets of information in the spectrogram like the separate feature bundles that make up letters in printed words.

Because of the overlap of successive phonemes, it is difficult, and we believe counterproductive, to try to divide the speech stream up into separate phoneme units in advance of identifying the units. A number of other researchers (e.g., Fowler, 1984; Klatt, 1980) have made much

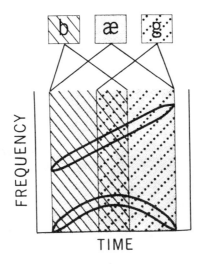

FIGURE 1. A schematic spectrogram for the syllable *bag*, indicating the overlap of the information specifying the different phonemes. (From "The Grammars of Speech and Language" by A. M. Liberman, 1970, *Cognitive Psychology*, *1*, p. 309. Copyright 1970 by Academic Press, Inc. Reprinted by permission.)

the same point. A superior approach seems to be to allow the phoneme identification process to examine the speech stream for characteristic patterns, without first segmenting the steam into separate units.

The problem of overlap is less severe for words than for phonemes, but it does not go away completely. In rapid speech, words run into each other, and there are no pauses between words. To be sure, there are often cues that signal the locations of boundaries between words— stop consonants are generally aspirated at the beginnings of stressed words in English, and word initial vowels are generally preceeded by glottal stops, for example. These cues have been studied by a number of investigators, particularly Lehiste (e.g., Lehiste, 1960, 1964) and Nakatani and collaborators. Nakatani and Dukes (1977) demonstrated that perceivers exploit some of these cues, but found that certain utterances do not provide sufficient cues to word boundaries to permit reliable perception of the intended utterance. Speech errors often involve errors of word segmentation (Bond & Garnes, 1980), and certain segmentation decisions are easily influenced by contextual factors (R. A. Cole & Jakimik, 1980). Thus, it is clear that word recognition cannot count on an accurate segmentation of the phoneme stream into separate word units, and in many cases such a segmentation would perforce exclude from one of the words a shared segment that is doing double duty in each of two successive words.

Context Sensitivity of Cues

A fourth major fact about speech is that the cues for a particular unit vary considerably with the context in which they occur. For example, the transition of the second formant carries a great deal of information about the identity of the stop consonant /b/ in Figure 1, but that formant would look quite different had the syllable been *big* or *bog* instead of *bag*. Thus, the context in which a phoneme occurs restructures the cues to the identity of that phoneme (Liberman, 1970).

Not only are the cues for each phoneme dramatically affected by preceding and following context, they are also altered by more global factors such as rate of speech (J. L. Miller, 1981), by morphological and prosodic factors such as position in the word and in the stress contour of the utterance, and by characteristics of the speaker such as size and shape of the vocal tract, fundamental frequency of the speaking voice, and dialectical variations (see Klatt, 1980, and Repp & Liberman, 1984, for discussions).

A number of different approaches to the problem have been tried by different investigators. One approach is to try to find relatively invariant—generally relational—features (e.g., Stevens & Blumstein, 1981). Another approach has been to redefine the unit so that it encompasses the context, and therefore becomes more invariant (Fujimura & Lovins, 1982; Klatt, 1980; Wickelgren, 1969). While these are both sensible and useful approaches, the first has not yet succeeded in establishing a sufficiently invariant set of cues, and the second may alleviate but does not eliminate the problem: Even units such as demisyllables (Fujimura & Lovins, 1982), context-sensitive allophones (Wickelgren, 1969), or even whole words (Klatt, 1980) are still influenced by context. We have chosen to focus instead on a third possibility: that the perceptual system uses information from the context in which an utterance occurs to alter connections dynamically, thereby effectively allowing the context to retune the perceptual mechanism in the course of processing.

Noise and Indeterminacy in the Speech Signal

To compound all the problems alluded to above, there is the additional fact that speech is often perceived under less than ideal circumstances. While a slow and careful speaker in a quiet room may produce sufficient cues to allow correct perception of all of the phonemes in an utterance without the aid of lexical or other higher-level constraints, these conditions do not always obtain. People can correctly perceive speech under quite impoverished conditions if it is semantically coherent and syntactically well-formed (G. A. Miller, Heise, & Lichten, 1951). This means that the speech mechanisms must be able to function, even with a highly degraded stimulus. In particular, as Grosjean and Gee (1984), Norris (1982), and Thompson (1984) have pointed out, the mechanisms of speech perception cannot count on accurate information about any part of a word. As we shall see, this fact poses a serious problem for one of the best current psychological models of the process of spoken word recognition, the COHORT model of Marslen-Wilson and Welsh (1978).

Many of the characteristics that we have reviewed differentiate speech from print—at least, from very high quality print on white paper—but it would be a mistake to think that similar problems are not encountered in other domains. Certainly, the sequential nature of spoken input sets speech apart from vision, in which there can be some degree of simultaneity of input. However, the problems of ill-defined boundaries, context sensitivity of cues, and noise and indeterminacy are central problems in vision just as much as they are in speech (cf. Ballard, Hinton, & Sejnowski, 1983; Barrow & Tenenbaum, 1978; Marr, 1982). Thus, though the model we present here is focused on speech perception, we would hope that the ways in which it deals with the challenges posed by the speech signal will be applicable in other domains.

The Importance of the Right Architecture

All of the considerations listed above played an important role in the formulation of the TRACE model. The model is an instance of a PDP model, but it is by no means the only instance of such a model that we have considered or that could be considered. Other formulations we considered simply did not appear to offer a satisfactory framework for dealing with these central aspects of speech (see Elman & McClelland, 1984, for discussion). Thus, the TRACE model hinges on the particular processing architecture it proposes for speech perception as well as on the PDP mechanisms that implement the interactive activation processes that occur within it.

Sources of TRACE's architecture. The inspiration for the architecture of TRACE goes back to the HEARSAY speech understanding system (Erman & Lesser, 1980; Reddy, Erman, Fennell, & Neely, 1973). HEARSAY introduced the notion of a Blackboard, a structure similar to the Trace in the TRACE model. The main difference is that the Trace is a dynamic processing structure that is self-updating, while the Blackboard in HEARSAY was a passive data structure through which autonomous processes shared information. The architecture of TRACE also bears a resemblance to the *neural spectrogram* proposed by Crowder (1978; 1981) to account for interference effects between successive items in short-term memory.

THE TRACE MODEL

The TRACE model consists primarily of a very large number of units, organized into three levels, the *feature*, *phoneme*, and *word* levels. Each unit stands for an hypothesis about a particular perceptual object—feature, phoneme, or word—occurring at a particular point in time defined relative to the beginning of the utterance. Thus, the TRACE model uses local representation.

A small subset of the units in TRACE II, the version of the model with which we will be mostly concerned, is illustrated in Figures 2, 3, and 4. Each of the three figures replicates the same set of units, illustrating a different property of the model in each case. In the figures, each rectangle corresponds to a separate processing unit. The labels on the units and along the side indicate the spoken object (feature, phoneme, or word) for which each unit stands. The left and right edges of each rectangle indicate the portion of the input the unit spans.

At the feature level, there are several banks of feature detectors, one for each of several dimensions of speech sounds. Each bank is replicated for each of several successive moments in time, or time slices. At the phoneme level, there are detectors for each of the phonemes. There is one copy of each phoneme detector centered over every three time-slices. Each unit spans six time slices, so units with adjacent centers span overlapping ranges of slices. At the word level, there are detectors for each word. There is one copy of each word detector centered over every three feature slices. Here, each detector spans a stretch of feature slices corresponding to the entire length of the word. Again, then, units with adjacent centers span overlapping ranges of slices.

Input to the model, in the form of a pattern of activation to be applied to the units at the feature level, is presented sequentially to the feature-level units in successive slices, as it would be if it were a real stream of speech. Mock-speech inputs on the three illustrated dimensions for the phrase *tea cup* (/tik^p/) are shown in Figure 2. At any instant, input is arriving only at the units in one slice at the feature

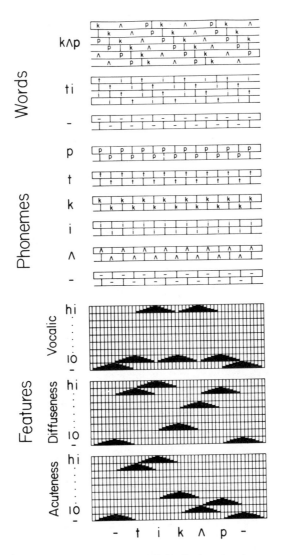

FIGURE 2. A subset of the units in TRACE II. Each rectangle represents a different unit. The labels indicate the item for which the unit stands, and the horizontal edges of the rectangle indicate the portion of the Trace spanned by each unit. The input feature specifications for the phrase *tea cup*, preceded and followed by silence, are indicated for the three illustrated dimensions by the blackening of the corresponding feature units.

level. In terms of the display in Figure 2, then, we can visualize the input being applied to successive slices of the network at successive moments in time. However, it is important to remember that all the units are continually involved in processing, and processing of the input

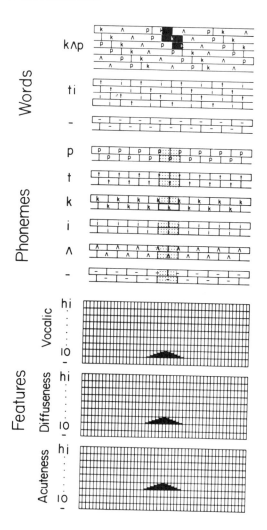

FIGURE 3. The connections of the unit for the phoneme /k/, centered over Time-Slice 24. The rectangle for this unit is highlighted with a bold outline. The /k/ unit has mutually excitatory connections to all the word- and feature-level units colored either partly or wholly in black. The more coloring on a unit's rectangle, the greater the strength of the connection. The /k/ unit has mutually inhibitory connections to all of the phoneme-level units colored partly or wholly in grey. Again, the relative amount of inhibition is indicated by the extent of the coloring of the unit; it is directly proportional to the extent of the temporal overlap of the units.

arriving at one time is just beginning as the input is moved along to the next time slice.

The entire network of units is called *the Trace*, because the pattern of activation left by a spoken input is a trace of the analysis of the input at

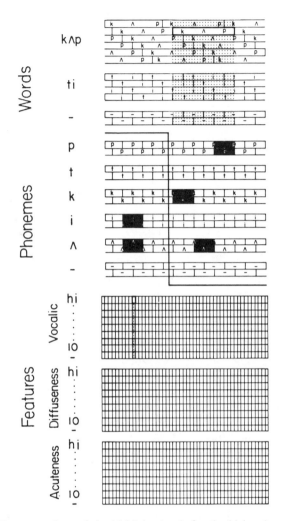

FIGURE 4. The connections of the highlighted unit for the high value on the vocalic feature dimension in Time-Slice 9 and for the highlighted unit for the word /kˆp/ starting in Slice 24. Excitatory connections are represented in black, inhibitory connections in gray, as in Figure 3.

each of the three processing levels. This trace is unlike many traces, though, in that it is active since it consists of activations of processing elements, and these processing elements continue to interact as time goes on. The distinction between perception and (primary) memory is completely blurred since the percept is unfolding in the same structures that serve as working memory, and perceptual processing of older portions of the input continues even as newer portions are coming into the system. These continuing interactions permit the model to incorporate right context effects and allow the model to account directly for certain aspects of short-term memory, such as the fact that more information can be retained for short periods of time if it hangs together to form a coherent whole.

Processing takes place through the excitatory and inhibitory interactions of the units in the Trace. Units on different levels that are mutually consistent have mutually excitatory connections, while units on the same level that are inconsistent have mutually inhibitory connections. All connections are bidirectional. Thus, the unit for the phoneme /k/ centered over Feature-Slice 24 (shown in Figure 3) has bidirectional excitatory connections to feature units that would be activated if the input contained that phoneme centered on Time-Slice 24. It also has bidirectional excitatory connections to all the units at the word level for words containing /k/ at Time-Slice 24. The connections of illustrative feature- and word-level units are shown in Figure 4. Units on the same level are mutually incompatible, and hence mutually inhibitory, to the extent that the input patterns they stand for would overlap with each other in time. That is to say, units on the same level inhibit each other in proportion to the extent of the overlap of their temporal spans, or windows. At the feature level, units stand for the content of only a single time slice, so they only compete with units standing for other values on the same dimension (see Figure 4). At the phoneme and word level, however, there can be different degrees of overlap, and hence of mutual inhibition. The extent of the mutual inhibition between the /k/ in Slice 24 and other phoneme-level units is indicated in Figure 3 by the amount of shading that falls over the rectangle for the other unit. Similarly, the extent of mutual inhibition between the unit for /k^p/ starting in Slice 24 and other word-level units is indicated in Figure 4.

Context-Sensitive Tuning of Phoneme Units

The connections between the feature and phoneme levels determine what pattern of activations over the feature units will most strongly activate the detector for each phoneme. To cope with the fact that the features representing each phoneme vary according to the phonemes surrounding them, the model uses multiplicative connections of the kind proposed by Hinton (1981b) and discussed in Chapters 4 and 16. These multiplicative connections essentially adjust the connections from units at the feature level to units at the phoneme level as a function of activations at the phoneme level in preceding and following time slices.

For example, when the phoneme /t/ is preceded or followed by the vowel /i/, the feature pattern corresponding to the /t/ is very different than it is when the /t/ is preceded or followed by another vowel, such as /a/. Accordingly, when the unit for /i/ in a particular slice is active, it changes the pattern of connections for units for /t/ in preceding and following slices.

TRACE I and TRACE II

In developing TRACE and in trying to test its computational and psychological adequacy, we found that we were sometimes led in rather different directions. We wanted to show that TRACE could process real speech, but to build a model that did so, it was necessary to worry about exactly what features must be extracted from the speech signal, about differences in duration of different features of different phonemes, and about how to cope with the ways in which features and feature durations vary as a function of context. Obviously, these are important problems, worthy of considerable attention. However, concern with these issues tended to obscure attention to the fundamental properties of the model and the model's ability to account for basic aspects of the psychological data obtained in many experiments.

To cope with these conflicting goals, we have developed two different versions of the model, called TRACE I and TRACE II. Both models spring from the same basic assumptions, but focused on different aspects of speech perception. TRACE I was designed to address some of the challenges posed by the task of recognizing phonemes from real speech. This version of the model is described in detail in Elman and McClelland (in press). With this version of the model, we have been able to show that the TRACE framework could indeed be used to process real speech—albeit from a single speaker uttering isolated monosyllables at this point. We have also demonstrated the efficacy of the idea of using multiplicative connections to adjust feature-to-phoneme connections on the basis of activations produced by surrounding context.

The second version of the model, TRACE II, will be the main focus of this chapter. We developed this version of the model to account for lexical influences on phoneme perception and for what is known about on-line recognition of words, though we will use it to illustrate how certain other aspects of phoneme perception fall out of the TRACE framework. This version of the model is actually a simplified version of TRACE I. Most importantly, we eliminated the connection-strength adjustment facility, and we replaced the real speech inputs to TRACE I with mock speech. This mock-speech input consisted of overlapping but contextually invariant specifications of the features of successive phonemes. Thus, TRACE II sidesteps many of the issues addressed by TRACE I, but it makes it much easier to see how the mechanism can account for a number of aspects of phoneme and word recognition. A number of further simplifying assumptions were made to facilitate examination of basic properties of the interactive activation processes taking place within the model.

Implementation Details

The material in this section is included for completeness, but the basic line of development may be followed without reading it. Readers uninterested in these details may wish to skip to the section on factors influencing phoneme identification.

Units and their dynamics. The dynamic properties of the units in TRACE are the same as those used in the interactive activation model of visual word perception; these are described in detail in Chapter 2. In brief, the model is a synchronous model, in that all the units update their activation at the same time, based on the activations computed in the previous update cycle. Each unit takes a sum of the excitatory and inhibitory influences impinging on it. Each influence is essentially the product of the output of the influencing unit and the weight on the connection between it and the receiver. If this net input is positive, it drives the activation of the unit upward in proportion to the distance left to the fixed maximum activation level; if the net input is negative, it drives the activation of the unit down in proportion to the distance left to the fixed minimum. Activations also tend to decay back to their resting activation level, which was fixed at 0 for all units. The output of a unit is 0 if the activation is less than or equal to 0; otherwise it is equal to its activation.

TRACE I. The inputs to TRACE I are sets of 15 parameter values extracted at 5 msec intervals from syllables spoken by a male native speaker of English. The bulk of the TRACE I simulations have been done with a set of CV syllables consisting of an unvoiced stop consonant ($/p/$, $/t/$, or $/k/$) followed by one of the vowels $/a/$, $/i/$, and $/u/$, as in the words *shah*, *tea*, and *who*. At the feature level, TRACE I consists of detectors for each of eight different value ranges on each of the 15 input parameters. There is a complete set of detectors for each 5 msec time slice of the input. Since there are 100 slices, the model is capable of processing 500 msec samples of speech.

There are no word-level units in TRACE I. However, there are phoneme-level units for each successive 15 msec time slice of the speech. The connections from the feature to the phoneme units were determined by using the perceptron convergence procedure (see Chapter 2) under two different conditions. First, in the invariant connections condition, a single set of connection strengths was found for each phoneme, using tokens of the phoneme spoken in all different contexts. In the context-sensitive connections condition, separate sets of connection strengths were found for each stop consonant in the context of each of the vowels.

TRACE I can be tested either using the invariant connections or using the multiplicative context-sensitive connections described above. In the latter case, the weights coming into a particular phoneme are weighted according to the relative activation of other phonemes in the surrounding context. Consider an arbitrary phoneme unit which we will designate, for now, the target unit. The strengths of the connections coming into this unit can be designated by the vector $\mathbf{w}$, where the elements of the vector are just the individual weights from each feature unit to the phoneme unit. This vector is the average over all context phonemes k of the context-specific weight vectors appropriate for the target phoneme in the context of k, where the contribution of each of these context-specific weight vectors is proportional to the exponential of the activation of phoneme k summed over the time slices adjacent to the target phoneme unit (see Elman & McClelland, 1986, in press, for further details).

TRACE II. Inputs to TRACE II are not real speech, but *mock speech* of the kind illustrated in Figure 2. The mock speech is a series of specifications for inputs to units at the feature level, one for each 25 msec time slice of the mock utterance. These specifications are generated by a simple computer program from a sequence of to-be-presented segments provided by the human user of the simulation program. The allowed segments consist of the stop consonants /b/, /p/, /d/, /t/, /g/, and /k/; the fricatives /s/ and /S/ (*sh* as in *ship*); the liquids /l/ and /r/; and the vowels /a/ (as in *pot*), /i/ (as in *beet*), /u/ (as in *boot*), and /ˆ/ (as in *but*). /ˆ/ is also used to represent reduced vowels such as the second vowel in *target*. There is also a "silence" segment represented by /-/. Special segments, such as a segment halfway between /b/ and /p/, can be constructed as well.

A set of seven dimensions is used in TRACE II to represent the feature-level inputs. Of course, these dimensions are intentional simplifications of the real acoustic structure of speech, in much the same way that the font used by McClelland and Rumelhart (1981) in the interactive activation model of visual word recognition was an intentional simplification of the real structure of print. Each dimension is divided into eight value ranges. Each phoneme has a value on each dimension; the values on the vocalic, diffuseness, and acuteness dimensions for the phonemes in the utterance /tik^p/ are shown in Figure 2. The dimensions and the values assigned to each phoneme on each dimension are indicated in Table 1. Numbers in the cells of the table indicate which value on the indicated dimension is most strongly activated by the feature pattern for the indicated phoneme. Values range from 1 (very low) to 8 (very high). The last two dimensions were altered for the categorical perception and trading relations simulations, as described below.

Values are assigned to approximate the values real phonemes would have on these dimensions and to make phonemes that fall into the same phonetic category have identical values on many of the dimensions. Thus, for example, all stop consonants are assigned the same values on the power, vocalic, and consonantal dimensions. We do not claim to have captured the details of phoneme similarity exactly. Indeed, one cannot do so in a fixed feature set because the similarities vary as a function of context. However, the feature sets do have the property that the feature pattern for one phoneme is more similar to the feature pattern for other phonemes in the same phonetic category (stop, fricative, liquid, or vowel) than it is to the patterns for phonemes

TABLE 1

PHONEME FEATURE VALUES USED IN TRACE II

PHONEME	POW	VOC	DIF	ACU	CON	VOI	BUR
p	4	1	7	2	8	1	8
b	4	1	7	2	8	7	7
t	4	1	7	7	8	1	6
d	4	1	7	7	8	7	5
k	4	1	2	3	8	1	4
g	4	1	2	3	8	7	3
s	6	4	7	8	5	1	-
S	6	4	6	4	5	1	-
r	7	7	1	2	3	8	-
l	7	7	2	4	3	8	-
a	8	8	2	1	1	8	-
i	8	8	8	8	1	8	-
u	8	8	6	2	1	8	-
^	7	8	5	1	1	8	-

POW = power, VOC = vocalicness, DIF = diffuseness, ACU = acuteness, CON = consonantal, VOI = voicing, BUR = burst amplitude. Only the stops have values on this last dimension.

in other categories. Among the stops, those phonemes sharing place of articulation or voicing are more similar than those sharing neither attribute.

The feature specification of each phoneme in the input stream extends over 11 time slices of the input. The strength of the pattern grows to a peak at the sixth slice and falls off again, as illustrated in Figure 2. Peaks of successive phonemes are separated by six slices. Thus, specifications of successive phonemes overlap, as they do in real speech (Fowler, 1984; Liberman, 1970).

Generally, there are no cues in the speech stream to word boundaries—the feature specification for the last phoneme of one word overlap with the first phoneme of the next in just the same way feature specifications of adjacent phonemes overlap within words. However, entire utterances presented to the model for processing—be they individual syllables, words, or strings of words—are preceded and followed by silence. Silence is not simply the absence of any input; rather, it is a pattern of feature values, just like the phonemes. Thus, a ninth value on each of the seven dimensions is associated with silence. These values are actually outside the range of values that occurred in the phonemes themselves so that the features of silence are completely uncorrelated with the features of any of the phonemes.

TRACE II contains a unit for each of the nine values on each of the seven dimensions, in each time slice of the Trace. At the phoneme level, each Trace contains a detector for each of the 15 phonemes and a detector for the presence of silence. The silence detectors are treated like all other phoneme detectors. Each member of the set of detectors for a particular phoneme is centered over a different time-slice at the feature level, and the centers are spaced three time-slices apart. The unit centered over a particular slice receives excitatory input from feature units in a range of 11 slices, extending both forward and backward from the slice in which the phoneme unit is located. It also sends excitatory feedback down to the same feature units in the same range of slices.

The connection strengths between the feature-level units and a particular phoneme-level unit exactly match the feature pattern the phoneme is given in its input specification. Thus, as illustrated in Figure 3, the strengths of the connections between the unit for /k/ centered over Time-Slice 24 and the units at the feature level are exactly proportional to the pattern of input to the feature level produced by an input specification containing the features of /k/ centered in the same time slice.

TRACE II also contains detectors for the 211 words found in a computerized phonetic word list that met all of following criteria: (a) The word consisted only of phonemes in the list above; (b) it was not an inflection of some other word that could be made by adding *ed*, *s*, or *ing*; and (c) the word together with its *ed*, *s*, and *ing* inflections occurred with a frequency of 20 or more per million in the Kucera and Francis (1967) word count. It is not claimed that the model's lexicon is an exhaustive list of words meeting these criteria since the computerized phonetic lexicon was not complete, but it is reasonably close to this. To make specific points about the behavior of the model, detectors for the following three words not in the main list were added: *blush*, *regal*, and *sleet*. The model also has detectors at the word level for silence (/-/), which is treated like a one-phoneme word.

Presentation and processing of an utterance. Before processing of an utterance begins, the activations of all of the units are set at their resting values. At the start of processing, the input to the initial slice of feature units is applied. Activations are then updated, ending the initial time cycle. On the next time cycle, the input to the next slice of feature units is applied, and excitatory and inhibitory inputs to each unit resulting from the pattern of activation left at the end of the previous time slice are computed.

It is important to remember that the input is applied, one slice at a time, proceeding from left to right as though it were an ongoing stream of speech "writing on" the successive time slices of the Trace. The interactive activation process is occurring throughout the Trace on each time slice, even though the external input is only coming in to the feature units one slice at a time. Processing interactions can continue even after the left to right sweep through the input reaches the end of the Trace. Once this happens, there are simply no new input specifications applied to the Trace; the continuing interactions are based on what has already been presented. This interaction process is assumed to continue indefinitely, though for practical purposes it is always terminated after some predetermined number of time cycles has elapsed.

Activations and overt responses. Activations of units in the Trace rise and fall as the input sweeps across the feature level. At any time, a decision can be made based on the pattern of activation as it stands at that moment. The decision mechanism can, we assume, be directed to consider the set of units located within a small window of adjacent slices within any level. The units in this set then constitute the set of response alternatives, designated by the identity of the item for which the unit stands (note that with several adjacent slices included in the set, several units in the alternative set may correspond to the same overt response). Word-identification responses are assumed to be based on readout from the word level, and phoneme-identification responses are assumed to be based on readout from the phoneme level.

As far as phoneme identification is concerned, then, a homogeneous mechanism is assumed to be used with both word and nonword stimuli. The decision mechanism can be asked to make a response either (a) at a critical time during processing, or (b) when a unit in the alternative set reaches a critical strength relative to the activation of other alternative units. Once a decision has been made to make a response, one of the alternatives is chosen from the members of the set. The probability of choosing a particular alternative i is then given by the Luce (1959) choice rule:

$$p\left(R_{i}\right) = \frac{S_{i}}{\sum\limits_{j} S_{j}}$$

where j indexes the members of the alternative set, and $S_{i} = e^{ka_{i}}$. The exponential transformation ensures that all activations are positive and gives great weight to stronger activations; the Luce rule ensures that the sum of all of the response probabilities adds up to 1.0. Substantially the same assumptions were used by McClelland and Rumelhart (1981).

Parameters. At the expense of considerable realism, we have tried to keep both TRACE I and TRACE II simple by using homogeneous parameters wherever possible. The strength of the total excitation coming into a particular phoneme unit from the feature units is normalized to the same value for all phonemes, thus making each phoneme equally excitable by its own canonical pattern. Other simplifying assumptions should be noted as well. For example, there are no differences in connections or resting levels for words of different frequency. It would have been a simple matter to incorporate frequency as McClelland and Rumelhart (1981) did, and a complete model would, of course, include some account for the ubiquitous effects of word frequency. We left it out here to facilitate an examination of the many other factors that appear to influence the process of word recognition in speech perception.

Even with all the simplifications described above, TRACE II still has 10 free parameters; these are listed in Table 2. There was some trial and error in finding the set of parameters used in the reported simulations, but, in general, the qualitative behavior of the model is remarkably robust under parameter variations, and no systematic search of the space of parameters is necessary.

In all the reported simulations using TRACE II, the parameters were held at the values given in Table 2. The only exception to this occurred in the simulations of categorical perception and trading relations. Since we were not explicitly concerned with the effects of feedback to the feature level in any of the other simulations, we set the

TABLE 2

PARAMETERS OF TRACE II

Parameter	Value
Feature-Phoneme Excitation	.02
Phoneme-Word Excitation	.05
Word-Phoneme Excitation	.03
Phoneme-Feature Excitation	.00
Feature-Level Inhibition	.04
Phoneme-Level Inhibition*	.04
Word-Level Inhibition*	.03
Feature-Level Decay	.01
Phoneme-Level Decay	.03
Word-Level Decay	.05

*Per 3 time slices of overlap.

feedback from the phoneme level to the feature level to zero to speed up the simulations in all other cases. In the categorical perception and trading relations simulations this parameter was set at 0.05. Phoneme-to-feature feedback tended to slow the effective rate of decay at the feature level and to increase the effective distinctiveness of different feature patterns. Rate of decay of feature level activations and strength of phoneme-to-phoneme competition were set to 0.03 and 0.05 to compensate for these effects. No lexicon was used in the categorical perception and trading relations simulations, which is equivalent to setting the phoneme-to-word excitation parameter to zero. In TRACE I, the parameters were tuned separately to compensate for the finer time scale of that version of the model.

FACTORS INFLUENCING PHONEME IDENTIFICATION

We are ready to examine the performance of TRACE, to see how well it can account for psychological data on the process of speech perception and, to determine how well it can cope with the computational challenges posed by speech. In this section we consider the process of phoneme identification. In the next section we examine several aspects of word recognition. The sections may be read independently, in either order.

In the introduction, we motivated the approach taken in the TRACE model in general terms. In this section, we will see that the simple concepts that lead to TRACE provide the basis for a coherent and synthetic account of a large number of different kinds of findings on the perception of phonemes. Previous models have been able to provide fairly accurate accounts of a number of these phenomena. For example, Massaro and Oden's feature integration model (Massaro, 1981; Massaro & Oden, 1980a, 1980b; Oden & Massaro, 1978) accounts in detail for a large body of data on the influences of multiple cues to phoneme identity, and the Pisoni/Fujisaki-Kawashima model of categorical perception (Fujisaki & Kawashima, 1968; Pisoni, 1973, 1975) accounts for a large body of data on the conditions under which subjects can discriminate sounds within the same phonetic category. Marslen-Wilson's COHORT model (Marslen-Wilson & Welsh, 1978) can account for the time course of certain aspects of lexical influences on phoneme identification. Recently Fowler (1984) has proposed an interesting account of the way listeners cope with coarticulatory influences on the acoustic parameters of speech sounds. Here we will show that TRACE brings these phenomena, and several others not considered by any of these other models, together into a coherent picture of the process of phoneme perception as it unfolds in time.

This section consists of four main parts. The first focuses on lexical effects on phoneme identification and the conditions under which these effects are obtained. The second part of this section focuses on the question of the role of phonotactic rules—that is, rules specifying which phonemes can occur together in English—in phoneme identification. Here, we see how TRACE mimics the apparently rule-governed behavior of human subjects, in terms of a "conspiracy" of the lexical items that instantiate the rule. The third part focuses on two aspects of phoneme identification often considered quite separately from lexical effects—namely, the contrasting phenomena of cue tradeoffs in phoneme perception and categorical perception. The simulations in the first three parts were all done using TRACE II. The fourth part describes our simulations with TRACE I, illustrating how the connection-modulation mechanisms embedded in that version of the model account for the fact that listeners appear to alter the cues they use to identify phonemes in different contexts.

Lexical Effects

You can tell a phoneme by the company that it keeps.[1] In this section, we describe a simple simulation of the basic lexical effect on

---

[1] This title is adapted from the title of a talk by David E. Rumelhart on related phenomena in letter perception. These findings are described in Rumelhart and McClelland (1982).

phoneme identification reported by Ganong (1980). We start with this phenomenon because it, and the related phonemic restoration effect, were among the primary reasons why we felt that the interactive activation mechanisms provided by PDP models would be appropriate for speech perception, as well as visual word recognition and reading.

For the first simulation, the input to the model consisted of a feature specification which activated /b/ and /p/ equally, followed by (and partially overlapping with) the feature specifications for /l/, then /ˆ/, then /g/. Figure 5 shows phoneme- and word-level activations at several points in the unfolding of this input specification. Each panel of the figure represents a different point in time during the presentation and concomitant processing of the input. The upper portion of each panel is used to display activations at the word level; the lower panel is used for activations at the phoneme level. Each unit is represented by a rectangle labeled with the identity of the item the unit stands for. The horizontal extension of the rectangle indicates the portion of the input spanned by the unit. The vertical position of the rectangle indicates the degree of activation of the unit. In this and subsequent figures, activations of the phoneme units located between the peaks of the input specifications of the phonemes (at Slices 3, 9, 15, etc.) have been deleted from the display for clarity. The input itself is indicated below each panel, with the successive phonemes positioned at the temporal positions of the centers of their input specifications. The "ˆ" along the x-axis represents the point in the presentation of the input stream at which the snapshot was taken.

The figure illustrates the gradual build-up of activation of the two interpretations of the first phoneme, followed by gradual build-ups in activation for subsequent phonemes. As these processes unfold, they begin to produce word-level activations. It is difficult to resolve any word-level activations in the first few frames, however, since in these frames, the information at the phoneme level simply has not evolved to the point where it provides enough constraint to select any one particular word. It is only after the /g/ has come in that the model has information telling it whether the input is closer to *plug*, *plus*, *blush*, or *blood* (TRACE's lexicon contains no other words beginning with /plˆ/ or /blˆ/). After that point, as illustrated in the fourth panel, *plug* wins the competition at the word level, and through feedback support to /p/, causes /p/ to dominate /b/ at the phoneme level. The model, then, provides an explicit account for the way in which lexical information can influence phoneme identification.

Factors influencing the lexical effect. There is now a reasonable body of literature on lexical effects on phoneme identification. One important property of this literature is the fact that the lexical effect is

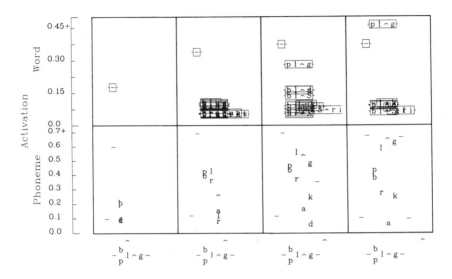

FIGURE 5. Phoneme- and word-level activations at several points in the unfolding of a segment ambiguous between /b/ and /p/, followed by /l/, /ˆ/, and /g/. See text for a full explanation.

often somewhat difficult to obtain. For example, Fox (1982, 1984) found that the lexical effect can be eliminated by time pressure. Ganong reported that the lexical effect only shows up with segments that are ambiguous; we know that in running speech, people often perceived as correctly pronounced words with deliberate errors (Marslen-Wilson & Welsh, 1978), but at the beginnings of isolated words lexical influences appear to lead to misperceptions of unambiguous tokens of phonemes. In reaction time studies, it has been observed by Foss and Blank (1980) that there is no lexical effect on the reaction time to detect word-initial phonemes.

Many of these findings have been taken as evidence against the view that top-down influences really play a role in normal perceptual processing (Foss & Gernsbacher, 1983), and only come into play in a post-perceptual stage of processing (Fox, 1982). However, we observe the same results in simulations with TRACE, where top-down influences are always at work. The reason why lexical effects do not emerge until late in processing for word-initial targets is simply that the contextual information is not available until then. The reason why lexical effects do not emerge with word-initial targets that are not ambiguous is simply that the bottom-up information is there to identify the target, long before the contextual information would be available. Simulations demonstrating the absence of lexical effects for word-initial segments under speeded conditions or when the segment is unambiguous are described in McClelland and Elman (in press).

The crucial observations concern what happens with lexical effects on word-final segments. It is well known that lexical effects are larger later in words than they are at the beginnings of words (Marslen-Wilson & Welsh, 1978) and can be obtained in reaction time studies even with unambiguous segments (Marslen-Wilson, 1980).

TRACE produces stronger lexical effects when the target comes late in the word, simply because the context is already providing top-down support for the target when it starts to come in under these circumstances. We illustrate by comparing response strength for the phoneme /t/ in /sikr^t/ (the word *secret*) and in the nonword /g^ld^t/ (*guldut*) in Figure 6. The figure shows the strength of the /t/ response as a function of processing cycles, relative to all other responses based on activations of phoneme units centered at Cycle 42, the peak of the input specification for the /t/. Clearly, response strength grows faster for the /t/ in /sikr^t/ than for the /t/ in /g^ld^t/; picking an arbitrary threshold of 0.9 for response initiation, we find that the /t/ in /sikr^t/ reaches criterion about 3 cycles or 75 msec sooner than the /t/ in /g^ld^t/. The size of the effect Marslen-Wilson (1980) obtains is quite comparable to the effect observed in Figure 6.

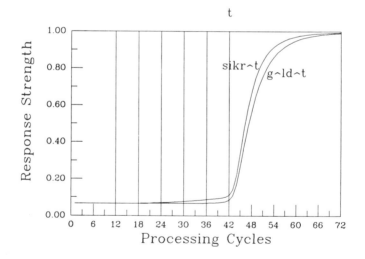

FIGURE 6.   Probability of the /t/ response as a function of processing cycles, based on activation of phoneme units at Cycle 42, for the stream /sikr^t/ (*secret*) and /g^ld^t/ (*guldut*). Vertical lines indicate the peaks of the input patterns corresponding to the successive phonemes in either stream.

Are Phonotactic Rule Effects the Result of a Conspiracy?

Recently, Massaro and M. M. Cohen (1983) have reported evidence they take as support for the use of phonotactic rules in phoneme identification. In one experiment, Massaro and Cohen's stimuli consisted of phonological segments ambiguous between /r/ and /l/ in different contexts. In one context (/t_i/), /r/ is permissible in English, but /l/ is not. In another context (/s_i/), /l/ is permissible in English but /r/ is not. In a third context (/f_i/), both are permissible, and in a fourth (/v_i/), neither is permissible. Massaro and Cohen found a bias to perceive ambiguous segments as /r/ when /r/ was permissible, or as /l/ when /l/ was permissible. No bias appeared in either of the other two conditions.

With most of these stimuli, phonotactic acceptability is confounded with the actual lexical status of the item; thus /fli/ and /fri/ (*flee* and *free*) are both words, as is /tri/ but not /tli/. In the /s_i/ context, however, neither /sli/ or /sri/ are words, yet Massaro and Cohen found a bias to hear the ambiguous segment as /l/, in accordance with phonotactic rules.

It turns out that TRACE produces the same effect, even though it lacks phonotactic rules. The reason is that the ambiguous stimulus produces partial activations of a number of words (*sleep* and *sleet* in the model's lexicon; it would also activate *sleeve*, *sleek*, and others in a model with a fuller lexicon). None of these word units gets as active as it would if the entire word had been presented. However, all of them (in the simulation, there are only two, but the principle still applies) are partially activated, and all conspire together and contribute to the activation of /l/. This feedback support for the /l/ allows it to dominate the /r/, just as it would if /sli/ were an actual word, as shown in Figure 7.

The hypothesis that phonotactic rule effects are really based on word activations leads to a prediction: We should be able to reverse these effects if we present items that are supported strongly by one or more lexical items even if they violate phonotactic rules. A recent experiment by Elman (1983) confirms this prediction. In this experiment, ambiguous phonemes (for example, halfway between /b/ and /d/) were presented in three different types of contexts. In all three types, one of the two (in this case, the /d/) was phonotactically acceptable, while the other (the /b/) was not. However, the contexts differed in their relation to words. In one case, the legal item actually occurred in a word (*bwindle/dwindle*). In a second case, neither item made a word, but the illegal item was very close to a word (*bwacelet/dwacelet*). In a third case, neither item was particularly close to a word (*bwiffle/dwiffle*).

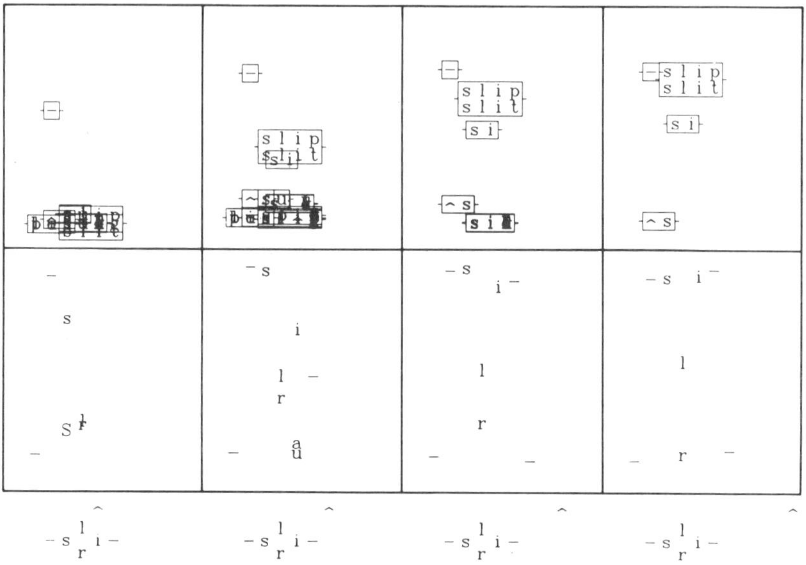

FIGURE 7. State of the Trace at several points in processing a segment ambiguous between /l/ and /r/ in the context /s_i/. The units for *sleep* (/slip/) and *sleet* (/slit/) are boxed together since they take on identical activation values.

Results of the experiment are shown in Table 3. The existence of a word identical to one of the two alternatives or differing from one of the alternatives by a single phonetic feature of one phoneme strongly influenced the subjects' choices between the two alternatives. Indeed, in the case where the phonotactically irregular alternative (*bwacelet*) was one feature away from a particular lexical item (*bracelet*), subjects tended to hear the ambiguous item in accord with the similar lexical item (that is, as a /b/) even though it was phonotactically incorrect.

TABLE 3

PERCENT CHOICE OF PHONOTACTICALLY IRREGULAR CONSONANT

Stimulus Type	Example	Percentage of Identifications as "Illegal" Phoneme (/b/)*
Legal word/illegal nonword	dwindle/bwindle	37%
Legal nonword/illegal nonword	dwiffle/bwiffle	46%
Legal nonword/illegal near-word	dwacelet/bwacelet	55%

*$F_{(2,34)} = 26.414$, $p < .001$

To determine whether the model would also produce such a reversal of the phonotactic rule effects with the appropriate kinds of stimuli, we ran a simulation using a simulated input ambiguous between /p/ and /t/ in the context /_luli/. /p/ is phonotactically acceptable in this context, but /t/ in this context makes an item that is very close to the word *truly*. The results of this run, at two different points during processing, are shown in Figure 8. Early on in processing, there is a slight bias in favor of the /p/ over the /t/ because at first a large number of /pl/ words are slightly more activated than any words beginning with /t/. Later, though, the /t/ gets the upper hand as the word *truly* comes to dominate at the word level. Thus, by the end of the word or shortly thereafter, the closest word has begun to play a dominating role, causing the model to prefer the phonotactically inappropriate interpretation of the ambiguous initial segment.

Of course, at the same time the word *truly* tends to support /r/ rather than /l/ for the second segment. Thus, even though this segment is not ambiguous and the /l/ would suppress the /r/ interpretation in a more neutral context, the /r/ stays quite active.

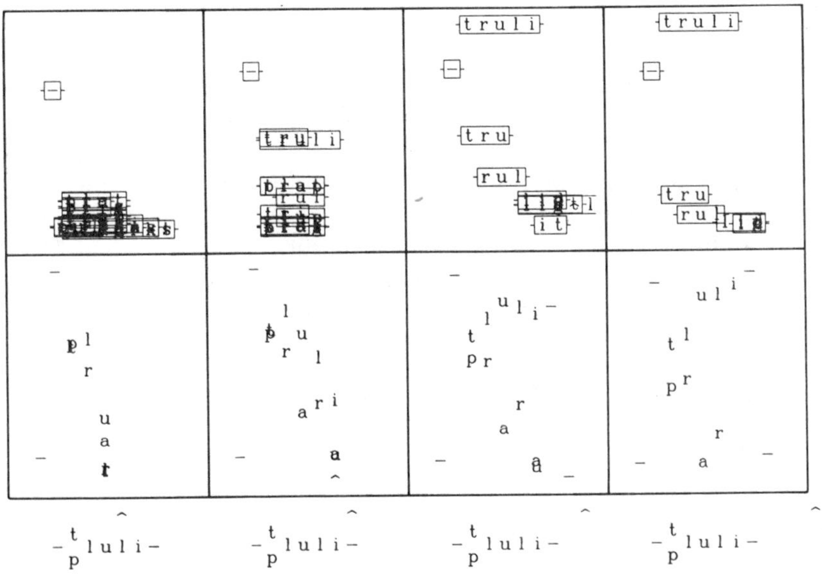

FIGURE 8. State of the Trace at several points in processing an ambiguous /p/-/t/ segment followed by /luli/.

Trading Relations and Categorical Perception

In the simulations considered thus far, phoneme identification is influenced by two different kinds of factors: featural and lexical. When one sort of information is lacking, the other can compensate for it. The image that emerges from these kinds of findings is of a system that exhibits great flexibility by being able to base identification decisions on different sources of information. It is, of course, well established that within the featural domain each phoneme is generally signaled by a number of different cues and that human subjects can trade these cues off against each other. The TRACE model exhibits this same flexibility, as we shall detail shortly.

But there is something of a paradox. While the perceptual mechanisms exhibit great flexibility in the cues that they rely on for phoneme identification, they also appear to be quite "categorical" in nature. That is, they produce much sharper boundaries between phonetic categories than we might expect based on their sensitivity to multiple cues; and they appear to treat acoustically distinct feature patterns as perceptually equivalent, as long as they are identified as instances of the same phoneme.

In this section, we illustrate that in TRACE, just as in human speech perception, flexibility in feature interpretation coexists with a strong tendency toward categorical perception.

For these simulations, the model was stripped down to the essential minimum necessary so that the basic mechanisms producing cue trade-offs and categorical perception could be brought to the fore. The word level was eliminated altogether, and at the phoneme level there were only three phonemes, /a/, /g/, and /k/, plus silence (/-/). From these four items, inputs and percepts of the form /-ga-/ and /-ka-/ could be constructed. The following additional constraints were imposed on the feature specifications of each of the phonemes: (a) the /a/ and /-/ had no featural overlap with either /g/ or /k/ so that neither /a/ nor /-/ would bias the activations of the /g/ and /k/ phoneme units where they overlapped with the consonant in time; (b) /g/ and /k/ were identical on five of the seven dimensions and differed only on the remaining two dimensions.

The two dimensions which differentiated /g/ and /k/ were voice onset time (VOT) and the onset frequency of the first formant (F1OF). These dimensions replaced the voicing and burst amplitude dimensions used in all of the other simulations. Figure 9 illustrates how F1OF tends to increase as voice onset time is delayed.

Trading relations. TRACE quite naturally tends to produce trading relations between features since it relies on the weighted sum of the

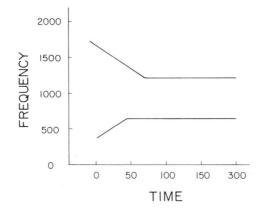

FIGURE 9. Schematic diagram of a syllable that will be heard as /ga/ or /ka/, depend-ing on the point is the syllable at which voicing begins. Before the start of voicing, F2 (top curve) and F3 are energized by aperiodic noise sources, and F1 is "cut back" (the noise source has little or no energy in this range). Because of the fact that F1 rises over time after syllable onset (as the vocal tract moves from a shape consistent with the con-sonant into a shape consistent with the vowel), its frequency at the onset of voicing is higher for later values of VOT. Parameters used in constructing this schematic syllable are derived from Kewley-Port (1982).

excitatory inputs to determine how strongly the input will activate a particular phoneme unit. All else being equal, the phoneme unit receiving the largest sum bottom-up excitation will be more strongly activated than any other and will therefore be the most likely response when a choice must be made between one phoneme and another. Since the net bottom-up input is just the sum of all of the inputs, no one input is necessarily decisive in this regard.

Generally, experiments demonstrating trading relations between two or more cues manipulate each of the cues over a number of values ranging between a value more typical of one of two phonemes and a value more typical of the other. Summerfield and Haggard (1977) did this for VOT and F1OF and found the typical result, namely, that the value of one cue that gives rise to 50% choices of /k/ was affected by the value of the other cue: The higher the value of F1OF, the shorter the value of VOT needed for 50% choices of /k/. Unfortunately, they did not present full curves relating phoneme identification to the values used on each of the two dimensions. In lieu of this, we present curves in Figure 10 from a classic trading relations experiment by Denes (1955). Similar patterns of results have been reported in other studies, using other cues (e.g., Massaro, 1981), though the transitions are often somewhat steeper.

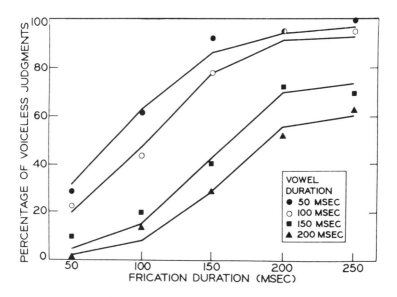

FIGURE 10. Results of an experiment demonstrating the trade-off between two cues to the identity of /s/ and /z/. Data from Denes, 1955, reprinted and fitted by the model of Massaro and Cohen. (From "The Contribution of Voice-Onset Time and Fundamental Frequency as Cues to the /zi/-/si/ Distinction" by D. W. Massaro and M. M. Cohen, 1977, *Perception & Psychophysics, 22*, p. 374. Copyright 1977 by the Psychonomic Society. Reprinted by permission.)

To demonstrate that TRACE would simulate the basic tradeoff effect, we generated a set of 25 intermediate phonetic segments made up by pairing each of five different intermediate patterns on the VOT dimension with each of five different intermediate patterns on the F1OF dimension. The different feature patterns used on each dimension are shown in Figure 11, along with the canonical feature patterns for /g/ and /k/ on each of the two dimensions. On the remaining five dimensions, the intermediate segments all had the common canonical feature values for /g/ and /k/.

The model was tested with each of the 25 stimuli, preceded by silence (/-/) and followed by /a-/. The peak on the intermediate phonetic segment occurred at Slice 12, the peak of the following vowel occurred at Slice 18, and the peak of the final silence occurred at Slice 24. For each input presented, the interactive activation process was allowed to continue through a total of 60 time slices, well past the end of the input. At the end of the 60th time slice, we recorded the activation of the units for /g/ and /k/ in Time-Slice 12 and the probability of choosing /g/ based on these activations. It makes no difference to the general pattern of the results if a different decision time is used.

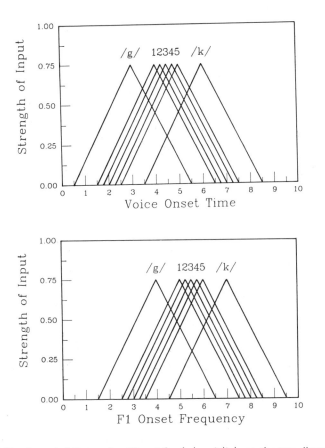

FIGURE 11. Canonical feature level input for /g/ and /k/, on the two dimensions that distinguish them, and the patterns used for the five intermediate values used in the trading relations simulation. Along the abscissa of each dimension, the nine units for the nine different value ranges of the dimension are arrayed. The curves labeled /g/ and /k/ indicate the relative strength of the excitatory input to each of these units produced by the indicated phoneme. The canonical curves also indicate the strengths of the feature to phoneme connections for /g/ and /k/ on these dimensions.

Response probabilities computed using the formulas given earlier are shown in Figure 12 for each of the 25 conditions of the experiment. The pattern of results is quite similar to that obtained in Denes' (1955) experiment on the /s/-/z/ continuum. The contribution of each cue is approximately linear and additive in the middle of the range, and the curves flatten out at the extremes, as in the Denes (1955) experiment. More importantly, the model's behavior exhibits the ability to trade one cue off against another. In terms of Summerfield and Haggard's measure, the value of VOT needed to achieve 50% probability of reporting

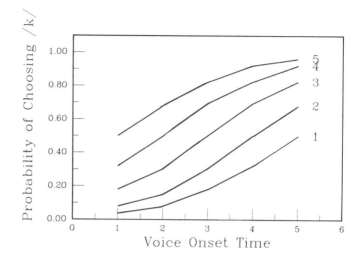

FIGURE 12. Simulated probability of choosing /k/ at Time-Slice 60, for each of the 25 stimuli used in the trading relations simulation experiment. Numbers next to each curve refer to the intermediate pattern on the F1OF continuum used in the 5 stimuli contributing to each curve. Higher numbers correspond to higher values of F1OF.

/k/, we can see that the VOT needed increases as the F1OF decreases, just as these investigators found.

Categorical perception. In spite of the fact that TRACE is quite flexible in the way it combines information from different features to determine the identity of a phoneme, the model is quite categorical in its overt responses. This is illustrated in two ways: First, the model shows a much sharper transition in its choices of responses as we move from /g/ to /k/ along the VOT and F1OF dimensions than we would expect from the slight changes in the relative excitation of the /g/ and /k/ units. Second, the model tends to obliterate differences between different inputs which it identifies as the same phoneme, while sharpening differences between inputs assigned to different categories. We will consider each of these two points in turn, after we describe the stimuli used in the simulations.

Eleven different consonant feature patterns were used, embedded in the same simulated /-_a-/ context as in the trading relations simulation. The stimuli varied from very low values of both VOT and F1OF, more extreme than the canonical /g/, through very high values on both dimensions, more extreme than the canonical /k/. All the stimuli were spaced equal distances apart on the VOT and F1OF dimensions. The locations of the peak activation values on each of these two continua are shown in Figure 13.

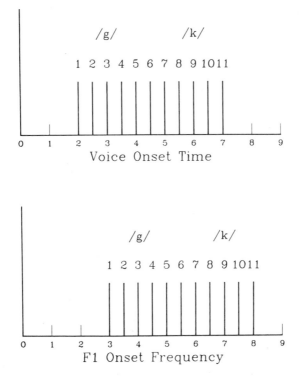

FIGURE 13. Locations of peak activations along the VOT and F1OF dimensions, for each of the 11 stimuli used in the categorical perception simulation.

Figure 14 indicates the relative initial bottom-up activation of the /g/ and /k/ phoneme units for each of the 11 stimuli used in the simulation. The first thing to note is that the relative bottom-up excitations of the two phoneme units differ only slightly. For example, the canonical feature pattern for /g/ sends 75% as much excitation to /g/ as it sends to /k/. The feature pattern two steps toward /g/ from /k/ (stimulus number 5), sends 88% as much activation to /g/ as to /k/.

The figure also indicates, in the second panel, the resulting activations of the units for /g/ and /k/ at the end of 60 cycles of processing. The slight differences in net input have been greatly amplified, and the activation curves exhibit a much steeper transition than the relative bottom-up excitation curves.

There are two reasons why the activation curves are so much sharper than the initial bottom-up excitation functions. The primary reason is *competitive inhibition*. The effect of the competitive inhibition at the phoneme level is to greatly magnify the slight difference in the excitatory inputs to the two phonemes. It is easy to see why this happens.

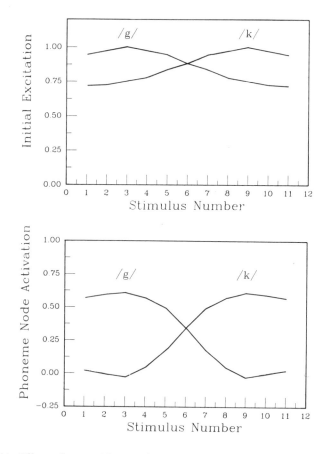

FIGURE 14. Effects of competition on phoneme activations. The first panel shows relative amounts of bottom-up excitatory input to /g/ and /k/ produced by each of the 11 stimuli used in the categorical perception simulation. The second panel shows the activations of units for /g/ and /k/ at Time-Cycle 60. Stimuli 3 and 9 correspond to the canonical /g/ and /k/, respectively.

Once one phoneme is slightly more active than the other, it exerts a stronger inhibitory influence on the other than the other can exert on it. The net result is that "the rich get richer." This general property of competitive inhibition mechanisms has been noted many times (Grossberg, 1976; Levin, 1976; McClelland & Rumelhart, 1981). A second cause of the sharpening of the activation curves is the phoneme-to-feature feedback, which we will consider in detail in a moment.

The identification functions that result from applying the Luce choice rule to the activation values shown in the second panel of Figure 14 are shown in Figure 15 along with the ABX discrimination function, which

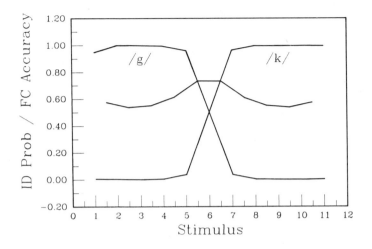

FIGURE 15. Simulated identification functions and forced-choice accuracy in the ABX task.

will be discussed later. The identification functions are even sharper than the activation curves; there is only a 4% chance that the model will choose /k/ instead of /g/ for Stimulus 5, for which /k/ receives 88% as much bottom-up support as /g/. The increased sharpness is due to the properties of the response strength assumptions. These assumptions essentially implement the notion that the sensitivity of the decision mechanism, in terms of d' for choosing the most strongly activated of two units, is a linear function of the difference in activation of the two units. When the activations are far enough apart, d' will be sufficient to ensure near-100% correct performance, even though both units have greater than zero activation. [2]

In TRACE, the categorical output of the model comes about only after an interactive competition process that greatly sharpens the differences in the activation of the detectors for the relevant units. This interactive process takes time. In the simulation results reported here, we assumed that subjects waited a fixed time before responding. But, if we assume that subjects are able to respond as soon as the response strength ratio reaches some critical level, we would find that subjects

---

[2] Many readers will note that the apparent sharpness of the identification functions shown in Figure 15 contrasts with the much shallower functions shown previously in the trading relations simulations. The reason for this is simply that the stimuli are spaced more closely together in the trading relations simulation than in the categorical perception case. This follows the standard experimental practice of emphasizing gradualness in trade-off experiments and sharpness in categorical perception experiments (Lane, 1965).

would be able to respond more quickly to stimuli near the prototype of each category than they can to stimuli near the boundary. This is exactly what was found by Pisoni and Tash (1974).

There is another aspect to categorical perception as exhibited by TRACE. This is the fact that feedback from the phoneme to the feature level tends to cause the model to obliterate the differences between input feature patterns that result in the identification of the same phoneme. This allows the model to account for poor within-category discrimination and good between-category discrimination—the second hallmark of categorical perception.[3] The way it works is this. When a feature pattern comes in, it sends more excitation to some phoneme units than others; as they become active, they begin to compete, and one gradually comes to dominate the others. This much we have already observed. But as this competition process is going on, there is also feedback from the phoneme level to the feature level. Thus, as a particular phoneme becomes active, it tends to impose its canonical pattern of activation on the feature level. The effect of the feedback becomes particularly strong as time goes on since the feature input only excites the feature units very briefly; the original pattern of activation produced by the phoneme units is, therefore, gradually replaced by the canonical pattern imposed by the feedback from the phoneme level. The result is that the pattern of activation remaining at the feature level after 60 cycles of processing has become assimilated to the prototype. In this way, feature patterns for different inputs assigned to the same category are rendered nearly indistinguishable.

This effect is illustrated in Figure 16, which shows how different pairs of patterns of activation at the feature level are at the end of 60 cycles of processing. The measure of difference is simply $1 - r_{ab}$, where r_{ab} stands for the correlation of the patterns produced by stimuli a and b. Only the two dimensions which actually differ between the canonical /g/ and /k/ are considered in the difference measure.

---

[3] Strictly speaking, at least as defined by Liberman, Cooper, Shankweiler, and Studdert-Kennedy (1967), true categorical perception is only exhibited when the ability to discriminate different sounds is no better than could be expected based on the assumption that the only basis a listener has for discrimination is the categorical assignment of the stimulus to a particular phonetic category. However, it is conceded that "true" categorical perception in this sense is never in fact observed (Studdert-Kennedy, Liberman, Harris, & Cooper, 1970). While it is true that the discrimination of sounds is much better for sounds that perceivers assign to different categories than for sounds they assign to the same category, there is also at least a tendency for discrimination to be somewhat better than predicted by the identification function, even between stimuli that are always assigned to the same category. TRACE II produces this kind of approximate categorical perception.

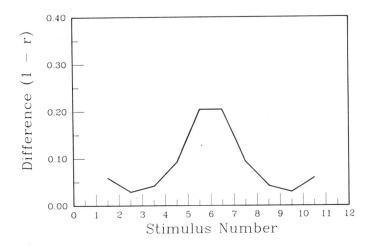

FIGURE 16. Differences between patterns of activation at the feature level at Cycle 60, for pairs of stimuli one step apart along the /g/-/k/ continuum used for producing the identification functions shown previously in Figure 15.

To relate the difference between two stimuli to probability correct choice performance in the ABX task generally used in categorical perception experiments, we once again use the Luce (1959) choice model. The probability of identifying stimulus x with alternative a is given by

$$p\,(R_{x=a}) = \frac{S_{ax}}{S_{ax} + S_{bx}},$$

where S_{ax} is the "strength" of the similarity between a and x. This is given simply by the exponential of the correlation of a and x,

$$S_{ax} = e^{k_r r_{ax}},$$

and similarly for S_{bx}. Here, k_r is the parameter that scales the relation between correlations and strengths. The resulting response probabilities are shown in Figure 15.

Basically, the figures show that the effect of feedback is to make the feature patterns for inputs well within each category more similar than those for inputs near the boundary between categories. Differences between stimuli near the prototype of the same phoneme are almost obliterated. When two stimuli straddle the boundary, the feature level patterns are much more distinct. As a result, the probability of correctly discriminating stimuli within a phoneme category is much lower than the probability of discriminating stimuli in different categories.

Like the completion process considered earlier, the process of "canonicalization" of the representation of a speech sound via the feedback mechanism takes time. During this time, two things are happening: One is that the activations initially produced by the speech input are decaying; another is that the feedback, which drives the representation toward the prototype, is building up. In the simulations, we allowed a considerable amount of time for these processes before computing similarities of different activation patterns to each other. Obviously, if we had left less time, there would not have been as much of an opportunity for these forces to operate. Thus, TRACE is in agreement with the finding that there tends to be an increase in within-category discrimination when a task is used that allows subjects to base their responses on judgments of the similarity of stimuli spaced closely together in time (Pisoni & Lazarus, 1974).

It should be noted that it would be possible to account for categorical perception in TRACE without invoking feedback from the phoneme level to the feature level. All we would need to do is assume that the feature information that gives rise to phoneme identification is inaccessible, as proposed by the motor theory of speech perception (Liberman, Cooper, Shankweiler, & Studdert-Kennedy, 1967), or is rapidly lost as proposed by the *dual code* model (Fujisaki & Kawashima, 1968; Massaro, 1975, 1981; Pisoni, 1973, 1975). The dual code model has had considerable success accounting for categorical perception data and accounts for all the aspects of categorical perception discussed thus far.

Both feedback models and dual code models can also accommodate the fact that vowels show less of a tendency toward categorical perception than consonants (Fry, Abramson, Eimas, & Liberman, 1962; Pisoni, 1973). It is simply necessary to assume that vowel features are more persistent than consonant features (Crowder, 1978, 1981; Fujisaki & Kawashima, 1968; Pisoni, 1973, 1975). However, the two classes of interpretations do differ in their predictions of performance in discriminating two stimuli, both away from the center of a category, but still within it. Here, TRACE tends to show greater discrimination than it shows between stimuli squarely in the middle of a category. Standard interpretations of categorical perception can account for increases in discriminability near the boundary between two categories (where identification may in fact be somewhat variable) by assuming that marginal stimuli are more likely to give rise to different category labels. But TRACE can account for increases in discriminability at extreme values of feature continua which would not give rise to different category labels. In TRACE, the reason for this increase in discriminability is that the activation of the appropriate item at the phoneme level is weaker, and therefore the feedback signal is weaker than it is when the input occurs near the center of the category. This results in less canonicalization of the extreme stimuli and produces a W-shaped discrimination function, as shown in Figure 16. Few studies of categorical perception use stimuli that extend far enough into the extreme ranges of each phonetic category to observe reliable W-shaped curves; however, Samuel (1977) did carry out such a study and obtained just such W-shaped discrimination curves.

In summary, TRACE appears to provide a fairly accurate account of the phenomena of cue trade-offs and categorical perception of speech sounds. It accounts for categorical perception without relying on the notion that the phenomenon depends on read-out from an abstract level of processing; it assumes instead that the feature level, like other levels of the system, is subject to feedback from higher levels which actually changes the representation as it is being retained in memory, pushing it toward a canonical representation of the phoneme most strongly activated by the input.

Retuning of Phoneme Detectors by Context

In our simulations of trading relations, we have shown that the boundary between phonetic categories on one dimension can be affected by inputs on other dimensions. Other factors also influence the phoneme perceived as a result of particular featural input. The identity of phonemes surrounding a target phoneme, the rate of speech of a syllable in which a particular feature value occurs, as well as characteristics of the speaker and the language being spoken all influence the interpretations of features. See Repp and Liberman (1984) for a review of these effects.

In TRACE, we account for local, coarticulatory influences on phoneme identification by assuming that activations of phoneme units can modulate the feature to phoneme connections among units in adjacent time slices. This idea provides one way of implementing what Fowler (1984) has called a *factoring* of coarticulatory influences out of the pattern of activation produced by a feature pattern at the phoneme level. The modulation of connections can also account for the fact that phoneme boundaries shift as a function of local phonetic context (Mann & Repp, 1980). In simulations using TRACE I, we were able to improve the performance of the model in identifying the correct consonant at the beginning of a CV syllable from 79% correct without using connection modulation to 90% correct with connection modulation in place. Interestingly, the model is capable of generalizing from the connection strengths appropriate for the vowels it has been trained on to other vowels, such as /e/, which fall between those on which it was trained. Figure 17 shows phoneme-level activations produced by the same token of the syllable /de/ with connection modulation turned on, in the upper panel, and off, in the lower panel. Though other units are activated in both cases, units for /d/ tend to dominate in the former case, but not the latter.

It has been suggested by J. L. Miller, Green, and Schermer (1984) that lexical effects and semantic and syntactic influences on phoneme identification may be due to a different mechanism than influences of such variables as speech rate and coarticulatory influences due to local phonetic context. The assumptions we have incorporated into TRACE

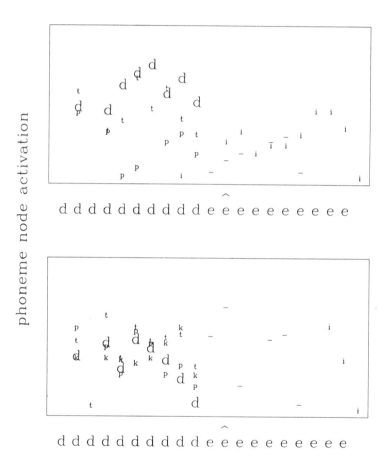

FIGURE 17. Activation of phoneme units resulting from the input /de/, with variable connection strengths enabled, in the upper panel, and disabled, in the lower panel. The /d/ units are depicted using a larger font just to increase their visibility.

make a similar distinction. Lexical effects are due to the additional source of input to the phoneme level provided by units at the word level. This is quite different from the connection modulation mechanism we have used to account for coarticulatory influences. In the discussion, we will consider ways of extending the connection modulation idea to accommodate effects of variations in rate and speaker parameters.

Summary of Phoneme Identification Simulations

We have considered a number of phenomena concerning the identification and perception of phonemes. These include lexical influences on phoneme identification and the lack thereof under certain circumstances; phonotactic rule effects on phoneme identification and the role of specific lexical items in influencing these effects; the integration of multiple cues to phoneme identity; and the categorical nature of the percept that results from this integration. We have also seen how connection modulation can be used to implement context-sensitive phoneme detectors, thereby allowing the model to improve its performance in identifying real speech and to account for effects of phonetic context on boundaries between phonemes. TRACE integrates all of these phenomena into a single account that incorporates aspects of the accounts offered for particular aspects of these results by other models. In the next section, we show how TRACE can also encompass a number of phenomena concerning the recognition of spoken words.

THE TIME COURSE OF SPOKEN WORD RECOGNITION

The study of spoken word recognition has a long history, and many models have been proposed. Morton's now-classic logogen model (Morton, 1969) was the first to provide an explicit account of the integration of contextual and sensory information in word recognition. Other models of this period (e.g., Broadbent, 1967) concentrated primarily on effects of word frequency. Until the midseventies, however, there was little explicit consideration of the time course of spoken word recognition. Several studies by Marslen-Wilson and his collaborators (Marslen-Wilson, 1973; Marslen-Wilson & Tyler, 1975), and by R. A. Cole and his collaborators (Cole, 1973; Cole & Jakimik, 1978, 1980) pioneered the investigation of this problem.

Marslen-Wilson's COHORT model (Marslen-Wilson & Tyler, 1980; Marslen-Wilson & Welsh, 1978) of speech perception was based on this early work on the time course of spoken word recognition. The COHORT model was one of the sources of inspiration for TRACE for two main reasons. First, it provided an explicit account of the way top-down and bottom-up information could be combined to produce a word recognition mechanism that actually worked in real time. Second, it accounted for the findings of a number of important experiments demonstrating the *on-line* character of the speech recognition process. However, several deficiencies of the COHORT model have been pointed out, as we shall see.

Because TRACE was motivated in large part by a desire to keep what is good about COHORT and improve upon its weaknesses, we begin this section by considering the COHORT model in some detail. First we review the basic assumptions of the model, then consider its strengths and weaknesses.

There appear to be four basic assumptions of the COHORT model:

1. The model uses the first sound (in Marslen-Wilson & Tyler, 1980, the initial consonant-cluster-plus-vowel) of the word to determine which words will be in an initial cohort or candidate set.

2. Once the candidate set is established, the model eliminates words from the cohort immediately, as each successive phoneme arrives, if the new phoneme fails to match the next phoneme in the word. Words can also be eliminated on the basis of semantic constraints, although the initial cohort is assumed to be determined by acoustic input alone.

3. Word recognition occurs immediately, as soon as the cohort has been reduced to a single member; in an auditory lexical decision task, the decision that an item is a nonword can be made as soon as there are no remaining members in the cohort.

4. Word recognition can influence the identification of phonemes in a word only after the word has been recognized.

There is a considerable body of data that supports various predictions of the COHORT model. It has been observed in a variety of paradigms that lexical influences on phoneme identification responses are much greater later in words than at their beginnings (Bagley, 1900; R. A. Cole & Jakimik, 1978, 1980; Marslen-Wilson, 1980; Marslen-Wilson

& Welsh, 1978). We considered some of this evidence in earlier sections. Another important finding supporting COHORT is the fact that the reaction time to decide that an item is a nonword is constant when measured from the occurrence of the first phoneme that rules out the last remaining word in the cohort (Marslen-Wilson, 1980).

Perhaps the most direct support for the basic word-recognition assumptions of COHORT comes from the gating paradigm, introduced first by Grosjean (1980). In this paradigm, subjects are required to guess the identity of a word after hearing successive presentations of the word. The first presentation is cut off so that the subject hears only the first N msec (N = 30 to 50 in different studies). Later presentations are successively lengthened in N msec increments until eventually the whole word is presented. The duration at which half the subjects correctly identify the word is called the *isolation point*. Considerably more input is required before subjects are reasonably sure of the identity of the word; that point is termed the *acceptance point*. Grosjean's initial study confirmed many basic predictions of COHORT, though it also raised a few difficulties for it (see below). In a more recent study using the same method, Tyler and Wessels (1983) carried out a very close analysis of the relation between the empirically determined isolation point and the point at which the input the subject has received is consistent with one and only one remaining item—the point at which recognition would be expected to occur in the COHORT model. They report that the isolation point falls very close to this theoretically derived recognition point, strongly supporting the basic immediacy assumptions of the COHORT model.

It should be noted that the gating task is not a timed task, and so it does not provide a direct measure of what the subject knows as the speech input is unfolding. However, it is now in fairly wide use, and Cotton and Grosjean (1984) have established that the basic patterns of results obtained in Grosjean's (1980) pioneering gating experiment do not depend on the presentation of successively longer and longer presentations of the same stimulus.

A dilemma for COHORT. Though the COHORT model accounts for a large body of data, there are several difficulties with it. We consider first the one that seems the most serious: as stated, COHORT requires accurate, undistorted information about the identity of the phonemes in a word up to the isolation point. Words cannot enter into consideration unless the initial consonant-cluster-plus-vowel is heard, and they are discarded from it as soon as a phoneme comes along that they fail to match. No explicit procedure is described for recovering words into the cohort once they have been excluded from it or when the beginning of the word is not accurately perceived due to noise or elision.

These aspects of COHORT make it very difficult for the model to explain recognition of words with distorted beginnings, such as *dwibble* (Norris, 1982), or words whose beginnings have been replaced by noise (Salasoo & Pisoni, 1985). From a computational point of view, this makes the model an extremely brittle one; in particular, it fails to deal with the problem of noise and underspecification which is so crucial for recognition of real speech (Thompson, 1984).

The recognizability of distorted items like *dwibble* might be taken as suggesting that all we need to do is liberalize the criterion for entering and retaining words in the cohort. Thus, the cohort could be defined as the set of words consistent with what has been heard or mild deviations (e.g., one or two features) from what has been heard. This would allow mild distortions like replacing /r/ with /w/ not to disqualify a word from the cohort. It would also allow the model to cope with cases where the beginning of the word is underspecified; in these cases, the initial cohort would simply be larger than in the case where the input clearly specified the initial phonemes.

However, there is still a problem. Sometimes we need to be able to rule out items that mismatch the input on one or two dimensions and sometimes we do not. Consider the items *pleasant* and *blacelet*. In the first case, we need to exclude *present* from the cohort, so the slight difference between /l/ and /r/ must be sufficient to rule it out; in the second case, we do not want to lose the word *bracelet*, since it provides the best fit overall to the input. Thus, in this case, the difference between /l/ and /r/ must not be allowed to rule a word candidate out.

Thus the dilemma: On the one hand, we want a mechanism that will be able to select the correct word as soon as an undistorted input specifies it uniquely, to account for the Tyler and Wessels results. On the other hand, we do not want the model to completely eliminate possibilities that might later turn out to be correct. We shall shortly see that TRACE provides a way out of this dilemma.

Another problem for COHORT. Grosjean (1985) has recently pointed out another problem for COHORT, namely, the possibility that the subject may be uncertain about the location of the beginning of each successive word. A tacit assumption of the model is that the subject goes into the beginning of each word knowing that it is the beginning. In the related model of R. A. Cole and Jakimik (1980), this assumption is made explicit. Unfortunately, it is not always possible to know in advance where one word starts and the next word ends. As we discussed in the introduction, acoustic cues to juncture are not always reliable, and in the absence of acoustic cues, even an optimally efficient mechanism cannot always know that it has heard the end of one word until it hears enough of the next to rule out the possible continuations of the first word.

What is needed, then, is a model that can account for COHORT's successes and overcome these two important deficiencies. The next two sections show that TRACE does quite well on both counts. The first of these sections examines TRACE's behavior in processing words whose beginnings and endings are clearly delineated for it by the presence of silence. The second considers the processing of multiword inputs, which the model must parse for itself.

One Word at a Time

In this section, we see how TRACE resolves the dilemma facing COHORT, in that it is immediately sensitive to new information but is still able to cope with underspecified or distorted word beginnings. We also consider how the model accounts for the preference for short-word responses early in processing a long word. The section concludes with a discussion of ways the model could be extended to account for word frequency and contextual influences.

Competition vs. bottom-up inhibition. TRACE deals with COHORT's dilemma by using competition rather than phoneme-to-word inhibition. The essence of the idea is simply this: Phoneme units have excitatory connections to all the word units they are consistent with. Thus, whenever a phoneme becomes active in a particular slice of the Trace, it sends excitation to all the word units consistent with that phoneme in that slice. The word units then compete with each other; items that contain each successive phoneme in a sequence dominate all others, but, if no word matches perfectly, a word that provides a close fit to the phoneme sequence can eventually win out over words that provide less adequate matches.

Consider, from this point of view, our two items *pleasant* and *blacelet* again. In the first instance, *pleasant* will receive more bottom-up excitation than *present* and so will win out in the competition. We have already seen, in our analysis of categorical perception at the phoneme level, how even slight differences in initial bottom-up excitation can be magnified by the joint effects of competition and feedback; but the real beauty of the competition mechanism is that this action is contingent on the activation of other word candidates. Thus, in the case of *blacelet*, since there is no word *blacelet*, *bracelet* will not be suppressed. Initially, it is true, words like *blame* and *blatant* will tend to dominate *bracelet*, but, since the input matches *bracelet* better than any other word, *bracelet* will eventually come to dominate the other possibilities.

This behavior of the model is illustrated using examples from its restricted lexicon in Figure 18. In one case, the input is *legal*, and the word *regal* is completely dominated by *legal*. In the other case, the input is *lugged*, and the word *rugged* eventually dominates because there is no word *lugged* (pronounced to rhyme with *rugged*—the word *lug* is not in the model's lexicon). Here, *rugged* must compete with other partial matches of *lugged*, of course, and it is less effective in this regard than it would be if the input exactly matched it, but it does win out in the end.

It should be noted that the details of what word will be most strongly activated in such cases depend on a number of factors, including, in particular, the distinctiveness of mismatching phonemes. Also, it is possible to find cases in which a word that correctly spans a part of a longer string dominates a longer word that spans the whole string but misses out on a phoneme in one place or another. An item like *vigorette* may or may not be a case in point. In such cases, though, the most important thing might not turn out to be winning or losing, but rather the fact that both tend to stay in the game. Such neologisms can suggest a poetic conjunction of meanings, if used just right: "He walked briskly down the street, puffing his vigorette."

Time course of word recognition in TRACE. So far we have shown how TRACE overcomes a difficulty with the COHORT model in cases where the beginning of a word has been distorted. In earlier sections on phoneme processing, some of the simulations illustrate that the model is capable of recognizing words with underspecified (i.e., ambiguous) initial phonemes. In this section, we examine how well TRACE emulates the COHORT model in cases where the input is an undistorted representation of some particular word. In particular, we wanted to see how close TRACE would come to behaving in accord with COHORT's assumption that incorrect words are dropped from the cohort of active candidates as soon as the input diverges from them.

To examine this process, we considered the processing of the word *product* (/ prad^ct/). Figure 19 shows the state of the Trace at various points in processing this word, and Figure 20 shows the response strengths of several units relative to the strength of the word *product* itself, as a function of time relative to the arrival of the successive phonemes in the input. In the latter figure, the response strength of *product* is simply given as 1.0 at each time slice and the response strengths of units for other words are given relative to the strength of *product*. The curves shown are for the words *trot, possible, priest, progress*, and *produce*; these words differ from the word *product* (according

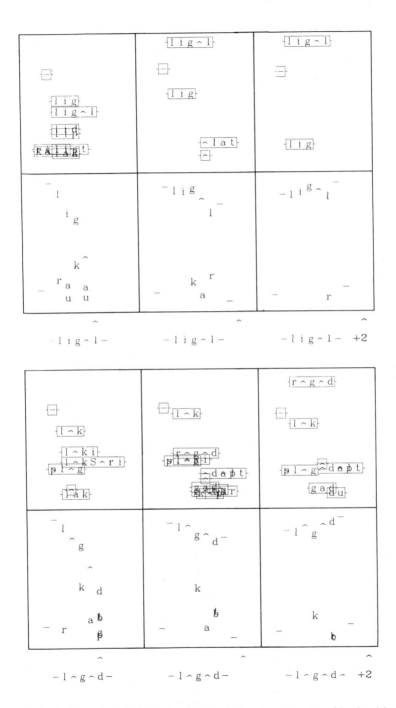

FIGURE 18. State of the Trace at three points during the processing of *legal* and *lugged*.

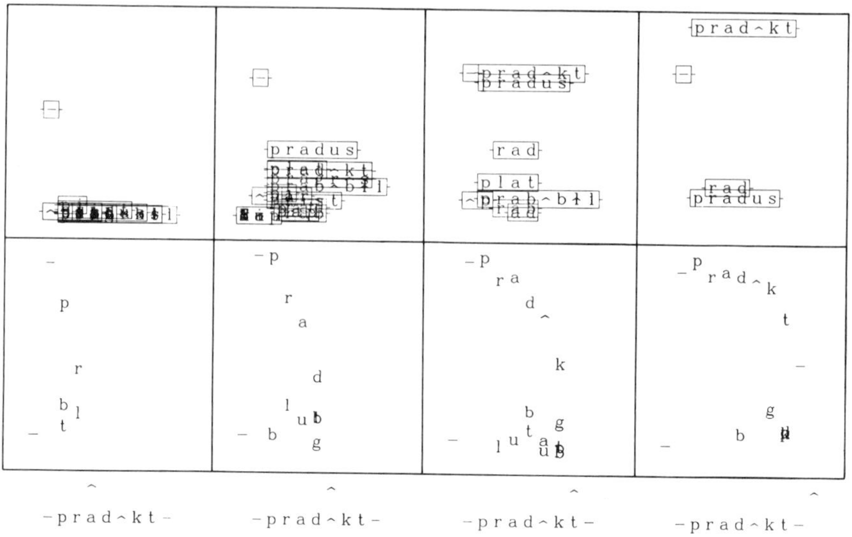

FIGURE 19. State of the Trace at various points in processing the word *product* (/prad·kt/).

to the simulation program's stressless encoding of them!) in the first, second, third, fourth, and fifth phonemes, respectively. Figure 20 shows that these items begin to drop out just after each successive phoneme comes in. Of course, there is nothing hard and fast or absolute about dropping a candidate in TRACE. What we see instead is that mismatching candidates simply begin to fade as the input diverges from them in favor of some other candidate. This is just the kind of behavior the COHORT model would produce in this case, though, of course, the drop-off would be assumed to be an abrupt, discrete event.[4]

There is one aspect of TRACE's behavior which differs from COHORT: Among those words that are consistent with the input up to a particular point in time, TRACE shows a bias in favor of shorter words over longer words. Thus, *priest* has a slight advantage before the /a/ comes in, and *produce* is well ahead of *product* until the /·/ comes in (in phonemes, *produce* is one shorter than *product*).

---

[4] The data reported by Tyler and Wessels actually appears to indicate an even more immediate drop-off than is seen in this simulation. However, it should be remembered that the curves shown in Figure 20 are on-line response strength curves, and thus reflect the lags inherent in the percolation of input from the feature to the word level. The gating task, on the other hand, does not require subjects to respond on-line. If the input is simply turned off at the peak of each phoneme's input specification and then allowed to run free for a few cycles, the dropout point shifts even earlier.

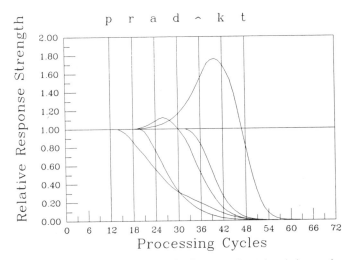

FIGURE 20. Response strengths of the units for several words relative to the response strength of the unit for *product* (/prad⌃kt/), as a function of time relative to the peak of the first phoneme that fails to match the word. The successive curves coming off of the horizontal line representing the normalized response strength of *product* are for the words *trot, possible, priest, progress,* and *produce,* respectively. In our lexicon they are rendered as /trat/, /pas⌃b⌃l/, /prist/, /pragr⌃s/, and /pradus/.

This advantage for shorter words is due to the competition mechanism. Recall that word units compete with each other in proportion to the overlap of the sets of time slices spanned by each of the words. Overlap is, of course, symmetrical, so long and short words inhibit each other to an equal extent; but longer words suffer more inhibition from other long words than short words do. For example, *progress* and *probable* inhibit *product* more than they inhibit *priest* and *produce.* Thus, units for longer words are generally subjected to extra inhibition, particularly early on in processing when many candidates are active, and so they tend to suffer in comparison to short words as a result.

We were at first somewhat disturbed by this aspect of the model's behavior, but it turns out to correspond quite closely with results obtained in experiments by Grosjean (1980) and Cotton and Grosjean (1984) using the gating paradigm. Both papers found that subjects hearing the beginnings of words like *captain* tended to report shorter words consistent with what they had heard (e.g., *cap*). However, we should observe that in the gating paradigm, when the word *captain* is truncated just after the /p/, it will sound quite a bit like *cap* followed by silence. In TRACE, this silence would activate silence units at the phoneme and word levels, and the word-level silence units would compete with units for words that extend into the silence. The silence will reinforce the preference of the model for short-word interpretations because the detection of the silence will inhibit the detector for the longer word. Thus, there are actually two reasons why TRACE might favor short-word interpretations over long-word interpretations in a gating experiment. Whether human subjects show a residual preference for shorter interpretations over longer ones in the absence of a following silence during the course of processing is not yet clear from available data.

We should point out that the experimental literature indicates that the advantage of shorter words over longer ones holds only under the special circumstances of gated presentation and then only with early gates, when shorter words are relatively more complete than longer ones would be. It has been known for a long time that longer words are generally more readily recognized than shorter ones when the whole word is presented for identification against a background of noise (Licklider & Miller, 1951). Presumably, the reason for this is simply that longer words generally provide a larger number of cues than shorter words do, and hence are simply less confusable.

Frequency and context effects. There are, of course, other factors that influence when word recognition will occur, beyond those we have considered thus far. Two very important ones are word frequency and contextual predictability. The literature on these two factors goes back to the turn of the century (Bagley, 1900). Morton's (1969) logogen model effectively deals with several important aspects of this huge literature, though not with the time course of these effects.

We have not yet included either word frequency or higher-level contextual influences in TRACE, though, of course, we believe they are important. Word frequency effects could be accommodated, as they were in the interactive activation model of word recognition, in terms of variation in the resting activation level of word units or in terms of variation in the strength of phoneme-to-word connections. Contextual influences can be thought of as supplying activation to word units from even higher levels of processing than the word level. In this way, basic aspects of these two kinds of influences can be captured. We leave it to future research, however, to determine to what extent these elaborations of TRACE would provide a detailed account of the data on the roles of these factors. For now, we turn to the problem of determining where one word ends and the next one begins.

Lexical Basis of Word Segmentation

How do we know when one word ends and the next word begins? This is by no means an easy task, as we noted in the introduction. To recap our earlier argument, there are some cues in the speech stream, but as several investigators have pointed out (R. A. Cole & Jakimik, 1980; Grosjean & Gee, 1984; Thompson, 1984), they are not always sufficient, particularly in fluent speech. It would appear that there is an important role for lexical knowledge to play in determining where one word ends and the next word begins, as well as in identifying the objects that result from the process of segmentation. Indeed, as Reddy (1976) has suggested, segmentation and identification may simply be joint results of the mechanisms of word recognition.

R. A. Cole and Jakimik (1980) discuss these points and present evidence that semantic and syntactic context can guide segmentation in cases where the lexicon is consistent with two readings (*car go* vs. *cargo*). Our present model lacks syntactic and semantic levels, so it cannot make use of these higher-level constraints; but it can make use of its knowledge about words, not only to identify individual words in isolation, but to pick out a sequence of words in continuous streams of phonemes. Word identification and segmentation emerge together from the interactive activation process as part and parcel of the process of word activation.

This section considers several aspects of the way in which word segmentation emerges from the interactive activation process, as observed in simulations with TRACE II. Before we consider these, it is worth recalling the details of some of the assumptions made about the bottom-up activation of word units and about competitive inhibition between word units. First, the extent to which a particular phoneme excites a particular word unit is independent of the length of the word. Second, the extent to which a particular word unit will inhibit another word unit is proportional to the temporal overlap of the two word units. This means that words which do not overlap in time will not inhibit each other but will gang up on other words that partially overlap each of them. These two assumptions form most of the basis of the effects we will observe in the simulations.

The boundary is in the ear of the "behearer." First, we consider the basic fact that the number of words we will hear in a sequence of phonemes can depend on our knowledge of the number of words the sequence makes. Consider the two utterances *she can't* and *secant*. Though we can say either item in a way that makes it sound like a single word or like two words, there is an intermediate way of saying them so that the first seems to be two words and the second seems like only one.

To see what TRACE II would do with single and multiple word inputs, we ran simulation experiments with each individual word in the main 211-word lexicon preceded and followed by silence, and then with

211 pairs of words, with a silence at the beginning and at the end of the entire stream. The pairs were made by simply permuting the lexicon twice and then abutting the two permutations so that each word occurred once as the first word and once as the second word in the entire set of 211 pairs. We stress, of course, that real speech would tend to contain cues that would mark word boundaries in many cases; the experiment is simply designed to show what TRACE can do in cases where these cues are lacking.

With the individual words, TRACE made no mistakes—that is, by a few slices after the end of the word, the word that spanned the entire input was more strongly activated than any other word. An example of this is shown using the item /parti/ in Figure 21. The stream /parti/ might be either one word (*party*) or two (*par tea*, or *par tee*—the model knows of only one word pronounced /ti/). At early points in processing the word, *par* dominates over *party* and other longer words for reasons discussed in the previous section. By the time the model has had a chance to process the end of the word, however, *party* comes to dominate.

Why does a single longer word eventually win out over two shorter ones in TRACE? There are two main reasons. First of all, a longer word eventually receives more bottom-up support than either shorter word, simply because there are more phonemes activating the longer word than the shorter word. The second reason has to do with the

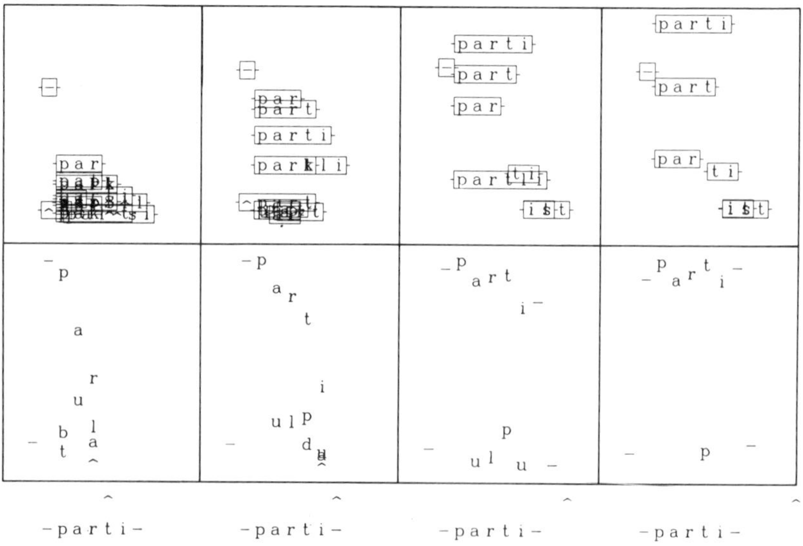

FIGURE 21. The state of the Trace at various points during processing of /parti/.

sequential nature of the input. In the case of /parti/, by the time the /ti/ is coming in, the word *party* is well enough established that it keeps /ti/ from getting as strongly activated as it would otherwise. This behavior of the model leads to the prediction that short words imbedded in the ends of longer words should not get as strongly activated as shorter words coming earlier in the longer word. This prediction could be tested using the gating paradigm or a cross-modal priming paradigm such as the one used by Swinney (1982).

However, it should be noted that this aspect of the behavior of the model can be overridden if there is bottom-up information favoring the two-word interpretation. Currently, this can only happen in TRACE through the insertion of a brief silence between the *par* and the *tea*. As shown in Figure 22, this results in *par* and *tea* dominating all other word candidates.

What happens when there is no long word that spans the entire stream, as in /barti/? In this case, the model settles on the two word interpretation *bar tea*, as shown in Figure 22. Note that other words, such as *art*, that span a portion of the input, are less successful than either *bar* or *tea*. The reason is that the interpretations *bar* and *art* overlap with each other, and *art* and *tea* overlap with each other, but *bar* and *tea* do not overlap. Thus, *art* receives inhibition from both *bar*

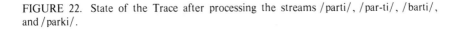

FIGURE 22. State of the Trace after processing the streams /parti/, /par-ti/, /barti/, and /parki/.

and *tea*, while *bar* and *tea* each receive inhibition only from *art*. Thus, two words that do not overlap with each other can gang up on a third word that each overlaps with partly and drive it out.

These remarkably simple mechanisms of activation and competition do a very good job of word segmentation without the aid of any syllabification, stress, phonetic word-boundary cues, or semantic and syntactic constraints. In 189 of the 211 word pairs tested in the simulation experiment, the model came up with the correct parse, in the sense that no other word was more active than either of the two words that had been presented. Some of the failures of the model occurred in cases where the input was actually consistent with two parses, either a longer spanning word rather than a single word (as in *party*) or a different parse into two words, as in *part rust* for *par trust*. In such cases TRACE tends to prefer parses in which the longer word comes first. There were, however, some cases in which the model did not come up with a valid parse, that is, a pattern than represents complete coverage of the input by a set of nonoverlapping words. For example, consider the input /parki/. Though this makes the two words *par* and *key*, the word *park* has a stronger activation than either *par* or *key*, as illustrated in Figure 22.

This aspect of TRACE II's behavior indicates that the present version of the model is far from the final word on word segmentation. A complete model would also exploit syllabification, stress, and other cues to word identity to help eliminate some of the possible interpretations of TRACE II's simple phoneme streams. The activation and competition mechanisms in TRACE II are sufficient to do quite a bit of the word segmentation work, but we do not expect them to do this perfectly in all cases without the aid of other cues.

Some readers may be troubled by a mechanism that does not insist upon a parse in which each phoneme is covered by one and only one word. Actually, though, this characteristic of the model is often a virtue, since in many cases, the last phoneme of a word must do double duty as the first phoneme of the next, as in *hound dog* or *brush shop*. While speakers tend to signal the doubling in careful speech, the cues to single vs. double consonants are not always sufficient for disambiguation, as is clear when strings with multiple interpretations are used as stimuli. For example, an utterance intended as *no notion* will sometimes be heard as *known notion* (Nakatani & Dukes, 1977). The model is not inclined to suppress activations of partially overlapping words, even when a nonoverlapping parse is available. This behavior of TRACE is illustrated with /b^stap/ (*bus top* or *bus stop*) in Figure 23. In this case, higher levels could provide an additional source of information that would help the model choose between overlapping and nonoverlapping interpretations.

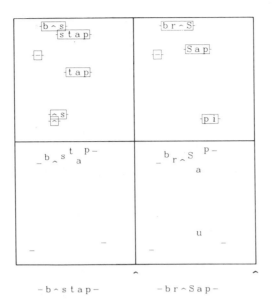

FIGURE 23. State of the Trace at the end of the streams /bustap/ (*bus stop* or *bus top*) and /bruSap/ (*brush shop*).

Thus far in this section, we have considered the general properties of the way in which TRACE uses lexical information to segment a speech stream into words, but we have not considered much in the way of empirical data that these aspects of the model shed light on. However, there are two findings in the literature which can be interpreted in accordance with TRACE's handling of multiword speech streams.

Where does a nonword end?   A number of investigators (e.g., R. A. Cole & Jakimik, 1980) have suggested that when one word is identified, its identity can be used to determine where it ends and, therefore, where the next word begins. In TRACE, the interactive activation process can often establish where a word will end even before it actually does end, particularly in the case of longer words or when activations at the word level are aided by syntactic and semantic constraints. However, it is much harder to establish the end of a nonword since the fact that it is a nonword means that we cannot exploit any knowledge of where it should end to do so.

This fact may account for the finding of Foss and Blank (1980) that subjects are much slower to respond to target phonemes at the beginning of a word preceded by a nonword than at the beginning of a word preceded by a word. For example, responses to detect word-initial /d/ were faster in stimuli such as the following:

At the end of last year, the government decided . . .

than they were when the word preceding the target phoneme (in this case, *government*) was replaced by a nonword such as *gatabont*. It should be noted that the targets were specified as word-initial segments. Therefore, the subjects had not only to identify the target phoneme; they had to determine that it fell at the beginning of a word as well. The fact that reaction times were faster when the target was preceded by a word suggests that subjects were able to use their knowledge of where the word *government* ends to help them determine where the next word begins.

An example of how TRACE allows one word to help establish where its successor begins is illustrated in Figure 24. In the example, the model receives the stream *possible target* or *pagusle target*, and we imagine that the target is word-initial /t/. In the first case, the word *possible* is clearly established, and competitors underneath it have been completely crushed by the time the initial /t/ in *target* becomes active at the phoneme level (second panel in the upper part of the figure), so there is no ambiguity about the fact that this /t/ is at the beginning of the next word. (The decision mechanism would, of course, be required to note that the model had established the location of the end of the preceding word. We have not yet incorporated explicit assumptions about how this would be done.) In the second case, words beginning and ending at a number of different places, including some that overlap with the location of the /t/, are partly activated. Thus, a listener would have to wait until the input is well into the word *target* before it becomes clear that the first /t/ in target is in fact a word-initial /t/.

In reality, the situation is probably not as bleak for the perceiver as it appears in this example because in many cases there will be cues in the manner of pronunciation and the syllabification of the input that will help to indicate the location of the word boundary. However, given the imprecision and frequent absence of such cues, it is not surprising that the lexical status of one part of a speech stream plays an important role in helping listeners determine where the beginning of the next word must be.

The long and short of word identification. One problematic feature of speech is the fact that it is not always possible to identify a word unambiguously until one has heard the word after it. Consider, for example, the word *tar*. If we are listening to an utterance and have gotten just to the /r/ in *The man saw the tar box*, though *tar* will tend to be

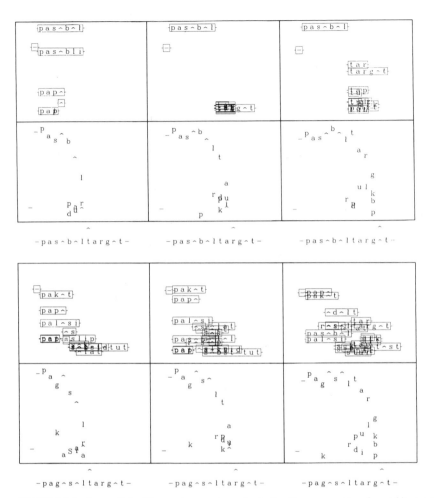

FIGURE 24. State of the Trace at several points during the processing of *possible target* and *pagusle target*.

the preferred hypothesis at this point, we do not have enough information to say unequivocally that the *tar* will not turn out to be *target* or *tarnished* or one of several other possibilities. It is only after more time has passed, and we have perceived either a silence or enough of the next word to rule out any of the continuations of /tar/, that we can decide we have heard the word *tar*. This situation, as it arises in TRACE with the simple utterance /tarbaks/ (*tar box*), is illustrated in Figure 25. Though *tar* is somewhat more active than the longer word *target* when the /r/ is coming in, it is only when the word *box* emerges as the interpretation of the phonemes following *tar* that the rival *target* finally fades as a serious contender.

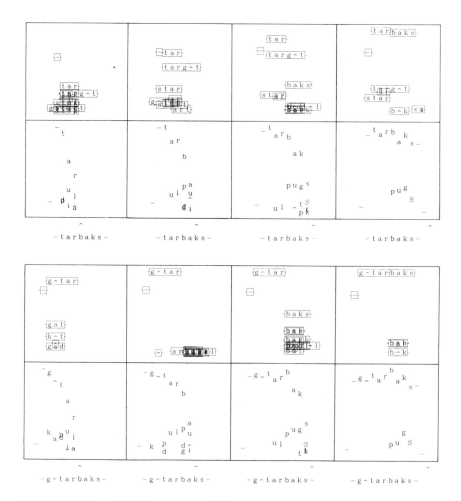

FIGURE 25. State of the Trace at several points in processing *tar box* and *guitar box*.

With longer words the situation is different. As we have already seen in another example, by the time the end of a longer word is reached it is much more likely that only one word candidate will remain. Indeed, with longer words it is often possible to have enough information to identify the word unambiguously well before the end of the word. An illustration of this situation is provided by a simulation using the utterance *guitar box* /gᵔtarbaks/. By the time the /r/ has registered, *guitar* is clearly dominant at the word level and can be unambiguously identified without further ado.

Recently, an experiment by Grosjean (1985) has demonstrated these same effects empirically. Grosjean presented subjects with long or short words followed by a second word and measured how much of the word and its successor the subject needed to hear to identify the target. With longer words, subjects could usually guess the word correctly well before the end of the word; and by the end of the word, they were quite sure of the word's identity. With monosyllabic words, on the other hand, many of the words could not be identified correctly until well into the next word. On average, subjects were not sure of the word's identity until about the end of the next word or the beginning of the one after. As Grosjean (1985) points out, a major reason for this is simply that the spoken input often does not uniquely specify the identity of a short word. In such cases, the perceptual system is often forced to process the short word and its successor at the same time.

Recognizing the words in a short sentence. One last example of TRACE II's performance in segmenting words is illustrated in Figure 26. The figure shows the state of the Trace at several points during the processing of the stream /SiSˆtˆbaks/. By the end, the words of the phrase *she shut a box*, which fits the input perfectly with no overlap, dominate all others.

This example illustrates how far it is sometimes possible to go in parsing a stream of phonemes into words without even considering syntactic and semantic constraints, or stress, syllabification, and juncture cues to word identification. The example also illustrates the difficulty the model has in perceiving short, unstressed words like *a*. This is, of course, just an extreme version of the difficulty the model has in processing monosyllabic words like *tar* and is consistent with Grosjean's data on the difficulty subjects have with identifying short words. In fact, Grosjean and Gee (1984) report pilot data indicating that these difficulties are even more severe with function words like *a* and *of*. It should be noted that TRACE makes no special distinction between content and function words per se, and neither do Grosjean and Gee. However, function words are usually unstressed and considerably shorter than content words. Thus, it is not necessary to point to any special mechanisms for closed versus open class morphemes to account for Grosjean and Gee's results.

Summary of Word Identification Simulations

While phoneme identification has been studied for many years, data from on-line studies of word recognition is just beginning to accumulate. There is an older literature on accuracy of word identification in noise, but it has only been quite recently that useful techniques have been developed for studying word recognition in real time.

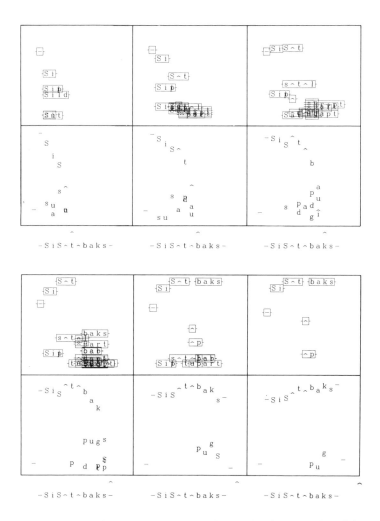

FIGURE 26. The state of the Trace at several points during the processing of the stream /SiSʿtˇbaks/ (*she shut a box*).

What evidence there is, though, indicates the complexity of the word identification process. While the word identification mechanism is sensitive to each new incoming phoneme as it arrives, it is nevertheless robust enough to recover from underspecification or distortion of word beginnings. And, it appears to be capable of some simultaneous processing of successive words in the input stream. TRACE appears to capture these aspects of the time course of word recognition. In these respects, it improves upon the COHORT model, the only previously extant model that provides an explicit account of the on-line process of word recognition. And, the mechanisms it uses to accomplish this are the same ones that it used for the simulations of the process of phoneme identification described in the preceding section.

GENERAL DISCUSSION

Summary of TRACE's Successes

In this chapter, we have seen that TRACE can account for a number of different aspects of human speech perception. We begin by listing the major correspondences between TRACE and what we know about the human speech understanding process.

1. TRACE, like humans, uses information from overlapping portions of the speech wave to identify successive phonemes.

2. The model shows a tendency toward categorical perception of phonemes, as do human subjects. The model's tendency toward categorical perception is affected by many of the same parameters that affect the degree of categorical perception shown by human subjects; in particular, the extent to which perception will be categorical increases with time between stimuli that must be compared.

3. The model combines feature information from a number of different dimensions and exhibits cue trade-offs in phoneme identification.

4. The model augments information from the speech stream with feedback from the lexical level in reaching decisions about the identity of phonemes. These lexical influences on phoneme identification occur in conditions similar to those in which lexical effects have been reported, but do not occur in conditions in which these effects have not been obtained.

5. Like human subjects, the model exhibits apparent phonotactic rule effects on phoneme identification, though it has no explicit representation of the phonotactic rules. The tendency to prefer phonotactically regular interpretations of ambiguous phonemes can be overridden by particular lexical items, just as it can in the human perceiver.

6. Our simulations with TRACE I show that the model is able to use activations of phoneme units in one part of the Trace to adjust the connection strengths determining which features will activate which phonemes in adjacent parts of the Trace. In this way the model can adjust as human subjects do to coarticulatory influences on the acoustic properties of phonemes (Fowler, 1984; Mann & Repp, 1980).

7. In processing unambiguous phoneme sequences preceded by silence, the model exhibits immediate sensitivity to information favoring one word interpretation over another. It shows an initial preference for shorter words relative to longer words, but eventually a sequence of phonemes that matches a long word perfectly will be identified as that word, overturning the initial preference for the short-word interpretation. These aspects of the model are consistent with human data from gating experiments.

8. Though the model is heavily influenced by word beginnings, it can recover from underspecification or distortion of a word's beginning.

9. The model can use its knowledge of the lexicon to parse sequences of phonemes into words and to establish where one word ends and the next one begins when cues to word boundaries are lacking.

10. Like human subjects, the model sometimes cannot identify a word until it has heard part of the next word. Also like human subjects, it can better determine where a word will begin when it is preceded by a word rather than a nonword.

11. The model does not demand a parse of a phoneme sequence that includes each phoneme in one and only one word. This allows it to cope gracefully with elision of phonemes at word boundaries. It will often permit several alternative parses to remain available for higher-level influences to choose among.

There is, of course, more data on some of these points than others. It will be very interesting to see how well TRACE will hold up against the data as further empirical studies are carried out.

Reasons for TRACE's Successes

We think there are two main reasons why TRACE has worked so well. The first is its use of massively parallel, interactive processing. The second is the Trace architecture. We do not believe that the model would have worked without both of these characteristics. The Trace provides a processing structure that lays out the hypotheses about the contents of an utterance in a way that captures directly the task to be performed—to find an interpretation of the utterance consisting of a sequence of units on each of several processing levels. Appropriate competition within levels and mutual facilitation between levels is quite naturally arranged in such a situation. This is important, but it would not work without the parallel, interactive processing that is provided by the PDP framework.

There is evidence to support both parts of this point. One source comes from our early attempts to model speech perception without adopting the Trace architecture. Our early model (described in Elman & McClelland, 1984) failed to provide a straightforward representation of the temporal structure of the speech stream.

It is a commonplace observation in the field of artificial intelligence that success in modeling depends on having the right representation, and Marr (1982) made very much of this point. But it is also true that one must have the right kind of processing system to exploit the architecture. The fact that the HEARSAY speech understanding system was not terribly successful attests to this. HEARSAY had the right architecture—a better one, in fact, in some ways, than we have in the current version of TRACE (see Chapter 16). But HEARSAY lacked the massively parallel, interactive capabilities of PDP models.

Some Deficiencies of TRACE

Although TRACE has had a number of important successes, it also has a number of equally important deficiencies. One fundamental deficiency is the fact that the model requires reduplication of units and connections, copying over and over again the connection patterns that determine which features activate which phonemes and which phonemes activate which words. One reason why this is a problem has to do with learning. Learning in PDP models involves tuning connections between pairs of units based on both of their states. This kind of learning is inherently *local* to the specific connections between the specific units involved. Given TRACE's architecture, such learning would not generalize from one part of the Trace to another and so would not be accessible for inputs arising at different locations in the Trace. A second problem is that the model, as is, is insensitive to variation in global parameters such as speaking rate, speaker characteristics and accent, and ambient acoustic characteristics. A third deficiency is that it fails to account for the fact that one presentation of a word has an effect on the perception of it a very short time later (Nusbaum & Slowiaczek, 1982). These two presentations, in the current version of the model, simply excite separate tokens for the same word in different parts of the Trace.

All these deficiencies reflect the fact that the Trace consists of a large set of independent tokens of each feature, phoneme, and word unit. What appears to be called for instead is a model in which there is a single stored representation of each phoneme and each word in some central representational structure. If this structure is accessed every time the word is presented, then we could account for repetition priming effects. Likewise, if there were a single central structure, learning could occur in just one set of units, as could dynamic retuning of feature-phoneme and phoneme-word connections to take account of changes in global parameters or speaker characteristics.

However, it remains necessary to keep straight the relative temporal location of different feature, phoneme, and word activations as we argued just above. Thus, it will not do to simply abandon the Trace in favor of a single set of units consisting of just one copy of each phoneme and one copy of each word.

It seems that we need to have things both ways: We need a central representation that plays a role in processing every phoneme and every word and that is subject to learning, retuning, and priming. We also need to keep a dynamic trace of the unfolding representation of the speech stream so that we can continue to accommodate both left and right contextual effects. The next chapter describes a model of reading that has some of these characteristics. It uses connection information stored in a central PDP network to set connections in a processing structure much like the Trace, thereby effectively programming this structure in the course of processing. The next step in the development of TRACE is to apply these ideas to speech perception. Some comments about how this might be done are included at the end of the next chapter.

CONCLUSION

Our aim in this chapter has been to show that parallel distributed processing mechanisms provide a natural framework for developing models capable of meeting the computational challenges posed by speech and of accounting for the data on human speech perception. The TRACE model does quite well on both counts. Though the architecture of the model is partially responsible for its successes, we have argued that the successes of the model depend at least as much on the parallel distributed processing operations that take place within the architecture. TRACE does have some limitations, but it successes so far have been quite encouraging. Just how easy it will be to overcome the limitations is a matter for future research.

ACKNOWLEDGMENTS

The work reported here was supported in part by a contract from the Office of Naval Research (N-00014-82-C-0374, NR 667-483), in part by a grant from the National Science Foundation (BNS-79-24062), and in part by a Research Scientist Career Development Award to the first author from the National Institute of Mental Health (5-K01-MH00385).

CHAPTER 16

The Programmable Blackboard Model of Reading

J. L. McCLELLAND

In 1975, Rumelhart outlined a model of reading called the *interactive model.* That model, inspired by the HEARSAY model of speech understanding (Reddy, Erman, Fennell, & Neely, 1973), supposed that reading involved simultaneous processing at a large number of levels, including visual feature, letter, word, syntactic, and semantic levels. Hypotheses at each level were activated when active hypotheses on adjacent levels suggested them and competed with alternative hypotheses at the same level. This model, of course, was a precursor of the interactive activation model of word recognition and of the approach that underlies this whole book.

In the interactive model of reading, the activation of hypotheses was guided by a set of structures called "knowledge sources," each of which had expertise with respect to a particular aspect of reading. For example, a lexical knowledge source that contained knowledge of the letter sequences that made up each word was proposed, along with a syntactic knowledge source, a semantic knowledge source, and an orthographic knowledge source containing knowledge of the appearance of the letters.

An important aspect of the interactive model of reading was parallel processing. Processing was supposed to occur in parallel both within and between levels, so that hypotheses could influence and be influenced by other hypotheses spanning large stretches of the input. However, no specific implementation of the mechanism was proposed. In HEARSAY, although the *conception* was parallel, the *reality* was quite sequential—each knowledge source could only be directed to a single small part of HEARSAY's BLACKBOARD at a time. The result was a model that was computationally very cumbersome and excruciatingly slow. Eventually, HEARSAY was abandoned in favor of a model in which the knowledge that guided processing was precompiled, by brute force, into a Markov chain.

PDP models such as the interactive activation model of word perception and the TRACE model of speech perception have tried to capture the parallelism inherent the conception of HEARSAY. (See Chapter 15.) However, these models differ from HEARSAY in a fundamental way. Instead of having a knowledge source that can be applied to an input, they build the knowledge into the connections between the units out of which the mechanisms are built. Thus, for example, in the word perception model, the knowledge that guides processing is built into the connections between the units. By making several copies of the same connection information, it is possible to allow parallel processing.

To make this point clear and to emphasize some of the properties of this aspect of the model, a sketch of the model is presented in Figure 1.

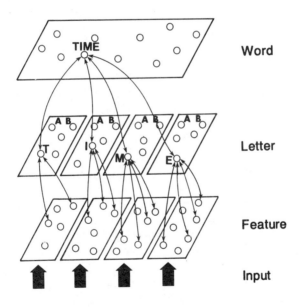

FIGURE 1. A sketch of the interactive activation model of word perception. Units within the same rectangle stand for incompatible alternative hypotheses about an input pattern, and are all mutually inhibitory. The bidirectional excitatory connections between levels are indicated for one word and its constituents. (From "Putting Knowledge in its Place: A Scheme for Programming Parallel Processing Structures on the Fly" by J. L. McClelland, 1985, *Cognitive Science*, 9, p. 115. Copyright 1985 by Ablex Publishing. Reprinted by permission.)

The figure brings out the fact that the model achieves parallel processing of all of the letters in a four-letter display by having four separate copies of the feature and letter units, and four separate copies of the connections between them. Parallel processing of more than one word at a time would require creating another copy of the entire network. In the TRACE model Elman and I did just this, creating a large number of copies of the entire connection network.

Now, the parallel processing permitted in models like the word perception model and the TRACE model are important to the functioning of these models, since parallel processing permits exploitation of mutual constraints, and to a large extent it is the exploitation of mutual constraints that parallel distributed processing is all about. But the massive reduplication of hardwired connection information leaves something to be desired as a way of providing a mechanism to exploit these mutual constraints. For one thing, it dedicates a lot of hardware to a single task. Also, even if we were willing to give up all the hardware, there would still be a serious problem with learning. In models of this sort, learning amounts to changing the strengths of the connections among units, based on their simultaneous activation. This kind of learning is, as was stressed in Chapter 1, *local* to the particular connections in which it occurred. Thus, any learning that occurred in one letter-processing channel of the word perception model would not be available for processing letters in other channels.

I was not pleased with this state of affairs. It seemed to me that if parallel distributed processing was going to prove viable, some way would have to be found to have a central knowledge representation, such as in HEARSAY, that could be made available for processing items occurring at different places in a visual display or at different points in time.

One obvious solution is just to "go sequential," and put the knowledge in a central location and map inputs into it one at a time. We have already reviewed a way that this can be done within the PDP framework in the introduction to this section of the book. The trouble with this solution is that it eliminates parallel processing, and thus the benefits thereof. Obviously, at some point we will have to go sequential—and I will start to do so at a later point in this chapter. But I was not happy with the possibility that the only way we could achieve parallel processing was through the reduplication of connection information. I sought, in short, a mechanism for achieving parallel processing without reduplication of connection information. This chapter reports on the results of my explorations in search of such a mechanism.

The organization of the chapter is as follows. The first section develops a model for processing two words at a time, using a single central representation of the knowledge of the letter patterns that make up words. The model is applied to some recent data collected by Mozer (1983) on the processing of two-word displays. The second section extends the ideas developed in the first part to a more complex processing structure called the *programmable blackboard*, a structure analogous to the Blackboard in HEARSAY. That model is applied to a number of phenomena in word recognition not covered by the original interactive activation model. It also gives an account of several important aspects of reading a line of print, including the integration of information over successive fixations in reading. The final section discusses the proposed mechanisms in more general terms, and describes how they might be extended to the processing of the syntactic and semantic content of sentences.

CONNECTION INFORMATION DISTRIBUTION

This section describes a simple system for programming parallel processing structures in response to ongoing task demands and applies it to the processing of words in one- and two-word displays. The system consists of a set of programmable modules. Each module is a network of processing units very similar to those in other PDP models, such as the interactive activation model of word perception. The difference is that these units are not dedicated permanently to stand for particular hypotheses, and the knowledge that determines the pattern of excitatory and inhibitory interactions is not hardwired into the connections between them. Rather, the connections in the network are programmable by inputs from a central network in which the knowledge that guides processing is stored.

The first part of this section describes an individual programmable network. Later parts describe the structures needed to program such networks in response to ongoing processing demands.

A Programmable Network

Figure 2 presents a very simple hardwired network. The task of this section is to see how we could replace this hardwired network with one that could be programmed to do the same work. The network shown in the figure is a very simple interactive activation system, consisting only of a letter and a word level. The figure is laid out differently from the previous figure to highlight the excitatory connections between the

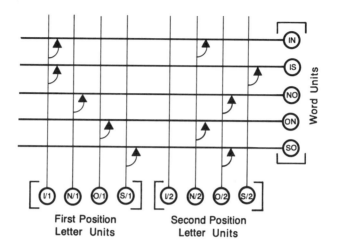

FIGURE 2. An extremely simple connectionist mechanism, capable of processing one two-letter string made up of the letters *I*, *N*, *O*, and *S*. The model knows only the five words that can be made of two of these letters, namely *IN*, *IS*, *NO*, *ON*, and *SO*. No top-down connections are included in this simple model. Units bracketed together are mutually inhibitory. (From "Putting Knowledge in its Place: A Scheme for Programming Parallel Processing Structures on the Fly" by J. L. McClelland, 1985, *Cognitive Science, 9*, p. 118. Copyright 1985 by Ablex Publishing. Reprinted by permission.)

units and lay them out in a way which will be convenient as we proceed.

In this simple network, there are detectors only for the letters *I*, *N*, *O*, and *S* in each of two letter positions. At the word level, there is a detector for each of the English words that can be made out of two of these letters. For simplicity, this model contains only letter-to-word connections; another matrix would be needed to capture word-to-letter feedback. Units that are in mutual competition are included in the same square brackets. This is just a shorthand for the bidirectional inhibitory connections, which could also be represented in another connection matrix.

In this diagram, letter units are shown having output lines that ascend from them. Word units are shown having input lines that run from left to right. Where the output line of each letter unit crosses the input line of each word unit, there is the possibility of a connection between them.

The knowledge built into the system, which lets it act as a processor for the words *IN*, *IS*, *NO*, *ON*, and *SO*, is contained in the excitatory connections between the letter and word units. These are represented by the filled triangles in the figure.

Now we are ready to see how we could build a programmable network, one that we could *instruct* to behave like the hardwired network shown in Figure 2. Suppose that instead of fixed connections from specific letter units to particular word units, there is a *potential* connection at the junction between the output line from each letter unit and the input line to each word unit. Then all we would need to do to "program" the network to process the words *IN, IS, NO, ON,* and *SO* correctly would be to send in signals from outside turning on the connections that are hardwired in Figure 2. This proposal is illustrated in Figure 3.

Multiplicative interactions yield programmable connections. At first glance, the notion of sending instructions to connections may seem to be adding a new kind of complexity to the basic processing elements out of which connectionist mechanisms are built. Actually, though, all we really need to do is to let each connection multiply two signals before passing along the result.

This point may be appreciated by considering the following equation. For the standard connections used in most connectionist models, the

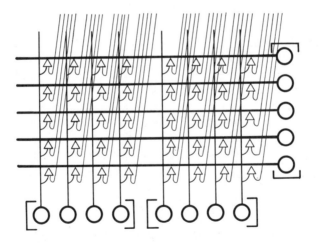

FIGURE 3. A programmable version of the simplified activation model shown in Figure 2. Each triangle represents a *programmable connection* that can be turned on by a signal coming from the central knowledge store, shown here as lying outside the figure to the upper right. If the triangular connections pass the product of the two signals arriving at their base along to the receiving unit, the lines coming into the matrix from above can be thought of as programming the network. (From "Putting Knowledge in its Place: A Scheme for Programming Parallel Processing Structures on the Fly" by J. L. McClelland, 1985, *Cognitive Science*, 9, p. 119. Copyright 1985 by Ablex Publishing. Reprinted by permission.)

time-varying signal to some unit i from some unit j is multiplied by the fixed weight or connection strength w_{ij} to determine the value of the input to i from j:

$$input_{ij}(t) = signal_j(t) \times w_{ij}.$$

All we are assuming now is that the signal from unit j is multiplied by a second time-varying signal, for example, the signal arising from some other unit k instead of the fixed connection strength w_{ij}:

$$input_{ij}(t) = signal_j(t) \times signal_k(t).$$

We can think of the signal from unit k as *setting the strength* of the connection to i from j. When the value of the second signal is greater than 0, we will say that the connection to i from j is *active*.

Function and implementation of programmable connections. Multiplicative connections of the kind proposed here were introduced in Chapter 2, and the increase in computational capability that their introduction affords is considered in Chapter 10. The specific idea of using a second signal to modulate connections has been used in other connectionist models. Hinton (1981b) used such a scheme to map inputs from local (retinocentric) feature detectors onto central (object-centered) feature detectors in a viewpoint-dependent way. My use of multiplicative connections here was inspired by Hinton's. J. A. Feldman and Ballard (1982) have also suggested the idea of making connections contingent on the activation of particular units. The general notion of using one set of signals to structure the way a network processes another set of signals has previously been proposed by Sejnowski (1981) and Hinton (1981a).

Let us briefly consider the functional significance of programmable connections. In essence, what connections do in PDP models of perceptual processing is specify *contingencies* between *hypotheses.*[1] A positive weight on the connection to unit i from unit j is like the conditional rule, "if j is active, excite i." Fixed connections establish such contingencies in a fixed, permanent way. Programmable connections allow us to specify what contingencies should be in force, in a way which is itself contingent on other signals. By using multiplicative interactions between signals, in place of fixed connections, we now have a way of setting from outside a network the functional connections or contingencies between the units inside the network. This means that we can dynamically program processing modules in response

---

[1] I would like to thank Geoff Hinton for pointing out the relation between connections and contingencies.

to expectations, task demands, etc. The little module shown in Figure 3 could be used for a variety of different processing tasks, if different connection patterns were sent into it at different times. For example, if we sent in different signals from outside, we could reprogram the module so that the word units would now respond to the two-letter words in some other language. In conjunction with reprogramming the connections from feature level units to the letter units, we could even assign the network the task of processing words in a language with a different alphabet or of processing completely different kinds of patterns.

At a neurophysiological level, multiplicative or quasi-multiplicative interactions between signals can be implemented in various ways. Neurons can implement multiplication-like interactions by allowing one signal to bring the unit's activation near threshold, thereby strongly increasing the extent to which another signal can make the unit fire (Sejnowski, 1981). There are other possibilities as well. A number of authors (e.g., Poggio & Torre, 1978) have suggested ways in which multiplication-like interactions could take place in subneuronal structures. Such interactions could also take place at individual synapses, though there is little evidence of this kind of interaction in cortex. For a fuller discussion of these issues, see Chapters 20 and 21.

The Connection Information Distribution Mechanism

We are now ready to move up to a complete Connection Information Distribution (CID) mechanism containing a number of programmable modules along with the structures required to program them. The basic parts of the mechanism are shown and labeled in Figure 4; they are shown again, with some of the interconnections, in Figure 5.

The CID mechanism consists of a central knowledge store, a set of programmable modules, and connections between them. The structure is set up in such a way that all of the connection information that is specific to recognition of words is stored in the central knowledge store. Incoming lines from the programmable modules allow information in each module to access the central knowledge, and output lines from the central knowledge store to the programmable modules allow connection activation information to be distributed back to the programmable modules.

The two programmable modules are just copies of the module shown in Figure 3. It is assumed that lower-level mechanisms, outside of the model itself, are responsible for aligning inputs with the two modules, so that when two words are presented, the left word activates

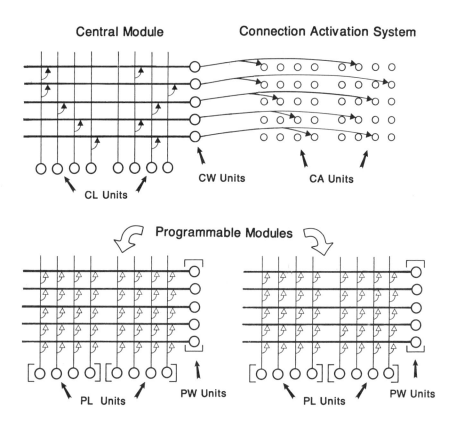

FIGURE 4. A simplified example of a connection information distribution (CID) mechanism, sufficient for simultaneous bottom-up processing of two two-letter words. The programmable modules consist of the programmable letter (PL) units, programmable word (PW) units, and programmable connections between them (open triangles). The central module consists of a set of central letter (CL) units, a set of central word (CW) units, and hardwired connections between them (filled triangles). The connection activation system includes the central word units, a set of connection activation (CA) units, and hardwired connections between them. Connections between the central knowledge system (central module plus connection activation system) and the programmable modules are shown in the next figure. (From "Putting Knowledge in its Place: A Scheme for Programming Parallel Processing Structures on the Fly" by J. L. McClelland, 1985, *Cognitive Science*, 9, p. 122. Copyright 1985 by Ablex Publishing. Adapted by permission.)

appropriate programmable letter units in the left module, and the right one activates appropriate programmable letter units in the right module.

The central knowledge store. The knowledge store in the CID mechanism is shown at the top of Figure 4. This is the part of the mechanism that contains the word-level knowledge needed to program

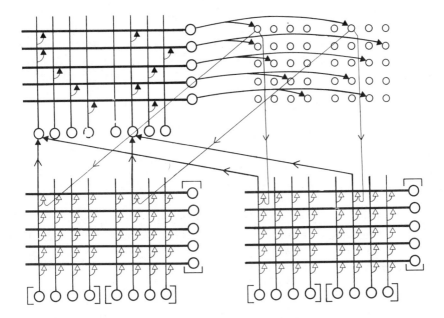

FIGURE 5. Each CA unit projects to the corresponding connection in both program-mable modules, and each central letter unit receives projections from the corresponding programmable letter unit in both programmable modules. The inputs to two central letter units and the outputs from two CA units are shown. (From "Putting Knowledge in its Place: A Scheme for Programming Parallel Processing Structures on the Fly" by J. L. McClelland, 1985, *Cognitive Science*, 9, p. 124. Copyright 1985 by Ablex Publishing. Reprinted by permission.)

the programmable modules. It consists of two parts. One part is called the *central module*, and the other part is called the *connection activation system*.

The central module consists of central letter units, central word units, and connections between the central letter units and the central word units. The letter units in the local modules project to the letter units in the central module, so that whenever a particular letter unit is active in either programmable module, the corresponding central letter unit is also (Figure 5). Note that the correspondence of local and central letter units is quite independent of what letters these units stand for.

The central letter units are connected to the central word units via connections of the standard, hardwired type. These connections allow patterns of activation at the letter level to produce corresponding activations at the word level, just as in the original interactive activation model. However, it should be noted that the central word unit activations are based on a superposition of the inputs to each of the two programmable modules. Thus, the activations in the central letter units do not specify which module the letters came from, though relative position within each module is preserved. Thus, activations in the central module do not distinguish between the input *IN SO* and the input *SO IN* or even *SN IO*. In short, the central module cannot correctly determine which aspects of its inputs belong together.

The second part of the central knowledge system, the connection activation system, also consists of two sets of units and their interconnections. One of these sets of units is the set of central word units—they belong both to the central module and to the connection activation system. The other set is the set of connection activation (CA) units. The purpose of the connection activation system is to translate activations of central word units into activations of connections appropriate for processing the corresponding words in the local modules. The CA units serve as a central map of the connections in each of the programmable modules, and provide a way to distribute connection information to all of the programmable modules at the same time. (The CA units are not strictly necessary computationally, but they serve to maintain the conceptual distinction between that part of the mechanism that contains the knowledge about words and the parts that simply distribute that knowledge to the local modules). There is one CA unit corresponding to the connection between a particular programmable letter unit and a particular programmable word unit. I have arranged the CA units in Figure 4 to bring out this correspondence. *Each* CA unit projects to the corresponding connection in *both* programmable modules. I have illustrated the projections of two of the CA units in Figure 5. For example, the top-left CA unit corresponds to the connection between the left-most programmable letter unit and the top-most programmable word unit. This CA unit projects to its corresponding connection in each of the programmable modules and provides one of that connection's two inputs. So, when a particular CA unit is active, it activates the corresponding connection in *all* of the programmable modules. In this way it acts as a master switch.

At a functional level, we can see each CA unit as standing for a contingency between two activations. Thus, if we index the programmable letter units by subscript i, and the programmable word units by j, the ijth CA unit stands for the contingency, "if letter unit i is active, excite word unit j." Thus, we can think of the CA units as contingency activation units, as much as connection activation units. When we activate a CA unit (to a certain degree) we are implementing the contingency it represents (with a corresponding strength) in both of the programmable modules at once.

The central word units, of course, are responsible for activating the CA units. There are excitatory connections from each word unit to each of the CA units for the connections needed to process the word. For example, the central word unit for *IN* activates two CA units. One is the CA unit for the connection between the left-most programmable letter unit and the top-most programmable word unit. The other is the CA unit for the connection from the sixth programmable letter unit from the left to the same programmable word unit. These connections effectively assign the top programmable word unit to be the detector for *IN* (assuming, of course, that lower levels of processing have been arranged so that *I* in the first position and *N* in the second position activate the appropriate programmable letter units).

In summary, the CID mechanism consists of (a) the programmable modules; (b) the central knowledge store, including the central module and the connection activation system; (c) converging inputs to the central knowledge store from the programmable modules; and (d) diverging outputs from the central knowledge store back to the programmable modules.

We can now see how this mechanism allows the programmable modules to be programmed dynamically in response to current inputs. When an input causes activations in some of the programmable letter units in one of the programmable modules (say, the programmable letter units for *I* in the first position and *N* in the second position of the left programmable module), these activations are passed to the corresponding central letter units. From the central letter units they activate the central word unit for *IN*. Central word units for patterns that overlap partially with the input (such as *IS* and *ON*) also receive excitation, but only in proportion to their overlap with the input. The central word units pass activation to the CA units, and these in turn pass activation back to the connections in both programmable modules. Connections are only turned on to the extent that they are consistent with the input. When different patterns are presented to each programmable module, connections appropriate for both patterns are turned on, thereby programming both programmable modules to process either pattern. Central word units—and therefore connections—are also turned on for any words that appear in the superimposed input from the two programmable modules. However, the results of processing in each programmable module still depend on the activations of the programmable letter units. Thus the appropriate programmable word unit will tend to be the most active in each local module. Although the output of the central module does not specify which word was presented to which local module, this information is represented (though with some tendencies to error, as we shall see) in the outputs of the local modules.

The correspondence between programmable and central units in the CID mechanism illustrated in Figures 4 and 5 may lead some readers to feel that the local units are really dedicated to process the particular word I have said that the mechanism programs it to process. This is an artifact of the use of one unit to represent each word. If distributed representations are used instead, each local output unit unit can be programmed in different ways on different occasions, depending on which of a large number of different distributed patterns the local modules are programmed to produce. The discussion that follows is generally applicable to both local and distributed CID mechanisms; I chose to use the local case because I thought it was generally easier to grasp intuitively. The main difference between local and distributed CID mechanisms is that the latter make much more efficient use of the units in the programmable modules, as discussed in Chapter 12.

Computer Simulation of Word Recognition Using the CID Mechanism

To examine the behavior of CID mechanisms in more detail, I implemented a CID version of the interactive activation model of word perception. The model, which I just call CID ("Sid"), is scaled-up from the example we have been considering so that it can process two strings of four letters each. Only three or four different letter alternatives were allowed in each position within each string. These were B, L, P, and S in the first position; A, E, I, and O in the second position; N, R, and V in the third position; and D, E, and T in the fourth position. The lexicon used in the simulation consisted of the 32 words shown in Table 1.

Like the smaller-scale version shown in the figures, the model consisted of two programmable modules, one for each of the two-letter strings, and a central knowledge store consisting of the central module and the connection activation system. Each programmable module had 16 programmable letter units and 32 programmable word units. The programmable letter units were grouped into four groups of four, with each group to be used for letters in one display location. The members of each group had mutual, hardwired, inhibitory connections. Similarly, all of the programmable word units in each module were mutually inhibitory. Each programmable module contained $16 \times 32 = 512$ programmable connections, and there were 512 CA units, one for each each programmable connection. The central module contained 16 letter and 32 word units, like the programmable modules. There were no inhibitory connections either between the central word units or between the central letter units. The connections between the central letter

TABLE 1

THE 32 WORDS USED IN THE SIMULATIONS

BAND	BARE	BEND	BIND
BIRD	BOND	BONE	BORE
LAND	LANE	LARD	LEND
LINE	LINT	LIVE	LONE
LORD	LOVE	PANE	PANT
PART	PINE	PINT	POND
PORE	PORT	SAND	SANE
SAVE	SEND	SORE	SORT

Note: From "Putting Knowledge in its Place: A Scheme for Programming Parallel Processing Structures on the Fly" by J. L. McClelland, 1985, *Cognitive Science*, 9, p. 126. Copyright 1985 by Ablex Publishing. Reprinted by permission.

units and the central word units and the connections from the central word units to the appropriate CA units were hardwired with the connection information needed to make the central letter units activate the right central word units and to make the central word units activate the right CA units.

Inputs to the simulation model were simply specifications of bottom-up activations to the programmable letter units in either or both programmable modules. Inputs were presented when all the units in the model were at their resting activation values and turned off after some fixed number of time cycles.

The only substantive difference between CID and the original interactive activation model is in the strengths of the excitatory connections between units. In CID, these strengths vary as a function of the current input, while in the original model they were fixed. Highly simplified activation rules are used to capture the essence of the connection activation process via the central letter units, central word units, and CA units. The activation of a particular central letter unit is simply the number of input units projecting to it that have activations greater than 0. Thus, the activation of a particular central letter unit just gives a count of the corresponding programmable letter units that are active. The activation of a central word unit is just the sum of the active central letter units that have hardwired connections to the central letter unit. The activation of a CA unit is just the activation of the central word unit that projects to it, and this value is transmitted unaltered to the corresponding programmable connection in each programmable module.

The net effect of these assumptions is to make the activation of the connections coming into a particular programmable word unit proportional to the number of active units for the letters of the word, summed over both modules. Active letter units count only if they stand for letters in appropriate positions within the programmable module of origin.

Output. So far we have said nothing about how the activations that arise in the programmable modules might give rise to overt responses. Following the original interactive activation model, I assume there is a readout mechanism of unspecified implementation which translates activations at either the letter or the word level into overt responses.[2] The readout mechanism can be directed to the word or the letter level of either module, and at the letter level it can be directed to a particular letter position within a module. In cases where more than one stimulus is to be identified on the same trial, the readout of each of the items is independent.

The probability of choosing a particular response depends on the strength of the unit corresponding to that response divided by the sum of the strengths of all the relevant alternatives (e.g., units for words in the same position) following the formulas introduced in Chapter 15.

The main import of these assumptions for present purposes is that the probability of a particular response is solely a function of the activations of units relevant to the response. All interactions between display items are thus attributed to the unit and connection activation mechanisms, and not to the readout mechanisms themselves.

RESULTS OF THE SIMULATIONS

Two principle findings emerged from working with the simulation model. First, when processing a single word, the CID mechanism causes the model to behave as though it were sharply tuned to its inputs, thereby mimicking the benefits of the bottom-up inhibition used in the original word perception model without actually incurring any of its deficiencies. Second, when processing two words at a time, the connection activation scheme causes the model to make errors similar to those made by human subjects viewing two-word displays. These

---

[2] There must be coordination between the readout mechanism and the CID mechanism. For example, it would not do for the system to program the top-most letter unit to represent the word *BAND*, say, if the readout mechanism took this unit to correspond to some other word. The problem is a general one and is no different in programmable networks than it is in standard ones.

errors arise as a result of the essential characteristics of the CID mechanism.

One Word at a Time: The Poor Get Poorer

In the original word perception model, bottom-up inhibition from the letter level to the word level was used to sharpen the net bottom-up input to word units. For example, consider a display containing the word *SAND*. Due to bottom-up inhibition, units for words matching only three of the four letters shown (e.g., *LAND*) would receive less than 3/4 as much net bottom-up excitation as the unit for the word *SAND* itself.

The CID version of the model closely emulates this feature of the original, even though it lacks these bottom-up inhibitory connections. In CID, the activation of the *connections* coming into a word unit varies with the number of letters of the word that are present in the input. At the same time, the number of inputs to these same connections from the programmable letter units also varies with the number of letters in the input that match the word. The result is that in the CID version of the model, the amount of bottom-up activation a programmable word unit receives varies as the *square* of the number of letters in common with the input. Poorer matches get penalized twice.

In working with the original model, Rumelhart and I picked values for the bottom-up excitation and inhibition parameters by trial and error as we searched for a set of parameters that allowed the model to fit the results of a large number of experiments. The values we hit upon put the strength of bottom-up inhibition at 4/7 the strength of bottom-up excitation. For words that share two, three, or all four letters in common with the input, this ratio produces almost exactly the same relative amounts of net bottom-up activation as is produced by the CID mechanism (Table 2). Words with less than two letters in common received net bottom-up inhibition in the old version, whereas in the CID version they receive little or no excitation. In both cases their activation stays below zero due to competition, and thus they have no effect on the behavior of the model.

This analysis shows that the CID version of the model can mimic the original, and even provides an unexpected explanation for the particular value of bottom-up inhibition that turned out to work best in our earlier simulations. As long as the bottom-up input to the letter level was unambiguous, the correspondence of the CID version and a no-feedback version of the original model is extremely close.

When the bottom-up input to the letter level is ambiguous, however, there is a slight difference in the performance of the two versions of

TABLE 2

ONE WORD AT A TIME: BOTTOM-UP ACTIVATIONS OF SEVERAL
WORD UNITS IN THE ORIGINAL AND CID VERSIONS OF THE
INTERACTIVE ACTIVATION MODEL

Input: SAND

Unit	Letters Shared w/Input	Original Relative Activation	Ratio	CID Version Relative Activation	Ratio
SAND	4	4	-	4×4	-
LAND	3	3 - 4/7	.61	3×3	.56
LANE	2	2 - 8/7	.21	2×2	.25

Note: Ratio is the net bottom-up activation of the unit, divided by the net bottom-up activation of the unit for SAND. From "Putting Knowledge in its Place: A Scheme for Programming Parallel Processing Structures on the Fly" by J. L. McClelland, 1985, *Cognitive Science*, 9, p. 129. Copyright 1985 by Ablex Publishing. Reprinted by permission.

the model. This actually reveals an advantage of eliminating bottom-up inhibition similar to some of the advantages discovered in the discussion of the TRACE model (see the previous chapter). Consider the input to a word unit from the letter units in a particular letter position. In the original model, if three or more letter candidates were active, two of them would always produce enough bottom-up inhibition to more than outweigh the excitatory effect any one of them might have on the word. For example, if *E*, *F*, and *C* are equally active in the second letter position, *F* and *C* together would inhibit the detectors for words with *E* in second position more than *E* will excite them. Thus, if three letters are active in all four letter positions, no word would ever receive a net excitatory input. This problem does not arise in the CID version because there is no bottom-up inhibition. Thus, the CID version can pull a word out of a highly degraded display in which several letters are equally compatible with the feature information presented, while the original model cannot. It thus appears that CID gives us the benefits of bottom-up inhibition without the costs.

Two Words at a Time: Interference and Crosstalk

So far we have seen how CID retains and even improves on some of the important aspects of the behavior of the original model. Now, I will show how CID captures important aspects of the data obtained in experiments in which subjects are shown two words at a time. Here the CID architecture becomes essential, since simultaneous processing of two patterns introduces considerations that do not arise in the processing of one pattern at a time.

When letters are presented to both modules, *all* of the letters are combined to turn on connections that are distributed to *both* of the programmable modules. The result is that the connections appropriate for the word presented in one module are turned on in the other module as well. This biases the resulting activations in each module. The programmable word unit for the word presented to a particular module will generally receive the most activation. However, the activation of programmable word units for words containing letters presented to the other module is enhanced. This increases the probability that incorrect responses to one of the words will contain letters presented in the other.

At first this aspect of the model disturbed me, for I had hoped to build a parallel processor that was less subject to crosstalk between simultaneously presented items. However, it turns out that human subjects make the same kinds of errors that CID makes. Thus, though CID may not be immune to crosstalk, its limitations in this regard seem to be shared by human subjects. I'll first consider some data on human performance, and then examine in detail why CID behaves the same way.

The data come from a recent experiment by Mozer (1983). In his paradigm, a pair of words (e.g, *SAND LANE*) is displayed, one to the left and one to the right of fixation. The display is followed by a patterned mask which occupies the same locations as the letters in the words that were presented. In addition, the mask display contains a row of underbars to indicate which of the two words the subject is to report. Subjects were told to say the word they thought they saw in the cued location or to say "blank" in case they had no idea.

In his first experiment, Mozer presented pairs of words that shared two letters in common. The pairs of words had the further property that either letter which differed between the two words could be transposed to the corresponding location in the other word and the result would still be a word. In our example *SAND-LANE*, *SAND* and *LANE* have two letters in common, and either the *L* or the *E* from *LANE* can be moved into the corresponding position in *SAND*, and the result would still be a word (*LAND* and *SANE*). Of course, it was also always true with these stimuli that the result would be a word if both letters "migrated." The duration of the two-word display was adjusted after each counterbalanced block of trials in an attempt to home in on a duration at which the subject would get approximately 70% of the whole-word responses correct. Thus, the overall error rate was fixed by design, though the pattern of errors was not.

The principal results of Mozer's experiment are shown in Table 3. Of the trials when subjects made errors, nearly half involved what Mozer called "migration errors" — errors in which a letter in the context word showed up in the report of the target. To demonstrate that these errors were truly due to the presentation of these letters in the context, Mozer showed that these same error responses occurred much less frequently when the context stimulus did not contain these letters. Such "control" errors are referred to in the table as pseudo-migration errors.

As I already suggested, migration errors of the type Mozer reported are a natural consequence of the CID mechanism. Since the letters from both words are superimposed as they project onto the central module, the connections for words whose letters are present (in the correct letter position) in either of the two input strings are strongly activated in both programmable modules. The result is that programmable units for words containing letters from the context are more easily activated than they would be in the absence of the input presented to the other module.

Table 4 compares relative programmable word unit activations for various words, for two different cases: In one case, the word *SAND* is

TABLE 3

METHOD AND RESULTS OF MOZER (1983), EXPERIMENT 1

Method:		
Example Display		SAND LANE
Target Cue		‾‾‾‾‾
Results:		
Response Type	Example	% of Total
Correct response	(SAND)	69.0
Single migration	(SANE or LAND)	13.3
Double migration	(LANE)	0.5
Other		17.2
Total		100.0
Pseudo-migration*		5.3

*Pseudo-migration rate is the percentage of reports of the given single migration responses (SANE,LAND) when a context word which does not contain these letters is presented. In this example, the context string might have been BANK. From "Putting Knowledge in its Place: A Scheme for Programming Parallel Processing Structures on the Fly" by J. L. McClelland, 1985, *Cognitive Science, 9*, p. 131. Copyright 1985 by Ablex Publishing. Reprinted by permission.

presented alone; in the other, it is presented in the context of the word *LANE*. When *SAND* is presented alone, all words that share three letters with it receive $(3/4)^2$ or $9/16$ as much bottom-up activation as the unit for *SAND* itself—we already explored this property of the CID model in the previous section. When *SAND* is presented with *LANE*, however, words fitting the pattern $(L$ or $S)$-A-N-$(D$ or $E)$ all have their connections activated to an equal degree because of the pooling of the input to the connection activation apparatus from both modules. These words are, of course, *SAND* and *LANE* themselves and the single migration error words *LAND* and *SANE*. Indeed, over both letter strings, there are 6 occurrences of the letters of each of these words (the *A* and the *N* each occur twice). The result is that the excitatory input to the programmable word units in the left module for *LAND* and *SANE* is $3/4$ of that for *SAND*, as opposed to $9/16$. Other words having three letters in common with the target have their connections less activated. Their bottom-up activation is either $5/8$ or $1/2$ that of *SAND*, depending on whether two of the letters they have in common with the target are shared with the context (as in *BAND*) or not (as in *SEND*). Thus, we expect *LAND* and *SANE* to be reported more often than other words sharing three letters in common with *SAND*.

The reader might imagine that the effect would be rather weak. The difference between $3/4$ and $5/8$ or $1/2$ does not seem strikingly large. However, a raw comparison of the relative bottom-up activation does not take into account the effects of within-level inhibition. Within-level inhibition greatly amplifies small differences in bottom-up activation.

TABLE 4

TWO WORDS AT A TIME: CROSSTALK.
RELATIVE BOTTOM-UP ACTIVATIONS PRODUCED BY SAND
PRESENTED EITHER ALONE OR WITH LANE AS CONTEXT

	alone		with LANE	
	activation	ratio	activation	ratio
SAND	4×4	-	4×6	-
LAND	3×3	.56	3×6	.75
BAND	3×3	.56	3×5	.62
SEND	3×3	.56	3×4	.50
LANE	2×2	.16	2×6	.50

Note: Ratio refers to the bottom-up activation of the unit, divided by bottom-up activation of SAND. From "Putting Knowledge in its Place: A Scheme for Programming Parallel Processing Structures on the Fly" by J. L. McClelland, 1985, *Cognitive Science*, 9, p. 132. Copyright 1985 by Ablex Publishing. Reprinted by permission.

This is especially true when two or more units are working together at the same level of activation. In this case, the units for *LAND* and *SANE* act together. Neither can beat out the other, and both "gang up" on those receiving slightly less bottom-up activation, thereby pushing these other alternatives out. This "gang effect" was also observed in the original version of the word perception model. The results of this feature of the model are illustrated in Figure 6. Through the mutual inhibition mechanism, *SAND* and *LANE* come to dominate other words that share three letters in common with the target. Some of these, in turn, dominate words that have only two letters in common with the target, including, for example, *LANE*, even though the connections for *LANE* are strongly activated. This result of the simulation agrees with the experimental result that double or "whole-word" migrations are quite rare in Mozer's experiment, as shown in Table 3.

Mozer (1983) reported several additional findings in his study of the processing of two-word displays. A full discussion of these effects can be found in McClelland (1985). Suffice it to say here that the major findings of Mozer's experiments are consistent with what we would expect from CID.

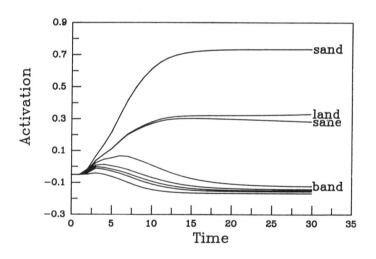

FIGURE 6. Activation curves for various programmable word units in the module to which *SAND* is shown when the input to the other module is *LANE*. The horizontal axis represents time cycles from the onset of the two-word display. (From "Putting Knowledge in its Place: A Scheme for Programming Parallel Processing Structures on the Fly" by J. L. McClelland, 1985, *Cognitive Science*, 9, p. 133. Copyright 1985 by Ablex Publishing. Reprinted by permission.)

THE PROGRAMMABLE BLACKBOARD MODEL

The model described in the previous section provides an illustrative example of the way in which knowledge in a central processing structure can be used to program local processing structures in response to ongoing processing demands, but it is a long way from a fully adequate model of the reading process, even excluding syntactic and semantic processing levels above the word. This section describes a more sophisticated model. The model is incomplete, but I believe it takes us several steps further toward a complete model of the reading process.

Overlapping the Programmable Processing Structures

In the preceding section, as in the original word perception model, we considered words of a fixed length, and we assumed that some mechanism not included in the model would solve the problem of aligning the inputs appropriately with the programmable letter units in each of the local modules.

Obviously, any plausible model of reading must be capable of accommodating arbitrary strings of text, including words of different lengths, without requiring prealignment of each word with a specific location in a module.

In the TRACE model described in the previous chapter, Elman and I dealt with this problem by allowing units in adjacent slots to have overlapping "receptive fields." The ideas from that model can be applied to reading as follows. Suppose we have several sets of letter units, one for each of a reasonably large number of letter positions in a string of text. Each of these sets of units will be the top end of a position-specific letter-processing channel, like the ones in the original word perception model. Let's suppose we can present letter strings so that the left letter projects to any one of the letter channels and adjacent letters in the string activate letter units in adjacent slots. Then the model would be able to process words starting in any slot if there were a number of overlapping word processing channels, one starting in every slot. If a string like *BINK*/4 was shown (*BINK* starting in the fourth letter position), units for *BLINK*/3, *INK*/5, *BIN*/4, and *SLINKY*/3 would all be activated, along with units for *BIND*/4, *SINK*/4 and others than would have been activated in the original model. These units would all produce feedback to the units for the letters they contain in the appropriate positions. This would allow conspiracies of partial activations of words of various lengths. The word units could also compete with each other to the extent that they spanned overlapping letter sequences, thereby allowing the model to settle on the best interpretation of each substring of letters separately. A hardwired version of this idea was implemented in the TRACE model of speech perception; the task now is to apply the same idea to a programmable processing structure.

The extension of the connection information distribution scheme to overlapping modules introduces a new problem. To see the problem, let's first look at a simple example of a hardwired word perception model with overlapping slots. Figure 7 illustrates such a case—a hardwired "blackboard" set up for processing the two-letter *I-N-O-S* words *IN*, *IS*, *NO*, *ON*, and *SO*, starting in any one of several letter positions. Each set of word units receives inputs from two adjacent sets of letter units. Members of each letter slot are mutually inhibitory. Members of word slots are mutually inhibitory to the extent that the words they stand for overlap at the letter level. As in the TRACE model, the idea is that word units should only compete in so far as they represent alternative interpretations of the contents of the same location.

A two-letter word can be presented to letter units in any two adjacent letter slots. The appropriate unit in the appropriate word slot will then become active. Thus, the second letter slot can represent the second letter of a two-letter word starting in the first position or the first letter of a word starting in the second position.

Let us consider what is involved in making a programmable version of this model. First, we would need to replace the hardwired connections in the figure with programmable connections. We would need a

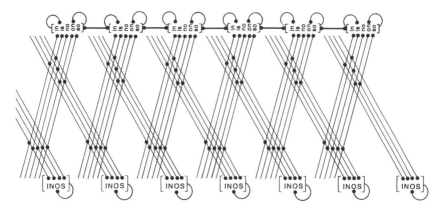

FIGURE 7. A hardwired processing structure for bottom-up processing of the words *IN*, *IS*, *NO*, *ON*, and *SO* presented to any two adjacent letter slots. Note that each letter slot participates in two different word slots, except at the edges.

lexical knowledge source, or central knowledge system, consisting of two sets of letter units, one for the first letter of the word and one for the second letter, and a set of word units. In addition, we would need a set of connection activation units, one for each programmable connection in each overlapping word-processing channel. Finally, we would need converging input lines coming into the central knowledge structure and diverging output lines from the CA units to the programmable connections.

A problem arises with the converging input lines. Since each letter slot can contain letters in either role within the word, the letter units in each letter slot would have to send outputs to the central letter units in each central letter channel. The problem with this would be that every occurrence of a particular letter, regardless of what role it turned out to play within a word, would turn on the connections for words which have that letter in them anywhere. Consider, for example, the display –ON– (dashes indicate blanks in the display). This will turn on the connections for all words having O in them anywhere, since the O will be treated as a potential first letter of a word beginning in the second position and as a potential second letter of a word beginning in the first position. Similarly, the N will turn on the weights for all words having N in them anywhere, since it will be treated as a potential first letter of a word beginning in the third position and a potential second letter of a word beginning in the second position. For this example, then, the connections for NO will be turned on just as much as the connections for ON. When multiple words are presented, this would tend to cause the model to make migration errors for nonhomologous positions. For example, in a version of the model with a forth overlapping module, the display IS-NO would activate the connections for the word IN just as much as the display IS-ON. We would therefore expect subjects to report IN as an incorrect report of the contents of the first two letter positions equally often in these two cases. More generally, migration errors would not be expected to honor relative position within the word. Analyses reported in Shallice and McGill (1978) reveal that nonhomologous migration errors are quite rare indeed.

Role-specific letter units. It appears that we need to have some way of treating letters differently depending on their role within the word. One poor way to do this would be to imagine that there is a separate detector in each letter channel for each letter in every role it might play. For our two-letter word case, letters can occur in either of two roles—as the first letter in a two-letter string, or as the last. So, if we went with role-specific letter detectors, each letter slot would contain eight letter units, one for each letter as the first letter in a word, and one for each letter as the second letter in a word.

Obviously this solution has serious weaknesses. For one thing, it duplicates units ad nauseam. If we wished to process longer words, we would need very large numbers of role-specific units in each letter slot.

Coarse coding. There is a better way. Suppose that each letter is represented, not as an activation of a single unit, but as a pattern of activation involving several active units. With such an encoding, we would imagine that a given letter—say, the letter *I*—produces a slightly different pattern when it occurs in different roles. All these different patterns will represent different versions of the letter *I*, and to the degree that they are similar they will have similar effects, but their differences can be used to produce a sufficient degree of differentiation of letters in different contexts to allow our model to work and to reduce the incidence of nonhomologous migration errors. This idea is an instance of the idea of coarse coding, which is described more fully in Chapter 3.

In the present application of the coarse coding idea, there are four detectors for each letter. Each detector serves as a partial specification of a letter/role combination. One of the detectors is activated when the letter is preceded by a blank; one when it is preceded by any other letter; one when it is followed by a blank; and one when it is followed by any other letter. With this scheme, the pattern of active letter/role units can indicate whether a particular letter is a separate string unto itself (blanks on both sides); the first letter in a string (blank on left, letter on right); the last letter in a string (letter on left, blank on right); or an internal letter in a string (letters on both sides).

Notationally, I will designate particular units by the letter they stand for in upper case, preceded or followed by an underbar to indicate a blank or by an *x* to indicate any letter. The code for a letter in a particular role would be given by two such specifications in the same parentheses, with a space between them. The code for an entire word is given by a series of letter codes in square brackets. For example, the word *EVE* would be coded [(_E Ex) (xV Vx) (xE E_)].

One property of this coding scheme is that similar letter roles produce overlapping patterns of activation. For example, the code for the letter *I* at the beginning of a word, (_I Ix), overlaps with the code for the same letter in medial position, (xI Ix). It also overlaps with the code for the same letter in isolation, (_I I_). Similarly, the code for a letter in final position, (xI I_), overlaps with the code for the same letter in medial position, (xI Ix), and with the code for the letter in isolation. However, there is no overlap of the codes for initial and final letters—disjoint subsets of units are activated in the two cases. Similarly, there is no overlap of codes for isolated letters and medial letters in strings of three letters or more.

Obviously this particular code could be improved upon to make all occurrences of a particular letter have something in common and to make nonterminal letters in different positions differ from each other. However, this scheme captures the essence of the coarse coding idea and works surprisingly well for words up to four letters long, so we will stick with it for now.

For our two-letter word example, this coding scheme does not buy us much. The advantages of the scheme only begin to emerge with longer words. However, for visualization, I represent a version sufficient for the two letter *I-N-O-S* words in Figure 8; some of the connections from the CA units to two of the programmable modules are shown in Figure 9. For two-letter words, we don't need all four different role specifications, so I've only included the $_x$ and $x_$ units where x stands for I, N, O, or S.

With these coarse-coded letter/role units in each letter channel, the units in each programmable letter channel project to a single central letter channel. Different role specifications are still preserved, so that the word *ON* will activate no central units in common with the word *NO*.

Other Enhancements in the Programmable Blackboard Model

Feedback. In describing CID, I did not discuss word-to-letter level feedback, although a version of that model has been implemented that incorporates this. For the programmable blackboard model, I wished to incorporate feedback to illustrate how the overlapping modules in CID allow feedback from activations of words of different lengths appropriately aligned with the input. I assume feedback is implemented by another set of programmable connections running from the programmable word units back down to the programmable letter units. (It is important to distinguish the connection activation process from word-to-letter level feedback. They are different things serving very different roles in the model). Because of the symmetry of the bottom-up and top-down activations, the very same set of CA units can be used to program both the bottom-up and the top-down connections. That is, the same CA unit that turns on the bottom-up connection from a particular letter unit to a particular word unit can also turn on the connection from that same word unit to the same letter unit.

Shifting the focus of attention. While there is some reason to believe we process more than one word at a time, it is clear that we do not process whole pages or even whole lines of text at a time. In reading, the eyes make saccadic jumps from word to word, stopping for

Central Knowledge System

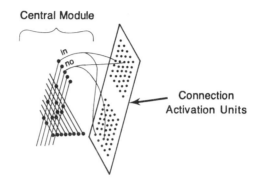

Programmable Blackboard

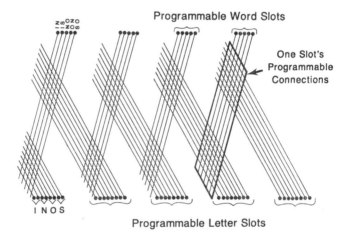

FIGURE 8. A simplified programmable blackboard sufficient for processing the two-letter *I-N-O-S* words, using coarse coded representations at the letter level. The hardwired connections between the central letter and word units are illustrated, as are the connections from the central word units to the CA units for the connections needed to process two of these words. The programmable letter units in each letter channel project to the central letter units, and the CA units project back to the programmable connections between the letter units and the word units. Some of these projections are shown in the next figure.

200-500 msec about once a word. Shorter words may be skipped and longer words may be fixated twice.

From these data, we might infer that the reading process is proceeding in a strictly sequential fashion, one word at a time. For certain

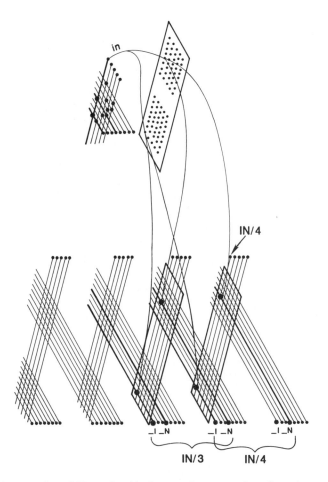

FIGURE 9. A replica of Figure 8, with the outgoing connections from the central word unit for *IN* to two of the programmable modules via the CA units. Also shown are the units for _*I* and *N*_ in the third, fourth, and fifth letter slots, and the unit for *IN* in the fourth programmable word slot. The programmable letter units in each letter channel project to the central letter units, and the CA units project back to the programmable connections between the letter and the word units, but these connections are not actually shown.

purposes, as an approximation, it may not be unreasonable to treat it so. But there is evidence that we can process more than one word per fixation, though there is some degradation of performance with an increase in the contents of the display. Of course, this finding is compatible with sequential processing within fixations, but I will assume that within a fixation, all processing activity is taking place in parallel. However, I will assume that the higher resolution of the fovea and the focusing of attention to the right of fixation causes some parts of the contents of the fixation to be processed more successfully and more completely than others.

A sketch that may help visualize these assumptions is shown in Figure 10. Several sets of programmable word and letter units are illustrated, with schematic illustrations of the connections between and within slots. Along the bottom, a fragment of a possible line of text is displayed with the letters lined up below the letter slots. The assumption is that the letters lying within the foveal region and the right penumbra are mapped into successive slots in the programmable blackboard. Converging inputs from these slots, and only these slots, are sent to the central letter units, and connection activations from the connection activation structures are projected back to the programmable connections in these slots only.

Connection activations are assumed to be "sticky": That is, it is assumed that once a connection has been programmed by the central knowledge structures, it stays that way. This allows interactive activation processes to continue in older parts of the programmable

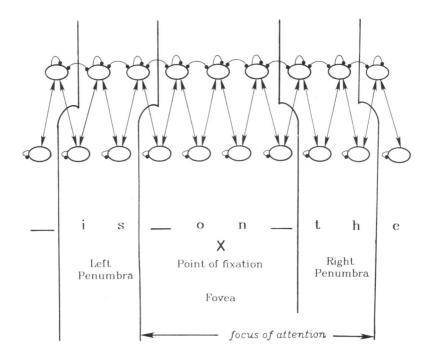

FIGURE 10. Schematic illustrations of the programmable blackboard, with illustrations of the fovea, the left and right penumbras, and the span of the focus of attention during a fixation in reading.

blackboard while they are being set up in newer parts as the eye moves along from left to right. So, what has already been processed in earlier fixations can continue to influence the processing of the contents of the current fixation, and, in case of unresolved ambiguities, can be influenced by the new information coming in.

To make this model work, we would need to imagine that when the eye moves, we change the mapping from a low-level, retinocentric representation of the contents of the current fixation into the programmable blackboard. The mapping mechanism proposed by Hinton (1981b) would be appropriate for doing this, assuming the activations of the "mapping units" could be updated in tandem with the commands to move the eyes. Hinton's scheme can also be used to map from the programmable letter units into the central letter units and from the connection activation units to the appropriate programmable connections.

Details of the Simulation Model

The simulation model developed to embody these assumptions is called PABLO, an approximate acronym for Programmable Blackboard. PABLO uses an alphabet of 9 letters, including blank (written as "_") and the letters A, C, D, E, H, I, N, and T. Its lexicon consists of all the words of one to four letters made up only of those letters that were found in a large computerized dictionary. Archaic forms, past tenses, plurals, proper nouns and abbreviations were deleted, leaving a lexicon of 92 words, as shown in Table 5. Also, blank was treated as a (special) one-letter word. As in TRACE (Chapter 15), blanks served to provide competition against words that would otherwise tend to invade the spaces between words.

PABLO contained a sufficient programmable blackboard to read lines of text 20 characters long. To accomplish this, it contained 20 overlapping programmable modules, each capable of being programmed for processing all of the words in the lexicon. Each module consisted of a set of programmable word units, four sets of programmable letter units shared with adjacent modules, and two sets of programmable connections between the letter units and the word units. One set of connections was used for bottom-up connections from the letter to the word level, the other for top-down connections from the word to the letter level.

There were 93 word units in each module, one for each of the 93 words (including blank). Each letter slot consisted of a complete set of letter/role specification units. Each set contained four units for each of

TABLE 5

THE WORDS USED IN PABLO

A	ACE	ACHE	ACID	ACNE	ACT
AD	ADD	AID	AIDE	AN	AND
ANT	ANTE	AT	CAD	CAN	CANE
CANT	CAT	CEDE	CENT	CHAT	CHIC
CHIN	CHIT	CITE	DAD	DATA	DATE
DEAD	DEAN	DEED	DEN	DENT	DICE
DIE	DIET	DIN	DINE	DINT	EACH
EAT	EDIT	END	ETCH	HAND	HAT
HATE	HE	HEAD	HEAT	HEED	HEN
HI	HIDE	HIND	HINT	HIT	I
ICE	IDEA	IN	INCH	INN	IT
ITCH	NEAT	NEED	NET	NICE	NINE
NIT	TACT	TAD	TAN	TEA	TEE
TEEN	TEN	TEND	TENT	THAN	THAT
THE	THEN	THIN	TIDE	TIE	TIN
TINE	TINT				

the nine letters, one for each partial role specification described above. Thus, there were 36 letter/role units in each letter slot.

Each complete programmable module included four letter slots. Thus, module i included letter slots i, $i + 1$, $i + 2$, and $i + 3$.

As in TRACE, competition between word units depends upon the extent of the overlap of the words the units represent. That is, each word unit inhibits every other word unit once for every letter position in which they overlap. Thus $CAT/1$ and $HAT/1$, which overlap in all three letter positions, inhibit each other three times. $CAT/1$ and $ATE/2$, which overlap in two letter positions, inhibit each other twice. Again, for simplicity, these inhibitory connections are hardwired.[3]

In addition to these programmable processing structures, there is a central knowledge structure consisting of a central module and a connection activation system. The central module consists of one set of central letter units, one set of 93 central word units, and one hardwired

---

[3] For these connections to be hardwired, knowledge of the length of the word the local word units will be allocated to must be built into the local modules. This state of affairs would obviously be unacceptable in the ultimate version of the model. These inhibitory interactions could, of course, be programmed, as suggested earlier. An alternative scheme, in which patterns of activation are coarse-coded at the word as well as the letter level, would get around the problem by using a kind of "natural competition" that occurs between patterns competing for activation over the same set of units.

connection matrix linking the central letter units to the central word units. The central unit for each word also has hardwired connections to the CA units appropriate for activating both bottom-up and top-down connections between the programmable letter units and the programmable word unit for this word.

An input to be presented to the model for processing could be a sequence of letters up to twenty letters long. The sequence could include blanks (represented by "_"). The blanks were treated as characters just as the letters were. The sequence of blanks and letters were used to specify which letter/role units should receive external excitatory input in each letter slot. The "_" character always activated the same two letter/role units no matter what preceded or followed it. So, for example, the input-specification _CAT_ caused external input to the letter/role units for _C and Cx in the second slot, xA and Ax in the third letter slot, xT and T_ in the fourth letter slot, and to the two letter/role units for "_" in the first and fifth slots.

Single and Multiple Fixations

The simulation model was studied in two modes: One involved short displays—one or at most two short words—which were presented for processing within a single fixation. The second mode involved whole lines of text. For the latter, I assumed that PABLO "read" them from left to right with one fixation per word. Fixation durations were fixed, for simplicity, though we know that they can be influenced by visual and linguistic factors (Just & Carpenter, 1980).

First I will describe the assumptions made for the multiple fixation case. It was assumed that, on each fixation, the spotlight of attention consisted of two regions: a *center* and a *penumbra*. The center of the spotlight consisted of several letter positions in the input centered around the point of fixation. The penumbra of the spotlight consisted of several more letter positions further to the right. Note that the parafoveal region to the left of the point of fixation is not included in the spotlight of attention. Evidence from a number of sources (e.g., Rayner, 1975) suggests that information in this region is not utilized.

Letters in both the center and the penumbra of the spotlight of attention are projected to the corresponding letter units. To keep things simple, I assume for present purposes that the letters in the penumbra simply produce weaker bottom-up activations than those in the center of the spotlight. A more realistic assumption would be that letters in the penumbra produce less precise activations (i.e., activations that encompass all the letters in the same shape-class) but I have foregone this assumption for now to keep things simple and as clear as possible.

The spotlight of attention falls on a particular place in the input text, and its contents are projected to the corresponding region in the programmable blackboard. This region is called the *illuminated region*. The programmable letter units in this part of the blackboard project to the central module, and the connection activation units project back to the programmable connections in the illuminated region.[4]

In single fixation mode, I assumed that both the display and the spotlight of attention were centered on the fovea. For simplicity, the entire display was assumed to fall within the center of the spotlight so that each letter in the display was given equal weight.

Details of activation dynamics. The rules for updating programmable letter and word unit activations were the same as in CID. However, the introduction of coarse coding and overlapping slots necessitated a number of changes in the quantitative parameters. The values I used were chosen so that PABLO would produce approximately the same behavior with single four-letter words as CID.

The activation assumptions for units in the connection programming loop (the central letter and word units, and the connection activator units), and for the programmable connections themselves, were kept simple as before. Two modifications were introduced, however. First, letter-level activations in the penumbra of attention were treated as weaker inputs to the central letter units than letter-level activations in the center: They only turned on the corresponding central letter units half as much as those in the center of attention. Second, activations of central word units were normalized by the length of the word. This was done so that that the connections for a word would be turned on to the same extent, regardless of word length, whenever all the letters of the word were present in the center of attention. For example, the activation of the central word unit for the word *CAT* is just a count of the active letter/role units for *CAT* (*_C, Cx, xA, Ax, xT*, and *T_*) in the center of the illuminated zone, plus .5 times a count of the active letter/role units in the penumbra, divided by 6. With the display *_CAT_*, centered in the fovea, this number would therefore be 1.0.

As before, each CA unit simply multiplies the activation value it receives from its CW unit by the appropriate global excitatory constant,

---

[4] An alternative conception would be to imagine that the contents of the fixation are always projected directly to the central module and to the illuminated region of the programmable blackboard. There are reasons for preferring this view, one of which is that it opens up the possibility that the information projected into the central module might be of higher grain than that projected to the programmable blackboard. This is desirable because higher fidelity is needed in the connection activation process than in the programmable blackboard once the correct connections have been activated. (See Chapter 12.)

depending on whether the connection is letter-to-word or word-to-letter, and passes the result on as the activation of the programmable connections it projects to in each local matrix. Projections into the penumbra were assumed to be half strength.

Finally, as noted above, the model assumes that programmable connection activations are "sticky"—they tend to retain the values programmed into them when the spotlight of attention moves on. For simplicity, I adopted the most radical version of this assumption. Input from the central module can increase the activation of a programmable connection, and, once activated to a level, it stays at that level while the rest of the line of text is processed. At the beginning of a new line, of course, the slate is wiped clean.

One Fixation at a Time

Simultaneous resolution of two tokens of the same ambiguous character. Consider the ambiguous character in *THE CHT* . Let us see what happens when this display is presented to PABLO. We assume that the whole display, with blanks on both sides, fits inside the center of attention, with the first *T* lined up with the second input position, so that each letter activates all of the appropriate letter/role units in the appropriate positions. For the two slots to which the *H* character is presented, we assume that *xH* and *Hx* and *xA* and *Ax* are all activated to equal degrees. PABLO will end up selecting *xH* and *Hx* for the first *H* and *xA* and *Ax* for the second *H* , in parallel.

Once the letter/role units have been activated in the appropriate letter level slots, they will activate the central word units for various words to various degrees. For example, the central word unit for the word *THE* will receive eight excitatory inputs: from the _*T* and *Tx* at the beginning of *THE*, from the *xH* and *Hx* in *THE*, from the *E*_ and *xE* at the end of *THE*, and from the *xH* and *Hx* in the middle of *CHT*. The central word unit for the word *CAT* will receive eight excitatory inputs as well: two from the initial *C* of *CHT*, two from the final *T* of *CHT*, and two from each (possible) medial *A*. Normalizing for length, the central word units for both words end up getting an activation value of 8/6 or 1.33.

The central word units for other words will also be turned on to some extent. For example, the unit for the word *HAT* will receive eight excitatory inputs as well. The two *Hx* units activated by the *H* figures will each contribute one input, the two *xA* and the two *Ax* activated by the two *H* figures will each contribute one input, and the *xT* and *T*_ occurring at the end of *CHT* will each contribute one input.

Note that the medial *H* turns on the weights for *HAT* half as much as an initial *H* would because it only has one letter-form feature in common with the initial *H* pattern of the word *HAT*. In any case, when we normalize again for length, the CW unit for *HAT* ends up getting an activation value of 8/6 or 1.33 as well.

The CW units for *THAT* and *CHAT* will be set at activation values of 1.25. The display contains a total of 10 tokens of the eight letter/role elements of each of these two words, and the denominator for each is 8. It is important to remember that the activation of central word units do not directly determine what is seen; they simply turn on connections in the programmable blackboard and thereby permit local letter units to interact with local word units. We now examine the activations that result from these interactions.

Letter and word activations resulting from *THE CHT* after 10 cycles of processing are shown in Figure 11. The model has successfully completed the first word as *THE* and the second as *CAT* by the end of 10

THE CHT

FIGURE 11. Word- and letter-level activations in PABLO in response to *THE CHT*. At the word level (upper panel), each rectangle represents a word unit. The vertical position of the rectangle indicates the activation of the unit, and the horizontal placement relative to the input shown at bottom indicates the portion of the input that the unit spans. At the letter level (lower panel), the height of each square represents the average activation of the two active letter/role units representing the letter inscribed in the square.

cycles and is in the process of reinforcing the correct interpretation of the *H* figure in each position through feedback. Words such as *HAT* whose weights are strongly activated play a stronger role in the competition than words such as *CAN*, which contain letters that do not occur anywhere in the input string, but even *HAT*, as well as *THAT* and *CHAT*, are eventually driven out. After 30 cycles (not shown) *THE*/2 and *CAT*/6 are the only word units left active, and, at the letter level, the *H* has beaten *A* in *THE* while *A* has beaten *H* in *CHT*.

Conspiracies of Words of Different Lengths

So far what we have illustrated could have been done in a three-letter word version of CID, though the details about what words got activated would have been somewhat different. Now, however, we can present the words starting in any letter position—as long as they do not fall off either end of our series of overlapped modules—and they can be of any length up to the maximum length allowed by the structural parameters of the model. An even more important advantage of PABLO is that, when nonwords are shown, they can profit from conspiracy effects arising from words of different lengths. The mechanism causes the activations of the conspirators to be aligned appropriately with the input.

PABLO has been tested with each of the words that it knows and with pseudowords made from each word by replacing either a vowel with a vowel or a consonant with a consonant. In all cases, the letter string was presented alone, surrounded by blanks. The model works well with either words or pseudowords. When a word of any length is shown surrounded appropriately by blanks, the appropriate word unit always wins the competition. When an nonword is shown, both shorter and longer strings can participate, as shown in Figure 12. In the figure, several representative conspiracies are shown for nonwords of four, three, and two letters in length. Though in some cases one word the model knows is a much better match than all the others, and therefore dominates the competition, in other cases there is a much more homogeneous conspiracy. Generally speaking, words both longer and shorter than the item shown get into the act—though of course no words longer than four letters are activated, because they have not been included in the model.

Within-word crosstalk. One idiosyncrasy of PABLO needs to be mentioned. It has a tendency to activate words that have repetitions of the same letter more strongly than words that do not, even when only one copy of the letter is shown. For example, *DAT* activated *DATA*

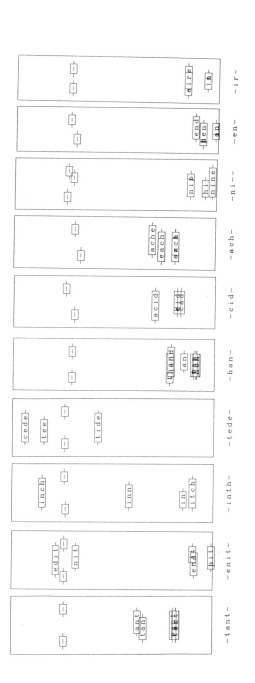

FIGURE 12. The sets of conspirators active at the word level 10 cycles after presentation of the display shown at the bottom of each column. Relative height corresponds to activation. There are several cases in which two or more words are overprinted on each other, particularaly in the cases of *HAN* (where *RAN, TAN, PAN, CAN, HAT, HAD,* and *HEN* are all active) and *TANT* (where *RANT, CANT, PANT, TART, TACT, TENT,* and *TINT* are all active).

more strongly than it activated *DATE*. The reason is that the *xA* unit activated by *DAT* turns on the weights for *DATA* two times—that is, there is a double excitatory link from this letter/role unit to the CW unit for *DATA*. At the present, I know of no data to indicate whether this characteristic is consistent with human behavior or not.

There is another kind of within-word crosstalk that occurs in PABLO for which there is evidence. Connections for words that are anagrams of the display are strongly activated, particularly if they involve switches of the two internal letters. For example, the display *CLAM* activates the connections for *CALM* as much as the connections for *CLAM*, and the display *BCAK* activates the weights for *BACK* as much as *BACK* itself does. These connection activations make *CALM* a reasonably likely error response when *CLAM* is shown and make *BACK* a very likely error response when *BCAK* is shown. Johnston, Hale, and van Santen (1983) have reported evidence of such "transposition" errors in word perception. Generally, such errors are attributed to imprecise positional specificity, so that letters in one position activate detectors for words having these same letters in adjacent positions. In PABLO, such effects occur for a slightly different reason: It is not that the model does not keep straight which letter occurred where, it is just that the description of a letter that is projected into the central module is very similar to the description that would be projected by the same letter in a similar role. The results of simulations illustrating these effects are shown in Figure 13. One feature of these data was originally puzzling when Johnston et al. first reported them. On the one hand, it seemed that subjects had very accurate letter position information, since they only rarely rearranged letters in real words to make other real words. On the other hand, they acted as though they had poor letter position information when the rearrangement of the string formed a word but the display itself did not. In PABLO, the reason for the difference is simply that in the first case, there is a word that matches the input better than the transposition word; due to competitive inhibition, the correct word dominates the transposition word. When there is no correct word, the transposition word is not dominated, so it may become quite strongly activated. This effect is clearly illustrated in the simulation shown in Figure 13.

Effects of Display Length on Amount of Feedback

One additional feature of PABLO is that shorter displays receive less benefit from feedback than longer displays. This is true, even though I have normalized the activation of central word units, and therefore of

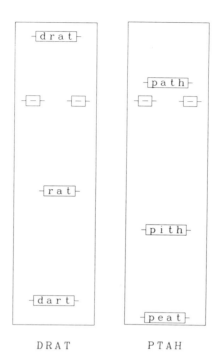

FIGURE 13. Word-level activations produced by two displays which can be made into words by rearranging the central two letters.

connection strengths, for word length. The reason for this is simply that shorter displays activate fewer letter/role units than longer ones, and therefore provide less excitatory input to the word level than longer displays. Word units are less strongly activated than they are by longer displays. Consequently, less feedback is generated. This occurs both for word and for pronounceable nonword displays. The activations shown in Figure 12 illustrate this aspect of the model in the case of pronounceable nonword displays.

At first I found this characteristic of the model undesirable. However, recent evidence collected by Samuel, van Santen, and Johnston (1982) is consistent with the model as it is. Samuel et al. examined accuracy of letter identification for letters in one-, two-, three-, and four-letter words and random letter strings and found that letter identification accuracy increased as a function of word-length for letters in words but not for letters in random strings. The advantage for letters in words over letters in random strings was either very small or nonexistent for one-letter words and grew as the words grew longer, up to about four or five letters. Thus, the data appear to support this aspect of the model's behavior. It should be noted that this aspect of the model does not depend on the connection activation aspects of PABLO but would occur in a hardwired version of the original model, as long as the net excitatory input to the word unit from a complete specification of the word varied linearly with the length of the word.

Sequences of Fixations

Up to now I have examined the behavior of PABLO within single fixations. Now, we will examine the model as it reads along a line of text. The present version of the model is fairly rudimentary, and there are a lot of important properties of human reading that it does not capture. However, it does read in a limited sense: It makes a series of fixations, and builds up a representation of the letters and words in its programmable blackboard as it moves along, as illustrated in Figure 14. PABLO's reading in this simulation is not affected by semantic and syntactic constraints since it lacks these higher levels. Furthermore, as it presently stands, PABLO does not even monitor its own performance. Instead, each fixation lasts a fixed number of cycles. PABLO does adjust the point of its fixation for the word it is viewing, however: It fixates on the center of words containing an odd number of letters, or the middle of the letter before the center for words containing an even number of letters. [5]

Although it lacks many of the features of Thibadeau, Just, and Carpenter's (1982) READER model, PABLO does show one property that READER does not: It integrates information over successive fixations. That is, while it is looking at one word in a particular fixation, it is also gathering some information about the next word. This is, of course, something that human readers do also (Rayner, 1975; for a dissenting view, see McConkie, Zola, Blanchard, & Wolverton, 1982).

We can see that PABLO is picking up information from peripheral vision in Figure 14, but it is hard to tell from the figure how much difference this makes. To get a handle on this matter, I repeated with PABLO the same experiment Rayner (1975) did to demonstrate the pickup of peripheral information in human readers. That is, I adjusted the display during PABLO's saccade, so that what the model saw in the periphery and what it saw when it later fixated the previously peripheral location was not necessarily the same. Rayner showed that when the

---

[5] The reader should be clear about the fact that PABLO's fixation shifting apparatus is *not* simulated using interactive activation processes, though I think it could be. A saccade for PABLO simply amounts to moving a pointer into the array of input letters and another into the programmable blackboard.

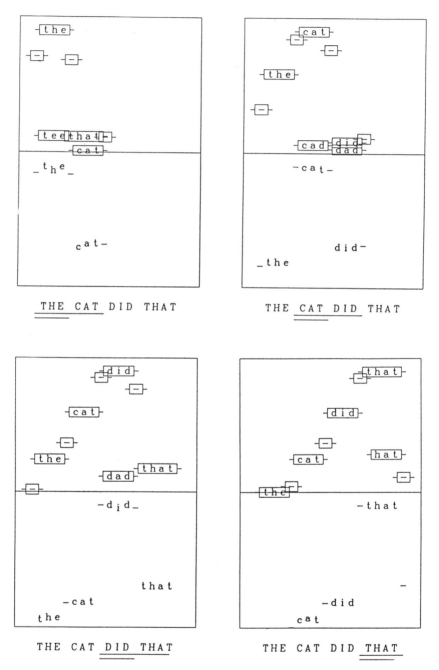

FIGURE 14. Pattern of activation at the letter and word level in PABLO, at the end of each fixation, as the model reads along a line of text. Location of the fovea and penumbra of the spotlight of attention is indicated with underbars.

information in peripheral vision differed from the information later fixated directly, subjects spent longer fixating the word than they did in the normal case in which the preview and the foveal view were the same. Other experiments have shown that subjects can name the fixated word more quickly if it matches the preview than if the preview is different (Rayner, McConkie, & Ehrlich, 1978).

Figure 15 illustrates PABLO's behavior in the following experiment. The model was given a two-word phrase and read it in two successive fixations. During the second fixation, when the fixation point was centered on the second word, the two-word phrase was always *CAR RIDE*. During the first fixation, however, there were different versions of the second word. In one case, the second word was intact on the first fixation. In the second case, the two internal letters were replaced with other letters having the same general shape. In the third case, all four letters were changed.

The figure shows that the activation of the word *RIDE* is greater at the end of the second fixation in the first case than in the other two cases. In the third case, the activation of *RIDE* is considerably less at the end of the second fixation than it would be without any preview of the second word at all (indicated by the curve marked *XXXX* in the figure). Generally, PABLO shows a facilitation, as it does in this example, when the preview matches the target. It also generally shows an interference effect when there is a gross mismatch between preview

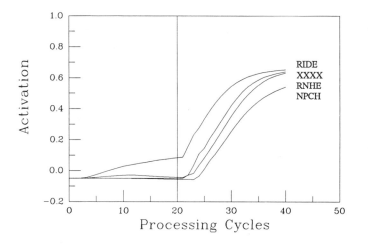

FIGURE 15. Time course of activation of the programmable word unit for the word *RIDE* in the sixth word channel of the programmable blackboard when the preview was *RIDE, XXXX, RNHE,* or *NPCH.* The *XXXX* marks the case of no preview information at all, as though there had never been a first display.

and target. For intermediate cases, it can show either a facilitation or an interference, depending on the target and the exact distortion of it. On the average, the intermediate condition shows a slight interference effect when two letters are changed, as in the example shown in the figure.

To relate these results to Rayner's, we would begin by assuming that human readers, unlike PABLO, move on to the next fixation when their certainty of the contents of one fixation reaches a satisfactory level. In the simulation, PABLO has a fixed duration for each fixation, but if the fixation was terminated when the activation of one of the words reached some criterion, then the duration of the fixation on the target would have been shortest in the case where the preview matches the target and longest in the case where the two differ the most, just as Rayner found.

Thus, PABLO integrates information over successive fixations, and conforms to the main result of Rayner's experiment. Though PABLO does not capture all the details, I would expect it to capture more of them if some of the simplifying assumptions were replaced with more realistic ones. For example, end letters appear to be more critical to human readers than to PABLO, and people are sensitive to the similarity of shape between original and replacement letters. A more elaborate version of PABLO, including a real feature level, would surely show these effects.

DISCUSSION

I have described two models that rely on the same basic principles of connection information distribution from a central knowledge source to local, programmable modules. CID, the first model, illustrated some basic properties of the connection information distribution scheme, but it had serious limitations because of its fixed-length, nonoverlapping programmable modules. PABLO overcame this deficiency of CID and provided a framework for beginning to think about the problem of reading by a series of fixations and integrating information from one fixation to the next.

Benefits of Connection Information Distribution

At this point, it is appropriate to step back from these particular models and take a look at the more general implications of the idea of connection information distribution. I believe this is quite a powerful idea. Connection information distribution allows us to instruct parallel processing structures from outside the network, making their behavior contingent on instructions originating elsewhere in the network. This means, for example, that the way a network responds to a particular input can be made contingent on the state of some other network in the system, thereby greatly increasing the flexibility of parallel processing mechanisms.

Perhaps the most general way of stating the benefit of connection information distribution is to note that it is analogous, in a way, to the invention of the stored program. [6] Before programs were invented, special-purpose logic circuits were built to carry out particular tasks; the invention of the computer program made it possible for a single, general-purpose machine to carry out any one of a huge number of tasks. The use of centrally stored connection information to program local processing structures is analogous. This allows the very same processing structures to be programmed to perform a very wide range of different tasks.

More specifically, there are two computational problems facing PDP models to which connection information distribution can be beneficially addressed. Connection information distribution allows multiple tokens of the same type to be active at once. PABLO, for example, would have no difficulty processing the same word twice if it occurred in two different locations in the programmable blackboard, and higher levels would be able to make different choices from alternative word activations in different places, just as the word level in PABLO was able to make different choices between A and H for the two instances of the H character in THE CHT. The ability to treat different tokens of the same type separately is, of course, very important in such domains as sentence processing, in which several tokens of the same kind of object (e.g., noun phrase, sentence, etc.) must be kept straight and assigned to appropriate roles in larger structures.

Connection information distribution also carries out a form of what is known in production systems as "resolution," binding the right tokens in the blackboard together into higher-order structural patterns. In PABLO, we have the following situation: Activations from everywhere in the programmable blackboard (analogous for present purposes to the working memory in production system models) are projected into the central knowledge system (analogous to the production memory). Central word units (analogous to productions) are activated to the extent the input from anywhere in the programmable blackboard activates them. The problem faced in production systems is to avoid spuriously

---

[6] I thank Gary Cottrell for pointing out this analogy.

firing productions when the activations in working memory lack the proper overall organization and to bind the correct pieces together, so that each production has the correct set of arguments rather than a mélange of arguments coming from different places. If spurious activations occur and the bindings are done improperly, the consequence will be that the wrong things will be added to the working memory. The ultimate problem faced in PABLO is analogous. We need to avoid allowing the wrong word units to get activated in the programmable blackboard by spurious conjunctions of letters. PABLO solves this problem, not by explicitly binding productions to particular elements and checking that they occur in the right conjunctions, but by distributing connection information so that when and if the right combination of elements occurs, the appropriate word unit will be activated. This allows different elements to be bound to different higher-order structures in parallel, as we saw in the case of *THE CHT*; one of the *H*'s was "bound" to an appropriate "instantiation" of *THE*, and the other to an appropriate instantiation of *CAT*. This scheme does involve some crosstalk, as we saw, but the crosstalk is of about the same kind and magnitude as the crosstalk shown by human subjects when they have to process more than one pattern in parallel. In short, it looks as though the CID model captured pretty well some of the features of the way elements are bound together in human information processing.

Costs of Connection Information Distribution

At first glance, it may seem that the ability to program parallel processing structures is being purchased at a very high price. For it appears that, wherever we had a connection in a hardwired PDP model, we now need a lot of extra circuitry to program that connection. The cost appears particularly high in terms of connection activation units and programmable connections. If there are $n \times n$ units in a programmable module, it appears that we need n^2 programmable connections and the same number of CA units to activate the programmable connections. This can get very expensive very quickly as processing structures get to realistic sizes.

Actually, however, the situation is not as bleak as this observation seems to suggest. It turns out that with programmable modules, it is possible to process patterns with far fewer units than we would need in a hardwired parallel processing structure. The reason is that the size of a module required to process a pattern without error depends on the number of patterns the module must be prepared to process at one time. To see this, consider a module which must either be hardwired to process three different languages or which can be programmed to process any one of them. Clearly, we can get by with a smaller module in the latter case. More generally, if we use distributed representations, simple mathematical analyses show that the more patterns a module is set up to process, the larger it must be to avoid errors in processing any one pattern that might be presented on a particular occasion. The reason why this fact is relevant is that in conventional, hardwired PDP mechanisms, the model is programmed to process all of the patterns that it knows, all of the time. With programmable modules, on the other hand, the module only gets loaded with the connection information relevant to processing the patterns whose central representations are active. Since this number is vastly smaller than the number of total patterns a person might know, the size of the programmable processing structure can be very much smaller than a hardwired processing structure would have to be to do the same work. A full discussion of these issues is presented in Chapter 12 on the resource requirements of PDP models.

One interesting result of the analysis of the resource requirements of connection information distribution is the fact that the required number of units and connections varies with the number of different patterns the network is being programmed to process at a particular time. This fact means that for a CID mechanism of a given size, there is a limit of the amount of parallelism it can handle (quite independently of the number of programmable modules). But the limit is in programming the modules, not in their actual parallel operation once they have been set up. Thus if we can program each module separately, we can let them run in parallel; though we are forced to go sequential at the level of programming the modules, they can still be allowed to process simultaneously and to mutually constrain each other though interactions with higher-level processing structures. We therefore get the main benefit we wanted parallel distributed processing for—simultaneous, mutual constraint—without reduplicating hardwired knowledge or paying an undue price in terms of programmable connections and connection activation units.

Extensions of Connection Information Distribution

A programmable version of TRACE. In PABLO, we programmed the programmable blackboard sequentially, in fixation-sized blocks. In extending the basic ideas inherent in PABLO to speech, we notice immediately that the speech signal forces seriality on us by its very sequential nature. We argued, though, in Chapter 15, for a dynamic parallel processing structure in which we could exploit both forward and backward interactions, simply allowing new constraints to be added from the input at the right edge of the Trace, as the speech input unfolded in time. Now, to make the Trace into a programmable parallel processing structure, we can use the seriality of speech to program the Trace sequentially, as the speech comes in, thereby minimizing the resource requirements of the connection information distribution apparatus.

There would be several advantages to implementing a programmable version of TRACE. One is that retuning, priming, and learning effects could be accounted for by adjustment of the strengths of connections in the central knowledge source. Another is that the rather high cost of the connection modulation scheme for tuning feature-phoneme connections based on phoneme activations in the context could be born just once in the central knowledge structures. A third is that the central knowledge source could be retuned by global variables such as rate, etc., thereby allowing it to create in the Trace just the right pattern of interconnections to fit the expected form different possible utterances might take in the given situation. Such a model might be seen as performing an analysis of the speech signal by synthesis of the appropriate connection information, capturing some of the flexibility and context sensitivity which Halle and Stevens (1964) noted that speech perception calls for in their seminal article on "Analysis by Synthesis."

Processing syntactic structures. Introducing programmable connections into the TRACE model of speech perception will be a challenging task. Equally challenging will be the problem of extending the approach described here to higher levels of language structure. At higher levels, we must come to grips with the recursive structure of sentences, and the unbounded dependencies among elements of sentences. Some ideas about how this might be done are described in Chapter 19.

CONCLUSION

In this chapter, I have tried to argue that connection information distribution provides a way of overcoming some apparent limitations of parallel distributed processing mechanisms. Using connection information distribution, we can create local copies of relevant portions of the contents of a central knowledge store. These local copies then serve as the basis for interactive processing among the conceptual entities they program local hardware units to represent. With this mechanism, PDP models can now be said to be able to create multiple *instantiations* of the same schema, bound appropriately to the correct local variables, though subject to just the kinds of binding errors human subjects seem to make.

Perhaps the main lesson is that some of the limitations of **PDP** mechanisms that connection information distribution has been proposed to overcome are more apparent than real. For I have not really done anything more than show how existing tools in the arsenal of parallel distributed processing mechanisms can be used to create local copies of networks. This is not to say that all the challenges facing **PDP** models have suddenly vanished. Obviously there is a long way still to go in the development of computationally adequate models of reading, speech perception, and higher levels of language processing. I hope this chapter helps to indicate that **PDP** mechanisms will help us get there in a way that captures the flexibility and interactive character of human processing capabilities.

ACKNOWLEDGMENTS

This work was supported by ONR Contract N00014-82-C-0374, NR 667-483, by a grant from the System Development Foundation, by a grant from the National Science Foundation (BNS-79-24062), and by a NIMH Research Scientist Career Development Award. Special thanks for useful ideas and suggestions are due to Gary Cottrell, Geoff Hinton, Dave Rumelhart, and David Zipser.

CHAPTER 17

A Distributed Model of
Human Learning and Memory

J. L. McCLELLAND and D. E. RUMELHART

The view that human memory is distributed has come and gone several times over the years. Hughlings-Jackson, the 19th century neurologist; Kurt Goldstein, the Gestalt neurologist of the early 20th century; and Karl Lashley, the physiological psychologist of the same era, all held to variants of a distributed model of the physiology of learning and memory.

While Lashley's and Goldstein's more radical views have not been borne out, the notion that memory is physiologically distributed within circumscribed regions of the brain seems to be quite a reasonable and plausible assumption (see Chapters 20 and 21). But given the rather loose coupling between a psychological or cognitive theory and a theory of physiological implementation, we can ask, does the notion of distributed memory have anything to offer us in terms of an understanding of human cognition?

In Chapter 3, several of the attractive properties of distributed models are described, primarily from a computational point of view. This chapter addresses the relevance of distributed models from the point of view of the psychology of memory and learning. We begin by

This chapter is a modified version of the article "Distributed Memory and the Representation of General and Specific Information" by J. L. McClelland and D. E. Rumelhart, which appeared in *Journal of Experimental Psychology: General*, 1985, *114*, 159-188. Copyright 1985 by the American Psychological Association, Inc. Adapted with permission.

considering a dilemma that faces cognitive theories of learning and memory, concerning the representations of general and specific information. Then we show how a simple version of a distributed model circumvents the dilemma. We go on to show how this model can account for a number of recent findings about the learning and representation of general and specific information. We end by considering some other phenomena that can be accounted for using a distributed model, and we consider a limitation of the model and ways this limitation might be overcome.

The Dilemma

One central dilemma for theories of memory has to do with the choice to represent *general* or *specific* information. On the one hand, human memory and human learning seem to rely on the formation of summary representations that generalize from the details of the specific experiences that gave rise to them. A large number of experiments, going back to the seminal findings of Posner and Keele (1968) indicate that we appear to extract what is common to a set of experiences. On the basis of this sort of evidence, a number of theorists have proposed that memory is largely a matter of generalized representations, either abstracted representations of concepts discarding irrelevant features, or prototypes—representations of typical exemplars (Rosch, 1975). On the other hand, specific events and experiences play a prominent role in memory. Experimental demonstrations of the importance of specific stimulus events even in tasks which have been thought to involve abstraction of a concept or rule are now legion. Responses in categorization tasks (Brooks, 1978; Brooks, Jacoby, & Whittlesea, 1985; Medin & Schaffer, 1978), perceptual identification tasks (Jacoby, 1983a, 1983b; Whittlesea, 1983), and pronunciation tasks (Glushko, 1979) all seem to be quite sensitive to the congruity between particular training stimuli and particular test stimuli, in ways which most abstraction or prototype formation models would not expect.

One response to this dual situation has been to propose models in which apparent rule-based or concept-based behavior is attributed to a process that makes use of stored traces of specific events or specific exemplars of the concepts or rules. According to this class of models, the apparently rule-based or concept-based behavior emerges from what might be called a conspiracy of individual memory traces or from a sampling of one from the set of such traces. Models of this class include the Medin and Schaffer (1978) context model, the Hintzman (1983) multiple trace model, and the Whittlesea (1983) episode model.

This trend is also exemplified by our interactive activation model of word perception and an extension of the interactive activation model to generalization from exemplars (McClelland, 1981). Both of these models were described in Chapter 1.

One feature of some of these exemplar models is very troublesome. Many of them are internally inconsistent with respect to the issue of abstraction. Thus, though our word perception model assumes that linguistic rules emerge from a conspiracy of partial activations of detectors for particular words, thereby eliminating the need for abstraction of rules, the assumption that there is a single detector for each word implicitly assumes that there is an abstraction process that lumps each occurrence of the same word into the same single detector unit. Thus, the model has its abstraction and creates it too, though at somewhat different levels.

One logically coherent response to this inconsistency is to simply say that each word or other representational object is itself the result of a conspiracy of the entire ensemble of memory traces of the different individual experiences we have had with the object. We will call this view the *enumeration of specific experiences* view. It is exemplified most clearly by Jacoby (1983a, 1983b), Hintzman (1983), and Whittlesea (1983).

As the papers just mentioned demonstrate, enumeration models can work quite well as an account of quite a number of empirical findings. However, there still seems to be a drawback. Enumeration models seem to require an unlimited amount of storage capacity, as well as mechanisms for searching an almost unlimited mass of data. This is especially true when we consider that the primitives out of which we normally assume each experience is built are themselves based on generalizations. For example, a word is a sequence of letters, and a sentence is a sequence of words. Are we to believe that all these units are mere notational conveniences for the theorist, and that every event is stored separately as an extremely rich (obviously structured) representation of the event, with no condensation of details?

Thus, the dilemma remains. It would appear that enumeration models cannot completely eliminate the need for some kind of abstractive representation in memory. Yet the evidence that the characteristics of particular events influence memory performance cannot be disposed of completely.

One response to this dilemma is to propose the explicit storage of both general and specific information. For example, Elio and J. R. Anderson (1981) have proposed just such a model. In this model, memory traces are stored in the form of productions, and mechanisms of production generalization and differentiation form summary representations and refine them as necessary during learning.

Distributed models suggest another alternative. As we shall demonstrate in this chapter, a very simple distributed model, relying on a very simple learning rule—the delta rule—is capable of accounting for empirical data that has been taken as suggesting that we store summary representations (e.g., prototypes) as well as data that has been taken as suggesting that memory consists of an enumeration of specific experiences.

The specific model we consider does have a limitation that it shares with enumeration models: It relies on a fixed set of representational primitives. However, we shall argue that natural extensions of the model which overcome this limitation are now possible. Though we do not explore these extensions in detail, we do consider some issues that will arise if these extensions are incorporated.

A DISTRIBUTED MODEL OF MEMORY

The model we shall describe is a member of the class of models discussed in Chapter 3. In developing the ideas presented here, we were strongly influenced by the work of J. A. Anderson (e.g., 1977, 1983; Anderson, Silverstein, Ritz, & Jones, 1977; Knapp & Anderson, 1984) and Hinton (1981a). We have adopted and synthesized what we found to be the most useful aspects of their distinct but related models, preserving (we hope) the basic spirit of both.

In keeping with the overall enterprise of the book, our distributed model is a model of the microstructure of memory. It specifies the internal workings of some of the components of information processing and memory, in particular those concerned with the retrieval and use of prior experience. The model does not specify in and of itself how these acts of retrieval and use are planned, sequenced, and organized into coherent patterns of behavior. Some thoughts about ways this might be done may be found in Chapters 8 and 14.

General Properties

Our model shares a number of basic assumptions about the nature of the processing and memory system with most other distributed models. In particular, the processing system is assumed to consist of a highly interconnected network of units that take on activation values and communicate with other units by sending signals modulated by weights associated with the connections between the units, according to the principles laid out in Chapter 2. Sometimes we may think of the units as corresponding to particular representational primitives, but they need not. For example, even what we might consider to be a primitive feature of something, like having a particular color, might be a pattern of activation over a collection of units.

Modular structure. We assume that the units are organized into modules. Each module receives inputs from other modules; the units within the module are richly interconnected with each other; and they send outputs to other modules. Figure 1 illustrates the internal structure of a very simple module, and Figure 2 illustrates some hypothetical interconnections between a number of modules. Both figures grossly under-represent our view of the numbers of units per module and the number of modules. We would imagine that there would be thousands to millions of units per module and many hundred or perhaps many thousand partially redundant modules in anything close to a complete memory system.

The state of each module represents a synthesis of the states of all of the modules it receives inputs from. Some of the inputs will be from relatively more sensory modules, closer to the sensory end-organs of

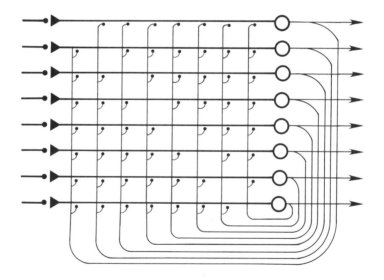

FIGURE 1. A simple information processing module, consisting of a small ensemble of eight processing units. Each unit receives inputs from other modules (indicated by the single input impinging on the input line of the unit from the left; this can stand for a number of converging input signals from several units outside the module) and sends outputs to other modules (indicated by the rightward output line from each unit). Each unit also has a modifiable connection to all the other units in the module, as indicated by the branches of the output lines that loop back onto the input lines leading into each unit.

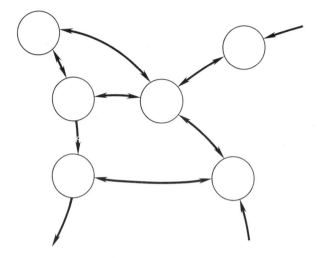

FIGURE 2. An illustrative diagram showing several modules and interconnections among them. Arrows between modules simply indicate that some of the units in one module send inputs to some of the units in the other. The exact number and organization of modules is, of course, unknown; the figure is simply intended to be suggestive.

one modality or another. Others will come from relatively more abstract modules, which themselves receive inputs from and send outputs to other modules placed at the abstract end of several different modalities. Thus, each module combines a number of different sources of information.

Units play specific roles within patterns. A pattern of activation only counts as the same as another if the same units are involved. The reason for this is that the knowledge built into the system for re-creating the patterns is built into the set of interconnections among the units, as we will explain below. For a pattern to access the right knowledge, it must arise on the appropriate units. In this sense the units play specific roles in the patterns.

Obviously, a system of this sort is useless without sophisticated perceptual processing mechanisms at the interface between memory and the outside world, so that input patterns arising at different locations in the world can be mapped into the same set of units internally. The scheme proposed by Hinton (1981b; see Chapter 4) is one such mechanism. Chapter 16 describes mechanisms that allow several different patterns to access the same set of units at the same time; in the present chapter, we restrict attention to the processing of one pattern at a time.

Relation to Basic Concepts in Memory

Several standard concepts in the memory literature map easily onto corresponding aspects of distributed memory models. Here we consider three key concepts in memory and their relation to our distributed memory model.

Mental state as pattern of activation. In a distributed memory system, a mental state is a pattern of activation over the units in some subset of the modules. The patterns in the different modules capture different aspects of the content of the mental states in a kind of a partially overlapping fashion. Alternative mental states are simply alternative patterns of activation over the modules. Information processing is the process of evolution in time of mental states.

Memory traces as changes in the weights. Patterns of activation come and go, leaving traces behind when they have passed. What are the traces? They are changes in the strengths or weights of the connections between the units in the modules.[1] As we already said, each memory trace is distributed over many different connections, and each connection participates in many different memory traces. The traces of different mental states are therefore superimposed in the same set of weights.

Retrieval as reinstatement of prior pattern of activation. Retrieval amounts to partial reinstatement of a mental state, using a cue which is a fragment of the original state. For any given module, we can see the cues as originating from outside of it. Some cues could arise ultimately from sensory input. Others would arise from the results of previous retrieval operations, fed back to the memory system under the control of a search or retrieval plan. It would be premature to speculate on how such schemes would be implemented in this kind of a model, but it is clear that they must exist.

Detailed Assumptions

In the rest of our presentation, we will be focusing on operations that take place within a single module. This obviously oversimplifies the behavior of a complete memory system since the modules are assumed

---

[1] These traces are not, of course, to be confused with the continuing pattern of activation stored in the Trace processing structure discussed in Chapter 15.

to be in continuous interaction. The simplification is justified, how-ever, in that it allows us to examine some of the basic properties of dis-tributed memory systems which are visible even without these interac-tions with other modules.

Let us look, therefore, at the internal structure of one very simple module, as shown in Figure 1. Again, our image is that in a system sufficient for real memories, there would be much larger numbers of units. We have restricted our analysis to small numbers simply to illus-trate basic principles as clearly as possible; this also helps to keep the running time of simulations in bounds.

Continuous activation values. The units used in the present model are fairly standard units, taking on continuous activation values in the range from −1 to +1. Zero is in this case a neutral resting value, toward which the activations of the units tend to decay.

Unlike the models described in some of the previous chapters in this section, there is no threshold activation value. This means that both positive and negative activations influence other units.

Inputs, outputs, and internal connections. Each unit receives input from other modules and sends output to other modules. For simplicity, we assume that the inputs from other modules occur at connections whose weights are fixed. In the simulations, we also treat the input from outside the module as a static pattern that comes on suddenly, ignoring for simplicity the fact that the input pattern evolves in time and might be affected by feedback from the module under study. While the input to each unit might arise from a combination of sources in other modules, we can lump the external input to each unit into a single real valued number representing the combined effects of all com-ponents of the external input. In addition to extramodular connections, each unit is connected to all other units in the module via a weighted connection. The weights on these connections are modifiable, as described below. The weights can take on any real values: positive, negative, or zero. There is no connection from a unit onto itself.

An input pattern is presented at some point in time over some or all of the input lines to the module and is then left on for several ticks, until the pattern of activation it produces settles down and stops changing.

The processing cycle. The model is a synchronous model, like all of the other models in this section. On each processing cycle, each unit determines its net input, based on the external input to the unit and the activations of all of the units at the end of the preceding cycle modulated by the weight coefficients which determine the strength and direction of each unit's effect on every other unit. Then the activations of all of the units are updated simultaneously.

The net input to a unit consists of two parts, the *external* input, arising from outside the module, and the *internal* input, arising from the other units in the module. The internal input to unit i, i_i, is then just the sum of all of these separate inputs, each weighted by the appropriate connection strength:

$$i_i = \sum_j a_j w_{ij}.$$

Here, j ranges over all units in the module other than i, a_j is the activation of unit j at the end of the previous cycle, and w_{ij} is the weight of the connection to unit i from unit j. This sum is then added to the *external* input to the unit, arising from outside the module, to obtain the net input to unit i, net_i:

$$net_i = i_i + e_i$$

where e_i is just the lumped external input to unit i.

Activations are updated according to a simple nonlinear "squashing" function like the one we have used in several other chapters. If the net input is positive, the activation of the unit is incremented by an amount proportional to the distance left to the ceiling activation level of $+1.0$. If the net input is negative, the activation is decremented by an amount proportional to the distance left to the floor activation level of -1.0. There is also a decay factor which tends to pull the activation of the unit back toward the resting level of 0. For unit i, if $net_i > 0$,

$$\Delta a_i = Enet_i (1 - a_i) - Da_i. \tag{1}$$

If $net_i \leqslant 0$,

$$\Delta a_i = Enet_i (a_i - (-1)) - Da_i. \tag{2}$$

In these equations, E and D are global parameters that apply to all units and set the rates of excitation and decay, respectively. The term a_i is the activation of unit i at the end of the previous cycle, and Δa_i is the change in a_i; that is, it is the amount added to (or, if negative, subtracted from) the old value a_i to determine its new value for the next cycle.

The details of these assumptions are quite unimportant to the behavior of the model. Many of the same results have been obtained using other sigmoid squashing functions. The fact that activations range from -1 to $+1$ is, likewise, not very relevant, though if activations ranged from 0 to 1 instead, it would be necessary to incorporate threshold terms in addition to the connection strengths to get the model to produce roughly the same results.

Given a fixed set of inputs to a particular unit, its activation level will be driven up or down in response until the activation reaches the point where the incremental effects of the input are balanced by the decay. In practice, of course, the situation is complicated by the fact that as each unit's activation is changing it alters the input to the others. Thus, it is necessary to run the simulation to see how the system will behave for any given set of inputs and any given set of weights. In all the simulations reported here, the model is allowed to run for 50 cycles, which is considerably more than enough for it to achieve a stable pattern of activation over all the units.

Memory traces. The memory trace of a particular pattern of activation is a set of changes in the entire set of weights in the module. We call the whole set of changes an *increment* to the weights. After a stable pattern of activation is achieved, weight adjustment takes place. This is thought of as occurring simultaneously for all of the connections in the module.

We use the delta rule to determine the size and direction of the change at each connection. This learning rule is explored extensively in Chapters 2, 8, and 11. Here, we consider it from the point of view of indicating how it implements, in a direct and simple way, the storage of the connection information that will allow a module to re-create complete patterns of activation originally produced by external input when only a part of the pattern is presented as a "retrieval" cue.

To achieve pattern completion, we want to set up the internal connections among the units in the module so that when part of the pattern is presented, the internal connections will tend to reproduce the rest. Consider, in this light, a particular pattern of activation, and assume for the moment that in this pattern the external input to a particular unit j is excitatory. Then we will want the internal input from other units in the module to tend to excite unit j when they take on the activation values appropriate for this particular pattern. In general, what we need to do for each unit is to adjust the weights so that the internal connections in the module will tend to reproduce the external input to that unit, given that the rest of the units have the activation values appropriate for this particular pattern.

The first step in weight adjustment is to see how well we are already doing. If the network is already doing what it should, the weights do not need to be changed. Therefore, for each unit i, we compute the difference δ_i between the external input to the unit and the net internal input to the unit from other units in the module:

$$\delta_i = e_i - i_i.$$

In determining the activation value of the unit, we added the external input together with the internal input. Now, in adjusting the weights, we are taking the difference between these two terms. This implies that the unit must be able to aggregate all inputs for purposes of determining its activation, but it must be able to distinguish between external and internal inputs for purposes of adjusting its weights.

Let us consider the term δ_i for a moment. If it is positive, the internal input is not activating the unit enough. If negative, it is activating the unit too much. If it is zero, everything is fine and we do not want to change anything. Thus, δ_i determines the magnitude and direction of the overall change that needs to be made in the internal input to unit i. To achieve this overall effect, the individual weights are then adjusted according to the following formula:

$$\Delta w_{ij} = \eta \delta_i a_j. \tag{3}$$

The parameter η is just a global strength parameter which regulates the overall magnitude of the adjustments of the weights; Δw_{ij} is the *change* in the weight to i from j.

The delta rule, given by Equation 3, has all the intended consequences. That is, it tends to drive the weights to the right values to make the internal inputs to a unit match the external inputs. For example, consider the case in which δ_i is positive and a_j is positive. In this case, the value of δ_i tells us that unit i is not receiving enough excitatory input, and the value of a_j tells us that unit j has positive activation. In this case, the delta rule will increase the weight from j to i. The result will be that the next time unit j has a positive activation, its excitatory effect on unit i will be increased, thereby reducing δ_i. Similar reasoning applies to cases where δ_i is negative, a_j is negative, or both are negative. Of course, when either δ_i or a_j is zero, w_{ij} is not changed. In the first case, there is no error to compensate for; in the second case, a change in the weight will have no effect the next time unit j has the same activation value.

What the delta rule can and cannot do. Several important points about the delta rule have been discussed in Chapters 2, 8, and 11. Here we consider just those that are most relevant to our present purposes. Basically, the most important result is that, for a set of patterns that we present repeatedly to a module, if there is a set of weights that will allow the system to reduce δ to 0 for each unit in each pattern, this rule will find it through repeated exposure to all of the members of the set of patterns. However, it is important to note that the existence of a set of weights that will allow δ to be reduced to 0 is not guaranteed, but depends on the structure inherent in the set of patterns that the model is given to learn. To be perfectly learnable by our model, the patterns must conform to the following *linear predictability constraint*:

> Over the entire set of patterns, the external input to each unit must be predictable from a linear combination of the activations of every other unit.

While this limitation is severe, it is important to realize that the extent to which a set of patterns satisfy the linear predictability constraint depends on the way the set of patterns is represented. From this point of view, we must distinguish clearly between the theoretical description of a set of stimuli presented to a human subject for processing and the patterns of activation produced in some module deeply embedded in the cognitive system.   As a rule of thumb, an encoding which treats each dimension or aspect of a stimulus separately is unlikely to be sufficient; what is required is a *context sensitive* or *conjunctive* encoding, such that the representation of each aspect is colored by other aspects. A fuller discussion of this issue is presented in Chapter 3; we also return to it below in considering some recent evidence obtained by Medin and Schwanenflugel (1981).

No hidden units. As explained in Chapters 7 and 8, the linear predictability constraint arises from the fact that the present model contains no hidden units. If hidden units were incorporated, the generalized delta rule described in Chapter 8 could be used to train the connections into these units, thereby allowing the model to form new representational primitives to overcome the linear predictability constraint when it arises. At the end of the chapter, we consider the introduction of hidden units and the effects that this would have on the behavior of the model.

Decay in the increments to the weights. We assume that each trace or increment undergoes a decay process, though the rate of decay of the increments is assumed to be much slower than the rate of decay of patterns of activation. Following a number of theorists (e.g., Wickelgren, 1979), we imagine that traces at first decay rapidly, but then the remaining portion becomes more and more resistant to further decay. Whether it ever reaches a point where it is no longer decaying at all, we do not know. The basic effect of this assumption is that individual inputs exert large short-term effects on the weights, but, after they decay, the residual effect is considerably smaller. The fact that each increment has its own temporal history increases the complexity of computer simulations enormously. In many of the simulations, therefore, we will specify simpler assumptions to keep the simulations tractable.

Parameters and details. In simulations using the model, it is important to keep the net input to each unit on each processing cycle relatively small, so that activations do not jump too suddenly and go out of range. For this purpose we set the parameters E and D equal to .15. It is also important to reduce η in proportion to the number of internal inputs to each unit. The number of internal inputs to each unit is just 1 minus the number of units, since each unit receives an input from every other unit. Thus, if we define $\eta = S/(n-1)$, where n is the number of units, then the rate of learning, in terms of the reduction in δ, will be about the same for all values of n. Instability can result if S is set greater than 1.0. Generally a value of .85 was used in the following simulations.

Illustrative Examples

In this section, we describe a number of simulations to illustrate several key aspects of the model's behavior. We wish to demonstrate several points:

1. The model can extract what appears to be the prototype or central tendency of a set of patterns, if the patterns are in fact random distortions of the same base or prototype pattern.

2. The model can do this for several *different* patterns, using the same set of connections to store its knowledge of all the prototypes.

3. This ability does not depend on the exemplars being presented with labels so that the model is given the where-with-all to keep them straight.

4. Representations of specific, repeated exemplars can coexist in the same set of connections with knowledge of the prototype.

Learning a prototype from exemplars. To illustrate the first point, we consider the following hypothetical situation. A little boy sees many different dogs, each only once and each with a different name. All the dogs are a little different from each other, but in general there is a pattern which represents the typical dog—each one is just a different distortion of this prototype. (We are not claiming that the dogs in the world have no more structure than this; we make this assumption for purposes of illustration only). For now we will assume that the names of the dogs are all completely different. We would expect, given this experience, that the boy would learn the prototype of the category, even without ever seeing any particular dog that matches the prototype directly (Posner & Keele, 1968, 1970; J. A. Anderson, 1977, applies an earlier version of a distributed model to this case). That is, the prototype will seem as familiar as any of the exemplars, and the boy will be able to complete the pattern corresponding to the prototype from any part of it. He will not, however, be very likely to remember the names of each of the individual dogs, though he may remember the most recent ones.

We model this situation with a module consisting of 24 units. We assume that the presentation of a dog produces a visual pattern of activation over 16 of the units in the hypothetical module (the 9th through 24th, counting from left to right). The name of the dog produces a pattern of activation over the other 8 units (Units 1 to 8, counting from left to right).

Each visual pattern, by assumption, is a distortion of a single prototype. The prototype used for the simulation simply had a random series of +1 and −1 values. Each distortion of the prototype was made by probabilistically flipping the sign of randomly selected elements of the prototype pattern. For each new distorted pattern, each element has an independent chance of being flipped, with probability .2. Each name pattern was simply a random sequence of +1s and −1s for the eight name units. Each encounter with a new dog is modeled as a presentation of a new name pattern with a new distortion of the prototype visual pattern. Fifty different trials were run, each with a new name-pattern/visual-pattern pair.

For each presentation, the pattern of activation is allowed to stabilize, and then the weights are adjusted as described above. The increment to the weights is then allowed to decay considerably before the next input is presented. For simplicity, we assume that before the next pattern is presented, the last increment decays to a fixed small proportion (5%) of its initial value and thereafter undergoes no further decay.

What does the model learn? The module acquires a set of weights which is continually buffeted about by the latest dog exemplar, but which captures the prototype dog quite well. Waiting for the last increment to decay to the fixed residual yields the weights shown in Figure 3.

These weights capture the correlations among the values in the prototype dog pattern quite well. The lack of exact uniformity is due to the more recent distortions presented, whose effects have not been corrected by subsequent distortions. This is one way in which the model gives priority to specific exemplars, especially recent ones. The

Prototype pattern:

```
                        + − + + − − − − + + + + + − − −
```

Weights acquired after learning:

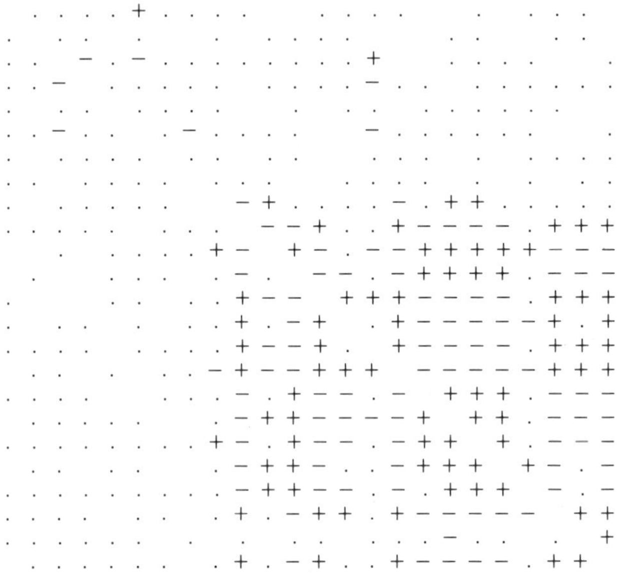

```
.  .  .  .  +  .  .  .  .      .  .  .  .           .  .  .  .
.  .  .  .         .  .  .  .  .      .  .  .  .           .  .  .  .
.  .  −  .  −  .  .  .  .  .  .  +           .  .  .  .      .
.  .  −  .  .  .  .  .  .  .  .  −  .  .      .  .  .  .  .
.  .  .  .  .  .  .  .  .  .  .  .  .  .  .  .  .  .  .  .
.  .  −  .  .  .  −  .  .  .  .  −  .  .  .  .  .  .  .
.  .  .  .  .  .  .  .  .  .  .  .  .  .  .  .  .  .  .
.  .  .  .  .  .  .  .  .  .  .  .  .  .  .  .  .  .  .
.  .  .  .  .  .  .  .  − +  .  .  .  .  −  .  + +  .  .  .  .  .
.  .  .  .  .  .  .  .  − − +  .  .  + − − − −  .  + + +
.  .  .  .  .  + −    + −  .  − − + + + + + + − − −
.  .  .  .  .  −  .    − − .  − + + + + .  − − −
.  .  .  .  .  + − −    + + + − − − − .  + + +
.  .  .  .  .  + .  − +    .  + − − − − − + .  +
.  .  .  .  .  + − − + .    + − − − − .  + + +
.  .  .  .  − + − − + + +    − − − − − + + +
.  .  .  .  .  − .  + − − .  −    + + + .  − − −
.  .  .  .  .  − + + − − − − +    + + .  − − −
.  .  .  .  + − .  + − − .  − + +    + − .  −
.  .  .  .  .  − + + − .  .  − + + +    + − .  −
.  .  .  .  .  − + + − − .  − .  + + +    − .  −
.  .  .  .  + .  − + + .  + − − − − −    + +
.  .  .  .  .  .  .  .  .  .  .  .  .  .  −  .  .  .      .  +
.  .  .  .  .  + .  − + .  .  + − − − − .  + +
```

FIGURE 3. Weights acquired in learning from distorted exemplars of a prototype. (The prototype pattern is shown above the weight matrix. Blank entries correspond to weights with absolute values less than .01; dots correspond to absolute values less than .06; plusses or minuses are used for weights with larger absolute values.)

effects of recent exemplars are particularly strong, of course, before they have had a chance to decay. The module can complete the prototype quite well, and it will respond more strongly to the prototype than to any distortion of it. It has, however, learned no particular relation between this prototype and any name pattern since a totally different random association was presented on each trial. If the pattern of activation on the name units had been the same in every case (say, each dog was just called "dog"), or even in just a reasonable fraction of the cases, then the module would have been able to retrieve this shared name pattern from the prototype of the visual pattern and the prototype pattern from the name; we will see cases of this kind of behavior in the next section.

Multiple, nonorthogonal prototypes. In the preceding simulation, we have seen how the distributed model acts as a sort of signal averager, finding the central tendency of a set of related patterns. In and of itself this is an important property of the model, but the importance of this property increases when we realize that the model can average several *different* patterns in the same composite memory trace. Thus, several different prototypes can be stored in the same set of weights. This property is important because it means that the model does not fall into the trap of needing to decide which category to put a pattern into before knowing which prototype to average it with. The acquisition of the different prototypes proceeds without any sort of explicit categorization. If the patterns are sufficiently dissimilar (i.e., orthogonal), there is no interference among them at all. Increasing similarity leads to increased confusability during learning, but eventually the delta rule finds a set of connection strengths that minimizes the confusability of similar patterns. These points are discussed mathematically in Chapter 11; here we illustrate this through a simulation of the following hypothetical situation.

Let us suppose that our little boy sees different dogs, different cats, and different bagels in the course of his day-to-day experience. First, let's consider the case in which each experience with a dog, a cat, or a bagel is accompanied by someone saying "dog," "cat," or "bagel," as appropriate.

The simulation analog of this situation involved forming three "visual" prototype patterns of 16 elements: two of them (the one for dog and the one for cat) somewhat similar to each other ($r = .5$) and the third (for the bagel) orthogonal to both of the other two. Paired with each visual pattern was a name pattern of eight elements. Each name pattern was orthogonal to both of the others. Thus, the prototype visual pattern for cat and the prototype visual pattern for dog were similar to each other, but their names were not related.

Stimulus presentations involved presentations of distorted exemplars of the name/visual pattern pairs to a module like the one used in the previous simulation. This time, both the name pattern and the visual pattern were distorted, with each element having its sign flipped with an independent probability of .1 on each presentation. Fifty different distortions of each name/visual pattern pair were presented in groups of three, consisting of one distortion of the dog pair, one distortion of the cat pair, and one distortion of the bagel pair. Weight adjustment occurred after each presentation, with decay to a fixed residual before each new presentation.

At the end of training, the module was tested by presenting each name pattern and observing the resulting pattern of activation over the visual units, and by presenting each visual pattern and observing the pattern of activation over the name units. The results are shown in Table 1. In each case, the model reproduces the correct completion for

TABLE 1

RESULTS OF TESTS AFTER LEARNING THE DOG, CAT, AND BAGEL PATTERNS

	Name Pattern	Visual Pattern
Pattern for dog prototype	+ − + − + − + −	+ − + + − + − − + + + + − − − −
Response to dog name	+5 −4 +4 −5 +5 −4 +4 −4	+3 −4 +4 −4 −4 −4 +4 +4 +4 +3 +4 −4 −4 −4 −3
Response to dog visual pattern	+5 −4 +4 −5 +5 −4 +4 −4	
Pattern for cat prototype	+ + − − + + − −	+ − + + − − + − + + − + + − + +
Response to cat name	+5 +4 −4 −5 +4 +4 −4 −4	+4 −3 +4 +4 −4 −3 −3 −4 −4 −4 +4 +4 +4 −4 +4 +4
Response to cat visual pattern	+5 +4 −4 −5 +4 +4 −4 −4	
Pattern for bagel prototype	+ − + + − − + −	+ + − + − − + − + + − + + + + −
Response to bagel name	+4 −4 −4 +4 −4 −4 +4 +4	+3 +4 −4 −4 +4 +4 −4 −4 −4 +4 +4 +3 +4 +4 −4
Response to bagel visual pattern	+4 −4 −4 +4 −4 −4 +4 +4	

Note: Decimal points have been suppressed for clarity; thus, an entry of +4 represents an activation value of +.4.

the probe, and there is no apparent contamination of the "cat" pattern by the "dog" pattern, even though the visual patterns are both similar.

In general, pattern completion is a matter of degree. Below, in simulating particular experimental results, we will introduce an explicit measure of the degree to which a particular pattern is active in the units of a module. For now, it is sufficient simply to note that the sign of all of the elements is correct; given this, the average magnitude of the elements gives an approximate measure of the "degree" of pattern reinstatement.

In a case like the present one, in which the patterns known to the model are not all orthogonal, the values of the connection strengths that the model produces do not necessarily have a simple interpretation. Though their sign always corresponds to the sign of the correlation between the activations of the two patterns, their magnitude is not a simple reflection of the magnitude of their correlation, but is influenced by the degree to which the model is relying on this particular correlation to predict the activation of one unit from the others. Thus, in a case where two units (call them i and j) are perfectly correlated, the strength of the connection from i to j will depend on the number of other units whose activations are correlated with j. If i is the only unit correlated with j, it will have to do all the work of "predicting" j, so the weight will be very strong; on the other hand, if many units besides i are correlated with j, then the work of exciting j will be spread around, and the weight between i and j will be considerably smaller. The situation is exactly the same as the one that arises in linear regression: if several variables predict another, they share the weight. The weight matrix acquired as a result of learning the dog, cat, and bagel patterns (Figure 4) reflects these effects. For example, across the set of three prototypes, Units 1 and 5 are perfectly correlated, as are Units 2 and 6. Yet the connection from Unit 2 to Unit 6 is stronger than the connection from Unit 1 to Unit 5 (these connections are *d in the figure). The reason for the difference is that Unit 2 is one of only three units which correlate perfectly with Unit 6, while Unit 1 is one of seven units which correlate perfectly with Unit 5.[2]

Thus far we have seen that several prototypes, not necessarily orthogonal, can be stored in the same module without difficulty. It is true, though we do not illustrate it, that the model has more trouble with the cat and dog visual patterns earlier on in training, before learning has essentially reached asymptotic levels, as it has by the end of 50

---

[2] In Figure 4, the weights do not reflect these contrasts perfectly in every case because the noise introduced into the learning happens, by chance, to alter some of the correlations present in the prototype patterns. Averaged over time, though, the weights will conform to their expected values.

Prototypes

Dog:

+ − + − + − + −   + − + + − − − − + + + + + − − −

Cat:

+ + − − + + − −   + − + + − − − − + − + − + + − +

Bagel:

+ − − + + − − +   + + − + − + + − + − − + + + + −

Weights

```
   −1 −1   +2*−1   −1      +3        +5 −2 −1    −3 +2 −1       +2        −1
         −2   −1 +4*         +1 −1          −1 −1 +1 −1 −1     −3 −1     −3 +3
   −1 −2   −1       +5          +1   +1 −1 −1    −1 +4    +1     −5    −1
            −1   −1   −1 +2    −1 +3 −3         +2 +3      −1 −2    −1 +1 +3
   +3 −1 −1              +1    −1    +3 −3       −2 +2      +1 +3 +1 −1 −1
   −1 +5 −3            −2     −1 −1    −1     −1        −1 −2 +1 −5 −1 +2 −1 +5
   −1    +3 −1 −1 −2            +1       −1    +1    +4      −1 −2
   −1 −1 −1 +2         −1    −1 +3 −2 −1    +3 +3    −1 −1 −3 +1 −1 +3 +3


   +3   −1 −1 +2 −1 −1                 +3 −2       −3 +2    +1    +3 +1    −1
   −1 −1 −1 +3    −1    +3          −3 −1    +3 +3 +1 −1 −2 −2 +1 −1    +2
   +1 +1 +1 −3           −2    +1 −3       −1 −3 −3       +1 +3 −1    −1 −2 +2
   +2 −1 −1 −1 +3    −1       +3 −1 +1    −2 −1    −2 +2 −1        +2 +1 −1
   −2 +1 +1    −2 +1 +1 +1    −3       −3       +1 +2 −3      −1 −2 −1 +1
      −1    +2 −1          +2    −1 +2 −2    +1    +2 +1 −1    −3    −1    +2
   −1 −1 −1 +2    −1 +2       +2 −3       +3    +1       −3    −1    +3
   −2 +1 +1 +1 −3    +1 +1    −2 +1 −1 −3 +3          −2 +1 −1 −1 −3 −1 +1
   +3    −1    +2    −1       +2       +3 −3       −3    −1 +1    +3 +1 −1 −1
      +5 −1 −1 −1 +4 −1                 +1    −3 +1       +1    −1 −5 −1 −1
      +1    −2    +1 +1 −3    +1 −2 +2 +1 −1 −3 −3 −1    +1    −2       −3 +1
   +1 −5 +2 +1 +1 −5 +3             −2 +1 −1 +1    +1 +1 −2    +1 −2 +1 −3
   +2 −1 −1    +3 −1          +2    +1 +2 −2       −3 +2    +1
      +3 −4 +1    +3 −4 +1    +1    −1       +1 +1    +1 −5 −1 −3       +1 +2
         +2 −1    −1          −1 +2 −3       +2 +2       −1 −3       +1
   −1 +3 −2 −1 −1 +5 −3       −1 +3    +1 −1       −1    +1 −2    +3 −1
```

FIGURE 4. Weights acquired in learning the three prototype patterns shown. (Blanks in the matrix of weights correspond to weights with absolute values less than or equal to +5 stands for a weight of +.25. The gap in the horizontal and vertical dimensions is used to separate the name field from the visual pattern field.)

cycles through the full set of patterns. And, of course, even at the end of learning, if we present as a probe a part of the visual pattern that does not differentiate between the dog and the cat, the model will produce a blended response. Both of these aspects of the model seem generally consistent with what we should expect from human subjects.

Category learning without labels. An important further fact about the model is that it can learn several different visual patterns, even without the benefit of distinct identifying name patterns during learning. To demonstrate this we repeated the previous simulation, simply replacing the name patterns with 0s. The model still learns about the internal structure of the visual patterns so that, after 50 cycles through the stimuli, any unique subpart of any one of the patterns is sufficient to reinstate the rest of the corresponding pattern correctly. This aspect of the model's behavior is illustrated in Table 2. Thus, we have a model that can, in effect, acquire a number of distinct categories, simply through a process of incrementing connection strengths in response to each new stimulus presentation. Noise, in the form of distortions in the patterns, is filtered out. The model does not require a name or other guide to distinguish the patterns belonging to different categories.

Coexistence of the prototype and repeated exemplars. One aspect of our discussion up to this point may have been slightly misleading. We may have given the impression that the model is simply a prototype extraction device. It is more than this, however; it is a device that captures whatever structure is present in a set of patterns (subject, of course, to the linear predictability constraint). When the set of patterns has a prototype structure, the model will act as though it is extracting prototypes; but when it has a different structure, the model will do its best to accommodate this as well. For example, the model permits the coexistence of representations of prototypes with representations of particular, repeated exemplars.

TABLE 2

RESULTS OF TESTS AFTER LEARNING
THE DOG, CAT, AND BAGEL PATTERNS WITHOUT NAMES

Dog visual pattern:	+	−	+	+	−	−	−	−	+	+	+	+	+	−	−	−
Probe:								+	+	+	+					
Response:	+3	−3	+3	+3	−3	−4	−3	−3	+6	+5	+6	+5	+3	−2	−3	−2
Cat visual pattern:	+	−	+	+	−	−	−	−	+	−	+	−	+	+	−	+
Probe:								+	−	+	−					
Response:	+3	−3	+3	+3	−3	−3	−3	−3	+6	−5	+6	−5	+3	+2	−3	+2
Bagel visual pattern:	+	+	−	+	−	+	+	−	+	−	−	+	+	+	+	−
Probe:								+	−	−	+					
Response:	+2	+3	−4	+3	−3	+3	+3	−3	+6	−6	−6	+6	+3	+3	+3	−3

As an illustration of this point, consider the following situation. Let us say that our little boy knows a dog next door named Rover and a dog at his grandma's house named Fido. And let's say that the little boy goes to the park from time to time and sees dogs, each of which his father tells him is a dog.

The simulation analog of this involved three different eight-element name patterns, one for Rover, one for Fido, and one for dog. The visual pattern for Rover was a particular randomly generated distortion of the dog prototype pattern, as was the visual pattern for Fido. For the dogs seen in the park, each one was simply a new random distortion of the prototype. The probability of flipping the sign of each element was again .2. The learning regime was otherwise the same as in the dog–cat–bagel example.

At the end of 50 learning cycles, the model was able to retrieve the visual pattern corresponding to either repeated exemplar (see Table 3) given the associated name as input. When given the dog name pattern as input, it retrieves the prototype visual pattern for dog. It can also retrieve the appropriate name from each of the three visual patterns. This is true, even though the visual pattern for Rover differs from the visual pattern for dog by only a single element. Because of the special importance of this particular element, the weights from this element to the units that distinguish Rover's name pattern from the the prototype name pattern are quite strong. Given part of a visual pattern, the model will complete it; if the part corresponds to the prototype, then that is what is completed, but if it corresponds to one of the repeated exemplars, then that exemplar is completed. The model, then, knows both the prototype and the repeated exemplars quite well. Several other sets of prototypes and their repeated exemplars could also be stored in the same module, as long as its capacity is not exceeded; given large numbers of units per module, a lot of different patterns can be stored.

Let us summarize the observations we have made in these several illustrative simulations. First, our distributed model is capable of storing not just one but a number of different patterns. It can pull the "central tendency" of a number of different patterns out of the noisy inputs; it can create the functional equivalent of perceptual categories with or without the benefit of labels; and it can allow representations of repeated exemplars to coexist with the representation of the prototype of the categories they exemplify in the same composite memory trace. The model is not simply a categorizer or "prototyping" device; rather, it captures the structure inherent in a set of patterns, whether it be characterizable by description in terms of prototypes or not, as long as the ensemble of patterns adheres to the linear predictability constraint.

TABLE 3

RESULTS OF TESTS WITH PROTOTYPE AND SPECIFIC EXEMPLAR PATTERNS

	Name Pattern	Visual Pattern
Pattern for dog prototype	+ − + − + − + −	+ − + + − − − + + + + + − − − −
Response to prototype name		+4 −5 +3 +3 −4 −3 −3 −3 +3 +4 +3 +4 −3 −4 −4
Response to prototype visual pattern	+5 −4 +4 −4 +5 −4 +4 −4	
Pattern for Fido exemplar	+ − − − + − − −	+ − (−)+ − − − − + + + + + + (+) − −
Response to Fido name		+4 −4 −4 +4 −4 −4 −4 +4 +4 +4 +4 +4 −4 −4 −4
Response to Fido visual pattern	+5 −5 −3 −5 +4 −5 −3 −5	
Pattern for Rover exemplar	+ − − + + − − +	+ (+)+ + − − − − + + + + + − − −
Response to Rover name		+4 +5 +4 +4 −4 −4 −4 −4 +4 +4 +4 +4 −4 −4 −4
Response to Rover visual pattern	+4 −4 −2 +4 +4 +4 −2 +4	

The ability to retrieve accurate completions of similar patterns is a property of the model that depends on the use of the delta learning rule. This allows both the storage of different prototypes that are not orthogonal and the coexistence of prototype representations and repeated exemplars.

SIMULATIONS OF EXPERIMENTAL RESULTS

Up to this point, we have discussed our distributed model in general terms and have outlined how it can accommodate both generalization and representation of specific information in the same network. We now consider, in the next two sections, how well the model does in accounting for some recent evidence about the details of the influence of specific experiences on performance and the conditions under which functional equivalents of summary representations such as logogens and prototypes emerge.

Repetition and Familiarity Effects

When we perceive an item—say a word, for example—this experience has effects on our later performance. If the word is presented again within a reasonable interval of time, the prior presentation makes it possible for us to recognize the word more quickly or from a briefer presentation.

Traditionally, this effect has been interpreted in terms of units that represent the presented items in memory. In the case of word perception, these units are called *word detectors* or *logogens* and a model of repetition effects for words has been constructed around the logogen concept (Morton, 1979). The idea is that the threshold for the logogen is reduced every time it "fires" (that is, every time the word is recognized), thereby making it easier to fire the logogen at a later time. There is supposed to be a decay of this priming effect, with time, so that eventually the effect of the first presentation wears off.

This traditional interpretation has come under serious question of late, for a number of reasons. Perhaps paramount among the reasons is the fact that the exact relation between the specific context in which the priming event occurs and the context in which the test event occurs makes a huge difference (Jacoby, 1983a, 1983b). Generally speaking, nearly any change in the stimulus—from spoken to printed, from male speaker to female speaker, etc.—tends to reduce the magnitude of the priming effect.

These facts might easily be taken to support the enumeration of specific experiences view, in which the logogen is replaced by the entire ensemble of experiences with the word, with each experience capturing aspects of the specific context in which it occurred. Such a view has been championed most strongly by Jacoby (1983a, 1983b).

Our distributed model offers an alternative interpretation. We see the traces laid down by the processing of each input as contributing to the composite, superimposed memory representation. Each time a stimulus is processed, it gives rise to a slightly different memory trace—either because the item itself is different or because it occurs in a different context that conditions its representation. The logogen is replaced by the set of specific traces, but *the traces are not kept separate.* Each trace contributes to the composite, but the characteristics of particular experiences tend nevertheless to be preserved, at least until they are overridden by canceling characteristics of other traces. Also, the traces of one stimulus pattern can coexist with the traces of other stimuli, within the same composite memory trace.

It should be noted that we are not faulting either the logogen model or models based on the enumeration of specific experiences for their physiological implausibility here, since these models are generally not stated in physiological terms, and their authors might reasonably argue that nothing in their models precludes distributed storage at a physiological level. What we are suggesting is that a model which proposes explicitly distributed, superpositional storage can account for the kinds of findings that logogen models have been proposed to account for, as well as other findings which strain the utility of the concept of the logogen as a psychological construct.

To illustrate the distributed model's account of repetition priming effects, we carried out the following simulation experiment. We made up a set of eight random vectors, each 24 elements long, each one to be thought of as the prototype of a different recurring stimulus pattern. Through a series of 10 training cycles using the set of eight vectors, we constructed a composite memory trace. During training, the model did not actually see the prototypes, however. On each training presentation it saw a new random distortion of one of the eight prototypes. In each of the distortions, each of the 24 elements had its value flipped with a probability of .1. Weights were adjusted after every presentation and were then allowed to decay to a fixed residual before the presentation of the next pattern.

The composite memory trace formed as a result of this experience plays the same role in our model that the set of logogens or detectors play in a model like Morton's or, indeed, the interactive activation model of word perception. That is, the trace contains information which allows the model to enhance perception of familiar patterns, relative to unfamiliar ones. We demonstrate this by comparing the activations resulting from the processing of subsequent presentations of new distortions of our eight familiar patterns, compared to other random patterns with which the model is not familiar. The pattern of activation that is the model's response to the input is stronger and grows to a particular level more quickly if the stimulus is a new distortion of an old pattern than if it is a new pattern. This effect is illustrated in Figure 5.

Pattern activation and response strength. The measure of activation shown in the figure is the dot product of the pattern of activation over the units of the module with the stimulus pattern itself, normalized for the number n of elements in the pattern: For the pattern p we call this expression α_p. In mathematical notation it is just

$$\alpha_p = \frac{1}{n}\Sigma_i a_i e_{pi}(t)$$

where i indexes the units in the module, and e_{pi} indexes the external input to unit i in pattern p. Essentially, α represents the degree to which the actual pattern of activation on the units captures the input pattern. It is an approximate analog of the activation of an individual unit in models that allocate a single unit to each whole pattern.

To relate these pattern activations to response probabilities, we must assume that mechanisms exist for translating patterns of activation into

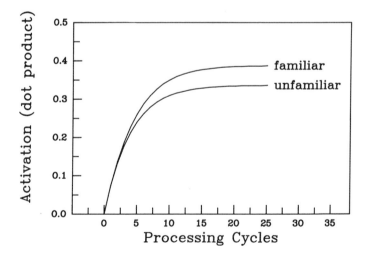

FIGURE 5. Growth of the pattern of activation for new distortions of familiar and unfamiliar patterns. The measure of the strength of the pattern of activation is the dot product of the response pattern with the input vector. See text for an explanation.

overt responses measurable by an experimenter. We will assume that these mechanisms obey the same principles discussed in Chapter 15 for relating activations to response probabilities, simply replacing the activations of particular units with the α measure of pattern activation.[3]

These assumptions finesse an important issue, namely, the mechanism by which a pattern of activation gives rise to a particular response. A specific mechanism for response generation is described in Chapter 18. For now, we wish only to capture basic properties any actual response selection mechanism must have: It must be sensitive to the input pattern, and it must approximate other basic aspects of response selection behavior captured by the Luce (1963) choice model.

Effects of experimental variables on time-accuracy curves. Applying the assumptions described above, we can calculate probability of correct response as a function of processing cycles for familiar and unfamiliar patterns. The result for a particular choice of scaling parameters is shown in Figure 6. If we assume that performance in a perceptual identification task is based on the height of the curve at the point where processing is cut off by masking (McClelland & Rumelhart, 1981), then familiarity would lead to greater accuracy of perceptual identification at a given exposure duration. In a reaction time task, if the response is emitted when its probability reaches a particular threshold activation value, familiarity would lead to speeded responses. Thus, the model is consistent with the ubiquitous influence of familiarity both on response accuracy and speed, in spite of the fact that it has no detectors for familiar stimuli.

But what about priming and the role of congruity between the prime event and the test event? To examine this issue, we carried out a second experiment. Following learning of eight patterns as in the previous experiment, new distortions of half of the random vectors previously learned by the model were presented as primes. For each of these primes, the pattern of activation was allowed to stabilize, and changes in the strengths of the connections in the model were then made. We then tested the model's response to (a) the same four distortions, (b) four new distortions of the same patterns, and (c) distortions of the four previously learned patterns that had not been

---

[3] One complication arises due to the fact that it is not, in general, possible to specify exactly what the set of alternative responses might be for the denominator of the Luce choice rule used in the word perception model. For this reason, the strengths of other responses are represented by a constant C (which stands for the competition). Thus, the expression for probability of choosing the response appropriate to pattern p is just $p(r_p) = (e^{k\bar{\alpha}_p})/(C + e^{k\bar{\alpha}_p})$ where $\bar{\alpha}_p$ represents the time average of α_p, and k is a scaling constant.

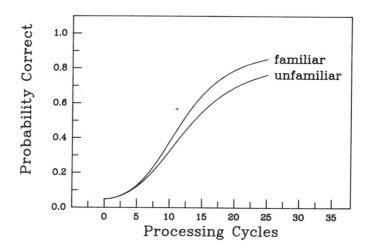

FIGURE 6. Simulated growth of response accuracy over the units in a 24-unit module, as a function of processing cycles, for new distortions of previously learned patterns compared to new distortions of patterns not previously learned.

presented as primes. There was no decay in the weights over the course of the priming experiment; if decay had been included, its main effect would have been to reduce the magnitude of the priming effects.

The results of the experiment are shown in Figure 7. The response of the model is greatest for the patterns preceded by identical primes, intermediate for patterns preceded by similar primes, and weakest for patterns not preceded by any related prime.

Our model, then, appears to provide an account, not only for the basic existence of priming effects, but also for the graded nature of priming effects as a function of congruity between prime event and test event. It avoids the problem of multiplication of context-specific detectors which logogen theories fall prey to, while at the same time avoiding enumeration of specific experiences. Congruity effects are captured in the composite memory trace.

The model also has another advantage over the logogen view. It accounts for repetition priming effects for unfamiliar as well as familiar stimuli. When a pattern is presented for the first time, a trace is produced just as it would be for stimuli that had previously been presented. The result is that, on a second presentation of the same pattern or a new distortion of it, processing is facilitated. The functional equivalent of a logogen begins to be established from the very first presentation.

To illustrate the repetition priming of unfamiliar patterns and to compare the results with the repetition priming we have already

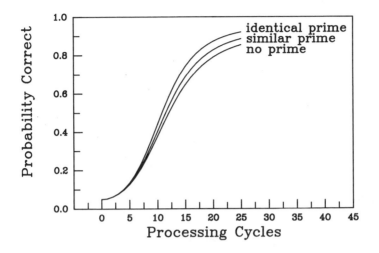

FIGURE 7. Response probability as a function of exposure time, for patterns preceded by identical primes, similar primes, or no related prime.

observed for familiar patterns, we carried out a third experiment. This time, after learning eight patterns as before, a priming session was run, in which new distortions of four of the familiar patterns and distortions of four new patterns were presented. Then, in the test phase, 16 stimuli were presented: New distortions of the primed, familiar patterns; new distortions of the unprimed, familiar patterns; new distortions of the primed, previously unfamiliar patterns; and finally, new distortions of four patterns that were neither primed nor familiar. The results are shown in Figure 8. What we find is that long-term familiarity and recent priming have approximately additive effects on the asymptotes of the time-accuracy curves. The time to reach any given activation level shows a mild interaction, with priming having slightly more of an effect for unfamiliar than for familiar stimuli.

These results are consistent with the bulk of the findings concerning the effects of pre-experimental familiarity and repetition in a recent series of experiments by Feustel, Shiffrin, and Salasoo (1983) and Salasoo, Shiffrin, and Feustel (1985). They found that pre-experimental familiarity of an item (word vs. nonword) and prior exposure had this very kind of interactive effect on exposure time required for accurate identification of all the letters of a string, at least when words and nonwords were mixed together in the same lists of materials.

A further aspect of the results reported by Salasoo, Shiffrin, and Feustel is also consistent with our approach. In one of their experiments, they examined the threshold for accurate identification as a function of number of prior presentations, for both words and

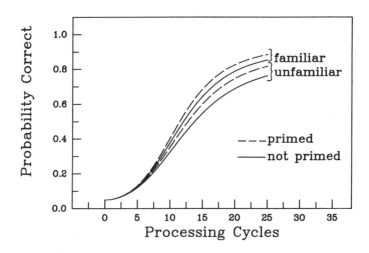

FIGURE 8. Response to new distortions of primed, familiar patterns; unprimed, familiar patterns; primed, unfamiliar patterns; and unprimed, unfamiliar patterns.

pseudowords. While thresholds were initially elevated for pseudo-words, relative to words, there was a rather rapid convergence of the thresholds over repeated presentations, with the point of convergence coming at about the same place on the curve for two different versions of their perceptual identification task (Salasoo et al., 1985). Our model, likewise, shows this kind of convergence effect, as illustrated in Figure 9.

There is one finding by Salasoo et al. (1985) that appears at first glance to support the view that there is some special process of unit formation that is distinct from the priming of old units. This is the fact that after a year between training and testing, performance with pseu-dowords used during training is indistinguishable from performance with words, but performance with words used during training shows no residual benefit compared to words not previously used. The data cer-tainly are consistent with the view that training experience made the pseudowords into lasting perceptual units at the same time that it pro-duced transitory priming of existing units. We have not attempted to account for this finding in detail, but we doubt that it is inconsistent with a distributed model. In support of this, we offer one reason why repetition effects might seem to persist longer for pseudowords rather than for words in the Salasoo et al. experiment. For pseudowords, a strong association would be built up between the item and the learning context during initial training. Such associations would be formed for words, but because these stimuli have been experienced many times before and have already been well learned, smaller increments in

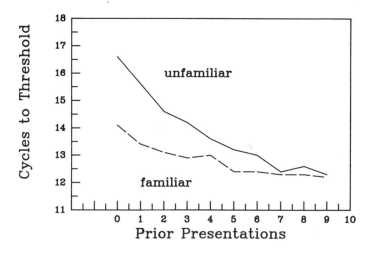

FIGURE 9. Time to reach a fixed accuracy criterion (60% correct) for previously familiar and unfamiliar patterns, as a function of repetitions.

connection strengths are formed for these stimuli during training, and thus the strength of the association between the item and the learning context would be less.

If our interpretation is correct, we would expect to see a disadvantage for pseudowords relative to words if the testing were carried out in a situation which did not reinstate the mental state associated with the original learning experience since for these stimuli much of what was learned would be tied to the specific learning context. Such a prediction would appear to differentiate our account from any view that postulated the formation of an abstract, context-independent logogen as the basis for the absence of a pseudoword decrement effect.

Representation of General and Specific Information

In the previous section, we cast our distributed model as an alternative to the view that familiar patterns are represented in memory either by separate detectors or by an enumeration of specific experiences. In this section, we show that the model provides alternatives to both abstraction and enumeration models of learning from exemplars of prototypes.

Abstraction models were originally motivated by the finding that subjects occasionally appeared to have learned better how to categorize the prototype of a set of distorted exemplars than the specific exemplars they experienced during learning (Posner & Keele, 1968). However, pure abstraction models have never fared very well since there is nearly always evidence of some superiority of the particular training stimuli over other stimuli equally far removed from the prototype. A favored model, then, is one in which there is both abstraction and memory for particular training stimuli.

Recently, proponents of models involving only enumeration of specific experiences have noted that such models can account for the basic fact that abstraction models are primarily designed to account for—enhanced response to the prototype, relative to particular previously seen exemplars, under some conditions—as well as failures to obtain such effects under other conditions (Hintzman, 1983; Medin & Schaffer, 1978). In evaluating distributed models, then, it is important to see if they can do as well. J. A. Anderson (1977) has made important steps in this direction, and Knapp and J. A. Anderson (1984) have shown how their distributed model can account for many of the details of the Posner-Keele experiments. Recently, however, two sets of findings have been put forward that appear to strongly favor the enumeration of specific experiences view, at least relative to pure abstraction models. It is important, therefore, to see how well our distributed model can do in accounting for these effects.

The first set of findings comes from a set of studies by Whittlesea (1983). In a large number of studies, Whittlesea demonstrated a role for specific exemplars in guiding performance on a perceptual identification task. We wanted to see whether our model would demonstrate a similar sensitivity to specific exemplars. We also wanted to see whether our model would account for the conditions under which such effects are not obtained.

Whittlesea used letter strings as stimuli. The learning experiences subjects received involved simply looking at the stimuli one at a time on a visual display and writing down the sequence of letters presented. Subjects were subsequently tested for the effect of this training on their ability to identify letter strings bearing various relationships to the training stimuli and to the prototypes from which the training stimuli were derived. The test was a perceptual identification task; the subject was simply required to try to identify the letters from a brief flash.

The stimuli Whittlesea used were all distortions of one of two prototype letter strings. Table 4 illustrates the essential properties of the sets of training and test stimuli he used. The stimuli in Set Ia were each one step away from the prototype. The Ib items were also one step from the prototype and one step from one of the Ia distortions. The Set IIa stimuli were each two steps from the prototype and one step from a particular Ia distortion. The Set IIb items were also two steps

TABLE 4

SCHEMATIC DESCRIPTION OF STIMULUS SETS
USED IN SIMULATIONS OF WHITTLESEA'S EXPERIMENTS

Prototype	Ia	Ib	IIa	IIb	IIc	III	V
PPPPP	APPPP	BPPPP	ABPPP	ACPPP	APCPP	ABCPP	CCCCC
	PAPPP	PBPPP	PABPP	PACPP	PAPCP	PABCP	CBCBC
	PPAPP	PPBPP	PPABP	PPACP	PPAPC	PPABC	BCACB
	PPPAP	PPPBP	PPPAB	PPPAC	CPPAP	CPPAB	ABCBA
	PPPPA	PPPPB	BPPPA	CPPPA	PCPPA	BCPPA	CACAC

Note: The actual stimuli used can be filled in by replacing P with +−+−; A with ++−−;
B with +−−+; and C with ++++. The model is not sensitive to the fact the same
subpattern was used in each of the 5 slots.

from the prototype, and each was one step from one of the IIa distortions. The Set IIc distortions were two steps from the prototype also, and each was two steps from the closest IIa distortion. Over the set of five IIc distortions, the A and B subpatterns each occurred once in each position, as they did in the case of the IIa distortions. The distortions in Set III were three steps from the prototype and one step from the closest member of Set IIa. The distortions in Set V were each five steps from the prototype.

Whittlesea ran seven experiments using different combinations of training and test stimuli. We carried out simulation analogs of all of these experiments plus one additional experiment that Whittlesea did not run. The main difference between the simulation experiments and Whittlesea's actual experiments was that he used two different prototypes in each experiment, while we only used one.

The simulation employed a simple 20-unit module. The set of 20 units was divided into five submodules, one for each letter in Whittlesea's letter strings. The prototype pattern and the different distortions used can be derived from the information provided in Table 4.

Each simulation experiment began with null connections between the units. The training phase involved presenting the set or sets of training stimuli analogous to those Whittlesea used, for the same number of presentations. To avoid idiosyncratic effects of particular orders of training stimuli, each experiment was run six times, each with a different random order of training stimuli. On each trial, activations were allowed to settle down through 50 processing cycles, and then connection strengths were adjusted. There was no decay of the increments to the weights over the course of an experiment.

In the test phase, the model was tested with the sets of test items analogous to the sets Whittlesea used. As a precaution against effects of prior test items on performance, we simply turned off the adjustment of weights during the test phase.

A summary of the training and test stimuli used in each of the experiments, of Whittlesea's findings, and of the simulation results are shown in Table 5. The numbers represent relative amounts of enhancement in performance as a result of the training experience,

TABLE 5

SUMMARY OF PERCEPTUAL IDENTIFICATION EXPERIMENTS
WITH EXPERIMENTAL AND SIMULATION RESULTS

Whittlesea's Experiment Number	Training Stimulus Set(s)	Test Stimulus Sets	Experimental Results	Simulation Results
1	Ia	Ia	.27	.24
		Ib	.16	.15
		V	.03	-.05
2	IIa	IIa	.30	.29
		IIc	.15	.12
		V	.03	-.08
3	IIa	IIa	.21	.29
		IIb	.16	.14
		IIc	.10	.12
4	IIa	P	-	.24
		Ia	.19	.21
		IIa	.23	.29
		III	.15	.15
4'	Ia	P	-	.28
		Ia	-	.24
		IIa	-	.12
5	IIa,b,c	P	-	.25
		Ia	.16	.21
		IIa	.16	.18
		III	.10	.09
6	III	Ia	.16	.14
		IIa	.16	.19
		III	.19	.30
7	IIa	IIa	.24	.29
		IIc	.13	.12
		III	.17	.15

relative to a pretest baseline. For Whittlesea's data, this is the per-letter increase in letter identification probability between a pre- and posttest. For the simulation, it is the increase in the size of the dot product for a pretest with null weights and a posttest after training. For comparability to the data, the dot-product difference scores have been doubled. This is simply a scaling operation to facilitate qualitative comparison of experimental and simulation results.

A comparison of the experimental and simulation results shows that wherever there is a within-experiment difference in Whittlesea's data, the simulation produced a difference in the same direction. (Between-experiment comparisons are not considered because of subject and material differences which render such differences unreliable). The next several paragraphs review some of the major findings in detail.

Some of the comparisons bring out the importance of congruity between particular test and training experiences. Experiments 1, 2, and 3 show that when distance of test stimuli from the prototype is controlled, similarity to particular training exemplars makes a difference both for the human subject and in the model. In Experiment 1, the relevant contrast was between Ia and Ib items. In Experiment 2, it was between IIa and IIc items. Experiment 3 shows that the subjects and the model both show a gradient in performance with increasing distance of the test items from the nearest old exemplar.

Experiments 4, 4', and 5 explore the status of the prototype and other test stimuli closer to the prototype than any stimuli actually shown during training. In Experiment 4, the training stimuli were fairly far away from the prototype, and there were only five different training stimuli (the members of the IIa set). In this case, controlling for distance from the nearest training stimuli, test stimuli closer to the prototype showed more enhancement than those farther away. (Ia vs. III comparison). However, the actual training stimuli nevertheless had an advantage over both other sets of test stimuli, including those that were closer to the prototype than the training stimuli themselves (IIa vs. Ia comparison).

In experiment 4' (not run by Whittlesea), the same number of training stimuli were used as in Experiment 4, but these were closer to the prototype. The result is that the simulation shows an advantage for the prototype over the old exemplars. The specific training stimuli used, even in this experiment, do influence performance, however, as Whittlesea's first experiment (which used the same training set) shows (Ia-Ib contrast). This effect holds both for the subjects and for the simulation. The pattern of results is similar to the findings of Posner and Keele (1968), in the condition where subjects learned six exemplars which were rather close to the prototype. In this condition, their subjects' categorization performance was most accurate for the prototype, but more accurate for old than for new distortions, just as in this simulation experiment.

In Experiment 5, Whittlesea demonstrated that a slight advantage for stimuli closer to the prototype than the training stimuli would emerge, even with high-level distortions, when a large number of different distortions were used once each in training, instead of a smaller number of distortions presented three times each. The effect was rather small in Whittlesea's case (falling in the third decimal place in the per-letter enhancement effect measure) but other experiments have produced similar results, and so does the simulation. In fact, since the prototype was tested in the simulation, we were able to demonstrate a monotonic drop in performance with distance from the prototype in this experiment.

Experiments 6 and 7, which used small numbers of training exemplars rather far from the prototype, both explore in different ways the relative influence of similarity to the prototype and similarity to the set of training exemplars. Both in the data and in the model, similarity to particular training stimuli is more important than similarity to the prototype, given the sets of training stimuli used in these experiments.

Taken together with other findings, Whittlesea's results show clearly that similarity of test items to particular stored exemplars is of paramount importance in predicting perceptual performance. Other experiments show the relevance of these same factors in other tasks, such as recognition memory, classification learning, etc. It is interesting to note, though, that performance does not honor the specific exemplars so strongly when the training items are closer to the prototype. Under such conditions, performance is superior on the prototype or stimuli closer to the prototype than the training stimuli. Even when the training stimuli are rather distant from the prototype, they produce a benefit for stimuli closer to the prototype if there are a large number of distinct training stimuli each shown only once. Thus, the dominance of specific training experiences is honored only when the training experiences are few and far between. Otherwise, an apparent advantage for the prototype, though with some residual benefit for particular training stimuli, is the result.

The congruity of the results of these simulations with experimental findings underscores the applicability of distributed models to the question of the nature of the representation of general and specific information. In fact, we were somewhat surprised by the ability of the model to account for Whittlesea's results, given the fact that we did not rely on context-sensitive encoding of the letter-string stimuli. That is, the distributed representation we assigned to each letter was independent of the other letters in the string. A context-sensitive encoding would, however, prove necessary to capture a larger ensemble of stimuli.

Whether a context-sensitive encoding would produce the same or slightly different results depends on the exact encoding. The exact degree of overlap of the patterns of activation produced by different distortions of the same prototype determines the extent to which the model will tend to favor the prototype relative to particular old exemplars. The degree of overlap, in turn, depends upon the specific assumptions made about the encoding of the stimuli. However, the general form of the results of the simulation would be unchanged: When all the distortions are close to the prototype, or when there is a very large number of different distortions, the central tendency will produce the strongest response; but when the distortions are fewer and farther from the prototype, the training exemplars themselves will produce the strongest activations. What the encoding would effect is the similarity metric.

In this regard, it is worth mentioning another finding that may appear to challenge our distributed account of what is learned through repeated experiences with exemplars. This is the finding of Medin and Schwanenflugel (1981). Their experiment compared learning of two different sets of stimuli in a categorization task. One set of stimuli could be categorized by a linear combination of weights assigned to particular values on each of four dimensions considered independently. The other set of stimuli could not be categorized in this way. The experiment demonstrated clearly that linear separability was not necessary for categorization learning. Linearly separable stimuli were less easily learned than a set of stimuli that were not linearly separable but had a higher degree of similarity between exemplars within categories.

At first glance, it may seem that Medin and Schwanenflugel's experiment is devastating to our distributed approach since our distributed model can only learn linear combinations of weights. However, whether a linear combination of weights can suffice in the Medin and Schwanenflugel experiments depends on how patterns of activation are assigned to stimuli. If each stimulus dimension is encoded separately in the representation of the stimulus, then the Medin and Schwanenflugel stimuli cannot be learned by our model. But if each stimulus dimension is encoded in a context sensitive way, then the patterns of activation associated with the different stimuli become linearly separable again.

One way of achieving context sensitivity is via separate enumeration of traces. But it is well known that there are other ways as well. Several different kinds of context-sensitive encodings which do not require separate enumeration of traces or the allocation of separate units to individual experiences are considered in Chapter 3 and in Chapters 18 and 19.

It should be noted that the motivation for context-sensitive encoding in the use of distributed representations is captured by, but by no means limited to, the kinds of observations reported in the experiment by Medin and Schwanenflugel. Context-sensitive encoding is required if distributed models are to be able to overcome the linear predictability constraint, and this constraint pervades both computational and psychological applications of the idea of distributed representations.

EXTENSIONS OF THE MODEL

There are a number of phenomena in the learning and memory literature that lend themselves to accounts in terms of distributed models of memory. Here we will give a brief list. In some cases, the phenomena are addressed in other chapters, and we merely give pointers. In other cases, we have not yet had an opportunity to carry out detailed simulations; in these cases we describe briefly how we might envision encompassing the phenomena in terms of distributed models.

The Emergence of Semantic Memory From Episodic Traces

A distributed model leads naturally to the suggestion that semantic memory may be just the residue of the superposition of episodic traces. Consider, for example, representation of a proposition encountered in several different contexts, and assume for the moment that the context and content are represented in separate parts of the same module. Over repeated experience with the same proposition in different contexts, the proposition will remain in the interconnections of the units in the proposition submodule, but the particular associations to particular contexts will wash out. However, material that is only encountered in one particular context will tend to be somewhat contextually bound. So we may not be able to retrieve what we learn in one context when we need it in other situations. Other authors (e.g., J. R. Anderson & Ross, 1980) have recently argued against a separation between episodic and semantic memory, pointing out interactions between traditionally episodic and semantic memory tasks. Such findings are generally consistent with the view we have taken here.

Emergence of Regularities in Behavior and Their Coexistence With Exceptions

Distributed models also influence our thinking about how human behavior might come to exhibit the kind of regularity that often leads linguists to postulate systems of rules. In Chapter 18 we describe a distributed model of a system that can learn the past-tense system of English, given as inputs pairs of patterns corresponding to the phonological structure of the present and past tense forms of actual English verbs. The model accounts for several interesting phenomena which have been taken as evidence for the acquisition of "linguistic rules," such as the fact that children at a certain stage overregularize the pronunciation of irregular forms. It also accounts for the coexistence of irregular pronunciations with productive use of the regular past tense in the mature language user.

In general, then, distributed models appear to provide alternatives to a variety of models that postulate abstract, summary representations such as prototypes, logogens, semantic memory representations, or even linguistic rules.

Spared Learning in Amnesia

Distributed models also provide a natural way of accounting for spared learning by amnesics—persons who have diminished ability to learn new information after some traumatic event. There are generally two types of spared learning effects observed. First, in domains where amnesics show deficits, they nevertheless show a residual ability to learn gradually from repeated experiences. Indeed, they generally appear to be relatively more spared in their ability to extract what is common about an ensemble of experiences than in their ability to remember the details of individual events. Second, there are some domains in which most amnesics show no deficits at all. These phenomena are taken up from the point of view of a distributed model of learning and memory in Chapter 25. Here, we briefly consider the main points of the arguments made in that chapter.

First, distributed models provide a natural way of explaining why there should be residual ability to learn gradually from repeated experience even within those domains where amnesics are grossly deficient in their memory for specific episodic experiences. For if we simply imagine that the effective size of the increments to the connections is reduced in amnesia, then the general properties of distributed models—the fact that they extract the central tendency from a set of similar experiences and build up a trace of the prototype from a series of repeated exemplars—automatically provide an account of the gradual accumulation of repeated information in the face of a profound deficit in remembering any specific episode in which the information was presented. Indeed, in some cases reduced increments in the sizes of connections can actually be an advantage for pulling out the central tendency of a set of experiences since reduction in the size of the increments means that the idiosyncratic properties of recent experiences will not unduly dominate the composite memory trace.

The fact that large changes in connection strengths are not universally beneficial in distributed models may also provide a basis for explaining why certain aspects of learning and memory are completely unaffected in amnesia. The domains in which learning is almost completely spared are just the ones in which learning is very gradual in normals and is independent of awareness of having learned. These kinds of learning, we suggest, are acquired gradually by normals because large changes in the strengths of connections do not facilitate learning in such cases.

Memory Blends

Loftus (1977) has observed that subjects often combine information from two episodes when they attempt to describe some aspect of just one of these events. Many observations of this kind of effect have been observed in the context of legal testimony, but Loftus has now begun to explore the same effects in more basic experimental paradigms. The general result is that subjects do not simply report the correct property from one or the other memory trace; rather, they report a property that results from averaging or blending the properties of the individual traces. In one study, for example, many subjects reported that the color of a vehicle was green. However, no subject had actually seen a green truck; instead, they had seen a yellow truck in the original scene and a blue one in a distorted version of it. Such blend errors, obviously, are the beginnings of the formation of a summary representation and fall naturally out of distributed models.

Interference and Fan Effects

One of the oldest phenomena in the memory literature is interference. That similar memories or memories acquired in similar contexts interfere with each other was a central topic in memory research for years (see Crowder, 1976, for a review). Recently, this sort of interference effect has been explored by John Anderson, under the name of the *fan effect* (see J. R. Anderson, 1983, for a review). The fan effect refers to the fact that reaction times to indicate recognition of propositions goes up with the number of propositions studied that share arguments with the proposition under test. For example, *The boy kissed the girl* is harder to recognize if the subject has studied other sentences involving *the boy* or *the girl* than if he has not.

Distributed models like the one we have described here exhibit interference of just this kind. Singley (personal communication, 1985) has in fact simulated a simple fan-effect experiment. He trained a module of 48 units on "propositions" consisting of four parts. The parts correspond to a subject, a verb, an object, and a fourth part that served as a conjunctive representation of the three other constituents (following Hinton, 1981a). Some of the propositions shared arguments (subpatterns) with others. After several exposures to the set of patterns, Singley tested the module's response to each of the learned patterns and found that the response was generally weaker and took longer to reach a strength criterion for those patterns which shared arguments with other patterns. Obviously there is a lot more to fan effects than this; how well distributed models will be able to do in accounting for the details of the fan effect literature and other aspects of interference is a matter for further research.

AUGMENTING THE MODEL WITH HIDDEN UNITS

For all its successes, it must be said that our model does suffer from the fact that it can only learn to respond appropriately to sets of patterns that obey the linear predictability constraint. While this constraint can be overcome in specific cases by providing a conjunctive coding capable of serving as a basis for learning any prespecified set of patterns, it cannot be overcome, in general, without great cost in terms of units to cover the space of possibly necessary conjunctive representations. To overcome this limitation, it is necessary to have a model that can create the right conjunctive representation to solve the problems it faces in a way that is efficient in terms of units and connections.

Both the Boltzmann learning rule (Chapter 7) and the generalized delta rule (Chapter 8) provide ways of constructing just the required set of representations. Here we briefly sketch the application of the generalized delta rule to the case of an auto-associator with hidden units.

The basic idea is simple. A module consists of a large network of units. Some units are designated as *input/output* or simply *input* units and the remaining are designated as *hidden* units. As in the case without hidden units, the goal of the system is to have the net input from internal sources impinging on a particular input/output unit equal the external input to that unit. In this case, however, input from the hidden units is added as part of the total input from internal sources. The basic difference between a hidden unit and an input/output unit is that hidden units receive no inputs from outside the module, nor do they send outputs outside of the module. Rather, they receive inputs from other hidden units or from the input/output units. The generalized delta rule allows the system to develop a set of connections to and among the hidden units which, in principle, allows the system to retrieve the correct input pattern from *any* unique subportion of the input pattern—provided there are enough hidden units.

The "classical" example of a set of vectors for which hidden units are required is due to Hinton (1981a). This is the *one-same-one* problem. Suppose that the input vectors are divided into three portions. The first portion represents a number (either one or zero); the second portion represents a relation (either same or different); and the final portion represents a number (again either one or zero). Further, suppose that we want to store four vectors in the system representing the four propositions, *one-same-one*, *one-different-zero*, *zero-same-zero*, and *zero-different-one*. Now, we probe the system with two of the three sections of a stored input vector and hope that the memory system reconstructs the missing portion. For example, we present the system with *one-same-?* or *zero-?-one* and see what values get filled in for the unspecified portion of the vector.

For this example, learning done with the standard delta rule will not succeed. These vectors violate the linear predictability constraint. For example, suppose we present the system with *one-same-?*. In this case, the pattern associated with *one* in the first position are as often associated with *zero* in the third position as they are with *one* in that position. Thus, this pattern will try to establish a blend of *zero* and *one* in the unspecified portion of the input vector. Similarly, the pattern representing *same* in the middle portion are as often associated with the *zero* as with *one* in both other slots. Each portion will singly try to produce a blend of the two possibilities in the missing portion and both together will simply produce the same blend. The same argument can be made for each slot. However, with hidden units, the problem can be solved. A given hidden unit can relate bits from two or more slots together, and in that way it can fill in the appropriate pattern when it is missing from the input.

We have tested a network on a very simple version of this problem, using the generalized delta rule (Chapter 8) to learn the appropriate weights to and from the hidden units. In this case there were only three input/output units, one for each of the three components of the *one–same–one* propositions. The number *one* and the predicate *same* were represented by 1s in the appropriate positions, and the number *zero* and the *predicate* different were represented by 0s in the appropriate positions. Thus, the proposition *one–same–one* was represented by the input pattern 111 and the proposition *zero–different–one* was represented by pattern 001. The other two propositions were analogously represented as 010 and 100.

The basic auto-associative memory paradigm was modified slightly from the ones presented earlier in this chapter to accommodate the requirements of hidden units. Most of the modifications are not substantive, but were made to accommodate the technical requirements of the generalized delta rule. In particular, a *logistic* sigmoid "squashing" function was substituted for the one described in Equations 1 and 2. This was done because our original squashing function was not differentiable as required by the generalized delta rule (see Chapter 8 for details). This is a minor difference and has little effect on the performance of the system. Similarly, output values of units were allowed to range between [0,1] rather than [−1,1]. This change again is minor. It essentially involves a scale change and the addition of a *bias* or threshold term (again, see Chapter 8 for details). In this case 0.5 is the resting level rather than 0.0. One other somewhat more significant change of paradigm was also required. In the standard auto-associator we did not allow units to connect to themselves. The technical reason for this is that they can "learn" the input (i.e., they can make the internal input equal to the external input) simply by learning to "predict" themselves. However, this does not facilitate reconstruction of the whole pattern from a part. If the strongest connections are from a unit to itself, when that unit is not present in the probe pattern, the missing element will not be well predicted. We wished to impose the same restriction to the model with hidden units, but in this case the restriction becomes harder to enforce. That is, the system can learn to get around a constraint on direct connections by using hidden units to "shadow" each of the input units. When a particular input unit is turned on, it turns on its "shadow" hidden unit which in turn feeds activation back to the input unit. Thus we again have a unit predicting itself. Since we are most interested in the case of pattern completion, we want to avoid this situation. One way to avoid this is to require that the input from all units *except* the unit in question match the external input. To accomplish this, we used the following procedure: For each pattern to be learned, for example, 111, we actually trained the module to associate all of the patterns 11?, 1?1 and ?11 with the whole 111 pattern.

There were three input units and three hidden units. A network was set up which allowed any input unit to connect to any other input unit or to any hidden unit. Hidden units were allowed to connect to any input unit and any hidden unit—including themselves. The model was repeatedly presented with each of the three training inputs of each of the four patterns as indicated above, and the network was allowed to learn until it became stable and correctly filled in the missing parts of the patterns. The system was taught to fill in the missing portion of the pattern in three cycles and then to hold this completed pattern. Table 6 shows the state achieved by the three hidden units for each of the four input patterns. We see a distributed pattern over the hidden units in which the first unit signals that the predicate was either different or the proposition was *one–same–one*. The second hidden unit indicates that the predicate was either *same* or the proposition was *one–different–zero*, and the final hidden unit indicates that the predicate was either *different* or the proposition was *one–same–one*. This provides a unique, linearly independent basis from which to reconstruct the required input patterns.

The final network is shown in Figure 10 and the weight matrix is shown in Table 7. A number of interesting features of the network should be observed. First, since none of the input units can predict any other of the input units, the potential connections among input units have all gone to zero. Second, the hidden units all connect positively to themselves and negatively to each other. This allows a unit once turned on to stay on and shut down any noise feeding into those units that were not turned on. Finally, the connections between the input units and the hidden units are roughly symmetric. That is, whenever a particular hidden unit is turned on by a given input unit, that hidden unit tries, in turn, to turn on that input unit. It is interesting that this general kind of architecture in which units are connected roughly symmetrically and in which there are pools of mutually inhibitory units which evolved following the presentation of these very difficult patterns is very much like the architecture we have assumed in a number of our

TABLE 6

HIDDEN UNIT REPRESENTATIONS FOR
THE ONE–SAME–ONE PROBLEM

Propositions	Input Patterns		Hidden Unit Patterns
one–same–one	111	→	111
one–different–zero	100	→	011
zero–same–zero	010	→	110
zero–different–one	001	→	101

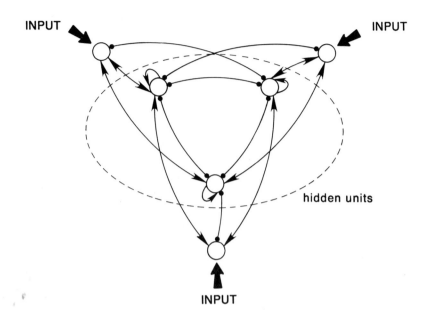

INPUT

INPUT

hidden units

INPUT

FIGURE 10. The network used in solving the *one–same–one* problem. See text for explanation.

models, including the interactive activation model of word perception and the models described in Chapters 15 and 16.

Finally, before leaving this section, it should be noted that the hidden-units version of an auto-associator will behave very much like the system we have described throughout this chapter if the situation does not require hidden units. In this case, the connections among the

TABLE 7

WEIGHTS ACQUIRED IN LEARNING THE ONE–SAME–ONE PROBLEM

		FROM					
		Input Units			Hidden Units		
TO	Input Units	0	0	0	-6	4	4
		0	0	0	4	4	-6
		0	0	0	4	-6	4
	Hidden Units	-5	6	6	6	-3	-3
		6	6	-5	-3	6	-3
		6	-5	6	-3	-3	6

input units can capture the structure of the problem, and the connections to and from the hidden units will tend not to play an important role. This is because direct connections are learned more quickly than those which propagate information through hidden units. These will lag behind, and if they are not required, their strength will stay relatively small.

It should also be emphasized that the use of hidden units overcomes a problem that our present model shares with enumeration models: the problem of having a fixed set of representational primitives. Hidden units really are new representational primitives that grow and develop to meet the representational needs of the particular set of patterns that have been encountered. They allow the model to change its representation and thereby to change the pattern of similarity among the input patterns. We have only begun our exploration of the generalized delta rule and its application to the case of the auto-associator. We are very optimistic that it can serve as a mechanism for building the kinds of internal representations required by allowing distributed memories the flexibility of adapting to arbitrary sets of stimulus patterns.

CONCLUSION

In this chapter, we have argued that a distributed model of memory provides a natural way of accounting for the fact that memory appears to extract the central tendencies of a set of experiences and for the fact that memory is sensitive to the details of specific events and experiences. We have seen how a simple distributed model, using the delta rule, provides a fairly direct and homogeneous account for a large body of evidence relevant to the representation of general and specific information. The delta rule—and now the generalized delta rule—provide very simple mechanisms for extracting regularities from an ensemble of inputs without the aid of sophisticated generalization or rule-formulating mechanisms that oversee the performance of the processing system. These learning rules are completely local, in the sense that they change the connection between one unit and another on the basis of information that is locally available to the connection rather than on the basis of global information about overall performance. The model thus stands as an alternative to the view that learning in cognitive systems involves the explicit formulation of rules and abstractions under the guidance of some explicit overseer. We do not wish to suggest that explicit rule formation has no place in a cognitive theory of learning and memory; we only wish to suggest that such mechanisms need not be invoked whenever behavior is observed that appears to be describable by some generalization or rule.

We emphasize that the basic properties of distributed models are largely independent of specific detailed assumptions about implementation, such as the dynamic range of the units involved or the exact form of the activation update rule. This points are underscored by the models considered in Chapters 18 and 19, in which some of the same general properties arise from models making use of somewhat different detailed assumptions.

ACKNOWLEDGMENTS

The work reported was supported in part by a grant from the System Development Foundation, in part by a contract from the Office of Naval Research (N00014-79-C-0323, NR667-437), and in part by a NIMH Career Development Award (MH00385) to the first author. Many of the ideas described here were developed and elaborated in the context of a conference organized by Lee Brooks and Larry Jacoby entitled "The Priority of the Specific," June, 1983, Elora, Ontario. The chapter owes a great deal to this conference and to stimulating discussions with the organizers and participants.

CHAPTER 18

On Learning the
Past Tenses of English Verbs

D. E. RUMELHART and J. L. McCLELLAND

THE ISSUE

Scholars of language and psycholinguistics have been among the first to stress the importance of rules in describing human behavior. The reason for this is obvious. Many aspects of language can be characterized by rules, and the speakers of natural languages speak the language correctly. Therefore, systems of rules are useful in characterizing what they will and will not say. Though we all make mistakes when we speak, we have a pretty good ear for what is right and what is wrong—and our judgments of correctness—or grammaticality—are generally even easier to characterize by rules than actual utterances.

On the evidence that what we will and won't say and what we will and won't accept can be characterized by rules, it has been argued that, in some sense, we "know" the rules of our language. The sense in which we know them is not the same as the sense in which we know such "rules" as "*i* before e except after *c*," however, since we need not necessarily be able to state the rules explicitly. We know them in a way that allows us to use them to make judgments of grammaticality, it is often said, or to speak and understand, but this knowledge is not in a form or location that permits it to be encoded into a communicable verbal statement. Because of this, this knowledge is said to be *implicit*.

A slight variant of this chapter will appear in B. MacWhinney (Ed.), *Mechanisms of language acquisition.* Hillsdale, NJ: Erlbaum (in press).

So far there is considerable agreement. However, the exact characterization of implicit knowledge is a matter of great controversy. One view, which is perhaps extreme but is nevertheless quite clear, holds that the rules of language are stored in explicit form as propositions, and are used by language production, comprehension, and judgment mechanisms. These propositions cannot be described verbally only because they are sequestered in a specialized subsystem which is used in language processing, or because they are written in a special code that only the language processing system can understand. This view we will call the *explicit inaccessible rule* view.

On the explicit inaccessible rule view, language acquisition is thought of as the process of inducing rules. The language mechanisms are thought to include a subsystem—often called the *language acquisition device* (LAD)—whose business it is to discover the rules. A considerable amount of effort has been expended on the attempt to describe how the LAD might operate, and there are a number of different proposals which have been laid out. Generally, though, they share three assumptions:

- The mechanism hypothesizes explicit inaccessible rules.

- Hypotheses are rejected and replaced as they prove inadequate to account for the utterances the learner hears.

- The LAD is presumed to have *innate* knowledge of the possible range of human languages and, therefore, is presumed to consider only hypotheses within the constraints imposed by a set of *linguistic universals*.

The recent book by Pinker (1984) contains a state-of-the-art example of a model based on this approach.

We propose an alternative to explicit inaccessible rules. We suggest that lawful behavior and judgments may be produced by a mechanism in which there is no explicit representation of the rule. Instead, we suggest that the mechanisms that process language and make judgments of grammaticality are constructed in such a way that their performance is characterizable by rules, but that the rules themselves are not written in explicit form anywhere in the mechanism. An illustration of this view, which we owe to Bates (1979), is provided by the honeycomb. The regular structure of the honeycomb arises from the interaction of forces that wax balls exert on each other when compressed. The honeycomb can be described by a rule, but the mechanism which produces it does not contain any statement of this rule.

In our earlier work with the interactive activation model of word perception (McClelland & Rumelhart, 1981; Rumelhart & McClelland,

1981, 1982), we noted that lawful behavior emerged from the interactions of a set of word and letter units. Each word unit stood for a particular word and had connections to units for the letters of the word. There were no separate units for common letter clusters and no explicit provision for dealing differently with orthographically regular letter sequences—strings that accorded with the rules of English—as opposed to irregular sequences. Yet the model did behave differently with orthographically regular nonwords than it behaved with words. In fact, the model simulated rather closely a number of results in the word perception literature relating to the finding that subjects perceive letters in orthographically regular letter strings more accurately than they perceive letters in irregular, random letter strings. Thus, the behavior of the model was lawful even though it contained no explicit rules.

It should be said that the pattern of perceptual facilitation shown by the model did not correspond exactly to any system of orthographic rules that we know of. The model produced as much facilitation, for example, for special nonwords like *SLNT*, which are clearly irregular, as it did for matched regular nonwords like *SLET*. Thus, it is not correct to say that the model exactly mimicked the behavior we would expect to emerge from a system which makes use of explicit orthographic rules. However, neither do human subjects. Just like the model, they showed equal facilitation for vowelless strings like *SLNT* as for regular nonwords like *SLET*. Thus, human perceptual performance seems, in this case at least, to be characterized only approximately by rules.

Some people have been tempted to argue that the behavior of the model shows that we can do without linguistic rules. We prefer, however, to put the matter in a slightly different light. There is no denying that rules still provide a fairly close characterization of the performance of our subjects. And we have no doubt that rules are even more useful in characterizations of sentence production, comprehension, and grammaticality judgments. We would only suggest that parallel distributed processing models may provide a mechanism sufficient to capture lawful behavior, without requiring the postulation of explicit but inaccessible rules. Put succinctly, our claim is that PDP models provide an alternative to the explicit but inaccessible rules account of implicit knowledge of rules.

We can anticipate two kinds of arguments against this kind of claim. The first kind would claim that although certain types of rule-guided behavior might emerge from PDP models, the models simply lack the computational power needed to carry out certain types of operations which can be easily handled by a system using explicit rules. We believe that this argument is simply mistaken. We discuss the issue of computational power of PDP models in Chapter 4. Some applications of PDP models to sentence processing are described in Chapter 19.

The second kind of argument would be that the details of language behavior, and, indeed, the details of the language acquisition process, would provide unequivocal evidence in favor of a system of explicit rules.

It is this latter kind of argument we wish to address in the present chapter. We have selected a phenomenon that is often thought of as demonstrating the acquisition of a linguistic rule. And we have developed a parallel distributed processing model that learns in a natural way to behave in accordance with the rule, mimicking the general trends seen in the acquisition data.

THE PHENOMENON

The phenomenon we wish to account for is actually a sequence of three stages in the acquisition of the use of past tense by children learning English as their native tongue. Descriptions of development of the use of the past tense may be found in Brown (1973), Ervin (1964), and Kuczaj (1977).

In Stage 1, children use only a small number of verbs in the past tense. Such verbs tend to be very high-frequency words, and the majority of these are irregular. At this stage, children tend to get the past tenses of these words correct if they use the past tense at all. For example, a child's lexicon of past-tense words at this stage might consist of *came, got, gave, looked, needed, took*, and *went*. Of these seven verbs, only two are regular—the other five are generally idiosyncratic examples of irregular verbs. In this stage, there is no evidence of the use of the rule—it appears that children simply know a small number of separate items.

In Stage 2, evidence of implicit knowledge of a linguistic rule emerges. At this stage, children use a much larger number of verbs in the past tense. These verbs include a few more irregular items, but it turns out that the majority of the words at this stage are examples of the *regular* past tense in English. Some examples are *wiped* and *pulled*.

The evidence that the Stage 2 child actually has a linguistic rule comes not from the mere fact that he or she knows a number of regular forms. There are two additional and crucial facts:

● The child can now generate a past tense for an invented word. For example, Berko (1958) has shown that if children can be convinced to use *rick* to describe an action, they will tend to say *ricked* when the occasion arises to use the word in the past tense.

- Children now *incorrectly* supply regular past-tense endings for words which they used correctly in Stage 1. These errors may involve either adding *ed* to the root as in *comed* /kˆmd/, or adding *ed* to the irregular past tense form as in *camed* /kʌmd/[1] (Ervin, 1964; Kuczaj, 1977).

Such findings have been taken as fairly strong support for the assertion that the child at this stage has acquired the past-tense "rule." To quote Berko (1958):

> If a child knows that the plural of *witch* is *witches*, he may simply have memorized the plural form. If, however, he tells us that the plural of *gutch* is *gutches*, we have evidence that he actually knows, albeit unconsciously, one of those rules which the descriptive linguist, too, would set forth in his grammar. (p. 151)

In Stage 3, the regular and irregular forms coexist. That is, children have regained the use of the correct irregular forms of the past tense, while they continue to apply the regular form to new words they learn. Regularizations persist into adulthood—in fact, there is a class of words for which either a regular or an irregular version are both considered acceptable—but for the commonest irregulars such as those the child acquired first, they tend to be rather rare. At this stage there are some clusters of exceptions to the basic, regular past-tense pattern of English. Each cluster includes a number of words that undergo identical changes from the present to the past tense. For example, there is a *ing/ang* cluster, an *ing/ung* cluster, an *eet/it* cluster, etc. There is also a group of words ending in /d/ or /t/ for which the present and past are identical.

Table 1 summarizes the major characteristics of the three stages.

Variability and Gradualness

The characterization of past-tense acquisition as a sequence of three stages is somewhat misleading. It may suggest that the stages are clearly demarcated and that performance in each stage is sharply distinguished from performance in other stages.

---

[1] The notation of phonemes used in this chapter is somewhat nonstandard. It is derived from the computer-readable dictionary containing phonetic transcriptions of the verbs used in the simulations. A key is given in Table 5.

TABLE 1

CHARACTERISTICS OF THE THREE STAGES
OF PAST TENSE ACQUISITION

Verb Type	Stage 1	Stage 2	Stage 3
Early Verbs	Correct	Regularized	Correct
Regular	—	Correct	Correct
Other Irregular	—	Regularized	Correct or Regularized
Novel	—	Regularized	Regularized

In fact, the acquisition process is quite gradual. Little detailed data exists on the transition from Stage 1 to Stage 2, but the transition from Stage 2 to Stage 3 is quite protracted and extends over several years (Kuczaj, 1977). Further, performance in Stage 2 is extremely variable. Correct use of irregular forms is never completely absent, and the same child may be observed to use the correct past of an irregular, the base + ed form, and the past + ed form, within the same conversation.

Other Facts About Past-Tense Acquisition

Beyond these points, there is now considerable data on the detailed types of errors children make throughout the acquisition process, both from Kuczaj (1977) and more recently from Bybee and Slobin (1982). We will consider aspects of these findings in more detail below. For now, we mention one intriguing fact: According to Kuczaj (1977), there is an interesting difference in the errors children make to irregular verbs at different points in Stage 2. Early on, regularizations are typically of the base + ed form, like *goed*; later on, there is a large increase in the frequency of past + ed errors, such as *wented*.

THE MODEL

The goal of our simulation of the acquisition of past tense was to simulate the three-stage performance summarized in Table 1, and to see whether we could capture other aspects of acquisition. In particular, we wanted to show that the kind of gradual change characteristic of normal acquisition was also a characteristic of our distributed model, and we wanted to see whether the model would capture detailed aspects of the phenomenon, such as the change in error type in later phases of development and the change in differences in error patterns observed for different types of words.

We were not prepared to produce a full-blown language processor that would learn the past tense from full sentences heard in everyday experience. Rather, we have explored a very simple past-tense learning environment designed to capture the essential characteristics necessary to produce the three stages of acquisition. In this environment, the model is presented, as learning experiences, with pairs of inputs—one capturing the phonological structure of the root form of a word and the other capturing the phonological structure of the correct past-tense version of that word. The behavior of the model can be tested by giving it just the root form of a word and examining what it generates as its "current guess" of the corresponding past-tense form.

Structure of the Model

The basic structure of the model is illustrated in Figure 1. The model consists of two basic parts: (a) a simple *pattern associator* network similar to those studied by Kohonen (1977; 1984; see Chapter 2) which learns the relationships between the base form and the past-tense

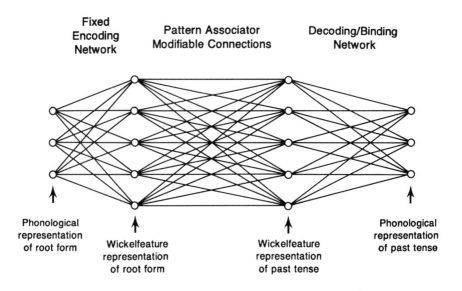

FIGURE 1. The basic structure of the model.

form, and (b) a decoding network that converts a featural representation of the past-tense form into a phonological representation. All learning occurs in the pattern associator; the decoding network is simply a mechanism for converting a featural representation which may be a near miss to any phonological pattern into a legitimate phonological representation. Our primary focus here is on the pattern associator. We discuss the details of the decoding network in the Appendix.

Units. The pattern associator contains two pools of units. One pool, called the input pool, is used to represent the input pattern corresponding to the root form of the verb to be learned. The other pool, called the output pool, is used to represent the output pattern generated by the model as its current guess as to the past tense corresponding to the root form represented in the inputs.

Each unit stands for a particular feature of the input or output string. The particular features we used are important to the behavior of the model, so they are described in a separate section below.

Connections. The pattern associator contains a modifiable connection linking each input unit to each output unit. Initially, these connections are all set to 0 so that there is no influence of the input units on the output units. Learning, as in other PDP models described in this book, involves modification of the strengths of these interconnections, as described below.

Operation of the Model

On test trials, the simulation is given a phoneme string corresponding to the root of a word. It then performs the following actions. First, it encodes the root string as a pattern of activation over the input units. The encoding scheme used is described below. Node activations are discrete in this model, so the activation values of all the units that should be on to represent this word are set to 1, and all the others are set to 0. Then, for each output unit, the model computes the net input to it from all of the weighted connections from the input units. The net input is simply the sum over all input units of the input unit activation times the corresponding weight. Thus, algebraically, the net input to output unit i is

$$net_i = \sum_j a_j w_{ij}$$

where a_j represents the activation of input unit j, and w_{ij} represents the weight from unit j to unit i.

Each unit has a threshold, θ, which is adjusted by the learning procedure that we will describe in a moment. The probability that the unit is turned on depends on the amount the net input exceeds the threshold. The *logistic* probability function is used here as in the Boltzmann machine (Chapter 7) and in harmony theory (Chapter 6) to determine whether the unit should be turned on. The probability is given by

$$p\,(a_i = 1) = \frac{1}{1 + e^{-(net_i - \theta_i)/T}} \tag{1}$$

where T represents the temperature of the system. The logistic function is shown in Figure 2. The use of this probabilistic response rule allows the system to produce different responses on different occasions with the same network. It also causes the system to learn more slowly so the effect of regular verbs on the irregulars continues over a much longer period of time. As discussed in Chapter 2, the temperature, T, can be manipulated so that at very high temperatures the response of the units is highly variable; with lower values of T, the units behave more like *linear threshold units*.

Since the pattern associator built into the model is a one-layer net with no feedback connections and no connections from one input unit to another or from one output unit to another, iterative computation is of no benefit. Therefore, the processing of an input pattern is a simple matter of first calculating the net input to each output unit and then

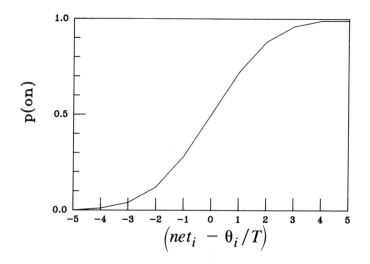

FIGURE 2. The logistic function used to calculate probability of activation. The x-axis shows values of $(net_i - \theta_i/T)$, and the y-axis indicates the corresponding probability that unit i will be activated.

setting its activation probabilistically on the basis of the logistic equation given above. The temperature T only enters in setting the variability of the output units; a fixed value of T was used throughout the simulations.

To determine how well the model did at producing the correct output, we simply compare the pattern of output Wickelphone activations to the pattern that the correct response would have generated. To do this, we first translate the correct response into a target pattern of activation for the output units, based on the same encoding scheme used for the input units. We then compare the obtained pattern with the target pattern on a unit-by-unit basis. If the output perfectly reproduces the target, then there should be a 1 in the output pattern wherever there is a 1 in the target. Such cases are called *hits*, following the conventions of signal detection theory (Green & Swets, 1966). There should also be a 0 in the output whenever there is a 0 in the target. Such cases are called *correct rejections*. Cases in which there are 1s in the output but not in the target are called *false alarms*, and cases in which there are 0s in the output that should be present in the input are called *misses*. A variety of measures of performance can be computed. We can measure the percentage of output units that match the correct past tense, or we can compare the output to the pattern for any other response alternative we might care to evaluate. This allows us to look at the output of the system independently of the decoding network. We can also employ the decoding network and have the system synthesize a phonological string. We can measure the performance of the system either at the featural level or at the level of strings of phonemes. We shall employ both of these mechanisms in the evaluation of different aspects of the overall model.

Learning

On a learning trial, the model is presented with both the root form of the verb and the target. As on a test trial, the pattern associator network computes the output it would generate from the input. Then, for each output unit, the model compares its answer with the target. Connection strengths are adjusted using the classic *perceptron convergence procedure* (Rosenblatt, 1962). The perceptron convergence procedure is simply a discrete variant of the delta rule presented in Chapter 2 and discussed in many places in this book. The exact procedure is as follows: We can think of the target as supplying a teaching input to each output unit, telling it what value it ought to have. When the actual output matches the target output, the model is doing the right thing and so none of the weights on the lines coming into the unit are adjusted. When the computed output is 0 and the target says it should be 1, we want to increase the probability that the unit will be active the next time the same input pattern is presented. To do this, we increase the weights from all of the input units that are active by a small amount η. At the same time, the threshold is also reduced by η. When the computed output is 1 and the target says it should be 0, we want to decrease the probability that the unit will be active the next time the same input pattern is presented. To do this, the weights from all of the input units that are active are reduced by η, and the threshold is increased by η. In all of our simulations, the value of η is simply set to 1. Thus, each change in a weight is a unit change, either up or down. For nonstochastic units, it is well known that the perceptron convergence procedure will find a set of weights that will allow the model to get each output unit correct, provided that such a set of weights exists. For the stochastic case, it is possible for the learning procedure to find a set of weights that will make the probability of error as low as desired. Such a set of weights exists if a set of weights exists that will always get the right answer for nonstochastic units.

Learning Regular and Exceptional Patterns in a Pattern Associator

In this section, we present an illustration of the behavior of a simple pattern associator model. The model is a scaled-down version of the main simulation described in the next section. We describe the scaled-down version first because in this model it is possible to actually examine the matrix of connection weights, and from this to see clearly how the model works and why it produces the basic three-stage learning phenomenon characteristic of acquisition of the past tense. Various aspects of pattern associator networks are described in a number of places in this book (Chapters 1, 2, 8, 9, 11, and 12, in particular) and elsewhere (J. A. Anderson, 1973, 1977; J. A. Anderson, Silverstein, Ritz, & Jones, 1977; Kohonen, 1977, 1984). Here we focus our attention on their application to the representation of rules for mapping one set of patterns into another.

For the illustration model, we use a simple network of eight input and eight output units and a set of connections from each input unit to each output unit. The network is illustrated in Figure 3. The network is shown with a set of connections sufficient for associating the pattern of activation illustrated on the input units with the pattern of activation illustrated on the output units. (Active units are darkened; positive

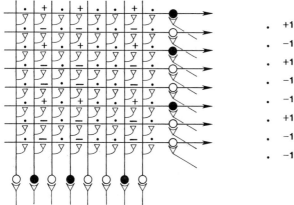

.	+15	.	+15	.	.	+15	.
.	−15	.	−15	.	.	−15	.
.	+15	.	+15	.	.	+15	.
.	−15	.	−15	.	.	−15	.
.	−15	.	−15	.	.	−15	.
.	+15	.	+15	.	.	+15	.
.	−15	.	−15	.	.	−15	.
.	−15	.	−15	.	.	−15	.

FIGURE 3. Simple network used in illustrating basic properties of pattern associator networks; excitatory and inhibitory connections needed to allow the active input pattern to produce the illustrated output pattern are indicated with + and −. Next to the network, the matrix of weights indicating the strengths of the connections from each input unit to each output unit. Input units are indexed by the column they appear in; output units are indexed by row.

and negative connections are indicated by numbers written on each connection). Next to the network is the matrix of connections abstracted from the actual network itself, with numerical values assigned to the positive and negative connections. Note that each weight is located in the matrix at the point where it occurred in the actual network diagram. Thus, the entry in the ith row of the jth column indicates the connection w_{ij} from the jth input unit to the ith output unit.

Using this diagram, it is easy to compute the net inputs that will arise on the output units when an input pattern is presented. For each output unit, one simply scans across its rows and adds up all the weights found in columns associated with active input units. (This is exactly what the simulation program does!) The reader can verify that when the input pattern illustrated in the left-hand panel is presented, each output unit that should be on in the output pattern receives a net input of +45; each output unit that should be off receives a net input of −45. [2] Plugging these values into Equation 1, using a temperature

---

[2] In the examples we will be considering in this section, the thresholds of the units are fixed at 0. Threshold terms add an extra degree of freedom for each output unit and allow the unit to come on in the absence of input, but they are otherwise inessential to the operation of the model. Computationally, they are equivalent to an adjustable weight to an extra input unit that is always on.

of 15,[3] we can compute that each output unit will take on the correct value about 95% of the time. The reader can check this in Figure 2; when the net input is +45, the exponent in the denominator of the logistic function is 3, and when the net input is −45, the exponent is −3. These correspond to activation probabilities of about .95 and .05, respectively.

One of the basic properties of the pattern associator is that it can store the connections appropriate for mapping a number of different input patterns to a number of different output patterns. The perceptron convergence procedure can accommodate a number of arbitrary associations between input patterns and output patterns, as long as the input patterns form a linearly independent set (see Chapters 9 and 11). Table 2 illustrates this aspect of the model. The first two cells of the table show the connections that the model learns when it is trained on each of the two indicated associations separately. The third cell shows connections learned by the model when it is trained on both patterns in alternation, first seeing one and then seeing the other of the two. Again, the reader can verify that if either input pattern is presented to a network with this set of connections, the correct corresponding output pattern is reconstructed with high probability; each output unit that should be on gets a net input of at least +45, and each output unit that should be off gets a net input below −45.

The restriction of networks such as this to linearly independent sets of patterns is a severe one since there are only N linearly independent patterns of length N. That means that we could store at most eight unrelated associations in the network and maintain accurate performance. However, if the patterns all conform to a general rule, the capacity of the network can be greatly enhanced. For example, the set of connections shown in Table 2D is capable of processing all of the patterns defined by what we call the *rule of 78*. The rule is described in Table 3. There are 18 different input/output pattern pairs corresponding to this rule, but they present no difficulty to the network. Through repeated presentations of examples of the rule, the perceptron convergence procedure learned the set of weights shown in cell D of Table 2. Again, the reader can verify that it works for any legal association fitting the rule of 78. (Note that for this example, the "regular" pairing

---

[3] For the actual simulations of verb learning, we used a value of T equal to 200. This means that for a fixed value of the weight on an input line, the effect of that line being active on the unit's probability of firing is much lower than it is in these illustrations. This is balanced by the fact that in the verb learning simulations, a much larger number of inputs contribute to the activation of each output unit. Responsibility for turning a unit on is simply more distributed when larger input patterns are used.

TABLE 2

WEIGHTS IN THE 8-UNIT NETWORK
AFTER VARIOUS LEARNING EXPERIENCES

A. Weights acquired in learning
(2 4 7) → (1 4 6)

```
.  15   .  15   .   .  15   .   .
. -16   . -16   .   . -16   .   .
. -17   . -17   .   . -17   .   .
.  16   .  16   .   .  16   .   .
. -16   . -16   .   . -16   .   .
.  17   .  17   .   .  17   .   .
. -16   . -16   .   . -16   .   .
. -17   . -17   .   . -17   .   .
```

B. Weights acquired in learning
(3 4 6) → (3 6 7)

```
.   .   . -16 -16   . -16   .   .
.   .   . -17 -17   . -17   .   .
.   .   .  17  17   .  17   .   .
.   .   . -16 -16   . -16   .   .
.   .   . -17 -17   . -17   .   .
.   .   .  16  16   .  16   .   .
.   .   .  17  17   .  17   .   .
.   .   . -17 -17   . -17   .   .
```

C. Weights acquired in learning
A and B together

```
.  24 -24   .   . -24  24   .
. -13 -13 -26   . -13 -13   .
. -23  24   1   .  24 -23   .
.  24 -25  -1   . -25  24   .
. -13 -13 -26   . -13 -13   .
.  13  13  26   .  13  13   .
. -25  24  -1   .  24 -25   .
. -12 -13 -25   . -13 -12   .
```

D. Weights acquired in learning
the rule of 78

```
 61 -37 -37  -5  -5  -3  -6  -7
-35  60 -38  -4  -6  -3  -5  -8
-39 -35  61  -4  -5  -4  -7  -6
 -6  -4  -5  59 -37 -37  -8  -7
 -5  -5  -4 -36  60 -38  -7  -7
 -5  -4  -6 -37 -38  60  -8  -7
  .   1   .   1   .   . -50  51
  .   .  -1  -2   1   .  49 -50
```

TABLE 3

THE RULE OF 78

Input patterns consist of one active unit from each of the following sets:	(1 2 3) (4 5 6) (7 8)
The output pattern paired with a given input pattern consists of:	The same unit from (1 2 3) The same unit from (4 5 6) The other unit from (7 8)
Examples:	2 4 7 → 2 4 8 1 6 8 → 1 6 7 3 5 7 → 3 5 8
An exception:	1 4 7 → 1 4 7

of (1 4 7) with (1 4 8) was used rather than the exceptional mapping illustrated in Table 3).

We have, then, observed an important property of the pattern associator: If there is some structure to a set of patterns, the network may be able to learn to respond appropriately to all of the members of the set. This is true, even though the input vectors most certainly do not form a linearly independent set. The model works anyway because the response that the model should make to some of the patterns can be predicted from the responses that it should make to others of the patterns.

Now let's consider a case more like the situation a young child faces in learning the past tenses of English verbs. Here, there is a regular pattern, similar to the rule of 78. In addition, however, there are exceptions. Among the first words the child learns are many exceptions, but as the child learns more and more verbs, the proportion that are regular increases steadily. For an adult, the vast majority of verbs are regular.

To examine what would happen in a pattern associator in this kind of a situation, we first presented the illustrative 8-unit model with two pattern pairs. One of these was a regular example of the 78 rule $[(2\ 5\ 8) \rightarrow (2\ 5\ 7)]$. The other was an exception to the rule $[(1\ 4\ 7) \rightarrow (1\ 4\ 7)]$. The simulation saw both pairs 20 times, and connection strengths were adjusted after each presentation. The resulting set of connections is shown in cell A of Table 4. This number of learning trials is not enough to lead to perfect performance; but after this much experience, the model tends to get the right answer for each output unit close to 90 percent of the time. At this point, the fact that one of the patterns is an example of a general rule and the other is an exception to that rule is irrelevant to the model. It learns a set of connections that can accommodate these two patterns, but it cannot generalize to new instances of the rule.

This situation, we suggest, characterizes the situation that the language learner faces early on in learning the past tense. The child knows, at this point, only a few high-frequency verbs, and these tend, by and large, to be irregular, as we shall see below. Thus each is treated by the network as a separate association, and very little generalization is possible.

But as the child learns more and more verbs, the proportion of regular verbs increases. This changes the situation for the learning model. Now the model is faced with a number of examples, all of which follow the rule, as well as a smattering of irregular forms. This new situation changes the experience of the network, and thus the pattern of interconnections it contains. Because of the predominance of the regular

TABLE 4

REPRESENTING EXCEPTIONS: WEIGHTS IN THE 8-UNIT NETWORK

A. After 20 exposures to
    (1 4 7)→(1 4 7), (2 5 8)→(2 5 7)

12	-12	.	12	-12	.	12	-12
-11	13	.	-11	13	.	-11	13
-11	-11	.	-11	-11	.	-11	-11
12	-12	.	12	-12	.	12	-12
-11	11	.	-11	11	.	-11	11
-11	-12	.	-11	-12	.	-11	-12
12	11	.	12	11	.	12	11
-11	-13	.	-11	-13	.	-11	-13

B. After 10 more exposures to
    all 18 associations

44	-34	-26	-2	-10	-4	-8	-8
-32	46	-27	-11	2	-4	-9	-4
-30	-24	43	-5	-5	-1	-2	-9
-1	-7	-7	45	-34	-26	-4	-11
-8	-3	-3	-31	44	-27	-7	-7
-6	-8	-3	-31	-28	42	-7	-10
11	-2	-6	11	-2	-6	-35	38
-9	-4	7	-13	1	6	36	-42

C. After 30 more exposures to
    all 18 associations

61	-38	-38	-6	-5	-4	-6	-9
-38	62	-39	-6	-5	-4	-8	-7
-37	-38	62	-5	-5	-3	-7	-6
-4	-6	-6	62	-40	-38	-8	-8
-5	-5	-4	-38	62	-38	-7	-7
-6	-4	-5	-38	-39	62	-8	-7
20	-5	-4	22	-5	-6	-50	61
-19	8	5	-18	5	7	54	-60

D. After a total of 500 exposures
    to all 18 associations

64	-39	-39	-5	-4	-5	-7	-7
-39	63	-39	-5	-5	-5	-7	-8
-39	-40	64	-5	-5	-5	-8	-7
-5	-5	-5	64	-40	-39	-8	-7
-5	-5	-5	-39	63	-39	-7	-8
-5	-5	-5	-39	-39	63	-8	-7
71	-28	-29	70	-28	-28	-92	106
-70	27	28	-70	27	28	91	-106

form in the input, the network learns the regular pattern, temporarily "overregularizing" exceptions that it may have previously learned.

Our illustration takes this situation to an extreme, perhaps, to illustrate the point. For the second stage of learning, we present the model with the entire set of eighteen input patterns consisting of one active unit from (1 2 3), one from (4 5 6), and one from (7 8). All of these patterns are regular except the one exception already used in the first stage of training.

At the end of 10 exposures to the full set of 18 patterns, the model has learned a set of connection strengths that predominantly captures the "regular pattern." At this point, its response to the exceptional pattern is *worse* than it was before the beginning of Phase 2; rather than getting the right output for Units 7 and 8, the network is now *regularizing* it.

The reason for this behavior is very simple. All that is happening is that the model is continually being bombarded with learning experiences directing it to learn the rule of 78. On only one learning trial out of 18 is it exposed to an exception to this rule.

In this example, the deck has been stacked very strongly against the exception. For several learning cycles, it is in fact quite difficult to tell from the connections that the model is being exposed to an exception mixed in with the regular pattern. At the end of 10 cycles, we can see that the model is building up extra excitatory connections from input Units 1 and 4 to output Unit 7 and extra inhibitory strength from Units 1 and 4 to Unit 8, but these are not strong enough to make the model get the right answer for output Units 7 and 8 when the (1 4 7) input pattern is shown. Even after 40 trials (panel C of Table 4), the model still gets the wrong answer on Units 7 and 8 for the (1 4 7) pattern more than half the time. (The reader can still be checking these assertions by computing the net input to each output unit that would result from presenting the (1 4 7) pattern.)

It is only after the model has reached the stage where it is making very few mistakes on the 17 regular patterns that it begins to accommodate to the exception. This amounts to making the connection from Units 1 and 4 to output Unit 7 strongly excitatory and making the connections from these units to output Unit 8 strongly inhibitory. The model must also make several adjustments to other connections so that the adjustments just mentioned do not cause errors on regular patterns similar to the exceptions, such as (1 5 7), (2 4 7), etc. Finally, in panel D, after a total of 500 cycles through the full set of 18 patterns, the weights are sufficient to get the right answer nearly all of the time. Further improvement would be very gradual since the network makes errors so infrequently at this stage that there is very little opportunity for change.

It is interesting to consider for a moment how an association is represented in a model like this. We might be tempted to think of the representation of an association as the difference between the set of connection strengths needed to represent a set of associations that includes the association and the set of strengths needed to represent the same set excluding the association of interest. Using this definition, we see that the representation of a particular association is far from invariant. What this means is that learning that occurs in one situation (e.g., in which there is a small set of unrelated associations) does not necessarily transfer to a new situation (e.g., in which there are a number of regular associations). This is essentially why the early learning our illustrative model exhibits of the (1 4 7) → (1 4 7) association in the context of just one other association can no longer support correct performance when the larger ensemble of regular patterns is introduced.

Obviously, the example we have considered in this section is highly simplified. However, it illustrates several basic facts about pattern associators. One is that they tend to exploit regularity that exists in the mapping from one set of patterns to another. Indeed, this is one of the main advantages of the use of distributed representations. Second, they allow exceptions and regular patterns to coexist in the same network. Third, if there is a predominant regularity in a set of patterns, this can swamp exceptional patterns until the set of connections has been acquired that captures the predominant regularity. Then further, gradual tuning can occur that adjusts these connections to accommodate both the regular patterns and the exception. These basic properties of the pattern associator model lie at the heart of the three-stage acquisition process, and account for the gradualness of the transition from Stage 2 to Stage 3.

Featural Representations of Phonological Patterns

The preceding section describes basic aspects of the behavior of the pattern associator model and captures fairly well what happens when a pattern associator is applied to the processing of English verbs, following a training schedule similar to the one we have just considered for the acquisition of the rule of 78. There is one caveat, however: The input and target patterns—the base forms of the verbs and the correct past tenses of these verbs—must be represented in the model in such a way that the features provide a convenient basis for capturing the regularities embodied in the past-tense forms of English verbs. Basically, there were two considerations:

- We needed a representation that permitted a differentiation of all of the root forms of English and their past tenses.

- We wanted a representation that would provide a natural basis for generalizations to emerge about what aspects of a present tense correspond to what aspects of the past tense.

A scheme which meets the first criterion, but not the second, is the scheme proposed by Wickelgren (1969). He suggested that words should be represented as sequences of context-sensitive phoneme units, which represent each phone in a word as a triple, consisting of the phone itself, its predecessor, and its successor. We call these triples *Wickelphones*. Notationally, we write each Wickelphone as a triple of phonemes, consisting of the central phoneme, subscripted on the left by its predecessor and on the right by its successor. A phoneme occurring at the beginning of a word is preceded by a special symbol (#) standing for the word boundary; likewise, a phoneme occurring at the end of a word is followed by #. The word /kat/, for example, would be represented as $_{\#}k_a$, $_ka_t$, and $_at_{\#}$. Though the Wickelphones in a word are not strictly position specific, it turns out that (a) few words contain more than one occurrence of any given Wickelphone, and (b) there are no two words we know of that consist of the same sequence of Wickelphones. For example, /slit/ and /silt/ contain no Wickelphones in common.

One nice property of Wickelphones is that they capture enough of the context in which a phoneme occurs to provide a sufficient basis for differentiating between the different cases of the past-tense rule and for characterizing the contextual variables that determine the subregularities among the irregular past-tense verbs. For example, the word-final phoneme that determines whether we should add /d/, /t/ or /$^\wedge$d/ in forming the regular past. And it is the sequence $_iN_{\#}$ which is transformed to $_aN_{\#}$ in the *ing* → *ang* pattern found in words like *sing*.

The trouble with the Wickelphone solution is that there are too many of them, and they are too specific. Assuming that we distinguish 35 different phonemes, the number of Wickelphones would be 35^3, or 42,875, not even counting the Wickelphones containing word boundaries. And, if we postulate one input unit and one output unit in our model for each Wickelphone, we require rather a large connection matrix (4.3×10^4 squared, or about 2×10^9) to represent all their possible connections.

Obviously, a more compact representation is required. This can be obtained by representing each Wickelphone as a distributed pattern of activation over a set of feature detectors. The basic idea is that we represent each phoneme, not by a single Wickelphone, but by a pattern of what we call *Wickelfeatures*. Each Wickelfeature is a conjunctive, or context-sensitive, feature, capturing a feature of the central phoneme, a feature of the predecessor, and a feature of the successor.

Details of the Wickelfeature representation. For concreteness, we will now describe the details of the feature coding scheme we used. It contains several arbitrary properties, but it also captures the basic principles of coarse, conjunctive coding described in Chapter 3. First, we will describe the simple feature representation scheme we used for coding a single phoneme as a pattern of features without regard to its predecessor and successor. Then we describe how this scheme can be extended to code whole Wickelphones. Finally, we show how we "blur" this representation, to promote generalization further.

To characterize each phoneme, we devised the highly simplified feature set illustrated in Table 5. The purpose of the scheme was (a) to give as many of the phonemes as possible a distinctive code, (b) to allow code similarity to reflect the similarity structure of the phonemes

TABLE 5

CATEGORIZATION OF PHONEMES ON FOUR SIMPLE DIMENSIONS

| | | Place | | | | | |
| | | Front | | Middle | | Back | |
		V/L	U/S	V/L	U/S	V/L	U/S
Interrupted	*Stop*	b	p	d	t	g	k
	Nasal	m	-	n	-	N	-
Cont. Consonant	*Fric.*	v/D	f/T	z	s	Z/j	S/C
	Liq/SV	w/l	-	r	-	y	h
Vowel	*High*	E	i	O	ˆ	U	u
	Low	A	e	I	a/α	W	*/o

Key: N = ng in *sing*; D = th in *the*; T = th in *with*; Z = z in *azure*; S = sh in *ship*;
C = ch in *chip*; E = ee in *beet*; i = i in *bit*; O = oa in *boat*; ˆ = u in *but* or schwa;
U = oo in *boot*; u = oo in *book*; A = ai in *bait*; e = e in *bet*; I = i_e in *bite*;
a = a in *bat*; α = a in *father*; W = ow in *cow*; * = aw in *saw*; o = o in *hot*.

in a way that seemed sufficient for our present purposes, and (c) to keep the number of different features as small as possible.

The coding scheme can be thought of as categorizing each phoneme on each of four dimensions. The first dimension divides the phonemes into three major types: interrupted consonants (stops and nasals), continuous consonants (fricatives, liquids, and semivowels), and vowels. The second dimension further subdivides these major classes. The interrupted consonants are divided into plain stops and nasals; the continuous consonants into fricatives and sonorants (liquids and semivowels are lumped together); and the vowels into high and low. The third dimension classifies the phonemes into three rough places of articulation—front, middle, and back. The fourth subcategorizes the consonants into voiced vs. voiceless categories and subcategorizes the vowels into long and short. As it stands, the coding scheme gives identical codes to six pairs of phonemes, as indicated by the duplicate entries in the cells of the table. A more adequate scheme could easily be constructed by increasing the number of dimensions and/or values on the dimensions.

Using the above code, each phoneme can be characterized by one value on each dimension. If we assigned a unit for each value on each dimension, we would need 10 units to represent the features of a single phoneme since two dimensions have three values and two have two values. We could then indicate the pattern of these features that corresponds to a particular phoneme as a pattern of activation over the 10 units.

Now, one way to represent each Wickelphone would simply be to use three sets of feature patterns: one for the phoneme itself, one for its predecessor, and one for its successor. To capture the word-boundary marker, we would need to introduce a special eleventh feature. Thus, the Wickelphone $_{\#}\mathbf{k}_{\mathbf{a}}$ can be represented by

$$[\ (000) \ (00) \ (000) \ (00) \ 1 \ ]$$
$$[\ (100) \ (10) \ (001) \ (01) \ 0 \ ]$$
$$[\ (001) \ (01) \ (010) \ (01) \ 0 \ ].$$

Using this scheme, a Wickelphone could be represented as a pattern of activation over a set of 33 units.

However, there is one drawback with this. The representation is not sufficient to capture more than one Wickelphone at a time. If we add another Wickelphone, the representation gives us no way of knowing which features belong together.

We need a representation, then, that provides us with a way of determining which features go together. This is just the job that can be done with detectors for Wickelfeatures—triples of features, one from the central phoneme, one from the predecessor phoneme, and one from the successor phoneme.

Using this scheme, each detector would be activated when the word contained a Wickelphone containing its particular combination of three features. Since each phoneme of a Wickelphone can be characterized by 11 features (including the word-boundary feature) and each Wickelphone contains three phonemes, there are $11 \times 11 \times 11$ possible Wickelfeature detectors. Actually, we are not interested in representing phonemes that cross word boundaries, so we only need 10 features for the center phoneme.

Though this leaves us with a fairly reasonable number of units $(11 \times 10 \times 11$ or $1,210)$, it is still large by the standards of what will easily fit in available computers. However, it is possible to cut the number down still further without much loss of representational capacity since a representation using all 1,210 units would be highly redundant; it would represent each feature of each of the three phonemes 16 different times, one for each of the conjunctions of that feature with one of the four features of one of the other phonemes and one of the four features of the other.

To cut down on this redundancy and on the number of units required, we simply eliminated all those Wickelfeatures specifying values on two different dimensions of the predecessor and the successor phonemes. We kept all the Wickelfeature detectors for all combinations of different values on the same dimension for the predecessor and successor phonemes. It turns out that there are 260 of these (ignoring the word-boundary feature), and each feature of each member of each phoneme triple is still represented four different times. In addition, we kept the 100 possible Wickelfeatures combining a preceding word-boundary feature with any feature of the main phoneme and any feature of the successor; and the 100 Wickelfeatures combining a following word boundary feature with any feature of the main phoneme and any feature of the successor. All in all then, we used only 460 of the 1,210 possible Wickelfeatures.

Using this representation, a verb is represented by a pattern of activation over a set of 460 Wickelfeature units. Each Wickelphone activates 16 Wickelfeature units. Table 6 shows the 16 Wickelfeature units activated by the Wickelphone $_k A_m$, the central Wickelphone in the word *came*. The first Wickelfeature is turned on whenever we have a Wickelphone in which the preceding contextual phoneme is an interrupted consonant, the central phoneme is a vowel, and the following phoneme is an interrupted consonant. This Wickelfeature is turned on for the Wickelphone $_k A_m$ since /k/ and /m/, the context phonemes, are both interrupted consonants and /A/, the central phoneme, is a vowel. This same Wickelfeature would be turned on in the

TABLE 6

THE SIXTEEN WICKELFEATURES FOR THE WICKELPHONE $_k A_m$

Feature	Preceding Context	Central Phoneme	Following Context
1	Interrupted	Vowel	Interrupted
2	Back	Vowel	Front
3	Stop	Vowel	Nasal
4	Unvoiced	Vowel	Voiced
5	Interrupted	Front	Vowel
6	Back	Front	Front
7	Stop	Front	Nasal
8	Unvoiced	Front	Voiced
9	Interrupted	Low	Interrupted
10	Back	Low	Front
11	Stop	Low	Nasal
12	Unvoiced	Low	Voiced
13	Interrupted	Long	Vowel
14	Back	Long	Front
15	Stop	Long	Nasal
16	Unvoiced	Long	Voiced

representation of $_\mathbf{b}\mathbf{i}_\mathbf{d}$, $_\mathbf{p}\overset{\wedge}{\mathbf{t}}$, $_\mathbf{m}\mathbf{a}_\mathbf{p}$, and many other Wickelfeatures. Similarly, the sixth Wickelfeature listed in the table will be turned on whenever the preceding phoneme is made in the back, and the central and following phonemes are both made in the front. Again, this is turned on because /k/ is made in the back and /A/ and /m/ are both made in the front. In addition to $_\mathbf{k}\mathbf{A}_\mathbf{m}$ this feature would be turned on for the Wickelphones $_\mathbf{g}\mathbf{i}_\mathbf{v}$, $_\mathbf{g}\mathbf{A}_\mathbf{p}$, $_\mathbf{k}\mathbf{A}_\mathbf{p}$, and others. Similarly, each of the sixteen Wickelfeatures stands for a conjunction of three phonetic features and occurs in the representation of a large number of Wickelphones.

Now, words are simply lists of Wickelphones. Thus, words can be represented by simply turning on all of the Wickelfeatures in any Wickelphone of a word. Thus, a word with three Wickelphones (such as *came*, which has the Wickelphones $_\#\mathbf{k}_\mathbf{A}$, $_\mathbf{k}\mathbf{A}_\mathbf{m}$, and $_\mathbf{a}\mathbf{m}_\#$) will have at most 48 Wickelfeatures turned on. Since the various Wickelphones may have some Wickelfeatures in common, typically there will be less than 16 times the number of Wickelfeatures turned on for most words. It is important to note the temporal order is entirely implicit in this representation. All words, no matter how many phonemes in the word, will be represented by a subset of the 460 Wickelfeatures.

Blurring the Wickelfeature representation. The representational scheme just outlined constitutes what we call the *primary* representation of a Wickelphone. In order to promote faster generalization, we further blurred the representation. This is accomplished by turning on, in addition to the 16 primary Wickelfeatures, a randomly selected subset of the similar Wickelfeatures, specifically, those having the same value for the central feature and one of the two context phonemes. That is, whenever the Wickelfeature for the conjunction of phonemic features f_1, f_2, and f_3 is turned on, each Wickelfeature of the form $<?f_2 f_3>$ and $<f_1 f_2?>$ may be turned on as well. Here "?" stands for "any feature." This causes each word to activate a larger set of Wickelfeatures, allowing what is learned about one sequence of phonemes to generalize more readily to other similar but not identical sequences.

To avoid having too much randomness in the representation of a particular Wickelphone, we turned on the same subset of additional Wickelfeatures each time a particular Wickelphone was to be represented. Based on subsequent experience with related models (see Chapter 19), we do not believe this makes very much difference.

There is a kind of trade-off between the discriminability among the base forms of verbs that the representation provides and the amount of generalization. We need a representation which allows for rapid generalization while at the same time maintains adequate discriminability. We can manipulate this factor by manipulating the probability p that any one of these similar Wickelfeatures will be turned on. In our simulations we found that turning on the additional features with fairly high probability (.9) led to adequate discriminability while also producing relatively rapid generalization.

Although the model is not completely immune to the possibility that two different words will be represented by the same pattern, we have encountered no difficulty decoding any of the verbs we have studied. However, we do not claim that Wickelfeatures necessarily capture all the information needed to support the generalizations we might need to make for this or other morphological processes. Some morphological processes might require the use of units that were further differentiated according to vowel stress or other potential distinguishing characteristics. All we claim for the present coding scheme is its sufficiency for the task of representing the past tenses of the 500 most frequent verbs in English and the importance of the basic principles of distributed, coarse (what we are calling blurred), conjunctive coding that it embodies (see Chapter 3).

Summary of the Structure of the Model

In summary, our model contained two sets of 460 Wickelfeature units, one set (the input units) to represent the base form of each verb and one set (the output units) to represent the past-tense form of each verb.

The model is tested by typing in an input phoneme string, which is translated by the fixed encoding network into a pattern of activation over the set of input units. Each active input unit contributes to the net input of each output unit, by an amount and direction (positive or negative) determined by the weight on the connection between the input unit and the output unit. The output units are then turned on or off probabilistically, with the probability increasing with the difference between the net input and the threshold, according to the logistic activation function. The output pattern generated in this way can be compared with various alternative possible output patterns, such as the correct past-tense form or some other possible response of interest, or can be used to drive the decoder network described in the Appendix.

The model is trained by providing it with pairs of patterns, consisting of the base pattern and the target, or correct, output. Thus, in accordance with common assumptions about the nature of the learning situation that faces the young child, the model receives only correct input from the outside world. However, it compares what it generates internally to the target output, and when it gets the wrong answer for a particular output unit, it adjusts the strength of the connection between the input and the output units so as to reduce the probability that it will make the same mistake the next time the same input pattern is presented. The adjustment of connections is an extremely simple and *local* procedure, but it appears to be sufficient to capture what we know about the acquisition of the past tense, as we shall see in the next section.

THE SIMULATIONS

The simulations described in this section are concerned with demonstrating three main points:

- That the model captures the basic three-stage pattern of acquisition.

- That the model captures most aspects of differences in performance on different types of regular and irregular verbs.

- That the model is capable of responding appropriately to verbs it has never seen before, as well as to regular and irregular verbs actually experienced during training.

In the sections that follow we will consider these three aspects of the model's performance in turn.

The corpus of verbs used in the simulations consisted of a set of 506 verbs. All verbs were chosen from the Kucera and Francis (1967) word list and were ordered according to frequency of their gerund form. We divided the verbs into three classes: 10 high-frequency verbs, 410 medium-frequency verbs, and 86 low-frequency verbs. The ten highest frequency verbs were: *come* (/kˆm/), *get* (/get/), *give* (/giv/), *look* (/luk/), *take* (/tʌk/), *go* (/go/), *have* (/hav/), *live* (/liv/), and *feel* (/fɛl/). There is a total of 8 irregular and 2 regular verbs among the top 10. Of the medium-frequency verbs, 334 were regular and 76 were irregular. Of the low-frequency verbs, 72 were regular and 14 were irregular.

The Three-Stage Learning Curve

The results described in this and the following sections were obtained from a single (long) simulation run. The run was intended to capture approximately the experience with past tenses of a young child picking up English from everyday conversation. Our conception of the nature of this experience is simply that the child learns first about the present and past tenses of the highest frequency verbs; later on, learning occurs for a much larger ensemble of verbs, including a much larger proportion of regular forms. Although the child would be hearing present and past tenses of all kinds of verbs throughout development, we assume that he is only able to learn past tenses for verbs that he has already mastered fairly well in the present tense.

To simulate the earliest phase of past-tense learning, the model was first trained on the 10 high-frequency verbs, receiving 10 cycles of training presentations through the set of 10 verbs. This was enough to produce quite good performance on these verbs. We take the performance of the model at this point to correspond to the performance of a child in Phase 1 of acquisition. To simulate later phases of learning, the 410 medium-frequency verbs were added to the first 10 verbs, and the system was given 190 more learning trials, with each trial consisting of one presentation of each of the 420 verbs. The responses of the model early on in this phase of training correspond to Phase 2 of the acquisition process; its ultimate performance at the end of 190 exposures to each of the 420 verbs corresponds to Phase 3. At this point, the model exhibits almost errorless performance on the basic 420 verbs. Finally, the set of 86 lower-frequency verbs were presented to the system and the transfer responses to these were recorded. During this phase, connection strengths were not adjusted. Performance of the model on these transfer verbs is considered in a later section.

We do not claim, of course, that this training experience exactly captures the learning experience of the young child. It should be perfectly clear that this training experience exaggerates the difference between early phases of learning and later phases, as well as the abruptness of the transition to a larger corpus of verbs. However, it is generally observed that the early, rather limited vocabulary of young children undergoes an explosive growth at some point in development (Brown, 1973). Thus, the actual transition in a child's vocabulary of verbs would appear quite abrupt on a time-scale of years so that our assumptions about abruptness of onset may not be too far off the mark.

Figure 4 shows the basic results for the high frequency verbs. What we see is that during the first 10 trials there is no difference between regular and irregular verbs. However, beginning on Trial 11 when the 410 midfrequency verbs were introduced, the regular verbs show better performance. It is important to notice that there is no interfering effect on the regular verbs as the midfrequency verbs are being learned. There is, however, substantial interference on the irregular verbs. This interference leads to a dip in performance on the irregular verbs.

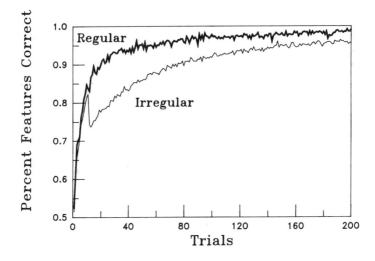

FIGURE 4. The percentage of correct features for regular and irregular high-frequency verbs as a function of trials.

Equality of performance between regular and irregular verbs is never again attained during the training period. This is the so-called U-shaped learning curve for the learning of the irregular past tense. Performance is high when only a few high-frequency, largely irregular verbs are learned, but then drops as the bulk of lower-frequency regular verbs are being learned.

We have thus far only shown that performance on high-frequency irregular verbs drops; we have not said anything about the nature of the errors. To examine this question, the response strength of various possible response alternatives must be compared. To do this, we compared the strength of response for several different response alternatives. We compared strengths for the correct past tense, the present, the base+ed and the past+ed. Thus, for example with the verb *give* we compared the response strength of /gʌv/, /giv/, /givd/, and /gʌvd/. We determined the response strengths by assuming that these response alternatives were competing to account for the features that were actually turned on in the output. The details of the competition mechanism, called a *binding network*, are described in the Appendix. For present purposes, suffice it to say that each alternative gets a score that represents the percentage of the total features that it accounts for. If two alternatives both account for a given feature, they divide the score for that feature in proportion to the number of features each accounts for uniquely. We take these response strengths to correspond roughly to relative response probabilities, though we imagine that the actual generation of overt responses is accomplished by a different version of the binding network, described below. In any case, the total strength of all the alternatives cannot be greater than 1, and if a number of features are accounted for by none of the alternatives, the total will be less than 1.

Figure 5 compares the response strengths for the correct alternative to the combined strength of the regularized alternatives.[4] Note in the figure that during the first 10 trials the response strength of the correct alternative grows rapidly to over .5 while that of the regularized alternative drops from about .2 to .1. After the midfrequency verbs are introduced, the response strength for the correct alternative drops rapidly while the strengths of regularized alternatives jump up. From about Trials 11 through 30, the regularized alternatives together are stronger than the correct response. After about Trial 30, the strength of the correct response again exceeds the regularized alternatives and continues to grow throughout the 200-trial learning phase. By the end, the correct response is much the strongest with all other alternatives below .1.

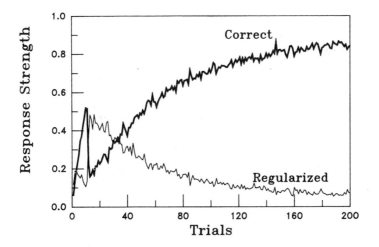

FIGURE 5. Response strengths for the high-frequency irregular verbs. The response strengths for the correct responses are compared with those for the regularized alternatives as a function of trials.

---

[4] Unless otherwise indicated, the regularized alternatives are considered the base+ed and past+ed alternatives. In a later section of the paper we shall discuss the pattern of differences between these alternatives. In most cases the base+ed alternative is much stronger than the past+ed alternative.

The rapidity of the growth of the regularized alternatives is due to the sudden influx of the medium-frequency verbs. In real life we would expect the medium-frequency verbs to come in somewhat more slowly so that the period of maximal regularization would have a somewhat slower onset.

Figure 6 shows the same data in a slightly different way. In this case, we have plotted the ratio of the correct response to the sum of the correct and regularized response strengths. Points on the curve below the .5 line are in the region where the regularized response is greater that the correct response. Here we see clearly the three stages. In the first stage, the first 10 trials of learning, performance on these high-frequency verbs is quite good. Virtually no regularization takes place. During the next 20 trials, the system regularizes and systematically makes errors on the verbs that it previously responded to correctly. Finally, during the remaining trials the model slowly eliminates the regularization responses as it approaches adult performance.

In summary, then, the model captures the three phases of learning quite well, as well as the gradual transition from Phase 2 to Phase 3. It does so without any explicit learning of rules. The regularization is the product of the gradual tuning of connection strengths in response

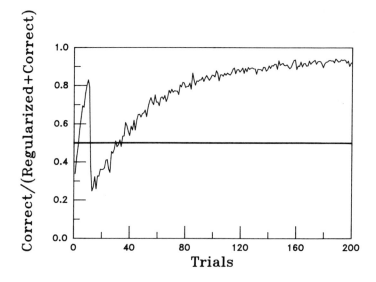

FIGURE 6. The ratio of the correct response to the sum of the correct and regularized response. Points on the curve below the .5 line are in the region where the regularized response is greater than the correct response.

to the predominantly regular correspondence exhibited by the medium-frequency words. It is not quite right to say that individual pairs are being stored in the network in any simple sense. The connection strengths the model builds up to handle the irregular forms do not represent these items in any separable way; they represent them in the way they must be represented to be stored along with the other verbs in the same set of connections.

Before discussing the implications of these kinds of results further, it is useful to look more closely at the kinds of errors made and at the learning rates of the medium-frequency regular and irregular verbs.

Learning the medium-frequency verbs. Figure 7A compares the learning curves for the regular verbs of high and medium frequency, and Figure 7B compares the learning curves for the corresponding groups of irregular verbs. Within only two or three trials the medium-frequency verbs catch up with their high-frequency counterparts. Indeed, in the case of the irregular verbs, the medium-frequency verbs seem to surpass the high-frequency ones. As we shall see in the following section, this results from the fact that the high-frequency verbs include some of the most difficult pairs to learn, including, for example, the *go/went* pair which is the most difficult to learn (aside from the verb *be*, this is the only verb in English in which the past and root form are completely unrelated). It should also be noted that even at this early stage of learning there is substantial generalization. Already, on Trial 11, the very first exposure to the medium-frequency verbs, between 65 and 75 percent of the features are produced correctly. Chance responding is only 50 percent. Moreover, on their first presentation, 10 percent more of the features of regular verbs are correctly responded to than irregular ones. Eventually, after 200 trials of learning, nearly all of the features are being correctly generated and the system is near asymptotic performance on this verb set. As we shall see below, during most of the learning period the difference between high- and medium-frequency verbs is not important. Rather, the differences between different classes of verbs is the primary determiner of performance. We now turn to a discussion of these different types.

Types of Regular and Irregular Verbs

To this point, we have treated regular and irregular verbs as two homogeneous classes. In fact, there are a number of distinguishable

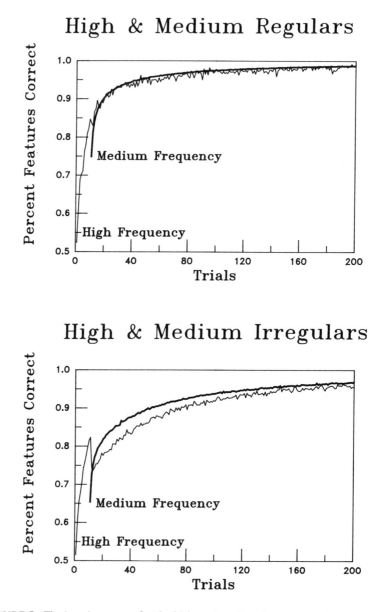

FIGURE 7. The learning curves for the high- and medium-frequency verbs.

types of regular and irregular verbs. Bybee and Slobin (1982) have studied the different acquisition patterns of the each type of verb. In this section we compare their results to the responses produced by our simulation model.

Bybee and Slobin divided the irregular verbs into nine classes, defined as follows: [5]

I. Verbs that do not change at all to form the past tense, e.g., *beat, cut, hit*.

II. Verbs that change a final /d/ to /t/ to form the past tense, e.g., *send/sent, build/built*.

III. Verbs that undergo an internal vowel change and also add a final /t/ or /d/, e.g., *feel/felt, lose/lost, say/said, tell/told*.

IV. Verbs that undergo an internal vowel change, delete a final consonant, and add a final /t/ or /d/, e.g., *bring/brought, catch/caught*.[6]

V. Verbs that undergo an internal vowel change whose stems end in a dental, e.g., *bite/bit, find/found, ride/rode*.

VIa. Verbs that undergo a vowel change of /i/ to /a/ e.g., *sing/sang, drink/drank*.

VIb. Verbs that undergo an internal vowel change of /i/ or /a/ to /ʌ/ e.g., *sting/stung, hang/hung*.[7]

VII. All other verbs that undergo an internal vowel change, e.g., *give/gave, break/broke*.

VIII. All verbs that undergo a vowel change and that end in a dipthongal sequence, e.g., *blow/blew, fly/flew*.

A complete listing by type of all of the irregular verbs used in our study is given in Table 7.

In addition to these types of irregular verbs, we distinguished three categories of regular verbs: (a) those ending in a vowel or voiced consonant, which take a /d/ to form the past tense; (b) those ending in a voiceless consonant, which take a /t/; and (c) those ending in /t/ or

---

[5] Criteria from Bybee and Slobin (1982, pp. 268-269).

[6] Following Bybee and Slobin, we included *buy/bought* in this class even though no final consonant is deleted.

[7] For many purposes we combine Classes VIa and VIb in our analyses.

TABLE 7

IRREGULAR VERBS

Type	High	Medium	Low
		Frequency	
I		beat fit set spread hit cut put	thrust bid
II		build send spend	bend lend
III	feel	deal do flee tell sell hear keep leave sleep lose mean say sweep	creep weep
IV	have make	think buy bring seek teach	catch
V	get	meet shoot write lead understand sit mislead bleed feed stand light find fight read meet hide hold ride	breed wind grind
VIa		drink ring sing swim	
VIb		drag hang swing	dig cling stick
VII	give take come	shake arise rise run become bear wear speak brake drive strike fall freeze choose	tear
VIII	go	throw blow grow draw fly know see	

/d/, which take a final /ˆd/ to form the past tense. The number of regular verbs in each category, for each of the three frequency levels, is given in Table 8.

Type I: No-change verbs. A small set of English verbs require no change between their present- and past-tense forms. One factor common to all such verbs is that they already end in /t/ or /d/. Thus, they superficially have the regular past-tense form—even in the present tense. Stemberger (1981) points out that it is common in inflectional languages not to add an additional inflection to base forms that already appear to have the inflection. Not all verbs ending in /t/ or /d/ show no change between present and past (in fact the majority of such verbs

TABLE 8

NUMBER OF REGULAR VERBS OF EACH TYPE

			Frequency		
Type	Suffix	Example	High	Medium	Low
End in dental	/ˆd/	start	0	94	13
End in voiceless consonant	/t/	look	1	64	30
End in voiced consonant or vowel	/d/	move	1	176	29

in English do show a change between present and past tense), but there is a reasonably large group—the Type I verbs of Bybee and Slobin—that do show this trend. Bybee and Slobin (1982) suggest that children learn relatively early on that past-tense verbs in English tend to end in /t/ or /d/ and thus are able to correctly respond to the no-change verbs rather early. Early in learning, they suggest, children also incorrectly generalize this "no-change rule" to verbs whose present and past tenses differ.

The pattern of performance just described shows up very clearly in data Bybee and Slobin (1982) report from an elicitation task with preschool children. In this task, preschoolers were given the present-tense form of each of several verbs and were asked to produce the corresponding past-tense form. They used the set of 33 verbs shown in Table 9.

The results were very interesting. Bybee and Slobin found that verbs not ending in t/d were predominately regularized and verbs ending in t/d were predominately used as no-change verbs. The number of occurrences of each kind is shown in Table 10. These preschool

TABLE 9

VERBS USED BY BYBEE & SLOBIN

Type of Verb	Verb List
Regular	walk smoke melt pat smile climb
Vowel change	drink break run swim throw meet shoot ride
Vowel change + t/d	do buy lose sell sleep help teach catch
No change	hit hurt set shut cut put beat
Other	go make build lend

TABLE 10

REGULAR AND NO CHANGE RESPONSES
TO t/d AND OTHER VERBS
(Data from Bybee & Slobin, 1982)

Verb Ending	Regular Suffix	No Change
Not t/d	203	34
t/d	42	157

children have, at this stage, both learned to regularize verbs not ending in t/d and, largely, to leave verbs ending in t/d without an additional ending.

Interestingly, our simulations show the same pattern of results. The system learns both to regularize and has a propensity *not* to add an additional ending to verbs already ending in t/d. In order to compare the simulation results to the human data we looked at the performance of the same verbs used by Bybee and Slobin in our simulations. Of the 33 verbs, 27 were in the high- and medium-frequency lists and thus were included in the training set used in the simulation. The other six verbs (*smoke, catch, lend, pat, hurt* and *shut*) were either in the low-frequency sample or did not appear in our sample at all. Therefore, we will report on 27 out of the 33 verbs that Bybee and Slobin tested.

It is not clear what span of learning trials in our simulation corresponds best to the level of the preschoolers in Bybee and Slobin's experiment. Presumably the period during which regularization is occurring is best. The combined strength of the regularized alternatives exceeds correct response strength for irregulars from about Trial 11 through Trials 20 to 30 depending on which particular irregular verbs we look at. We therefore have tabulated our results over three different time ranges—Trials 11 through 15, Trials 16 through 20, and Trials 21 through 30. In each case we calculated the average strength of the regularized response alternatives and of the no-change response alternatives. Table 11 gives these strengths for each of the different time periods.

The simulation results show clearly the same patterns evident in the Bybee and Slobin data. Verbs ending in t/d always show a stronger no-change response and a weaker regularized response than those not ending in t/d. During the very early stages of learning, however, the regularized response is stronger than the no-change response—even if the verb does end with t/d. This suggests that the generalization that the past tense of t/d verbs is formed by adding /ˆd/ is stronger than the generalization that verbs ending in t/d should not have an ending

TABLE 11

AVERAGE SIMULATED STRENGTHS OF
REGULARIZED AND NO-CHANGE RESPONSES

Time Period	Verb Ending	Regularized	No Change
11-15	not t/d	0.44	0.10
	t/d	0.35	0.27
16-20	not t/d	0.32	0.12
	t/d	0.25	0.35
21-30	not t/d	0.52	0.11
	t/d	0.32	0.41

added. However, as learning proceeds, this secondary generalization is made (though for only a subset of the t/d verbs, as we shall see), and the simulation shows the same interaction that Bybee and Slobin (1982) found in their preschoolers.

The data and the simulations results just described conflate two aspects of performance, namely, the tendency to make no-change *errors* with t/d verbs that are not no-change verbs and the tendency to make *correct* no-change responses to the t/d verbs that are no-change verbs. Though Bybee and Slobin did not report their data broken down by this factor, we can examine the results of the simulation to see whether in fact the model is making more no-change errors with t/d verbs for which this response is incorrect. To examine this issue, we return to the full corpus of verbs and consider the tendency to make no-change errors separately for irregular verbs other than Type I verbs and for regular verbs.

Erroneous no-change responses are clearly stronger for both regular and irregular t/d verbs. Figure 8A compares the strength of the erroneous no-change responses for irregular verbs ending in t/d (Types II and V) versus those not ending in t/d (Types III, IV, VI, VII, and VIII). The no-change response is erroneous in all of these cases. Note, however, that the erroneous no-change responses are stronger for the t/d verbs than for the other types of irregular verbs. Figure 8B shows the strength of erroneous no-change responses for regular verbs ending in t/d versus those not ending in t/d. Again, the response strength for the no-change response is clearly greater when the regular verb ends in a dental.

We also compared the regularization responses for irregular verbs whose stems end in t/d with irregulars not ending in t/d. The results are shown in Figure 8C. In this case, the regularization responses are initially stronger for verbs that do not end in t/d than for those that do.

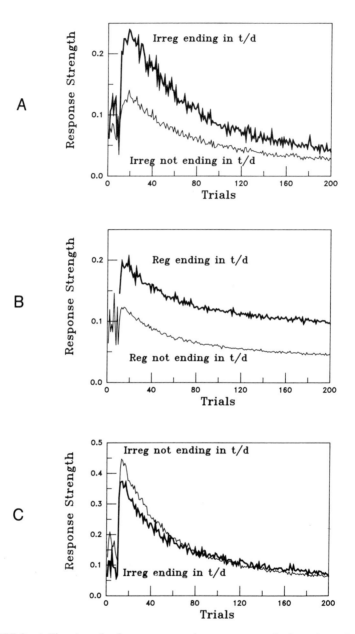

FIGURE 8. *A*: The strength of erroneous no-change responses for irregular verbs ending in a dental versus those not ending in a dental. *B*: The strength of erroneous no-change responses for regular verbs ending in a dental versus those not ending in a dental. *C*: The strength of erroneous regularization responses for irregular verbs ending in a dental versus those not ending in a dental.

Thus, we see that even when focusing only on erroneous responses, the system shows a greater propensity to respond with no change to *t/d* verbs, whether or not the verb is regular, and a somewhat greater tendency to regularize irregulars not ending in *t/d*.

There is some evidence in the literature on language acquisition that performance on Type I verbs is better sooner than for irregular verbs involving vowel changes—Types III through VIII. Kuczaj (1978) reports an experiment in which children were to judge the grammaticality of sentences involving past tenses. The children were given sentences involving words like *hit* or *hitted* or *ate* or *eated* and asked whether the sentences sounded "silly." The results, averaged over three age groups from 3;4 to 9;0 years, showed that 70 percent of the responses to the no-change verbs were correct whereas only 31 percent of the responses to vowel-change irregular verbs were correct. Most of the errors involved incorrect acceptance of a regularized form. Thus, the results show a clear difference between the verb types, with performance on the Type I verbs superior to that on Type III through VIII verbs.

The simulation model too shows better performance on Type I verbs than on any of the other types. These verbs show fewer errors than any of the other irregular verbs. Indeed the error rate on Type I verbs is equal to that on the most difficult of the regular verbs. Table 12 gives the average number of Wickelfeatures incorrectly generated (out of 460) at different periods during the learning processes for no-change (i.e., Type I) irregular verbs, vowel-change (i.e., Type III-VIII) irregular verbs, regular verbs ending in *t/d,* regular verbs not ending in *t/d,* and regular verbs ending in *t/d* whose stem is a CVC (consonant-vowel-consonant) monosyllable. The table clearly shows that throughout learning, fewer incorrect Wickelfeatures are generated for no-change verbs than for vowel-change verbs. Interestingly, the table

TABLE 12

AVERAGE NUMBER OF WICKELFEATURES INCORRECTLY GENERATED

Trial Number	Irregular Verbs		Regular Verbs		
	Type I	Types III-VIII	Ending in *t/d*	Not Ending in *t/d*	CV*t/d*
11-15	89.8	123.9	74.1	82.8	87.3
16-20	57.6	93.7	45.3	51.2	60.5
21-30	45.5	78.2	32.9	37.4	47.9
31-50	34.4	61.3	22.9	26.0	37.3
51-100	18.8	39.0	11.4	12.9	21.5
101-200	11.8	21.5	6.4	7.4	12.7

also shows that one subset of regulars are no easier than the Type I irregulars. These are the regular verbs which look on the surface most like Type I verbs, namely, the monosyllabic CVC regular verbs ending in *t/d*. These include such verbs as *bat*, *wait*, *shout*, *head*, etc. Although we know of no data indicating that people make more no-change errors on these verbs than on multisyllabic verbs ending in *t/d*, this is a clear prediction of our model. Essentially what is happening is that the model is learning that monosyllables ending in *t/d* sometimes take no additional inflection.[8] This leads to quicker learning of the no-change verbs relative to other irregular verbs and slower learning of regular verbs which otherwise look like no-change verbs. It should be noted that the two regular verbs employed by Bybee and Slobin which behaved like no-change verbs were both monosyllables. It would be interesting to see if whether no-change errors actually occur with verbs like *decide* or *devote*.

Types III-VIII: Vowel-change verbs. To look at error patterns on *vowel-change* verbs (Types III-VIII), Bybee and Slobin (1982) analyzed data from the spontaneous speech of preschoolers ranging from 1½ to 5 years of age. The data came from independent sets of data collected by Susan Ervin-Tripp and Wick Miller, by Dan Slobin, and by Zell Greenberg. In all, speech from 31 children involving the use of 69 irregular verbs was studied. Bybee and Slobin recorded the percentages of regularizations for each of the various types of vowel-change verbs. Table 13 gives the percentages of regularization by preschoolers, ranked from most to fewest erroneous regularizations. The results show that the two verb types which involve adding a *t/d* plus a vowel change (Types III and IV) show the least regularizations, whereas the verb type in which the present tense ends in a diphthong (Type VIII) shows by far the most regularization.

It is not entirely clear what statistic in our model best corresponds to the percentage of regularizations. It will be recalled that we collected response strength measures for four different response types for irregular verbs. These were the correct response, the no-change response, the base+ed regularization response, and the past+ed regularization response. If we imagine that no-change responses are, in general, difficult to observe in spontaneous speech, perhaps the measure that would be most closely related to the percentage of regularizations would be the ratio of the sum of the strengths of the regularization responses to

---

[8] Though the model does not explicitly encode number of syllables, monosyllabic words are distinguished from multisyllabic words by the fact that the former contain no Wickelphones of the form $_vC_v$. There are no no-change verbs in English containing such Wickelphones.

TABLE 13

PERCENTAGE OF REGULARIZATION
BY PRESCHOOLERS
(Data from Bybee & Slobin, 1982)

Verb Type	Example	Percentage Regularizations
VIII	blew	80
VI	sang	55
V	bit	34
VII	broke	32
III	felt	13
IV	caught	10

the sum of the strengths of regularization responses and the correct response—that is,

$$\frac{(base+ed\ +\ past+ed)}{(base+ed\ +\ past+ed\ +\ correct)}.$$

As with our previous simulation, it is not entirely clear what portion of the learning curve corresponds to the developmental level of the children in this group. We therefore calculated this ratio for several different time periods around the period of maximal overgeneralization. Table 14 shows the results of these simulations.

The spread between different verb classes is not as great in the simulation as in the children's data, but the simulated rank orders show a

TABLE 14

STRENGTH OF REGULARIZATION RESPONSES
RELATIVE TO CORRECT RESPONSES

Rank Order	Data		Trials 11-15		Trials 16-20		Trials 21-30		Average Trials 11-30	
	Type	Percent	Type	Ratio	Type	Ratio	Type	Ratio	Type	Ratio
1	VIII	80	VIII	.86	VIII	.76	VIII	.61	VIII	.71
2	VI	55	VII	.80	VII	.74	VII	.61	VII	.69
3	V	34	VI	.76	V	.60	IV	.48	V	.56
4	VII	32	V	.72	IV	.59	V	.46	IV	.56
5	III	13	IV	.69	III	.57	III	.44	III	.53
6	IV	10	III	.67	VI	.52	VI	.40	VI	.52

remarkable similarity to the results from the spontaneous speech of the preschoolers, especially in the earliest time period. Type VIII verbs show uniformly strong patterns of regularization whereas Type III and Type IV verbs, those whose past tense involves adding a t/d at the end, show relatively weak regularization responses. Type VI and Type VII verbs produce somewhat disparate results. For Type VI verbs, the simulation conforms fairly closely to the children's speech data in the earliest time period, but it shows rather less strength for regularizations of these verbs in the later time periods and in the average over Trials 11-30. For Type VII verbs, the model errs in the opposite direction: Here it tends to show rather greater strength for regularizations of these verbs than we see in the children's speech. One possible reason for these discrepancies may be the model's insensitivity to word frequency. Type VI verbs are, in fact, relatively low-frequency verbs, and thus, in the children's speech these verbs may actually be at a relatively earlier stage in acquisition than some of the more frequent irregular verbs. Type VII verbs are, in general, much more frequent—in fact, on the average they occur more than twice as often (in the gerund form) in the Kucera-Francis count than the Type VI verbs. In our simulations, all medium-frequency verbs were presented equally often and the distinction was not made. A higher-fidelity simulation including finer gradations of frequency variations among the verb types might lead to a closer correspondence with the empirical results. In any case, these verbs aside, the simulation seems to capture the major features of the data very nicely.

Bybee and Slobin attribute the pattern of results they found to factors that would not be relevant to our model. They proposed, for example, that Type III and IV verbs were more easily learned because the final t/d signaled to the child that they were in fact past tenses so the child would not have to rely on context as much in order to determine that these were past-tense forms. In our simulations, we found these verbs to be easy to learn, but it must have been for a different reason since the learning system was always informed as to what the correct past tense really was. Similarly, Bybee and Slobin argued that Type VIII verbs were the most difficult because the past and present tenses were so phonologically different that the child could not easily determine that the past and present tenses of these verbs actually go together. Again, our simulation showed Type VIII verbs to be the most difficult, but this had nothing to do with putting the past and present tense together since the model was always given the present and past tenses together.

Our model, then, must offer a different interpretation of Bybee and Slobin's findings. The main factor appears to be the degree to which the relation between the present and past tense of the verb is idiosyncratic. Type VIII verbs are most difficult because the relationship between base form and past tense is most idiosyncratic for these verbs. Thus, the natural generalizations implicit in the population of verbs must be overcome for these verbs, and they must be overcome in a different way for each of them. A very basic aspect of the mapping from present to past tense is that most of the word, and in particular everything up to the final vowel, is unchanged. For regular verbs, all of the phonemes present in the base form are preserved in the past tense. Thus, verbs that make changes to the base form are going against the grain more than those that do not; the larger the changes, the harder they will be to learn. Another factor is that past tenses of verbs generally end in /t/ or /d/.

Verbs that violate the basic past-tense pattern are all at a disadvantage in the model, of course, but some suffer less than others because there are other verbs that deviate from the basic pattern in the same way. Thus, these verbs are less idiosyncratic than verbs such as *go/went*, *see/saw*, and *draw/drew* which represent completely idiosyncratic vowel changes. The difficulty with Type VIII verbs, then, is simply that, as a class, they are simply more idiosyncratic than other verbs. Type III and IV verbs (e.g., *feel/felt*, *catch/caught*), on the other hand, share with the vast bulk of the verbs in English the feature that the past tense involves the addition of a *t/d*. The addition of the *t/d* makes these verbs easier than, say, Type VII verbs (e.g., *come/came*) because in Type VII verbs the system must not only learn that there *is* a vowel change, but it must also learn that there *is not* an addition of *t/d* to the end of the verb.

Type VI verbs (*sing/sang*, *drag/drug*) are interesting from this point of view, because they involve fairly common subregularities not found in other classes of verbs such as those in Type V. In the model, the Type VI verbs may be learned relatively quickly because of this subregularity.

Types of regularization. We have mentioned that there are two distinct ways in which a child can regularize an irregular verb: The child can use the base+ed form or the past+ed form. Kuczaj (1977) has provided evidence that the proportion of past+ed forms increases, relative to the number of base+ed forms, as the child gets older. He found, for example, that the nine youngest children he studied had more base+ed regularizations than past+ed regularizations whereas four out of the five oldest children showed more past+ed than base+ed regularizations. In this section, we consider whether our model exhibits this same general pattern. Since the base form and the past-tense form are identical for Type I verbs, we restrict our analysis of this issue to Types II through VIII.

Figure 9 compares the average response strengths for base+ed and past+ed regularizations as a function of amount of training. The results of this analysis are more or less consistent with Kuczaj's findings. Early in learning, the base+ed response alternative is clearly the stronger of the two. As the system learns, however, the two come together so that by about 100 trials the base+ed and the past+ed response alternatives are roughly equally strong. Clearly, the simulations show that the percentage of regularizations that are past+ed increases with experience—just as Kuczaj found in children. In addition, the two curves come together rather late, consistent with the fact, reported by Kuczaj (1977), that these past+ed forms predominate for the most part in children who are exhibiting rather few regularization errors of either type. Of the four children exhibiting more past+ed regularizations, three were regularizing less than 12% of the time.

A closer look at the various types of irregular verbs shows that this curve is the average of two quite different patterns. Table 15 shows the overall percentage of regularization strength due to the base+ed alternative. It is clear from the table that the verbs fall into two general categories, those of Types III, IV, and VIII which have an overall preponderance of base+ed strength (the percentages are all above .5) and Types II, VII, V, and VI which show an overall preponderance of past+ed strength (the percentages are all well below .5). The major variable which seems to account for the ordering shown in the table is the amount the ending is changed in going from the base form to the

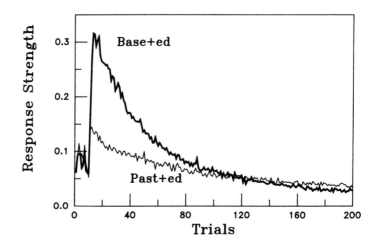

FIGURE 9. Average response strength for base+ed and past+ed responses for verb Types II through VIII.

TABLE 15

PERCENTAGE OF REGULARIZATION
STRENGTH DUE TO BASE+ED

Verb Type	Percent base+ed	Examples
III	0.77	sleep/slept
IV	0.69	catch/caught
VIII	0.68	see/saw
II	0.38	spend/spent
VII	0.38	come/came
V	0.37	bite/bit
VI	0.26	sing/sang

past-tense form. If the ending is changed little, as in *sing/sang* or *come/came*, the past+ed response is relatively stronger. If the past tense involves a greater change of the ending, such as *see/saw,* or *sleep/slept,* then the past+ed form is much weaker. Roughly, the idea is this: To form the past+ed for these verbs *two operations* must occur. The normal past tense must be created, and the regular ending must be appended. When these two operations involve very different parts of the verb, they can occur somewhat independently and both can readily occur. When, on the other hand, both changes occur to the same portion of the verb, they conflict with one another and a clear past+ed response is difficult to generate. The Type II verbs, which do show an overall preponderance of past+ed regularization strength, might seem to violate this pattern since it involves some change to the end in its past-tense form. Note, however, that the change is only a one feature change from /d/ to /t/ and thus is closer to the pattern of the verbs involving no change to the final phonemes of the verb. Figure 10A shows the pattern of response strengths to base+ed and past+ed regularizations for verb Types II, VII, V, and VI which involve relatively little change of the final phonemes from base to past form. Figure 10B shows the pattern of response strengths to base+ed and past+ed for verb Types III, IV, and VIII. Figure 10A shows very clearly the pattern expected from Kuczaj's results. Early in learning, base+ed responses are by far the strongest. With experience the past+ed response becomes stronger and stronger relative to the base+ed regularizations until, at about Trial 40, it begins to exceed it. Figure 10B shows a different pattern. For these verbs the past+ed form is weak throughout learning and never comes close to the base+ed regularization response. Unfortunately, Kuczaj did not present data on the relative frequency of the two types of regularizations separately for different verb types.

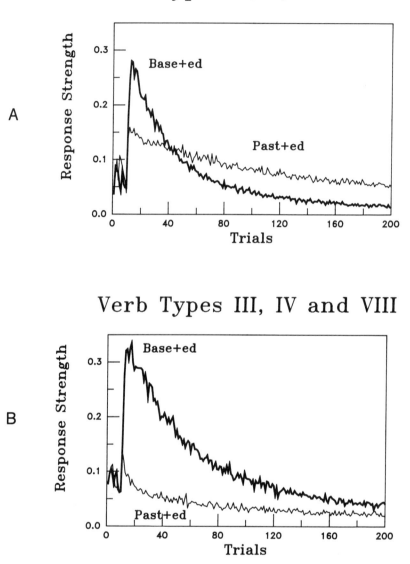

FIGURE 10. *A*: The pattern of response strengths to base + ed and past + ed regularizations for verb Types II, V, VI, and VII. *B*: The pattern of response strengths to base + ed and past + ed for verb Types III, IV, and VIII.

Thus for the present, this difference in type of regularization responses remains an untested prediction of the model.

Transfer to Novel Verbs

To this point we have only reported on the behavior of the system on verbs that it was actually taught. In this section, we consider the response of the model to the set of 86 low-frequency verbs which it never saw during training. This test allows us to examine how well the behavior of the model generalizes to novel verbs. In this section we also consider responses to different types of regular verbs, and we examine the model's performance in generating unconstrained responses.

Overall degree of transfer. Perhaps the first question to ask is how accurately the model generates the correct features of the new verbs. Table 16 shows the percentage of Wickelfeatures correctly generated, averaged over the regular and irregular verbs. Overall, the performance is quite good. Over 90 percent of the Wickelfeatures are correctly generated without any experience whatsoever with these verbs. Performance is, of course, poorer on the irregular verbs, in which the actual past tense is relatively idiosyncratic. But even there, almost 85 percent of the Wickelfeatures are correctly generated.

Unconstrained responses. Up until this point we have always proceeded by giving the model a set of response alternatives and letting it assign a response strength to each one. This allows us to get relative response strengths among the set of response alternatives we have provided. Of course, we chose as response alternatives those which we had reason to believe were among the strongest. There is the possibility, however, that the output of the model might actually favor some other, untested alternative some of the time. To see how well the output of the model is really doing at specifying correct past tenses or errors of the kind that children actually make, we must allow the model to choose among all possible strings of phonemes.

To do this, we implemented a second version of the binding network. This version is also described in the Appendix. Instead of a

TABLE 16

PROPORTION OF WICKELFEATURES
CORRECTLY GENERATED

Regular	.92
Irregular	.84
Overall	.91

competition among alternative strings, it involves a competition among individual Wickelphone alternatives, coupled with mutual facilitation between mutually compatible Wickelphones such as $_\#k_A$ and $_kA_m$.[9]

The results from the free-generation test are quite consistent with our expectations from the constrained alternative phase, though they did uncover a few interesting aspects of the model's performance that we had not anticipated. In our analysis of these results we have considered only responses with a strength of at least .2. Of the 86 test verbs, There were 65 cases in which exactly one of the alternatives exceeded .2. Of these, 55 were simple regularization responses, four were no-change responses, three involved double marking of regular verbs, (e.g., *type* was responded to with /tɪptˆd/), and there was one case of a vowel change (e.g., *slip/slept*). There were 14 cases in which two alternatives exceeded threshold and one case in which three exceeded threshold. Finally, in six cases, no response alternative exceeded threshold. This occurred with the regular verbs *jump, pump, soak, warm, trail,* and *glare*. In this case there were a number of alternatives, including the correct past-tense form of each of these verbs, competing with a response strength of about .1.

Table 17 shows the responses generated for the 14 irregular verbs. The responses here are very clear. All of the above-threshold responses made to an irregular verb were either regularization responses, no-change responses (to Type I and V verbs as expected) or correct vowel-change generalizations. The fact that *bid* is correctly generated as the past for *bid,* that *wept* is correctly generated as the past for *weep,* and that *clung* is correctly generated as a past tense for *cling* illustrates that the system is not only sensitive to the major regular past-tense pattern, but is sensitive to the subregularities as well. It should also be noted that the no-change responses to the verbs *grind* and *wind* occurs on monosyllabic Type V verbs ending in t/d, again showing evidence of a role for this subregularity in English past-tense formation.

Of the 72 regular verbs in our low-frequency sample, the six verbs mentioned above did not have any response alternatives above threshold. On 48 of the remaining 66 regular verbs, the only response exceeding threshold was the correct one. The threshold responses to the remaining 18 verbs are shown in Table 18.

---

[9] The major problem with this method of generating responses is that it is tremendously computer intensive. Had we used this method to generate responses throughout the learning phase, we estimate that it would have taken *over three years* of computer time to complete the learning phase alone! This compares to the 260 hours of computer time the learning phase took with the response alternatives supplied. It took about 28 hours to complete the response generation process in testing just the 86 low-frequency verbs used in this section of the study. Of course, in biological hardware, this would not be a problem since the processing would actually occur in parallel.

TABLE 17

THE MODEL'S RESPONSES TO UNFAMILIAR
LOW-FREQUENCY IRREGULAR VERBS

Verb Type	Presented Word	Phonetic Input	Phonetic Response	English Rendition	Response Strength
I	bid	/bid/	/bid/	(bid)	0.55
	thrust	/trˆst/	/trˆstˆd/	(thrusted)	0.57
II	bend	/bend/	/bendˆd/	(bended)	0.28
	lend	/lend/	/lendˆd/	(lended)	0.70
III	creep	/krɛp/	/krɛpt/	(creeped)	0.51
	weep	/wɛp/	/wɛpt/	(weeped)	0.34
			/wept/	(wept)	0.33
IV	catch	/kac/	/kact/	(catched)	0.67
V	breed	/brɛd/	/brɛdˆd/	(breeded)	0.48
	grind	/grɪnd/	/grɪnd/	(grind)	0.44
	wind	/wɪnd/	/wɪnd/	(wind)	0.37
VI	cling	/kliɴ/	/kliɴd/	(clinged)	0.28
			/klˆɴ/	(clung)	0.23
	dig	/dig/	/digd/	(digged)	0.22
	stick	/stik/	/stikt/	(sticked)	0.53
VII	tear	/ter/	/terd/	(teared)	0.90

Note that for 12 of the 18 verbs listed in the table, the correct response is above threshold. That means that of the 66 regular verbs to which any response at all exceeded threshold, the correct response exceeded threshold in 60 cases. It is interesting to note, also, that the model never chooses the incorrect variant of the regular past tense. As shown in Table 8, verbs ending in a /t/ or /d/ take /ˆd/ in the past tense; verbs ending in unvoiced consonants take /t/, and verbs ending in vowels or voiced consonants take /d/. On no occasion does the model assign a strength greater than .2 an incorrect variant of the past tense. Thus, the model has clearly learned the substructure of the regular correspondence and adds the correct variant to all different types of base forms. These results clearly demonstrate that the model acts in accordance with the regular pattern for English verbs and that it can apply this pattern with a high level of success to novel as well as familiar verbs.

In addition to the regular responses, five of the responses were no-change responses. In three cases the no-change response was to a verb ending in *t/d*. Four of the responses followed the pattern of Type III verbs, modifying the vowel and adding a final /t/. Thus, for example,

TABLE 18

SYSTEM RESPONSES TO UNFAMILIAR LOW-FREQUENCY REGULAR VERBS

Verb Type	Presented Word	Phonetic Input	Phonetic Response	English Rendition	Response Rendition
End in t/d	guard	/gɑrd/	/gɑrd/ /gɑrdˆd/	(guard) (guarded)	0.29 0.26
	kid	/kid/	/kid/ /kidˆd/	(kid) (kidded)	0.39 0.24
	mate	/mʌt/	/mʌtˆd/ /mʌdˆd/	(mated) (maded)	0.43 0.23
	squat	/skw*t/	/skw*tˆd/ /skw*t/ /skw*kt/	(squated) (squat) (squawked)	0.27 0.22 0.21
End in unvoiced consonant	carp	/kɑrp/	/kɑrpt/ /kɑrptˆd/	(carped) (carpted)	0.28 0.21
	drip	/drip/	/driptˆd/ /dript/	(dripted) (dripped)	0.28 0.22
	map	/map/	/maptˆd/ /mapt/	(mapted) (mapped)	0.24 0.22
	shape	/sʌp/	/sʌpt/ /sipt/	(shaped) (shipped)	0.43 0.27
	sip	/sip/	/sipt/ /sept/	(sipped) (sepped)	0.42 0.28
	slip	/slip/	/slept/	(slept)	0.40
	smoke	/smOk/	/smOktˆd/ /smOk/	(smokted) (smoke)	0.29 0.22
	snap	/snap/	/snaptˆd/	(snapted)	0.40
	step	/step/	/steptˆd/	(stepted)	0.59
	type	/tɪp/	/tɪptˆd/	(typted)	0.33
End in voiced consonant or vowel	brown	/brwn/	/brwnd/ /br*nd/	(browned) (brawned)	0.46 0.39
	hug	/hˆg/	/hˆg/	(hug)	0.59
	mail	/mʌˆl/	/mʌˆld/ /membˆld/	(mailed) (membled)	0.38 0.23
	tour	/tʊr/	/tʊrdˆr/ /tʊrd/	(toureder) (toured)	0.31 0.25

we have the past of *sip* rendered as *sept*, presumably on the model of *sleep/slept*, *keep/kept*, *sweep/swept*, etc. Interestingly, three of the four cases involved verbs whose base form ended in /p/ just as in the models listed above. Even though these last responses are, strictly speaking, incorrect, they all indicate a sensitivity to the regular and subregular patterns of the English past tense.

Perhaps the most surprising result evident from the table is the occurrence of a double past marker on the responses to seven of the verbs. Although we know of no references to this phenomenon in the literature, we expect that children (and adults) do occasionally make this kind of error. It is interesting, and possibly significant, that all seven of these responses occurred to verbs whose correct past tense is the addition of a /t/. It would be interesting to see whether children's errors of this type follow a similar pattern.

Finally, there were just four responses that involved the addition or modification of consonants. These were *maded* as a past tense of *mate*, *squawked* as a past tense for *squat, membled* as a past tense for *mail*, and *toureder* as a past tense for *tour*. It is unlikely that humans would make these errors, especially the last two, but these responses are, for the most part, near threshold. Furthermore, it seems likely that many of these responses could be filtered out if the model incorporated an auto-associative network of connections among the output units. Such a network could be used to clean up the output pattern and would probably increase the tendency of the model to avoid bizarre responses. Unfortunately, we have not yet had the chance to implement this suggestion.

Summary. The system has clearly learned the essential characteristics of the past tense of English. Not only can it respond correctly to the 460 verbs that it was taught, but it is able to generalize and transfer rather well to the unfamiliar low-frequency verbs that had never been presented during training. The system has learned about the conditions in which each of the three regular past-tense endings are to be applied, and it has learned not only the dominant, regular form of the past tense, but many of the subregularities as well.

It is true that the model does not act as a perfect rule-applying machine with novel past-tense forms. However, it must be noted that people—or at least children, even in early grade-school years—are not perfect rule-applying machines either. For example, in Berko's classic (1958) study, though her kindergarten and first-grade subjects did often produce the correct past forms of novel verbs like *spow, mott,* and *rick*, they did not do so invariably. In fact, the rate of regular past-tense forms given to Berko's novel verbs was only 51 percent.[10] Thus, we see

---

[10] Unfortunately, Berko included only one regular verb to compare to her novel verbs. The verb was *melt*. Children were 73 percent correct on this verb. The two novel verbs that required the same treatment as *melt* (*mott* and *bodd*) each received only 33 percent correct responses.

little reason to believe that our model's "deficiencies" are significantly greater than those of native speakers of comparable experience.

CONCLUSIONS

We have shown that our simple learning model shows, to a remarkable degree, the characteristics of young children learning the morphology of the past tense in English. We have shown how our model generates the so-called U-shaped learning curve for irregular verbs and that it exhibits a tendency to overgeneralize that is quite similar to the pattern exhibited by young children. Both in children and in our model, the verb forms showing the most regularization are pairs such as *know/knew* and *see/saw*, whereas those showing the least regularization are pairs such as *feel/felt* and *catch/caught*. Early in learning, our model shows the pattern of more no-change responses to verbs ending in t/d whether or not they are regular verbs, just as young children do. The model, like children, can generate the appropriate regular past-tense form to unfamiliar verbs whose base form ends in various consonants or vowels. Thus, the model generates an $/\hat{}d/$ suffix for verbs ending in $t/d,$ a $/t/$ suffix for verbs ending in an unvoiced consonant, and a $/d/$ suffix for verbs ending in a voiced consonant or vowel.

In the model, as in children, different past-tense forms for the same word can coexist at the same time. On rule accounts, such *transitional* behavior is puzzling and difficult explain. Our model, like human children, shows an relatively larger proportion of past+ed regularizations later in learning. Our model, like learners of English, will sometimes generate past-tense forms to novel verbs which show sensitivities to the subregularities of English as well as the major regularities. Thus, the past of *cring* can sometimes be rendered *crang* or *crung*. In short, our simple learning model accounts for all of the major features of the acquisition of the morphology of the English past tense.

In addition to our ability to account for the major *known* features of the acquisition process, there are also a number of predictions that the model makes which have yet to be reported. These include:

- We expect relatively more past+ed regularizations to irregulars whose correct past form *does not* involve a modification of the final phoneme of the base form.

- We expect that early in learning, a no-change response will occur more frequently to a CVC monosyllable ending in t/d than to a more complex base verb form.

- We expect that the double inflection responses (/driptˆd/) will occasionally be made by native speakers and that they will occur more frequently to verbs whose stem is ends in /p/ or /k/.

The model is very rich and there are many other more specific predictions which can be derived from it and evaluated by a careful analysis of acquisition data.

We have, we believe, provided a distinct alternative to the view that children learn the rules of English past-tense formation in any explicit sense. We have shown that a reasonable account of the acquisition of past tense can be provided without recourse to the notion of a "rule" as anything more than a *description* of the language. We have shown that, for this case, there is no *induction problem*. The child need not figure out what the rules are, nor even that there are rules. The child need not decide whether a verb is regular or irregular. There is no question as to whether the inflected form should be stored directly in the lexicon or derived from more general principles. There isn't even a question (as far as generating the past-tense form is concerned) as to whether a verb form is one encountered many times or one that is being generated for the first time. A uniform procedure is applied for producing the past-tense form in every case. The base form is supplied as input to the past-tense network and the resulting pattern of activation is interpreted as a phonological representation of the past form of that verb. This is the procedure whether the verb is regular or irregular, familiar or novel.

In one sense, every form must be considered as being derived. In this sense, the network can be considered to be one large rule for generating past tenses from base forms. In another sense, it is possible to imagine that the system simply stores a set of rote associations between base and past-tense forms with novel responses generated by "on-line" generalizations from the stored exemplars.

Neither of these descriptions is quite right, we believe. Associations are simply stored in the network, but because we have a *superpositional* memory, similar patterns blend into one another and reinforce each other. If there were no similar patterns (i.e., if the featural representations of the base forms of verbs were orthogonal to one another) there would be no generalization. The system would be unable to generalize and there would be no regularization. It is statistical relationships among the base forms themselves that determine the pattern of responding. The network merely reflects the statistics of the featural representations of the verb forms.

We chose the study of acquisition of past tense in part because the phenomenon of regularization is an example often cited in support of the view that children do respond according to general rules of language. Why otherwise, it is sometimes asked, should they generate forms that they have never heard? The answer we offer is that they do so because the past tenses of similar verbs they are learning show such a consistent pattern that the generalization from these similar verbs outweighs the relatively small amount of learning that has occurred on the irregular verb in question. We suspect that essentially similar ideas will prove useful in accounting for other aspects of language acquisition. We view this work on past-tense morphology as a step toward a revised understanding of language knowledge, language acquisition, and linguistic information processing in general.

ACKNOWLEDGMENTS

This research was supported by ONR Contracts N00014-82-C-0374, NR 667-483 and N00014-79-C-0323, NR 667-437, by a grant from the System Development Foundation, and by a Research Scientist Career Development Award MH00385 to the second author from the National Institute of Mental Health.

APPENDIX

One important aspect of the Wickelfeature representation is that it completely suppressed the temporal dimension. Temporal information is stored implicitly in the feature pattern. This gives us a representational format in which phonological forms of arbitrary length can be represented. It also avoids an a priori decision as to which part of the verb (beginning, end, center, etc.) contains the past-tense inflection. This grows out of the learning process. Unfortunately, it has its negative side as well. Since phonological forms *do* contain temporal information, we need to have a method of converting from the Wickelfeature representation into the time domain—in short, we need a decoding network which converts from the Wickelfeature representation to either the Wickelphone or a phonological representational format. Since we have probabilistic units, this decoding process must be able to work in the face of substantial noise. To do this we devised a special sort of decoding network which we call a *binding network*. Roughly speaking, a binding network is a scheme whereby a number of units *compete* for a set of available features—finally attaining a strength that is proportional to the number of features the units account for. We proceed by first describing the idea behind the binding network, then describing its application to produce the set of Wickelphones implicit in the Wickelfeature representation, and finally to produce the set of phonological strings implicit in the Wickelfeatures.

Binding Networks

The basic idea is simple. Imagine that there are a set of input features and a set of output features. Each output feature is consistent with certain of the input features, inconsistent with certain other of the input features, and neutral about still other of the input features. The idea is to find a set of output features that accounts for as many as possible of the output features while minimizing the number of input features accounted for by more than one output feature. Thus, we want each of the output features to *compete* for input features. The more input features it *captures*, the stronger its position in the competition and the more claim it has on the features it accounts for. Thus consider the case in which the input features are Wickelfeatures and the output features are Wickelphones. The Wickelphones compete among one another for the available Wickelfeatures. Every time a particular Wickelphone "captures" a particular Wickelfeature, that input feature no longer provides support for other Wickelphones. In this way, the system comes up with a set of more or less nonoverlapping Wickelphones which account for as many as possible of the available Wickelfeatures. This means that if two Wickelphones have many Wickelfeatures in common (e.g., $_k\hat{m}$ and $_kA_m$) but one of them accounts for more features than the other, the one that accounts for the most features will remove nearly all of the support for the very similar output feature which accounts for few if any input features uniquely. The binding network described below has the property that if two output units are competing for a set of input features, each will attain a strength proportional to the number of input features uniquely accounted for by that output feature divided by the total number of input features uniquely accounted for by any output feature.

This is accomplished by a network in which each input unit has a fixed amount of activation (in our case we assumed that it had a total activation value of 1) to be distributed among the output units consistent with that input feature. It distributes its activation in proportion to the strength of the output feature to which it is connected. This is thus a network with a dynamic weight. The weight from input unit j to output unit i is thus given by

$$w_{ij} = \frac{a_i}{\sum_{k_j} a_{k_j}}$$

where k_j ranges over the set of output units consistent with input units j. The total strength of output unit k at time t is a linear function of its inputs at time $t - 1$ and is thus given by

$$a_k(t) = \sum_{j_k} i_{j_k} w_{kj_k}(t) = \frac{\sum_{j_k} i_{j_k} a_k(t-1)}{\sum_{l_{j_k}} a_{l_{j_k}}(t-1)}$$

where j_k ranges over the set of input features consistent with output feature k, l_{j_k} ranges over the set of output features consistent with input feature j_k, and i_j takes on value 1 if input feature j is present and is 0 otherwise.

We used the binding network described above to find the set of Wickelphones which gave optimal coverage to the Wickelfeatures in the input. The procedure was quite effective. We used as the set of output all of the Wickelphones that occurred anywhere in any of the 500 or so verbs we studied. We found that the actual Wickelphones were always the strongest when we had 80 percent or more of the correct Wickelfeatures. Performance dropped off as the percentage of correct

Wickelfeatures dropped. Still when as few as 50 percent of the Wickel-features were correct, the correct Wickelphones were still the strongest most of the time. Sometimes, however, a Wickelphone not actually in the input would become strong and push out the "correct" Wickel-phones. If we added the constraint that the Wickelphones must fit together to form an entire string (by having output features activate features that are consistent neighbors), we found that more than 60 percent of correct Wickelfeatures lead to the correct output string more than 90 percent of the time.

The binding network described above is designed for a situation in which there is a set of input features that is to be divided up among a set of output features. In this case, features that are present, but not required for a particular output feature play no role in the evaluation of the output feature. Suppose, however, that we have a set of alternative output features, one of which is supposed to account for the entire pattern. In this case, input features that are present, but not consistent, with a given output feature must count against that output feature. One solution to this is to have input units *excite* consistent output units according the the rule given above and to *inhibit* inconsistent output units. In the case in which we tried to construct the entire phonological string directly from a set of Wickelfeatures we used the following activation rule:

$$a_k(t) = \sum_{j_k} i_{j_k} w_{kj_k}(t) - \sum_{l_k} i_{l_k}$$

where l_k indexes the input features that are inconsistent with output feature k. In this case, we used as output features all of the strings of less than 20 phonemes which could be generated from the set of Wick-elphones present in the entire corpus of verbs. This is the procedure employed to produce responses to the lowest frequency verbs as shown in Tables 17 and 18.

CHAPTER 19

Mechanisms of Sentence Processing:
Assigning Roles to Constituents of Sentences

J. L. McCLELLAND and A. H. KAWAMOTO

MULTIPLE CONSTRAINTS ON ROLE ASSIGNMENT

Like many natural cognitive processes, the process of sentence comprehension involves the simultaneous consideration of a large number of different sources of information. In this chapter, we consider one aspect of sentence comprehension: the assignment of the constituents of a sentence to the correct thematic case roles. Case role assignment is not, of course, all there is to comprehension, but it reflects one important aspect of the comprehension process, namely, the specification of who did what to whom.

Case role assignment is not at all a trivial matter either, as we can see by considering some sentences and the case roles we assign to their constituents. We begin with several sentences using the verb *break*:

(1) The boy broke the window.
(2) The rock broke the window.
(3) The window broke.
(4) The boy broke the window with the rock.
(5) The boy broke the window with the curtain.

We can see that the assignment of case roles here is quite complex. The first noun phrase (NP) of the sentence can be the Agent (Sentences 1, 4, and 5), the Instrument (Sentence 2), or the Patient

(Sentence 3). The NP in the prepositional phrase (PP) could be the Instrument (Sentence 4), or it could be a Modifier of the second NP, as it is in at least one reading of Sentence 5. Another example again brings out the ambiguity of the role assignment of with-NPs:

(6) The boy ate the pasta with the sauce.
(7) The boy ate the pasta with the fork.

In (6) the with-NP clearly does not specify an Instrument, but in (7) it clearly does.

Before we go much further, it should be said that there is no universally accepted set of case roles, nor universal agreement as to the correct assignment of constituents to roles. We have adopted conventions close to those originally introduced by Fillmore (1968) in "The Case for Case," but we do not think the details are of crucial importance to the behavior of our model. Later we will suggest ways in which an extension of our model might circumvent certain of the difficulties involved in specifying the correct assignment of cases.

These complications aside, it appears from the examples that the meaning of the words in these sentences influences the assignment of arguments to roles. However, the placement of NPs within the sentences is also very important. Consider these two cases:

(8) The vase broke the window.
(9) The window broke the vase.

Here we must rely on word-order constraints. That such constraints are very strong in English can be seen from sentences like:

(10) The pencil kicked the cow.

Even though semantic constraints clearly would indicate that the cow is a much more likely Agent and the pencil a much more likely Patient, Sentence 10 simply is not given this interpretation by adult readers who are native speakers of English.

Word-order constraints like those illustrated by (10) are very strong in English, but it is important to realize that such heavy reliance on such constraints is not universal. Bates and MacWhinney (in press; MacWhinney, Bates, & Kliegl, 1984) have shown that adult speakers of Italian will assign roles to sentences like (10) based predominantly on semantic constraints; [1] word order plays a very limited role and

---

[1] We use the phrase "semantic constraints" to refer to the constraints language users impose on the co-occurrence of constituents in particular roles in case-level representations. In the model, as we shall see, these constraints arise from the co-occurrences of constituents in the experiences the model is exposed to.

determines assignment only when semantics and case-marking inflections give no information.

As the work of Bates and MacWhinney amply demonstrates, case role assignment is influenced by at least three different kinds of factors: word order, semantic constraints, and (when available) inflectional morphology. Reliance on any one of these constraints is a matter of degree, and varies from language to language. In addition to these factors, there is one more that cannot be ignored, namely, the more global context in which the sentence is presented. Consider, for example, Sentence 11:

(11) The boy saw the girl with the binoculars.

We get one reading if prior context tells us "A boy was looking out the window, trying to see how much he could see with various optical instruments." We get quite a different one if it says "Two girls were trying to identify some birds when a boy came along. One girl had a pair of binoculars and the other did not." Crain and Steedman (1985) have experimentally demonstrated contextual influences on parsing decisions.

While the fact that word order and semantic constraints both influence role assignment has often been acknowledged (Bever, 1970; Fodor, Bever, & Garrett, 1974), there are few existing models that go very far toward proposing a mechanism to account for these effects. However, there are some researchers in language processing who have tried to find ways of bringing semantic considerations into syntactic processing in one way or another. One recent approach has been to rely on the lexicon to influence both syntactic processing and the construction of underlying functional representations (Ford, Bresnan, & Kaplan, 1982; Kaplan & Bresnan, 1982; MacWhinney & Sokolov, in press). Ford et al. (1982) considered cases like the following:

(12) The woman wanted the dress on the rack.
(13) The woman positioned the dress on the rack.

They noted that the preferred reading of the first of these had *on the rack* as a modifier of *the dress*, while the preferred reading of the second had *on the rack* as a locative argument of *positioned*. To account for this difference in role assignment, they proposed two principles: (a) *lexical preference* and (b) *final arguments*. Basically, lexical preference establishes an expected argument structure (e.g., Subject–Verb–Object in the case of *want*; Subject–Verb–Object–Prepositional Object in the case of *positioned*) by consulting an ordered list of possible argument structures associated with each verb. If a constituent is encountered that could fill a slot in the expected argument structure, the constituent is treated as an argument of the verb. However, if a constituent is encountered that appears to satisfy the conditions on the final argument of the expected argument structure, its attachment is delayed to allow for the incorporation into the constituent of subsequent constituents. Thus, with *want*, the NP *the dress* is a candidate for final argument and is not attached directly as a constituent of the VP; rather, a superordinate NP structure containing *the dress on the rack* is ultimately attached to the VP. With *position*, however, *the dress* would not be the final argument, and so is attached directly to the VP and closed. *On the rack* is then available for attachment as the final argument to the VP.

While this scheme certainly does some of the work that needs to be done in allowing the constraints imposed by the words in a sentence to influence role assignment, we do not think it goes nearly far enough. For as we saw in Sentences 4–7, the NPs of a sentence also influence syntactic decisions. Oden (1978) has verified that all three NPs in sentences like these influence subjects' role-assignment decisions.

In the literature on sentence processing, no one disputes that various factors influence the final reading that is assigned to a sentence. However, there are various views of the way in which these factors are taken into account on-line. Kurtzman (1985) argues that the parsing process is directly guided by an ongoing plausibility analysis; Marslen-Wilson and Tyler (1981) have pioneered this sort of view, and they stress the immediacy with which syntactic, semantic, and pragmatic considerations can all be brought to bear on the course of sentence processing. On the other hand, Frazier and her colleagues (e.g., Frazier & Rayner, 1982; Rayner, Carlson, & Frazier, 1983) argue that the syntactic parser imposes its preferred structuring on the sentence based only on syntactic considerations, passing the results of this processing on quickly to a thematic interpreter that can reject the syntactic parse in favor of a thematically more appropriate reading.

Whichever view one holds, it is clear that a mechanism is needed in which all the constituents of a sentence can work simultaneously to influence the assignment of roles to constituents. While we ourselves tend to favor a highly interactive view, the model we will describe here takes as its input a partial surface parse (though it is one that leaves certain attachment decisions unspecified) and generates from it a case-level representation. Intended extensions of the model, which we will describe below, would incorporate feedback to the syntactic structure level; but most of the model's behavior is not dependent on this feedback, and so readers committed to a less interactive view of the relation between syntactic and thematic analyses may yet find the model to be of interest.

GOALS

The primary goal of our model is to provide a mechanism that can begin to account for the joint role of word order and semantic constraints on role assignment. We wanted the model to be able to *learn* to do this based on experience with sentences and their case representations. We wanted the model to be able to *generalize* what it learned to new sentences made up of novel combinations of words.

In addition, we had several other goals for the model:

- We wanted the model to be able to select contextually appropriate readings of ambiguous words.

- We wanted the model to select the appropriate verb frame based on the pattern of arguments and their semantic features.

- We wanted the model to fill in missing arguments in incomplete sentences with plausible default values.

- We wanted the model to be able to generalize its knowledge of correct role assignment to sentences containing a word it has never seen before, given only a specification of some of the semantic properties of the word.

The model succeeded in meeting all these goals, as we shall see.

The model also exhibits an additional property that we had not actually anticipated, even though it is a central characteristic of language understanding: The model exhibits an uncanny tendency to shade its representation of the constituents of a sentence in ways that are contextually appropriate. It does this without any explicit training to do so; in fact, it does this in spite of the fact that the training inputs it receives are not contextually shaded as they would be in reality. We will examine this aspect of the model's behavior through examples, and observe how it emerges naturally from the model's structure.

The model is, of course, very far from a complete or final model of sentence processing or even case role assignment. Perhaps it is best seen as a partial instantiation of one view of what some properties of the interface between syntactic and more conceptual levels of language representation might be like. We offer the model not because it "solves the problem of sentence comprehension." Rather, we offer it because it suggests new ways of thinking about several aspects of language and language representation. The simulation model that embodies these ideas will undoubtedly require substantial development and elaboration.

It is our belief, though, that the basic principles that it embodies will prove extremely valuable as cognitive science continues to try to come to grips with the problem of understanding natural language.

We have limited the model in several ways. Most importantly, we have considered only single clause sentences. We have also considered only a limited set of roles and a limited vocabulary. Since we have restricted the analysis to English, case inflectional morphology does not arise. Within these bounds, we will see that we have been able to meet the goals of the model quite successfully, using a very simple PDP architecture.

Previous, Related Work

Both Cottrell (1985; Cottrell & Small, 1983) and Waltz and Pollack (1985) have preceded us in noting the appeal of connectionism as a means of exploiting the multiple constraints that appear to influence both case role assignment and the contextual disambiguation of ambiguous noun phrases. Their models differ from ours in several ways, most notably in that both rely primarily on local representations (one-unit–one-concept) as opposed to distributed representations, although Waltz and Pollack (1985) do suggest ways that a distributed representation could be used to represent global contextual influences on word meaning disambiguation. Within the context of distributed models, ours builds on the work of J. A. Anderson (1983) and Kawamoto (1985): Both models show how context can be used to select the appropriate reading of an ambiguous word. Our work incorporates mechanisms quite like theirs to accomplish this and other goals. Finally, Hinton's (1981a) early discussion of the use of distributed representations to represent propositions played an important role in the development of the ideas described here.

ARCHITECTURE OF THE MODEL

The role-assignment model is a distributed model, and has many properties in common with the verb learning model described in Chapter 18. The model consists of two sets of units: one for representing the surface structure of the sentence and one for representing its case structure. The model learns through presentations of correct surface-structure/case-structure pairs; during testing, we simply present the surface-structure input and examine the output the model generates at the case-structure level.

Sentences. The sentences processed by the model consist of a verb and from one to three NPs. There is always a Subject NP, and optionally there may be an Object NP. If this is present, there may also be a *with-NP*; that is, a NP in a sentence-final prepositional phrase beginning with the word *with*. All of the numbered sentences considered in the introduction are examples of sentence types that might be presented to the model.

Input format of sentences. What the model actually sees as input is not the raw sentence but a canonical representation of the constituent structure of the sentence, in a form that could be produced by a simple surface parser and a simple lexicon. Such a parser and lexicon are not, in fact, parts of the model in its present form—the sentences are simply presented to the model in this canonical format. We discuss ways such a parser could be implemented in a PDP model in the discussion section.

Semantic Microfeatures

In the canonical input format, words are represented as lists of semantic microfeatures (Hinton, 1981a; see Chapter 3; Waltz & Pollack, 1985, also make some use of a microfeature representation). For both nouns and verbs, the features are grouped into several dimensions. Each dimension consists of a set of mutually exclusive values, and, in general, each word is represented by a vector in which one and only one value on each dimension is ON for the word and all of the other values are OFF. Values that are set to be ON are represented in the feature vectors as 1s. Values that are set to be OFF are represented as dots (".").

We chose the dimensions and the values on each dimension to capture what we felt were important dimensions of semantic variation in the meanings of words that had implications for the role assignments of the words. We should be very clear about one point, though, which is that we do not want to suggest that the full range of the phenomena that are described under the rubric of the "meanings" of the words are captured by these semantic microfeatures. Indeed, we do not think of words as actually having some fixed meaning at all. Exactly how we do think of meanings will become clear after we examine the behavior of the model, so we postpone a fuller consideration of this issue until the discussion.

The full set of dimensions used in the feature sets are given in Table 1. The noun dimensions are largely self-explanatory, but the

TABLE 1

FEATURE DIMENSIONS AND VALUES

Nouns	
HUMAN	human nonhuman
SOFTNESS	soft hard
GENDER	male female neuter
VOLUME	small medium large
FORM	compact 1-D 2-D 3-D
POINTINESS	pointed rounded
BREAKABILITY	fragile unbreakable
OBJ-TYPE	food toy tool utensil furniture animate nat-inan

Verbs	
DOER	yes no
CAUSE	yes no-cause no-change
TOUCH	agent inst both none AisP
NAT_CHNG	pieces shreds chemical none unused
AGT_MVMT	trans part none NA
PT_MVMT	trans part none NA
INTENSITY	low high

Note: nat-inan = natural inanimate, AisP = Agent is Patient, NA = not applicable.

different dimensions of the verbs may need some explication. Basically, these dimensions are seen as capturing properties of the scenario specified by the verb. Thus, the DOER dimension indicates whether there is an Agent instigating the event. The CAUSE dimension specifies whether the verb is causal. If not, it indicates whether this is because there is no cause specified (as in the case of *the window broke*) or whether it is because there is no change (as in the case of *the boy touched the girl*). The TOUCH dimension indicates whether the Agent, the Instrument, both, or neither touches the Patient; the "AisP" value simply indicates that the Agent and the Patient are the same (as in *the cat moved*). The NAT_CHNG dimension specifies the nature of the change that takes place in the Patient. The AGT_MVMT and PT_MVMT specify the movement of the Agent and the Patient, respectively; and INTENSITY simply indicates the forcefulness of the action. The labels given to the dimensions are, of course, only for reference; they were chosen so that each noun or verb dimension would have a unique first letter that could be used to designate the dimension.

It must be stressed that we are not strongly wedded to this particular choice of features, and that other features would need to be included to extend the model to larger sets of nouns and verbs. On the other hand, the features that we did include were carefully chosen because they seemed highly relevant to determining the case role assignments. For example, the DOER dimension directly specifies whether there is or is not an Agent. Thus, the features of the verb, in particular, often have direct case-structural implications. (We would prefer a model that constructed its own semantic microfeatures using back propagation [Chapter 8] or a related method for learning, but this extension has not yet been implemented.)

Figures 1 and 2 give the vectors that we assigned to each of the words used in the model. It will be immediately noted that some of our encoding decisions were arbitrary, and that sometimes we seem to be forcing words into molds that they do not perfectly fit. Further, each feature has the same weight as all the others, and is as definite as all the others. Reality is not nearly so definite or evenhanded, of course. Balls are round, but may be soft or hard; paperweights are generally compact in shape but need not be, etc. The definiteness of the input used in the simulations is a simplification that we have adopted to make the initial coding of the input patterns as straightforward as possible. A more realistic coding would allow some features to be more definite than others. We will see that the model tends to correct this deficiency on its own accord.

One of our goals for the model is to show how it can select the contextually appropriate meaning for an ambiguous word. For ambiguous words (*bat*, flying or baseball, and *chicken*, living or cooked) the input pattern is the average of the feature patterns of each of the two readings of the word. This means that in cases where the two agree on the value of a particular input dimension, that dimension has the agreed value in the input representation. In cases where the two disagree, the feature has a value of .5 (represented by "?") in the input representation. A goal of the simulations is to see if the model can correctly fill in these unspecified values, effectively retrieving the contextually

	HU	SO	GND	VOL	FORM	PO	BR	OBJ_TYP
ball	. 1	1 .	. . 1	1 . .	1 . . .	. 1	. 1	. 1 . . . . .
fl-bat	. 1	1 .	1 . .	1 . .	. . . 1	1 .	. 1	. . . . . 1 .
bb-bat	. 1	. 1	. . 1	1 . .	. 1 . .	. 1	. 1	. 1 . . . . .
bat	. 1	? ?	? . ?	1 . .	. ? . ?	? ?	. 1	. ? . . . ? .
boy	1 .	1 .	. . 1	. 1 .	. . . 1	. 1	. 1	. . . . . 1 .
paperwt	. 1	. 1	. . 1	1 . .	1 . . .	1 .	. 1	. . . . 1 . .
cheese	. 1	1 .	. . 1	1 . .	. . 1 .	. 1	1 .	1 . . . . . .
li-chicken	. 1	1 .	. 1 .	1 . .	. . . 1	. 1	1 .	. . . . . 1 .
co-chicken	. 1	1 .	. . 1	1 . .	1 . . .	. 1	1 .	1 . . . . . .
chicken	. 1	1 .	. ? ?	1 . .	? . . ?	. 1	? ?	? . . . . ? .
curtain	. 1	1 .	. . 1	. 1 .	. . 1 .	. 1	1 .	. . . . 1 . .
desk	. 1	. 1	. . 1	. . 1	. . . 1	1 .	. 1	. . . . 1 . .
doll	. 1	1 .	. 1 .	1 . .	. . . 1	. 1	1 .	. 1 . . . . .
food	. 1	1 .	. . 1	1 . .	? ? ? ?	. 1	1 .	1 . . . . . .
fork	. 1	. 1	. . 1	1 . .	. 1 . .	1 .	. 1	. . . 1 . . .
girl	1 .	1 .	. 1 .	. 1 .	. . . 1	. 1	. 1	. . . . . 1 .
hatchet	. 1	. 1	. . 1	1 . .	. 1 . .	1 .	. 1	. . 1 . . . .
hammer	. 1	. 1	. . 1	1 . .	. 1 . .	. 1	. 1	. . 1 . . . .
man	1 .	1 .	1 . .	. . 1	. . . 1	. 1	. 1	. . . . . 1 .
woman	1 .	1 .	. 1 .	. . 1	. . . 1	. 1	. 1	. . . . . 1 .
plate	. 1	. 1	. . 1	1 . .	. . 1 .	. 1	1 .	. . . . 1 . .
rock	. 1	. 1	. . 1	1 . .	. . . 1	1 .	. 1	 1
potato	. 1	1 .	. . 1	1 . .	1 . . .	. 1	1 .	1 . . . . . .
pasta	. 1	1 .	. . 1	1 . .	. 1 . .	. 1	1 .	1 . . . . . .
spoon	. 1	. 1	. . 1	1 . .	. 1 . .	. 1	. 1	. . . 1 . . .
carrot	. 1	. 1	. . 1	1 . .	. 1 . .	1 .	1 .	1 . . . . . .
vase	. 1	. 1	. . 1	1 . .	. 1 . .	. 1	1 .	. . . . 1 . .
window	. 1	. 1	. . 1	. 1 .	. . 1 .	1 .	1 .	. . . . 1 . .
dog	. 1	1 .	1 . .	. 1 .	. . . 1	. 1	. 1	. . . . . 1 .
wolf	. 1	1 .	1 . .	. 1 .	. . . 1	1 .	. 1	. . . . . 1 .
sheep	. 1	1 .	. 1 .	. 1 .	. . . 1	. 1	1 .	. . . . . 1 .
lion	. 1	1 .	1 . .	. . 1	. . . 1	1 .	. 1	. . . . . 1 .

FIGURE 1. The nouns used in the model and their features. For ambiguous noun constituents, the correct, fully specified reading was used in specifying what the case role representation of the constituent should be, but the underspecified, ambiguous forms were used in the sentence-level input representation. See text for a full discussion.

appropriate missing values in the process of assigning the word to the appropriate case role. Figure 1 indicates both "full" readings of *bat* and *chicken*, as well as the ambiguous forms used as inputs.[2]

Another goal for the model is to show how it can select the contextually appropriate reading of a verb. This is handled in much the same

---

[2] For the concept *food*, which is taken to be the implied Patient in sentences like *The boy ate*, no particular shape seems appropriate. Therefore the intended output representation is assumed to be unspecified (as indicated by the "?") for all values on the shape dimension. For all other dimensions, *food* has what we take to be the typical values for foods.

	DO	CAU	TOUCH	N_CHG	A_MV	P_MV	IN
ate	1 .	1 . .	. . 1 . .	. . 1 . .	. 1 . .	1 . . .	1 .
ateAVP	1 .	1 . .	. . 1 . .	. . 1. .	. 1 . .	1 . . .	1 .
ateAVPI	1 .	1 . .	. . 1 . .	. . 1 . .	. 1 . .	1 . . .	1 .
ateAVF	·1 .	1 . .	. . 1 . .	. . 1 . .	. 1 . .	1 . . .	1 .
broke	1 .	1 . .	. 1 . . .	1	. 1 . .	. . 1 .	. 1
brokeAVPI	1 .	1 . .	. 1 . . .	1	. 1 . .	. . 1 .	. 1
brokeAVP	1 .	1 . .	1 . . .	1	. 1 . .	. . 1 .	. 1
brokeIVP	1 .	. 1 .	. 1 . . .	1	. . . 1	. . 1 .	. 1
brokePV	1 .	. 1 .	. . . 1 .	1	. 1 . .	. . 1 .	. 1
hit	1 .	. . : 1	. 1 . . .	. . . 1 .	. 1 . .	. . 1 .	. 1
hitAVPI	1 .	. . 1	. 1 . . .	. . . 1 .	. 1 . .	. . 1 .	. 1
hitAVP	1 .	. . 1	1	. . . 1 .	. 1 . .	. . 1 .	. 1
hitIVP	. 1	. . 1	. 1 . . .	. . . 1 .	. . . 1	. . 1 .	. 1
moved	1 .	1 . .	1	. . . 1 .	1 . . .	1 . . .	1 .
movedAVP	1 .	1 . .	1	. . . 1 .	1 . . .	1 . . .	1 .
movedAVS	1 .	1 . .	 1	. . . 1 .	1 . . .	1 . . .	1 .
movedPV	. 1	. 1 .	. . . 1 .	. . . 1 .	. . . 1	1 . . .	1 .
touched	1 .	. . 1	. 1 . . .	. . . 1 .	. 1 . .	. . 1 .	1 .
touchedAVPI	1 .	. . 1	. 1 . . .	. . . 1 .	. 1 . .	. . 1 .	1 .
touchedAVP	1 .	. . 1	1	. . . 1 .	. 1 . .	. . 1 .	1 .
touchedIVP	. 1	. . 1	. 1 . . .	. . . 1 .	. . . 1	. . 1 .	1 .

FIGURE 2. The verbs used in the model and their microfeature representations. The forms followed by strings of uppercase letters (e.g., AVPI) represent the alternative feature patterns that the model must choose between as its way of specifying the contextually appropriate reading of the verb. These alternative feature patterns correspond to the semantic features of the verb appropriate for particular configurations of case roles, as indicated by the uppercase letters: A = Agent, V = Verb, P = Patient, I = Instrument, M = Modifier, S = Self, F = implied Food. The position of the letter indicates the position of the corresponding constituent in the input sentence. The patterns given with the generic verb unadorned by uppercase letters were used in the sentence-level, input representations.

way as noun ambiguity resolution. The different readings are represented by (potentially) different sets of semantic microfeatures; for example, the Agent/No-Instrument reading of *broke* (brokeAVP) involves contact between the Agent and the Patient, while the Instrument/No-Agent version (brokeIVP) and the Agent/Instrument version (brokeAVPI) involve contact between the Instrument and the Patient. The input representation of the features of a given verb is the same, regardless of context, and the task given to the model is to activate the set of features for the sentence-appropriate version. Rather than use the average pattern based on all of the different possible readings of the verb, we used a "generic" pattern for each verb, which is the pattern for what we took to be the verb's most typical case frame.

This is indicated in Figure 2 by the pattern of features next to the plain verb.[3]

The feature patterns corresponding to the different case frames the model must choose among are indicated on the lines in the table following its generic pattern. (The labels on these lines are used simply to designate the feature patterns. They indicate the roles the various arguments in the surface structure of the sentence play. Thus, brokeAVPI specifies the case frame in which the surface subject is the Agent, the surface object is the Patient, and the with-NP is the Instrument.) Note that the microfeatures of two different readings of the same verb may or may not differ, depending on whether the features of the scenario do or do not change in different case frames.

The feature vectors for the constituents of the sentence *The boy broke the window with the hammer* are shown just below the corresponding constituents at the top of Figure 3. Note that these are displayed in the order: Verb, Subject NP, Object NP, and With-NP. The row of letters below the feature vectors indicates the first letter of the name of the dimension on which each feature represents a value. For example, the first two elements of the verb feature vector are labeled *d* for the DOER dimension; the first two values of each of the three noun feature vectors are labeled *h* for the HUMAN dimension.

Sentence-structure units. The sentence-structure level representation of an input sentence is not actually the set of constituent feature vectors; rather, it is the pattern of activation these vectors produce over units that correspond to *pairs* of features. These units are called sentence-structure (SS) units.[4]

---

[3] The different handling of nouns and verbs is not a principled distinction, but an exploration of two endpoints on a continuum ranging from underspecification of the input for ambiguous words to complete specification of an input representation, regardless of the fact that the features used in the case role representation will differ as a function of context. Perhaps the idea that the features will be altered as a function of context is the best way of putting things in this case. We imagine that the true state of affairs is intermediate between these two extremes, for both nouns and verbs. In any case, the model does not have any prior commitment to the idea that the features in the input representation should be preserved in the output representation; the full prespecification simply gives the model a fuller description to work from, thereby allowing greater differentiation of the different verbs.

[4] An alternative name for these units would be "surface-structure" units, to indicate that they do not capture the notion of underlying subject, object, etc. However, we have chosen the term "sentence-structure" because, for present purposes, the information they capture is not even a full surface-structure parse of the sentence; in particular, it does not specify the attachment of the with-NP.

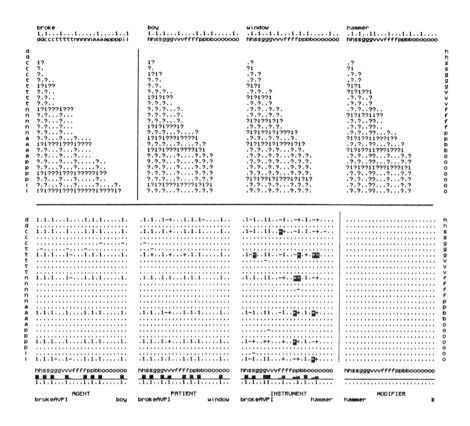

FIGURE 3. The top line of this figure displays the constituents of the sentence *The boy broke the window with the hammer* in the order: Verb, Subject NP, Object NP, and With-NP. Below these are the microfeatures of each constituent, and below these are conjunctive sentence-structure units for each constituent. Below the horizontal line are the blocks of case-structure units for the Agent, Patient, Instrument, and Modifier roles. Below these is an indication of the pattern of noun features the model is activating for each slot (represented by the vertical black bars), followed by a representation of the microfeatures of the correct filler of each slot. The last line gives the label of the correct head (verb frame or modified NP) and tail (slot filler) for each slot. See text for further explanation.

Each SS unit represents the conjunction of two microfeatures of the filler of a particular surface role. Since there are four sentence-structure roles, there are four sets of SS units. Within each set there is a unit that stands for the conjunction of every microfeature value on each dimension with every microfeature value on every other dimension. For example, for nouns there are units for:

HUMAN = yes / GENDER = male
SOLIDITY = hard / BREAKABILITY = fragile

among many others; for the verb, one of the units corresponds to

DOER = yes / TOUCH = instrument

(i.e., there is a doer—the Instrument touches the Patient).

The sentence-structure units are displayed in Figure 3 in four roughly triangular arrays. The verb array is separated from the arrays for the three NPs to indicate that different features are conjoined in the verb and NP representations.

Each array contains the conjunctive units for the constituent immediately above it. There is a unit wherever there is a 1, a "?", or a ".". Within each array, the units are laid out in such a way that the column a unit is in indicates one of the microfeatures that it stands for, and the row it is in indicates the other microfeature. Rows and columns are both ordered in the same way as the microfeature vectors at the top of the figure. The dimensions are indicated by the row of letters across the top of each array and along the left (for the verb units) or right (for the three sets of NP units). Note that the set of units in each array fills less than half of each block for two reasons. First, there are only $[n\,(n-1)]/2$ distinct pairs of n features; second, pairs of values on the same dimension are not included.

We considered various schemes for activating the sentence-structure units. One possible scheme would be to use a strict deterministic activation rule, so that a particular SS unit would be turned on only if both of the features the unit stands for were on in the feature vector. This use of the SS units would allow the model to learn to respond in a finely tuned way to particular conjunctions of microfeatures. However, we wished to see how well the model could function using an inherently noisy input representation. Furthermore, as discussed in Chapter 18, we knew that generalization is facilitated when units that only partially match the input have some chance of being activated. In the present case, we considered it important to be able to generalize to words with similar meanings. Therefore, the SS units were treated as stochastic binary units, like the units used in Chapter 18. Each SS unit received excitatory input from each of the two features that it stands for, and we set the bias and variance of the units so that when both of a SS unit's features were active, the unit came on with probability .85; and when neither was active, it came on with probability .15. These cases are represented in the figure by "1" and ".", respectively. Units receiving one excitatory input came on with probability .5; these units are represented in Figure 3 by "?".

The use of the SS units in conjunction with these particular activation assumptions means that the input representation the model must use as the basis for assigning words to case roles is both noisy and redundant. Each feature of the input is represented in the activation of many of the SS units, and no one of these is crucial to the representation. A drawback of these particular activation assumptions, however, is that they do not allow the model to learn to respond to specific conjunctions of inputs. While the model does well in our present simulations, we presume that simulations using a larger lexicon would require greater differentiation of some of the noun and verb representations. To handle such cases, we believe it would be necessary to allow tuning of the input connections to the SS units via back propagation (Chapter 8) so that greater differentiation can be obtained when necessary. In principle, also, higher-order conjunctions of microfeatures might sometimes be required. Our use of broadly tuned, pair-wise conjunctive units illustrates the *style* of representation that we think is appropriate for the input, but the present version is only an approximation to what we would expect a model with a tunable input representation to build for itself.

Case role representation. The case role representation takes a slightly different form than the sentence-structure representation. To understand this representation, it is useful to drop back to a more abstract viewpoint, and consider more generally how we might represent a structural description in a distributed representation. In general, a structural description can be represented by a set of triples of the form (A R B) where A and B correspond to nodes in the structural description, and R stands for the relation between the nodes. For example, a class-inclusion hierarchy can be represented by triples of the form (X IS-A Y), where X and Y are category names. Any other structural description, be it a syntactic constituent structure, a semantic constituent structure, or anything else, can be represented in just this way. Specifically, the case role assignment of the constituents of the sentence *The boy broke the window with the hammer* can be represented as:

> Broke Agent Boy
> Broke Patient Window
> Broke Instrument Hammer

The constituent structure of a sentence such as *The boy ate the pasta with the sauce* would be represented by:

> Ate Agent Boy
> Ate Patient Pasta
> Pasta Modifier Sauce

In a localist representation, we might represent each of these triples by a single unit. Each such unit would then represent the conjunction of a particular head or left-hand side of a triple, a particular relation, and a particular tail or right-hand side. Our more distributed approach is to allocate groups of units to stand for each of the possible relations (or roles), namely, Agent, Patient, Instrument, and Modifier, and to have units within each group stand for conjunctions of microfeatures of the first and third arguments (the head and the tail) of the triple. Thus, the triple is represented not by a single active unit, but by a pattern of activation over a set of units.

In our implementation, there is a group of units for each of the four relations allowed in the case structure. In Figure 3, the Agent, Patient, Instrument, and Modifier groups are laid out from left to right. Within each group, individual units stand for conjunctions of one microfeature of the head of each relation with a microfeature of the tail of each relation. Thus, for example, Broke–Agent–Boy is represented by a pattern of activation over the left-most square block of units. The unit in the ith row and jth column stands for the conjunction of feature i of the verb with feature j of the noun. Thus all the units with the same verb feature are lined up together on the same row, while all the units with the same noun feature are lined up together in the same column. For the Modifier group, the unit in the ith row and jth column stands for the conjunction of feature i of the modified NP and feature j of the modifier NP. Letters indicating the dimension specifications of the units are provided along the side and bottom edges.

The figure indicates the net input to each case role unit produced at the end of the training described below, in response to the sentence *The boy broke the window with the hammer.* (We will see very shortly how these net inputs are produced.) As before, a 1 indicates that the net input would tend to turn the unit on with probability (p) greater than or equal to .85, and a "." indicates that the net input would tend to turn it on with probability of .15 or less. A "+" indicates that the net input has a tendency to turn the unit on ($.85 > p > .5$), and a "−" indicates that the net input has a tendency to turn the unit off ($.5 > p > .15$).

The correct case-frame interpretation of the sentence is provided to the model by a specification that lists, for each of the four possible case roles, the label corresponding to the head and tail of the role. These are shown below each of the four blocks of case role units. The "#" is used to indicate a null slot filler, as in the Modifier role in the present example. From this it is possible to compute which units should be on in the case role representation. Here we simply assume that all the correct conjunctions should be turned on and all other units should be off.

In this example, the pattern of net inputs to the case role units corresponds quite closely to the correct case role representation of the sentence. The features of *boy* may be seen in the columns of the block of Agent units; the features of *window* in the columns of the block of Patient units; and the features of *hammer* in the columns of the block of Instrument units. The features of the Agent–Verb–Patient reading of the verb *broke* can be seen in the rows of each of these three sets of units. There are no features active in the fourth set of units, the Modifier units, because there is no Modifier in this case. In both the Agent and the Patient slots, the model tends to turn on ($p > .5$) all the units that should be on, and tends to turn off ($p < .5$) all the units that should be off. In the Instrument slot, there are some discrepancies; these are indicated by blackening the background for the offending units. All of the discrepancies are relatively mild in that the unit has either a weak tendency to go on when it should not (+ on a black background) or to go off when it should be on (− on a black background).

Several things should be said about the case-frame representations. The first thing is that the slots should not be seen as containing lexical items. Rather, they should be seen as containing patterns that specify some of the semantic properties assigned by the model to the *entities* designated by the words in the sentences. Thus, the pattern of feature values for the verb *break* specifies that in this instance there is contact between the Instrument and the Patient. This would also be the case in a sentence like *The hammer broke the window*. However, in a sentence like *The boy broke the window*, with no Instrument specified, the pattern of feature values specifies contact between the Agent and the Patient. Thus, the verb features provide a partial description of the scenario described by the sentence. The noun features, likewise, provide a partial description of the players (to use Fillmore's analogy) in the scenario, and these descriptions, as we will see later on, may actually be modulated by the model to take on attributes appropriate for the scenario in question.

Details of Sentence Processing and Learning

The model is very much like the verb learning model (Chapter 18). When a sentence is presented, a conventional computer program front-end determines the net input to each of the sentence-structure units, based on the feature vectors of the words. Each of these units is then turned on probabilistically, as described above. Each surface-structure unit has a modifiable connection to each of the case-structure units. In addition, each case-structure unit has a modifiable bias (equivalent to a connection from a special unit that is always on). Based on the sentence-structure pattern and the current values of the weights, a net input to each case-structure unit is computed; this is just the sum of the weights of the active inputs to each unit plus the bias term. Case-structure units take on activation values of 0 and 1, and activation is a probabilistic function of the net input, as in the verb learning model.

During learning, the resulting activation of each case-structure unit is compared to the value it should have in the correct reading of the sentence. The correct reading is supplied as a "teaching input" specifying which of the case role units should be on. The idea is that this teaching input is analogous to the representation a real language learner would construct of the situation in which the sentence might have occurred. Learning simply amounts to adjusting connection strengths to make the output generated by the model correspond more closely to the teaching input. As in the verb learning model, if a unit should be active and it is not, the weights on all the active input lines are incremented and the threshold is decremented. If a unit should not be active but it is, the weights on all the active output lines are decremented and the threshold is incremented. This is, of course, just the perceptron convergence procedure (Rosenblatt, 1962), whose strengths and weaknesses have been examined and relied upon throughout the book.

SIMULATION EXPERIMENTS

The most important thing about the model is the fact that its response to new inputs is strictly dependent upon its experience. In evaluating its behavior, then, it is important to have a clear understanding of what it has been exposed to during learning. We have done a number of different experiments with the model, but we will focus primarily on one main experiment.

The main experiment consisted of generating a corpus of sentences derived from the sentence frames listed in Table 2. It must be emphasized that these sentence frames were simply used to generate a set of legal sentences. Each frame specifies a verb, a set of roles, and a list of possible fillers of each role. Thus, the sentence frame *The human broke the fragile_object with the breaker* is simply a generator for all the sentences in which *human* is replaced with one of the words on the list of humans in Table 3, *fragile_object* is replaced with one of the words on the list of fragile objects in Table 3, and *breaker* is replaced with one of the words on the list of breakers in the table. It is clear that these generators do not capture all of the subtle distributional

TABLE 2

GENERATORS FOR SENTENCES USED IN TRAINING AND TESTS

Sentence Frame	Argument Assignment
The human ate.	AVF
The human ate the food.	AVP
The human ate the food with the food.	AVPM
The human ate the food with the utensil.	AVPI
The animal ate.	AVF
The predator ate the prey.	AVP
The human broke the fragile_object.	AVP
The human broke the fragile_object with the breaker.	AVPI
The breaker broke the fragile_object.	IVP
The animal broke the fragile_object.	AVP
The fragile_object broke.	PV
The human hit the thing.	AVP
The human hit the human with the possession.	AVPM
The human hit the thing with the hitter.	AVPI
The hitter hit the thing.	IVP
The human moved.	AVS
The human moved the object.	AVP
The animal moved.	AVS
The object moved.	PV

Note: Argument assignments specify the case role assignment of the constituents of a sentence from left to right. A = Agent, V = Verb, P = Patient, I = Instrument, M = Modifier, F = (implied) Food, S = Self.

properties of referents in real scenarios (e.g., the model is completely sex and age neutral when it comes to hitting and breaking things, contrary to reality), and so we cannot expect the model to capture all these subtleties. However, there are certain distributional facts implicit in the full ensemble of sentences encompassed by the generators. For example, all the breakers but one are hard, not soft (only *ball* is coded as *soft* in the feature patterns); only the humans enter as Agents into scenarios involving Instrument use; etc.

The "target" case-frame representations of the sentences were generated along with the sentences themselves. The case role assignments

TABLE 3

NOUN CATEGORIES

human	man woman boy girl
animal	fl-bat li-chicken dog wolf sheep lion
object	ball bb-bat paperwt cheese co-chicken curtain desk doll fork hatchet hammer plate rock pasta spoon carrot vase window
thing	human animal object
predator	wolf lion
prey	li-chicken sheep
food	co-chicken cheese spaghetti carrot
utensil	fork spoon
fragile_object	plate window vase
hitter	ball bb-bat paperwt hatchet hammer rock vase
breaker	paperwt ball bb-bat hatchet hammer rock
possession	ball dog bb-bat doll hatchet hammer vase

are indicated in Table 2 by the sequence of capital letters. These indicate the assignment of arguments from the sentences to the roles of Agent, Verb, Patient, Instrument, and Modifier (of the Patient).[5] Note that there are some sentences that could be generated by more than one generator. Thus, *The boy hit the girl with the ball* can be generated by the generator *The human hit the human with the possession*, in which case the ball is treated as a Modifier of the Patient. Alternatively, it may be generated by the generator *The human hit the thing with the hitter*. In this case, the ball is treated as the Instrument. Similarly, *The bat broke the vase* can be generated by *The breaker broke the fragile_object*, in which case its case-frame representation contains a

---

[5] Two special cases should be noted: For *The human ate*, the case frame contains a specification (F) that designates an implied Patient that is the generic food with unspecified shape, as indicated in the feature patterns displayed in Figure 1. For *The human moved* and *The animal moved*, the case frame contains a specification (S) that indicates that there is an implied Patient who is the same as the Agent (note that the sense of the verb *move* used here involves moving oneself and not one's possessions).

baseball bat serving as Instrument. The same sentence can also be generated by *The animal broke the fragile_object*, in which case, of course, its case-frame representation contains a flying bat serving as Agent.

For the main experiment, we generated all the sentences covered by the generators and then selected eight of each type to use as training sentences. Of these we selected two to be *familiar* test sentences. In addition, we selected two additional sentences from each generator to be used as *novel* test sentences. These sentences were never used to train the model.[6]

The model was given 50 cycles of training with the set of training sentences. On each cycle, each sentence was presented, the model's response to it was generated, and connection strengths were adjusted according to the perceptron convergence procedure.

After the 5th, 10th, 20th, 30th, 40th, and 50th training cycles, the model was tested on both the familiar and the novel test sentences. No learning occurred during the tests, so that the response to the novel test sentences always represented generalization from the training materials rather than the effects of direct experience.

Basic Results

Figure 4 gives a very global indication of the model's performance on both the familiar and the novel test sentences at each phase of testing. The figure indicates the average number of incorrect microfeatures produced by the model as a function of learning trials, for both the familiar and unfamiliar test sentences. There are a total of 2500 case role units, so on the average the model is getting over 95% correct, even with unfamiliar sentences, after the first 5 learning cycles, and is down to about 1% error at Cycle 50. However, these statistics are somewhat misleading, since on the whole, most of the case role units should be off. A more realistic indication of the absolute performance level is provided by the observation that between 56 and 168 of the units should be on in the correct case role representation of each sentence. In general, the errors that the model does make are about evenly distributed between sins of commission (false alarms) and sins of

---

6 Some of the generators (e.g., *The human hit the thing with the hitter*) generate rather large numbers of different sentences (in this case, 756), but others (e.g., *The human ate*, *The predator ate the prey*) generate only very small numbers of sentences (four in each of these cases). The training materials contained four copies of each of two of these sentences so that even here, there were two familiar test sentences and two unfamiliar test sentences.

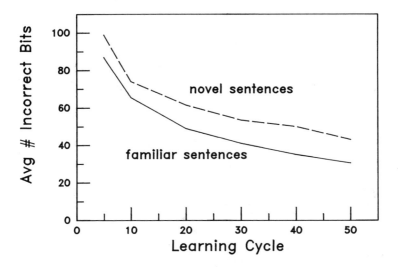

FIGURE 4. Average number of incorrect microfeatures produced as a function of amount of learning experience (in number of cycles through the full list of training sentences).

omission (incorrect rejections). Thus, on average, at the end of 50 cycles, the model is turning on about 85 of the approximately 100 microfeatures that should be on, and is turning on about 15 out of 2400 microfeatures that should be off. This corresponds to a d' of about 3.5.

Two things are apparent in the graph. First, there is a smooth and continuing improvement in performance for both familiar and unfamiliar sentences. Second, there is an advantage for sentences that the model has actually seen before, but it is not particularly great. As we shall see, there is considerable variability in the model's ability to deal with particular sentences; only a part of it is accounted for by whether a particular sentence happens to be old or new.

Figures 5 and 6 indicate the model's performance on each of the 38 familiar and unfamiliar test sentences, at each of the six tests.

In what follows, we will focus attention on the performance of the model at the end of 50 learning cycles.

Use of Semantic and Word-Order Cues

To assess the model's ability to make use of word-order and semantic cues to role assignment, we examined its performance on the verbs *hit*

FAMILIAR SENTENCES

Number of Cycles of Training

Input			Frame	5	10	20	30	40	50
man	ate		AVF	72	38	23	30	20	10
girl	ate		AVF	54	36	23	11	13	9
woman	ate	cheese	AVP	49	32	26	15	12	27
woman	ate	pasta	AVP	62	33	22	14	11	9
woman	ate	co_chic pasta	AVPM	90	66	49	37	38	29
man	ate	pasta co_chic	AVPM	78	47	31	27	25	10
girl	ate	pasta spoon	AVPI	72	63	55	26	19	23
boy	ate	co_chic fork	AVPI	73	47	45	20	26	8
dog	ate		AVF	58	27	13	22	26	12
sheep	ate		AVF	75	44	25	19	15	19
lion	ate	li_chic	AVP	60	48	31	17	19	10
lion	ate	sheep	AVP	52	34	11	21	10	9
woman	broke	window	AVP	65	38	33	21	14	13
boy	broke	plate	AVP	59	56	17	31	20	18
man	broke	window bb_bat	AVPI	86	70	69	36	47	25
boy	broke	plate hatchet	AVPI	101	52	37	35	30	14
paperwt	broke	vase	IVP	76	60	47	42	40	35
bb_bat	broke	plate	IVP*	108	91	94	111	83	47
fl_bat	broke	window	AVP*	107	149	119	105	111	108
wolf	broke	plate	AVP	91	62	44	47	26	30
vase	broke		PV	70	55	33	31	14	21
window	broke		PV	73	41	24	27	16	18
man	hit	pasta	AVP	85	55	34	30	19	16
girl	hit	boy	AVP	77	76	38	41	23	33
man	hit	girl hatchet	AVPM*	138	126	95	83	60	83
man	hit	woman hammer	AVPM*	130	112	79	63	59	69
woman	hit	fl_bat hammer	AVPI*	114	67	87	46	41	33
girl	hit	vase bb_bat	AVPI	147	102	81	56	54	49
hatchet	hit	pasta	IVP	100	70	49	41	34	24
hammer	hit	vase	IVP	103	75	57	55	49	32
man	moved		AVS	83	77	41	35	36	20
woman	moved		AVS	104	68	33	40	24	37
woman	moved	plate	AVP	79	64	49	32	36	31
girl	moved	pasta	AVP	66	57	45	29	29	18
fl_bat	moved		AVS*	122	109	103	71	50	52
dog	moved		AVS	97	64	46	64	45	42
doll	moved		PV	148	103	82	84	90	69
desk	moved		PV	93	86	48	44	50	42

FIGURE 5. Number of microfeature errors for each familiar sentence, after 5, 10, 20, 30, 40, and 50 cycles through the set of training stimuli. Sentences are written in SVOW (W = with-NP) order to facilitate reading; the column labeled "Verb Frame" indicates the role assignments of these arguments. The * indicates a sentence that is ambiguous in terms of the set of generators used, in that it could have been generated in two different ways. Recall that ambiguous words (*bat, chicken*) are ambiguously specified in the input; it is the job of the model to select the contextually appropriate reading.

and *break*. In the case of *break*, constraints are very important; in the case of *hit*, role assignment is sometimes determined by word order alone.

The dog broke the plate, the hammer broke the vase, and the window broke. The verb *break* can take its Instrument, its Agent, or even its Patient as its Subject. In the first two cases (as in *The dog broke the plate* and *The hammer broke the vase*), it is only the semantic properties of the subject that can tell us whether it is Agent or Instrument.

UNFAMILIAR SENTENCES

Number of Cycles of Training

Input				Frame	5	10	20	30	40	50
boy	ate			AV	67	26	34	19	18	6
woman	ate			AV	80	41	26	18	19	18
woman	ate	co_chic		AVP	64	43	31	25	23	19
man	ate	co_chic		AVP	51	49	34	29	21	21
woman	ate	co_chic	carrot	AVPM	121	99	101	70	60	64
boy	ate	carrot	pasta	AVPM	116	86	83	69	64	59
man	ate	co_chic	fork	AVPI	86	66	39	43	32	15
woman	ate	carrot	fork	AVPI	89	68	44	40	32	28
fl_bat	ate			AVF	79	81	43	38	45	43
li_chic	ate			AVF	104	57	62	56	39	46
wolf	ate	sheep		AVP	58	45	44	24	18	14
wolf	ate	li_chic		AVP	63	45	32	27	34	29
fl_bat	broke	vase		AVP*	113	128	133	109	119	136
dog	broke	plate		AVP	68	52	35	28	28	23
girl	broke	plate		AVP	82	52	21	16	14	14
woman	broke	plate		AVP	81	45	27	30	26	15
girl	broke	vase	hatchet	AVPI	96	67	55	32	24	28
man	broke	vase	ball	AVPI	147	91	80	74	58	60
hammer	broke	vase		IVP	86	61	55	44	62	32
ball	broke	vase		IVP	115	83	80	85	72	57
plate	broke			PV	73	60	34	36	29	26
plate	broke			PV	62	47	35	27	23	25
boy	hit	girl		AVP	84	62	31	44	31	36
girl	hit	carrot		AVP	86	68	45	42	35	34
man	hit	boy	hammer	AVPM*	159	126	121	101	96	71
boy	hit	woman	doll	AVPM	127	127	121	93	96	80
girl	hit	spoon	rock	AVPI	152	112	99	94	94	73
girl	hit	curtain	ball	AVPI	178	143	112	117	86	94
paperwt	hit	co_chic		IVP*	115	81	71	74	59	48
rock	hit	plate		IVP	117	97	65	58	58	57
girl	moved			AVS	115	95	73	70	74	57
boy	moved			AVS	101	64	67	53	55	50
girl	moved	hammer		AVP	103	62	56	46	43	42
man	moved	window		AVP	95	78	77	58	58	43
wolf	moved			AVS	101	74	61	60	54	35
sheep	moved			AVS	117	108	79	77	60	64
hatchet	moved			PV	95	76	60	60	67	40
paperwt	moved			PV	101	70	79	59	68	45

FIGURE 6. Number of missed features in output for each unfamiliar sentence, after 5, 10, 20, 30, 40, and 50 cycles of training. See previous caption for conventions.

These two sentences happened to be among the unfamiliar test sentences generated by the generators *The animal broke the fragile_object* and *The breaker broke the fragile_object*.[7] The patterns of activity produced by the model in response to each of these two sentences are shown in Figures 7 and 8. We can see that in the case of *The dog broke the plate*, there is a strong pattern of activation over the Agent units, while in the case of *The hammer broke the vase*, there is a strong pattern of activation over the Instrument units. In fact, the pattern of activation over the Agent units in the first case corresponds closely to the expected target pattern for brokeAVP–agent–dog, and the pattern of

---

[7] The others were *The bat broke the vase* and *The ball broke the vase*. Each of these sentences will be discussed later.

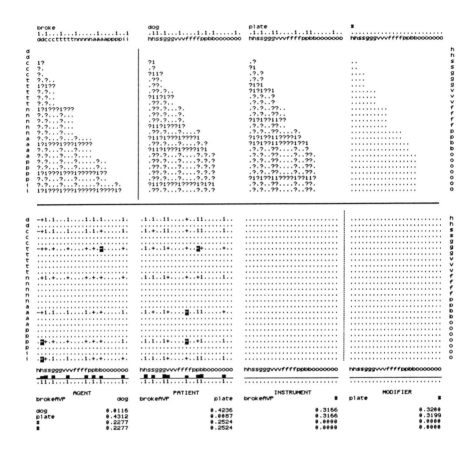

FIGURE 7. Display of the state of the model after processing the sentence *The dog broke the plate.*

activation over the Patient units in the second case corresponds closely to the expected target pattern for brokeIVP–instrument–hammer. Thus, the model has correctly assigned the word *dog* to be Agent in the first case, and the word *hammer* to be Instrument in the second. A summary display that makes the same point is shown below the case role representations in Figures 7 and 8, and is repeated, for these and other sentences, in Figure 9.

The summary display indicates in histogram form the features of the right-hand side of the triple stored in the corresponding block of units. The summary pattern for the Agent role in the sentence *The dog broke the plate* is shown below the block of Agent units in Figure 7. The pattern just below this indicates the "correct" pattern; in this instance, this

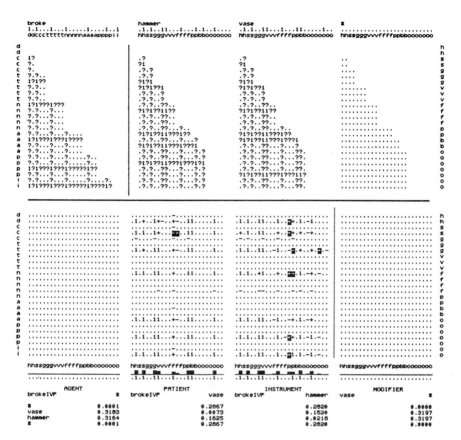

FIGURE 8. Display of the state of the model after processing the sentence *The hammer broke the vase*.

pattern is just the feature pattern for *dog*. In this case we can see that the pattern the model produced is very close to the correct answer.[8]

---

[8] The summaries are based on average probability of activation, with the average only taken over the subset of units associated with a particular noun feature that is also associated with the features of the correct verb frame. Thus, for example, there are 25 units for the feature *nonhuman* in each of the four case slots. These units are displayed as the second column of units in each slot; the ones that indicate whether this feature holds for the Agent are shown in the second column of the Agent units, the left-most block. Seven of these units conjoin this feature with a feature of the verb frame brokeAVP. In the case of *The dog broke the plate*, illustrated in Figure 7, most of these units have a tendency to come on ($.5 < p < .85$), while two (the last two) have a tendency to stay off ($.15 < p < .5$)—a tendency that is considered to be erroneous and results in these units having black backgrounds. The average of these probabilities is about .65. This value is translated into the height of the bar in the second position of the Agent slot for the sentence *The dog broke the plate*.

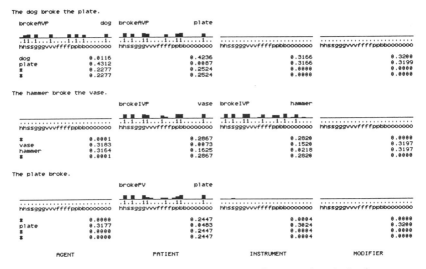

FIGURE 9. Summary of the activations produced on the case role units by the sentences *The dog broke the plate*, *The hammer broke the vase*, and *The plate broke*.

Below each feature-pattern summary, we present in quantitative form a measure of the similarity of all possible nouns from the sentence with the summary pattern for each slot. We also compare the contents of each slot to the "#" pattern, corresponding to "all units off." The numbers (called *deviation scores*) indicate the mean squared deviation of the indicated noun pattern from the summary representation of the contents of each slot. A small deviation score indicates a close correspondence. Thus, if we look at the row of numbers labeled *dog*, we can see again that the pattern over the Agent units (first column) is very similar to the pattern for *dog*. Further, we can see that the pattern for *dog* is not similar to the patterns in any of the other slots (second through fourth columns). The pattern on the Patient units is similar to the pattern for *plate*, and the patterns on the Instrument and Modifier units are similar to the null pattern. For the sentence *The hammer broke the vase*, the pattern on the Instrument units is similar to the pattern for *hammer*, the pattern on the Patient units is similar to the pattern for *vase*, and the patterns on the Agent and Modifier units are similar to the blank pattern. Thus, each argument has effectively been assigned the correct role, and each role has effectively been assigned the correct filler.[9]

---

[9] With respect to the summary pattern on the Instrument slot for *The hammer broke the window*, we can see that the model is trying to activate both values on the POINTINESS dimension. Because *hammer* and *hatchet* are identical except for this feature, the representation is really describing a somewhat pointy hammer (or perhaps blunt hatchet).

This example has been described in some detail, in part to explicate our displays of the model's responses to particular sentences. In the process we have seen clearly that *dog* and *hammer* trigger the appropriate bindings of arguments to slots. This is also the case for examples of the form *The plate broke*. There, Figure 9 indicates that the only slot in which there is appreciable activity is the Patient slot.

These examples illustrate that the model has learned quite a bit about assigning fillers to slots on the basis of the microfeatures of the slot fillers involved. For the word *break*, animate surface subjects are treated as Agents and inanimate surface subjects are treated as Instruments if an Object is specified; if not, the inanimate surface subject is treated as Patient. The model seems to capture this fact pretty well in its behavior.[10]

The boy hit the girl and the girl hit the boy. For the verb *hit*, there is a possibility that a sentence describing an instance of hitting will have two animate arguments, which may be equally plausible candidates to serve as Agent. The only way to tell which is the Agent (in the absence of other context) is to rely on word-order information. We know that *boy* is the Agent in the *The boy hit the girl* only because it occurs in the preverbal position. The model has no difficulty coping with this fact. Figure 10 shows in summary form the features activated by the sentences *The girl hit the boy* (a sentence the model actually experienced during learning) and *The boy hit the girl* (a sentence not experienced during learning). In both cases, the model activates the feature pattern for the correct argument in the correct slot. This is so, even though the feature patterns for *boy* and *girl* differ by a single feature.[11] As a more formal test of the model's ability to assign slot

---

[10] One thing that we see illustrated here is that with certain novel sentences, the model may have a tendency to misgenerate some of their features when assigning them to underlying slots. Thus, for example, in the case of *The plate broke*, only some of the features are produced faithfully. These are, in fact, the features associated with the slot fillers that the model actually learned to deal with in this sentence frame (*vase*, *window*). The ones that are poorly reproduced are the features that the familiar exemplars differ on or which differ between the familiar examples and *plate*. Such errors would be greatly reduced if a more disparate range of Patient-intransitive verbs with a more disparate range of subjects had been used in the learning. Such errors could also be cleaned up quite a bit by an auto-associative network of connections among the case role units. The virtues of augmenting the model with such a network are considered in more detail later.

[11] Though the model did not actually learn the sentence *The boy hit the girl* or any other sentence containing *boy* and *girl* as Subject and Object, it did learn several sentences in which *boy* was the subject of *hit* and several others in which *girl* was the object. As it happened, several of these involved modifiers of *girl*, hence the rather diffuse pattern of activation over the Modifier units.

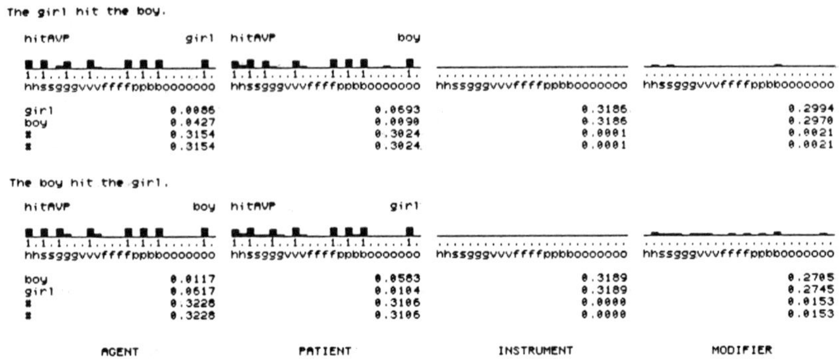

The girl hit the boy.

| hitAVP | girl | hitAVP | boy |

```
■.■..1..1....1.1.■.1......1.   1.1.1....1....1.1.1......1.   ..........................   ..........................
hhssgggvvvffffppbboooooooo    hhssgggvvvffffppbboooooooo   hhssgggvvvffffppbboooooooo   hhssgggvvvffffppbboooooooo

girl    0.0086        0.0693        0.3186        0.2994
boy     0.0427        0.0898        0.3186        0.2970
■       0.3154        0.3024        0.0001        0.0021
■       0.3154        0.3024        0.0001        0.0021
```

The boy hit the girl.

| hitAVP | boy | hitAVP | girl |

```
■■■.■..1..1....1.1.■.1......1.   1.1.1....1.1....1.1.1......1.   ..........................   ..........................
hhssgggvvvffffppbboooooooo     hhssgggvvvffffppbboooooooo    hhssgggvvvffffppbboooooooo   hhssgggvvvffffppbboooooooo

boy     0.0117        0.0583        0.3189        0.2705
girl    0.0617        0.0104        0.3189        0.2745
■       0.3228        0.3106        0.0000        0.0153
■       0.3228        0.3106        0.0000        0.0153
```

| AGENT | PATIENT | INSTRUMENT | MODIFIER |

FIGURE 10. Summaries of of the role assignment patterns produced by *The girl hit the boy* and *The boy hit the girl*.

fillers to slots based only on sentence-structure information, we tested the model on the full set of 12 different *human–hit–human* sentences. In all cases, the preverbal argument more closely matched the pattern of activation in the Agent role, and the postverbal argument more closely matched the pattern in the Patient role.

Verb-Frame Selection

Part of the task of the model is to put slot fillers in the correct places, but there is more that it must do than this. It must also determine from the ensemble of arguments what reading of the verb is intended. By this, we mean which of the possible scenarios the verb might describe is actually being described in this particular case. For example, the sentences *The boy broke the window with the hammer* generates quite a different mental scenario than *The dog broke the window* or *The window broke*. Our model captures the differences between these scenarios in two ways: one is simply in terms of the set of underlying roles and the assignment of sentence constituents to these roles. We have already seen in our earlier discussion of *break* that the model produced a different set of slots and assigned the preverbal noun phrase to a different role in each of the sentences *The dog broke the window*, *The hammer broke the vase*, and *The plate broke*. The other way the model captures the differences between scenarios is in terms of the pattern of activation of verb features in the Agent, Patient, and Instrument slots. Thus in *The boy hit the girl*, we visualize physical contact between the boy and the girl; in the case of *The boy hit the girl with the rock*, we visualize physical contact between the rock and the girl. These and other related distinctions are captured (admittedly, imperfectly) in the different feature patterns associated with the verb frames. As with the feature patterns for nouns, the patterns that we used do not adequately encode the differential flexibility of different aspects of the different scenarios. Nevertheless they capture the essence of the difference, say, between one or the other kind of hitting.

The features of the scenario are captured in the pattern of activation of the case role units. To this point, we have been summing along columns of units to determine the features of the object assigned to a particular role by the model. To determine the features of the action or scenario assigned to the sentence, we need to look along the rows. Figures 3, 7, and 8 indicate that features in somewhat different patterns of rows are activated by the sentences *The boy broke the window with the hammer*, *The dog broke the plate*, and *The hammer broke the vase*. These are, indeed, the correct verb features in each one. Thus, we see that the model has learned to successfully assign a different scenario, depending on the arguments supplied along with the verb.

In general, the model did quite well selecting the correct scenario. For every unambiguous test sentence, familiar or unfamiliar, the value on each scenario dimension that was most active was the correct value.

Filling in Missing Arguments

The model does a very good job filling in plausible default values for missing arguments. To demonstrate this, we tested the model on the sentence fragment *The boy broke*. The results are shown in Figure 11. The model fills in, fairly strongly, a plausible but underspecified fragile object—something that is nonhuman, neuter, nonpointed, fragile, and has object-type *furniture* (*plate*, *vase*, and *window* are all classified as furniture in the model). Values on the size (VOLUME) and shape (FORM) dimensions are very weakly specified.[12]

We see similar kinds of things happening with *The girl ate*, though in this case, the model is actually taught to fill in *food* of unspecified form, so this performance is not surprising. Something slightly

---

12 While the model's response to *The boy broke* clearly illustrates default assignment, it differs from the way many people appear to process this input; several people have commented to us that they read this fragment as a complete sentence specifying that it was the boy that broke, even though this produces a somewhat anomalous reading. The model's response to *The boy broke* is closer to the kind of thing most people get with a verb like *hit*, which does not have an intransitive reading.

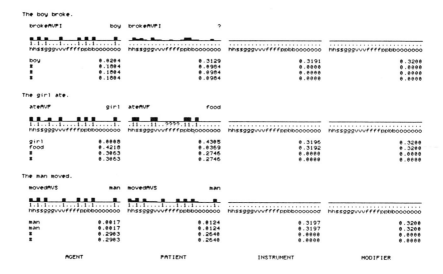

FIGURE 11. Summaries of the model's response to *The boy broke*, *The girl ate*, and *The man moved*.

different happens with *The man moved*: The model is taught to treat *man* as both Agent and Patient in such sentences; and indeed, the pattern for *man* predominates in the Patient slot. (Note that if an object is specified, as in *The man moved the piano*, it is handled correctly.)

These additional examples make two points: First, that the model can be explicitly trained to fill in implied arguments; and second, that the filling in of implied arguments is clearly specific to the particular verb. What is filled in on the Patient slot is quite different for each of these three examples.

Lexical Ambiguity Resolution

In general, the model does very well with ambiguous words. That is, it has little difficulty determining which reading to assign to an ambiguous word based on its context of occurrence—as long as the context is itself sufficient to disambiguate the meaning of the word.

To demonstrate this point, we carried out a number of analyses of the model's responses to sentences containing the ambiguous nouns *chicken* (live or cooked) and *bat* (flying or baseball). We divided the sentences containing these ambiguous words into those that had only one case-frame representation derivable from the generators and those that could have been generated in two different ways. An example of the former kind of sentence is *The chicken ate the carrot*, since only a live chicken could occur as the preverbal NP with *ate*. An example of the latter kind of sentence is *The bat broke the window*, which could be generated either from *The animal broke the fragile_object* or from *The breaker broke the fragile_object*. Given the set of generators, the context specifies which reading is correct for the first kind of sentence but does not specify which reading is correct for the latter. Since our present interest is to see how well the model can do at using context to resolve ambiguity, we focus here on the former type.

Our first analysis simply compared the model's performance with these sentences to its performance with sentences containing no ambiguous words. Since a large number of factors contribute to the number-of-bits-incorrect measure, we focused on a set of seven matched pairs of sentences that were generated from the same generators and were also either both old or both new, but differed in that one contained an ambiguous word (e.g., *The woman ate the chicken with the carrot*) and the other did not (*The boy ate the carrot with the pasta*). The average number of features missed was 28.4 for the items containing ambiguous words and 29.1 for the control items, $F(1,6) < 1$.

Another way to examine the data is to examine the relative strengths of the features of the two readings of an ambiguous word in the output summary representation. Figure 12 indicates a typical case and one interesting exception to the general pattern. In the typical case (*The man ate the chicken with the fork*), we see that the features of the cooked chicken have been strongly activated in the Patient slot, and though there are some weak traces of the features of the live chicken, they are not stronger than the weak extraneous activation we often see

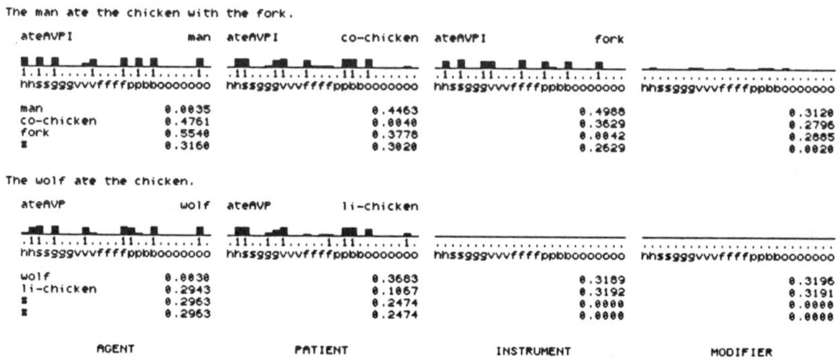

FIGURE 12. Summaries of the case role activations produced by *The man ate the chicken with the fork* and *The wolf ate the chicken*.

of incorrect features for unambiguous words. In the atypical case (*The wolf ate the chicken*), the pattern is really much closer to the cooked chicken than the live one that the model was "supposed" to have retrieved. It is difficult to fault the model too much in this case, however. Though it was never given sentences of the form *The animal ate the food*, it was given sentences of the form *The animal ate*, where the underlying case frame included (implied) *food*. Thus the model had considerable reason to treat the object of *The animal ate* as *food*. Though it had learned *The lion ate chicken* referred to a live chicken, it appears to prefer to treat *chicken* in *The wolf ate chicken* more as cooked food than as a living animal. Some of the properties of the live chicken are weakly present—e.g., its female sex and its natural, animate object-type classification—but the predominant pattern is that of *food*.

With this exception, the model has shown itself quite capable of handling sentences containing ambiguous words as well as sentences containing only unambiguous words. This is, of course, not really surprising in view of the fact that all the words in the sentence are used to help constrain the features assigned to the fillers of every role in the case frame. When one word does not provide the information to constrain the output, the model can exploit information contained in other words.

Structural Ambiguity

In this section, we briefly consider another type of ambiguity that sometimes arises even when the words in the sentence are unambiguous. This is the structural ambiguity of sentences such as *The man hit the woman with the hammer*. In such cases, *hammer* can either be the Instrument or simply a Modifier of *woman*. This case-structure ambiguity parallels an ambiguity in the syntactic structure of the sentence as well; if *hammer* is an Instrument, it is dominated directly by the VP; whereas if it is a Modifier, it is dominated by the NP *the window*. Because *hammer* was included both in the list of possible possession-modifiers of human objects and in the list of possible instruments of hitting (designated as *possessions* and *hitters* in the sentence frames in Table 2 and in the noun categories in Table 3), either of these readings is equally possible. Thus, it is not so surprising that the model has considerable difficulty with such sentences, generating a blend of the two readings.

A particularly interesting case of case-structure ambiguity occurs with the sentence *The bat broke the window*. As already mentioned, the sentence is both lexically ambiguous and ambiguous in case structure, and the structural ambiguity hinges on the lexical ambiguity. If *bat* is a baseball bat then the case frame specifies an Instrument and a Patient, but if it is a flying bat then the case frame specifies an Agent and a Patient.

Figure 13 illustrates the pattern of activation generated in this case. What the model does, quite sensibly, is activate one kind of *bat*—the flying bat—on the Agent units and the other—the baseball bat—on the Instrument units (see the figure caption for a detailed explanation of the use of the black background in this figure).

People generally get only one reading of a sentence at a time, even when (as in this case) the sentence is easy to interpret in either of two ways. In a later section of this chapter we explain how cross-connections among the case role units and back-connections to sentence-level units would tend to cause the model to choose a single interpretation, even in these ambiguous cases.

Shades of Meaning

Another property of the model, related to its handling of ambiguity, is the fact that it can shade the feature patterns it assigns, in ways that often seem quite appropriate. One example of this tendency arises in the model's response to the sentence *The ball broke the vase.* A summary of the pattern produced is shown in Figure 14. The feature pattern on the Instrument units matches the features of *ball* fairly closely. The value on each dimension that is most strongly activated is the correct value, except on one dimension—the hard/soft dimension. On this dimension, we can see that the model has gotten *ball* completely wrong—it has strongly activated *hard*, instead of *soft*.

In one sense, this is clearly an "error" on the model's part; all the balls that it learned about were soft, not hard balls. But in another sense, it is a perfectly sensible response for the model to make. All of the other instruments of breaking (called *breakers* in Table 3) were, in fact, hard. The model picked up on this fact, and shaded its interpretation of the meaning of the word *ball* as a result. As far as this model is concerned, balls that are used for breaking are hard, not soft.

This kind of shading of meaning is just another manifestation of the process that fills in missing arguments, chooses appropriate verb frames, and selects contextually appropriate meanings of words. It is part and parcel of the mechanism that generally results in the activation of the nominally correct feature pattern. It is a mechanism that naturally blends together what it learns into a representation that regularizes slot fillers, in the same way that the verb learning model discussed in Chapter 18 regularizes verbs.

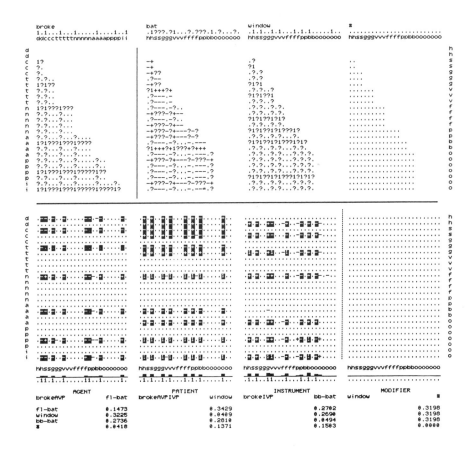

FIGURE 13. State of the model after processing the sentence *The bat broke the window*. For this sentence, the black background on the Agent units shows which units would be active in the pattern for *flying bat* as Agent of the agentive (AVP) reading of *broke*. The black background on the Instrument units indicates the units that would be active in the pattern for *baseball bat* as Instrument of the instrumental (IVP) reading of *broke*. The black background on the Patient units indicates the units that would be active in *either* the pattern appropriate for *window* as Patient of the agentive reading of *broke or* in the instrumental reading of *broke*.

Other Creative Errors

The model made a number of other interesting "errors." Its response to *The doll moved* was a particularly striking example. Recall that the training stimuli contained sentences from the frames *The animal moved*, *The human moved*, and *The object moved*. In the first two cases, the case frame contained the subject as both Agent and Patient, as in *The animal*

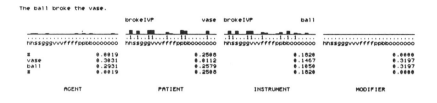

FIGURE 14. Summary of the activations on the case role units for the sentence *The ball broke the vase*.

moved itself. In the third case, the case frame contained the subject only as Patient. The model had some difficulty with these constructions, but generally put inanimate subjects into the Patient role only (as in *The desk moved*), and animate subjects into both the Agent and Instrument role. With *The doll moved*, however, the case-frame representation shows considerable activation in the Agent slot. The pattern of activation there (Figure 15) seems to be pretty much that of a small (though fragile and nonhuman) girl—or perhaps it is simply an animate doll.

Generalization to Novel Words

We have already seen that the model can generalize quite well from particular sentences that it has learned about to new ones that it has not seen before. It is, obviously, important to be able to generalize in this way, since we cannot expect the training set of sentences to cover all possible word combinations in the language, even if they contain only a single clause. Thus far, however, we have only considered generalization to sentences made up of familiar words. What would happen, we wondered, if we tested the model using new words that it had never seen before? To examine this issue, we tested the model on the verb *touch*. *Touch* differs from *hit* in only one feature, namely, intensity; otherwise all of the same verb features are appropriate to it. We assumed that the model already knew the meaning of *touch*—that is, we assumed that the correct input microfeatures were activated on presentation of a sentence containing it. We then took all of the test sentences containing the word *hit*, replaced *hit* with *touched*, and tested the model on each of these sentences. Overall, performance was somewhat worse than with *hit*, but the model was still able to assign arguments to the correct underlying roles. In particular, it had no difficulty assigning animate Subject NPs to the role of Agent and the Object NP to the role

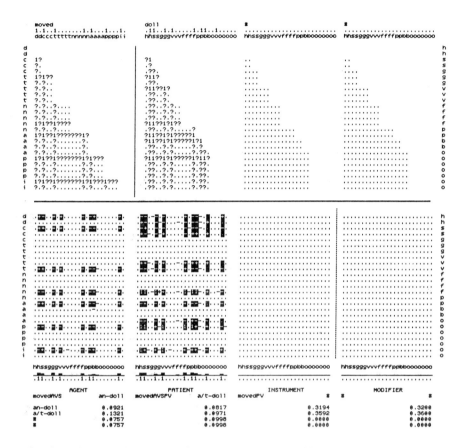

FIGURE 15. State of the model after processing the sentence *The doll moved*. The units with the black background in the Agent slot indicate the units appropriate for the pattern corresponding to an "animate doll" as Agent of the Agent-Verb-Self (AVS) reading of *move*. The units with the black background in the Patient slot indicate the units appropriate for either an animate or (normal) toy doll as Patient of either the AVS or Patient-Verb (PV) reading of *move*.

of Patient, nor did it have any problem assigning inanimate subjects to the role of Instrument, as illustrated in Figure 16.

Two characteristics of the model's performance with this novel verb should be noted. The first is that it does not activate all the correct verb features: The verb features captured in the case role representation are appropriate for *hit* rather than *touch*. There are two reasons for this. One is the redundancy of the representation of the verbs. The input representation we are using is an extremely rich and redundant representation for capturing the very small number of verbs that we have actually used, so the model learns to predict features from other

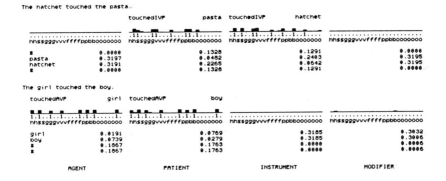

FIGURE 16. Summary of the pattern of activation produced by the model in response to *The hatchet touched the pasta* and *The girl touched the boy*.

features. In a much less redundant set of verbs, the model would be forced to learn to rely on just the right features of the sentence-level pattern to predict the features of the case role representation correctly. The second reason is that the deviant feature on the word *touch* does not fall inside a subspace delineated by a set of related exemplars that behave in similar ways. What we are seeing here, essentially, is assimilation to a single familiar verb that is near the novel verb in feature space. When a novel word is presented that more nearly falls inside the subspace delineated by a set of related exemplars, the model is better able to correctly represent all the features of the novel word in the case role representation. To show this, we defined a new fragile object, a *lamp*, that was nonhuman, hard, neuter, medium-sized, one-dimensional, not pointed, fragile, and furniture. This set of features matches at least one of the breakable objects on every dimension, but matches none of them on all of the dimensions. We then tested the model by taking all of the familiar test sentences containing any of the familiar breakable objects (*window*, *vase*, and *plate*) and then testing the model on these sentences with the word *lamp* in the place of the familiar breakable object. In many cases, as in the first example shown in Figure 17, the model activated all the features correctly. That is, the correct feature for *lamp* on each dimension was more strongly activated than any other. The model was most vulnerable to error on the VOLUME and FORM dimensions, as in the second example shown, where it differed from two of the three familiar fragile objects in each case. The example indicates, then, that the desirable tendency of the model to shade the meanings of the words it encounters to fit the typical features it expects in certain contexts need not lead to severe distortions of novel words, as long as their feature patterns fall inside the

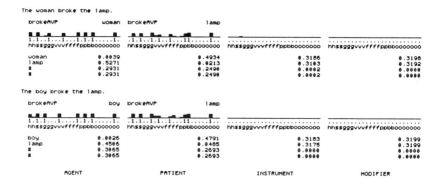

FIGURE 17. Summary of the response of the model to *The woman broke the lamp* and *The boy broke the lamp.*

space spanned by a set of similar words that are treated by the model in a consistent way.

Straightforward Extensions

The model does many things well, but it is really just the first step toward a PDP model of sentence processing. In the general discussion we will consider some of the more general issues that confront the attempt to extend a model of this type further. Here, we will just mention some straightforward extensions that should be fairly easy to implement. We have not yet implemented these changes, so our discussion is somewhat speculative.

One straightforward set of extensions to the model would involve adding cross-connections among the case role units. The cross-connections would allow argument co-occurrence constraints that are verb-frame specific to be captured more effectively than they can now. In the present version of the model, the influence of one argument on another can only essentially add together with the influence of the verb on that argument. This will ultimately prove insufficient, we believe, because different verbs carry with them radically different interdependencies among their arguments. For example, *see* imposes some restrictions between the Instrument and the Object (as in *Bill saw the man with the telescope* and *Bill saw the man with the revolver;* the first is ambiguous, the second is not) that are different from those posed by *hit* (*Bill hit the man with the telescope* and *Bill hit the man with the revolver* seem equally ambiguous). The present version of the model would not easily capture these kinds of verb-contingent argument constraints because each argument is encoded separately from the verb at the sentence level, and it is only connections from units at the sentence level that determine the input to the case role units. This means that conjunctions of noun characteristics with verb characteristics are not directly at work in determining case role unit activations. The case role units, however, do provide just such a conjunctive encoding. It seems likely, therefore, that connections among these units would be able to capture verb-specific (or, more exactly, verb-feature specific) contingencies between slot fillers and their assignments to roles.

Cross-connections among the case role units would add a number of other advantages, as well. They would allow competition among alternative interpretations of the same word at the case-frame level, so that the stronger of two competing interpretations of the same word could effectively suppress the weaker.

A second straightforward extension would be the addition of back-connections from the case role units to the sentence-structure units. This, too, could have several beneficial effects on the performance of the model. In an extended version of the model with cross-connections and back-connections, the computation performed by the present version of the model would be just the first step in an iterative settling process. This settling process could be used to fill in the features of one reading or another of an ambiguous word at the sentence level, based on the emerging pattern at the case role level. Once filled in, these features could then add to the support for the "dominant" reading over other, initially partially activated readings—the whole network would, in effect, drive itself into a stable "corner" that would tend to represent a coherent interpretation at both the sentence and the case role level. Kawamoto (1985) has observed just such effects in a simulation of word disambiguation based on the brain-state-in-a-box model of J. A. Anderson, Silverstein, Ritz, and Jones (1977). (Cottrell, 1985, and Waltz & Pollack, 1985, have also observed such effects in their more localist sentence-processing models.)

Back connections would also allow case role activations to actually specify the semantic features of novel or unfamiliar words occurring in constraining contexts. Consider, for example, the sentence, *The girl broke the shrafe with the feather*. The context provides a considerable amount of constraint on the properties of *shrafe*. The existing version of the model is able to fill in a plausible interpretation at the case level, but with feedback it would be able to pass this information back to the sentence level.

Another way of passing information back to the surface-structure level would be to use the back-propagation learning algorithm. The use of back propagation to train the sentence-structure units could allow the right features to be constructed at the surface level with only a phono-logical representation of the words as the predefined input. Back propa-gation might also allow us to cut down on the rather excessive numbers of units currently used in the surface-structure level. Right now there are far more SS units than are strictly necessary to do the work that the model is doing. While many units could be eliminated in a straightfor-ward way (e.g., many of the sentence units could be eliminated because they stand for unlikely conjunctions of features across dimensions, such as *human and tool*), many more are simply redundant ways of encoding the same information and so could be consolidated into fewer units. On the other hand, for a larger vocabulary, some conjunctive units will turn out to be necessary, and pair-wise conjunctive units ultimately will probably not suffice. Indeed, we feel quite sure that no predefined coding scheme of the kind we have used could provide a sufficient basis for learning all the sentences in any real language without being immensely wasteful, so it will become crucial to train the sentence and case role units to represent just the needed conjunctions of features.

Distributed Representation of Roles

We mention a final "straightforward" extension of the model under a separate heading because it is both more speculative and perhaps some-what more difficult to understand than the previous suggestions. This is the idea of using a distributed representation of the roles. The idea was first suggested by Hinton (1981a), and is currently under explora-tion by Derthick and Hinton at Carnegie-Mellon University. The essence of the idea is to think of roles not as strictly separate, mono-lithic objects, but as sets of role descriptors. Thus the role of Agent has certain properties: It specifies an active participant in the scenario, one that may be volitional; it specifies the instigator of the action. The role of Patient, on the other hand, specifies a passive participant, one whose volitional involvement is (pretty much) irrelevant, but who is the one that experiences the effects of the action.

Various problems arise with treating these roles as unitary objects. One is that some but not all of the Patient properties generally hold for the role nominally identified as Patient. Similarly, some but not all of the Agent properties generally hold for the role nominally identified as Agent. In certain cases, as with sentences like *The boy moved*, enough of these properties hold that we were led to assign *the boy* to both roles at once.

One suggestion that has often been made is to proliferate separate roles to deal separately with each of the slight variants of each of the traditional cases. This leads, of course, to a proliferation of roles that is ungainly, unwieldy, and inelegant, and that detracts considerably from the utility of the idea of roles as useful descriptive constructs.

Here, distributed representation can provide an elegant solution, just as it has in other instances where there appears to be a temptation to proliferate individualized, unitary representational constructs (see Chapter 17). If each role is represented by a conjunction of role properties, then far more distinct roles can be represented on the same set of role units. Furthermore, what the Agent roles of two verbs have in common is captured by the overlap of the role features in the representations of their roles, and how they differ is captured by their differences. The notion of a role that represents a combined Agent/Patient as in *The boy moved* is no longer a special case, and we get out of assigning the same argument to two different slots.

So far, the vision outstrips the implementation of this idea, but we will sketch briefly one very rudimentary instantiation of it, in what really amounts to a rather slight modification of the model we have described above. Currently, our case role units stand for conjunctions of a role, a feature of the verb, and a feature of the filler of the role. The suggestion, quite simply, is to replace these units with units that stand for a feature of a role, a feature of the verb, and a feature of the filler of the role. The first NP in *The boy broke the window with the hammer* will produce a pattern of activation over one set of these triples (corresponding pretty much to the canonical features of agenthood), while *the boy* in *The boy moved* would activate some units from other role feature sets, as well as many of the typical agent feature units.

Again, we stress that we do not yet have much experience using this kind of distributed representations of roles. However, Derthick and Hinton (personal communication, 1985) are exploring these ideas in the context of a PDP implementation of the representation language KL-TWO (Brachman & Schmolze, 1985; Moser, 1983). They have already shown that at least one version of the idea can be made to work and that the coarse coding of roles can be used to allow inheritance of constraints on role fillers.

DISCUSSION

Now that we have examined the model in some detail and considered some possible extensions of it, we turn to more general considerations. We consider three issues. First, we examine the basic principles of operation of the model and mention briefly why they are important and useful principles for a sentence-processing model to embody. Second, we consider some of the implications of the model for thinking about language and the representation of language. Third, we address the limitations of the present model. This part of the discussion focuses on a key question concerning the feasibility of our approach, namely, the requirement that any plausible model of language processing must be able to handle sentences containing embedded clauses.

Basic Features of the Model

We emphasize before we begin that the basic features of the present model are shared with a number of other distributed models, especially those of Hinton (1981a) and those described in Chapters 3, 14, 17, and 18 of this book. The two most important properties of the model are its ability to exploit the constraints imposed by all the arguments in a sentence simultaneously and its ability to represent *shades* of meaning. These aspects are basic, we believe, to any attempt to capture the flexibility and context-sensitivity of comprehension.

The first of these properties is, perhaps, just as easily capturable using local rather than distributed connectionist networks. These local connectionist models capture this property much more effectively than they have been captured in nonconnectionist mechanisms (e.g., Small's, 1980, word expert parser; cf. Cottrell & Small, 1983). Such networks have been applied to sentence processing, particularly to the problems of ambiguity resolution and role assignment (Cottrell, 1985; Waltz & Pollack, 1985). Both models use single units to stand for alternative meanings of words or as "binders" to tie words to alternative roles, and use mutual activation and inhibition to select between alternative meanings and alternative role assignments.

The present model exhibits many of these same properties, but uses distributed representations rather than local ones. What the distributed representations have that local representations lack is the natural ability to represent a huge palette of shades of meaning. With distributed representations, it is quite natural to represent a blend of familiar concepts or a shaded version of a familiar concept that fits a scenario (Waltz & Pollack, 1985, do this with their context microfeatures). Perhaps this is the paramount reason why the distributed approach appeals to us. To be sure, it is possible to represent different shades of meaning in a localist network. One can, for example, have different units for each significantly different variation of the meaning of a word. A problem arises, though, in specifying the meaning of "significantly different." We will probably all agree that there are different readings of the word *bat* in the sentences *The bat hit the ball* and *The bat flew round the cave*. But what about the word *chicken* in the sentences *The woman ate the chicken* and *The wolf ate the chicken*? Or what about the word *ball* in *The baby kicked the ball* and *The ball broke the window*? There is no doubt that we think of different balls in these cases; but do we really want to have a separate unit in memory for the soft, squishy, rubber ball the baby kicks and the small, hard ball that can break a window?

With distributed representations, we do not have to choose. Different readings of the same word are just different patterns of activation; really different readings, ones that are totally unrelated, such as the two readings of *bat* simply have very little in common. Readings that are nearly identical with just a shading of a difference are simply represented by nearly identical patterns of activation.

These properties of distributed representations are extremely general, of course, and they have come up before, particularly in the chapter on schemata (Chapter 14). We also just invoked them in suggesting that we might be able to use distributed representations instead of some fixed set of case roles. In both of these other cases, as in the case of distributed representation of word senses, the use of distributed representation allows for all shades and degrees of similarity and difference in two representations to be captured in a totally seamless way.

A final basic feature of the model is the gradualness of acquisition it exhibits. We have not stressed the time course of acquisition, but it was, of course, a crucial property of the verb learning model, described in Chapter 18, and it is quite evident that acquisition is gradual from Figure 4. As with the verb learning model, our model also seems to pick up on the strongest regularities first. This is seen most easily in Figures 5 and 6 by comparing NVN sentences from the *hit* and *broke* generators. Those with animate preverbal NPs, which are Agents, are learned more quickly than those with inanimate preverbal NPs, which are Instruments. This is because a far greater number of constructions have animate, Agent subjects than have inanimate, Instrument subjects.

These three basic properties—exploitation of multiple, simultaneous constraints, the ability to represent continuous gradations in meaning, and the ability to learn gradually, without formulating explicit rules, picking up first on the major regularities, are hallmarks of parallel distributed models, and they are no less applicable to comprehension of language than they are to any other aspect of cognitive processes.

Do Words Have Literal Meanings?

There is one further aspect of the distributed approach to representation of meaning that should be mentioned briefly. This is the stand our model takes on the issue of whether words have literal meanings. It is normal and natural to think of words as having literal meanings, but it is very difficult to say what these meanings really are. For, as we have noted throughout this chapter, the apparent meanings of words are infinitely malleable and very difficult to pin down. An alternative view is that words are clues to scenarios. This view, which has been proposed by Rumelhart (1979) among others, never made very much impression on us until we began to study the present model. However, in exploring the model, we have found that it embodies Rumelhart's idea exactly. A sentence assembles some words in a particular order, and each provides a set of clues that constrains the characteristics of the scenario, each in its own way. The verb, in and of itself, may specify a range of related scenarios and certain constraints on the players. The nouns further restrict the scenario and further constrain the players. But the words themselves are no longer present in the scenario, nor is there necessarily anything in the scenario that corresponds to the literal meaning of any of the words. Thus in the case of *The doll moved*, the (partially activated) Agent is not a copy of the standard doll pattern, but a pattern appropriate for a doll that can move under its own steam.

The crucial point, here, is that *all* the words work together to provide clues to the case frame representation of the sentence, and *none* of the words uniquely or completely determine the representation that is assigned to any of the constituents of the underlying scenario. Certainly, the word *hammer* most strongly constrains the filler of the Instrument role in *The boy broke the vase with the hammer*, but the other words contribute to the specification of the filler of this role, and *hammer* contributes to. the specification of the fillers of the other roles. Compare *The prisoner struck the rock with the hammer* and *The boy broke the vase with the feather*: The former suggests a heavier hammer; the latter suggests an extremely fragile vase (if we give an instrumental reading to the with-NP).

Toward a More Complete Model of Sentence Processing

As we have already made clear, the model that we have described in this chapter is far from a complete model of the psychological processes involved in sentence processing. It does not deal with the fact that sentence processing is an on-line process, a process that unfolds in real time as each word is heard. It does not deal with the integration of processed sentences into larger contextual frameworks. It does not handle anaphora and other referential phenomena, or tense, aspect, or number. No attempt is made to deal with quantification or scoping issues. The model even lacks a way of distinguishing different tokens with identical featural descriptions. Thus it does not explicitly designate separate dogs in *dog eat dog* and only one dog in *The dog chased himself.* Finally, the model completely ignores the complexities of syntax. For the present model, a sentence can come in only one rigidly structured form, and no embedded clauses, cleft sentences, or passive constructions are allowed.

Clearly, we have much work to do before we can claim to have a model that is in any sense complete or adequate. The question is, can it be done at all? Is there any fundamental limitation in the PDP approach that will prevent the successful development of a full-scale model of language processing that preserves the positive aspects of the distributed approach?

Obviously, the proof is in the pudding. But we think the enterprise can succeed. Rather than discuss all of the issues raised above, we will discuss one that seems quite central, namely, the application of PDP models to the processing of sentences with embedded clauses. We consider several different ways that PDP models could be applied to the processing of such sentences.

Interfacing PDP mechanisms with conventional parsers. We start with what might be the simplest view, or at any rate the most conventional: the idea that a model such as ours might be interfaced with a conventional parser. For example, we might imagine that a parser similar to Marcus's PARSIFAL (1980) might pass off the arguments of completed (or possibly even incomplete) clauses to a mechanism such as the one we have proposed for case role assignment and PP attachment. In this way, the "role-assignment module" could be used with any given sentoid and could be called repeatedly during the processing of a sentence containing embedded clauses.

Interfacing our model with a conventional parser would perhaps provide a way of combining the best of both conventional symbol processing and parallel distributed processing. We are not, however, particularly inclined to follow this route ourselves. For it appears that it will be possible to implement the parser itself as a PDP mechanism. As we shall see, there are at least three ways this might be done. One involves implementing a true recursive automaton in a PDP network. We describe this method first, even though we suspect that the human parser is not in fact such a machine. After describing the mechanism, we will explain our objections to this view of the human sentence-processing mechanism. This will lead us to suggest two other mechanisms. One relies on the connection information distribution mechanism described in Chapter 16 to program a parallel net to process sentences of indefinite length and embeddedness; the other operates iteratively rather than recursively. It is more computationally limited in some respects than the other mechanisms, but the limitations appear to conform to those of the human parser, as we shall see.

A PDP model that does recursion. It turns out that it is not difficult to construct a parallel network that does true recursive processing. Hinton (personal communication) worked out the scheme we will describe here and implemented a rudimentary version of it in 1973. While such a mechanism has not been applied directly to parsing, the fact that recursive processing is possible suggests that there is no reason, in principle, why it should not provide a sufficient basis for implementing some kind of parser.

The mechanism consists of a large network of units. Patterns of activation on these units are distributed representations of a particular state in a processing sequence. Processing occurs through the succession of states. The units are divided up into subnets that represent particular parts of the state. One important subnet is a set of units that provides a distributed pattern corresponding to a stack-level counter.

The connections in the network are set up so that successive states of a routine are associated with their predecessors. Thus, when one state is in place, it causes another to follow it. States may also be used to drive actions, such as output of a line segment, say, if the automaton is a mechanism for recursively drawing figures, as Hinton's was. Processing, then, amounts to carrying out a sequence of states, emitting actions (and possibly reading input) along the way.

Calling subroutines in such a mechanism is not particularly difficult, since all that is required is to associate a particular state (the calling state) with the start state of some routine. Passing parameters to the called routine is likewise not particularly difficult; in the simplest case they can be parts of the calling state that are carried along when the routine is called.

To implement recursion in such a network, all that is required is a way to reinstate the calling state when a routine is done. To do this, the mechanism associates the state of the stack-level units with the state that is in place over the rest of the units, using an associative learning rule to adjust connection strengths while processing is taking place. These associations are implemented by rapidly changing a short-term component of the weights whose long-term values implement the associations that allow the model to cycle through a sequence of states. The temporary associations stored in these short-term weights are not strong enough to overrule the long-term weights, but they are sufficiently strong to determine the next state of the network when the long-term weights leave several possibilities. So, at the end of a routine at stack level N, the network associatively reinstates stack level $N-1$, with the rest of the state cleared. This associative reinstatement of the previous stack-level state would be based on long-term, relatively permanent associations between states corresponding to adjacent depths in the stack. This reinstated stack-level state, which was associated in the short-term weights with the calling state just before the subroutine call occurred, would then simply use this short-term association to reinstate the pattern that existed before the call. (A "done" bit would have to be set to keep the process from doing the call again at this point.)

There is no apparent a priori limit to the number of embedded calls that could be carried out by such a network, though for any fixed size of the stack-level subnet, there will be a corresponding maximum number of associations that can be learned without error. Of course, similar limitations also occur with all other stack machines; the stack is always of finite depth. There would also likely be interference of previous calling states when returning from any particular level, unless the learning were carefully tuned so that earlier associations with a particular stack level were almost completely wiped out or decayed by the time a new one must be used. Care would also be necessary to avoid crosstalk between stack-level representations. However, these problems can be overcome by using enough units so that very different states are used to represent each level of the stack.

Drawbacks of true recursion. The scheme described in the previous section has several fairly nice properties and deserves considerably more exploration than it or some obvious variants have received to date. However, it does have one drawback from our point of view—one that it shares with other, more conventional implementations of truly recursive automata. The drawback is that the calling state is not present and active during the subroutine call; it is effectively inaccessible until it is reinstated after the return.

This property of truly recursive schemes limits their ability to simultaneously consider binding a prepositional phrase at each of two levels of embedding. Consider the sentences:

(14) The boy put the cake the woman made in the kitchen.
(15) The boy saw the cake the woman made in the kitchen.

Our preferred reading of the first of these two sentences has the boy putting the cake in the kitchen, rather than the woman preparing it there; while in the second case, the preferred interpretation appears to be that the woman made the cake in the kitchen, and the boy saw it at some unspecified location. Since the material is the same from the beginning of the embedding in both cases, it appears that the demand the matrix clause material (*The boy put the cake . . .*) makes for a locative argument influences the decision about whether *in the kitchen* should be bound into the subordinate clause. While it may be possible to arrive at these two different readings in a conventional parser by backtracking or by passing aspects of the calling state along when the subroutine is called, it would seem to be more natural to suppose that the matrix clause is actively seeking its missing locative argument as the embedded material is being processed, and so is prepared to steal *in the kitchen* from the verb in the embedded clause. Thus, it appears that a mechanism capable of processing at different levels of embedding at the same time is needed.

A fixed-length sentence processor. A connectionist parser that can, in principle, handle this kind of competition among alternative attachment decisions at different levels of embedding has recently been developed and implemented by Fanty (1985). He describes a mechanism that effectively parses a sentence at many levels at the same time. The parser consists of a fixed network of units. Some of the units represent the terminal and nonterminal symbols of the grammar; other units, called *match units*, represent the different possible expansions of each nonterminal. The symbol units are easily represented in a table in which the columns represent starting positions in the input string and the rows represent lengths. There is a unit for each terminal symbol in each position of the bottom row, and there is a unit for each nonterminal symbol at every position in the table; that is, there is a copy of each nonterminal unit for every possible portion of the input that it could cover. For each of these there is a set of *binder units*, one for each possible expansion of each nonterminal unit.

A simple version of the table, for the indicated three-rule grammar, is shown in Figure 18. Only a subset of the units—the ones that would become active in the parse of *aabbb*—are shown.

The parser can only process strings up to a predefined maximal length. Essentially, it processes the entire sentence in one two-pass processing sweep. In the first, bottom-up pass, all possible constituents are identified, and in the second, top-down pass, the constituents that fit together with the top S and the subconstituents of the top S are reinforced. These active units represent the parse tree.

A very nice feature of Fanty's parse is that it takes into account all levels of the parse tree simultaneously. This allows it to find a globally satisfactory parse in one pair of sweeps, eliminating possible constit%uents that do not fit together with others to make a globally acceptable structure. With some modifications (different degrees of strength for different rules; continuous, interactive processing as opposed to a single pair of sweeps), it would probably be possible to implement a mechanism that could choose the "better" of two alternative acceptable parses, as people seem to do with many ambiguous

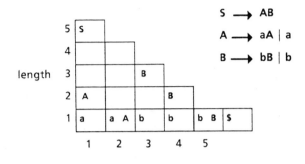

starting position

FIGURE 18. Parsing table used in Fanty's multilevel parser. Only the units active in the correct parse of the string *aabbb* are illustrated. (From *Context-Free Parsing in Connectionist Networks* [TR-174, p. 3] by M. Fanty, 1985, Rochester, NY: University of Rochester, Department of Computer Science. Copyright 1985 by M. Fanty. Reprinted by permission.)

sentences. The parser of Selman and Hirst (1985; Selman, 1985), which is similar to the Fanty parser in structure but uses simulated annealing, appears to have just the right characteristics in this regard.

However, this kind of parser does have some drawbacks. Most importantly, it is limited to sentences of a prespecified length. To expand it, one needs not only to add more units, one needs to program these units with the connections that allow them to do the jobs they are needed or potentially needed to do. Second, the size does grow rather quickly with the allowable length (see Fanty, 1985, for details).

Fanty's model is, in fact, somewhat reminiscent of the TRACE model of speech perception (Chapter 15) in its reduplication of dedicated units and connections. As with TRACE, it may be possible to use connection information distribution (Chapter 16) to program the necessary connections in the course of processing from a single, central network containing the system's long-term knowledge of the rules of English. Indeed, Fanty (1985) has explored an off-line variant of the connection information distribution scheme; his version learns new productions locally, and sends the results to a central network which then distributes the results to the rest of the net, off-line. If the programming of the connections could be made to work on-line in the course of processing, as with the programmable blackboard model of reading, we would have a mechanism that still needs large numbers of units and connections, but crucially, these would be *uncommitted* units and *programmable* connections. The computational capability of the machine would be expandable, simply by increasing the size of the programmable network.

Myths about the limitations of PDP models. We hope that these last two sections will help to dispel some widely held beliefs about computation and PDP models. We will briefly consider two variants of these beliefs. The first, quite simply, is that PDP models cannot do recursion. Hinton's recursive processor needs considerably more development before we will have a working simulation that proves that this belief is wrong, but it seems clear enough that a recursive PDP processor will be available fairly soon. In fact, it is likely that a slightly different approach will be explored more fully first: Touretzky and Hinton (1985) have recently developed a PDP implementation of a production system that can do rudimentary variable binding, and at present it appears that they may be able to extend it to perform recursive computations.

The second belief seems to follow from the first: It is that PDP models are inherently incapable of processing a full range of sentences. We say it only *seems* to follow because it depends upon accepting the assumption that in order to process sentences it is necessary to be able to do recursion. Most people who have thought computationally about sentence processing are familiar with sentence-processing mechanisms that are in fact recursive, such as the ATN (Woods, 1970) or the Marcus parser. Since sentences are recursively defined structures, the (implicit) argument goes, the mechanisms for parsing them must themselves operate recursively.

Fanty's (1985) parser, or an extension of it incorporating connection information distribution, begins to suggest that this may not be so. In a programmable version of Fanty's parser, we would have captured the essential property of a recursive automaton—that the same procedural knowledge be applicable at any point in a parse tree. But we would have captured it in a very exciting new way, a way that would free the mechanism from the serial processing constraint that prevents conventional recursive mechanisms from being able to exploit constraints from many levels at the same time. Connection information distribution may actually permit us to reap the benefits of simultaneous mutual constraints while at the same time enjoying the benefits of being able to apply the same bit of knowledge at many points in processing the same sentence.

There are a couple of caveats, however. One is that connection information distribution is very expensive computationally; a considerable number of units and connections are required to handle the input to and output from the central knowledge structures (see Chapter 12 for further discussion), and the resource demands made by even a fixed version of Fanty's parser are quite high already. There may turn out to be ways of reducing the resource demands made by the Fanty parser. In the meantime, it is worth asking whether some other approach might not succeed nearly as well with fewer resources.

Context-sensitive coding, iteration, and center embedding. There is one more belief about sentences, this one even more deeply ingrained than the last, that we have tacitly accepted up to this point. This is the belief that sentences are indeed recursively defined structures. Clearly, sentences are recursively definable, but there is one proviso: Only one level of center embedding is allowed. It may be controversial whether the "true" competence grammar of English accepts multiple center-embedded sentences, but people cannot parse such sentences without the use of very special strategies, and do not even judge them to be acceptable (G. A. Miller, 1962). Consider, for example, the "sentence":

(16) The man who the girl who the dog chased liked laughed.

The unparsability of sentences such as (16) has usually been explained by an appeal to adjunct assumptions about performance limitations (e.g., working-memory limitations), but it may be, instead, that they are unparsable because the parser, by the general nature of its design, is simply incapable of processing such sentences. [13]

It should be noted that parsers like the ATN and the Marcus parser are most certainly *not* intrinsically incapable of processing such sentences. Indeed, such parsers are especially well suited for handling an indefinite number of center embeddings. Such mechanisms are clearly necessary for processing such things as Lisp expressions, such as this one taken from a program (actually, itself a parser) written by Jeff Sokolov:

```
((and (eq (car (explode arg)) '/ )
      (eq (car (reverse (explode arg))) '/ ))
 (implode (reverse (cdr (reverse (cdr (explode arg)))))))
```

---

[13] This argument has been made previously by a number of other authors, including Reich (1969). In fact, Reich proposed an iterative approach to sentence processing that is similar, in some respects, to the one we consider here.

But sentences in natural language are simply not structured in this way. Perhaps, then, the search for a model of natural language processing has gone down the garden path, chasing a recursive white rabbit.

One approach to sentence processing holds that we should ask much less of a syntactic parser and leave most of the work of sentence processing to a more conceptual level of processing. This position is most strongly associated with Schank (1973), and Charniak (1983) is among the recent proponents of this view.

Let us assume that the job of the parser is to spit out phrases encoded in a form that captures their local context, in a way that is similar to the way the verb learning model, described in Chapter 18, in a form that captures their local context.[14] Such a representation may prove sufficient to allow us to reconstruct the correct bindings of noun phrases to verbs and prepositional phrases to nearby nouns and verbs. In fact, we suspect that this kind of local context-sensitive encoding can capture the attachments of NPs to the right verbs in "tail-recursive" sentences like

(17)   This is the farmer that kept the cock that waked the priest that married the man that milked the cow that tossed the dog that worried the cat that killed the rat that ate the malt that lay in the house that Jack built.[15]

Locally context-sensitive coding will begin to break down, however, when there is center embedding—specifically, more than one level of center embedding. Local context can be used to signal both the beginning and the end of an embedding, but cannot be used to specify which beginning of an embedding the material just after the end of an embedding should be bound to. Thus if we read to the last word of

(18)   The man who the girl who the dog chased laughed

we may not know whether to bind *laughed* (just after the end of the embedding) to *the man* (NP before an embedding) or to *the girl* (a different NP before an embedding). If we bind it to *the man*, we may

---

[14] For this to work it will be necessary to code units, not in terms of the adjacent words, but in terms of neighboring constituents more abstractly defined. Thus, in *The girl in the hat saw the mouse on the floor*, we will want to encode the complex NP *the girl in the hat* as adjacent to the verb *saw*. Thus, local context will have to be defined, as it is in Marcus (1980), in terms of constituents, not merely in terms of words. Getting this to work will be one of the major challenges facing this approach.

[15] Adapted from E. Johnson, E. R. Sickels, & F. C. Sayers (Eds.), *Anthology of Children's Literature* (4th ed., p. 16), 1970, Boston: Houghton-Mifflin.

experience a false sense of closure—this sentence is ungrammatical because it has only two verbs for three clauses.

These suggestions lead us to the following conjecture: It may be possible to build a PDP language-processing mechanism that works iteratively along a sentence, building constituents represented by distributed patterns conjunctively encoded with their local context of occurrence. The mechanism would need a way of unifying constituents on two sides of an intervening embedding. Exactly how this would be done remains to be established, but as long as it is done in terms of a mechanism sensitive only to the local context of the constituents before and after the embedding, it may succeed where there is a single embedding but fail in multiply embedded sentences where there are two suspended, incomplete constituents that the mechanism must choose between completing.

We hope it is clear that these ideas are speculative and that they are but pointers to directions for further research. Indeed, all three of the directions we have described in this section are only just beginning to be explored systematically, and it is unclear which of them will prove most attractive on closer scrutiny. We mention them because they suggest that ways of overcoming some of the apparent limitations of PDP mechanisms may not be very far beyond our present grasp, and that it may soon be possible to retain the benefits of parallel distributed processing in mechanisms that can cope with the structural complexity and semantic nuance of natural language.

ACKNOWLEDGMENTS

This work is supported by ONR Contract N00014-82-C-0374, NR 667-483, by a Research Scientist Career Development Award (MH00385) to the first author, and by grants from the Sloan Foundation and the System Development Foundation. We would like to thank Liz Bates, Gene Charniak, Gary Cottrell, Gary Dell, Mark Derthick, Mark Fanty, Geoff Hinton, Ron Kaplan, Brian MacWhinney, Dave Rumelhart, Bart Selman, and Michael Tanenhaus for useful comments and discussions.

PART V

BIOLOGICAL MECHANISMS

One of the appeals of parallel distributed processing is the fact that it seems closer to the neural basis of cognition than most other approaches to cognitive processes. The idea that intelligent processing can emerge from the interactions of a large number of simple computational units and their interactions is, of course, directly inspired by what we know about the way the brain works. Further, several of the specific assumptions embodied in particular models have been inspired by observed characteristics of neural function.

This part of the book explores the neural mechanisms underlying parallel distributed processing in three different, though interrelated ways. First, it examines what we know and do not know about relevant aspects of the brain. Second, it considers the precise nature of the relation between concepts at a neural level of analysis, and concepts at higher, more cognitive levels. Third, it presents three specific attempts to bring PDP and neuroscience closer together by capturing aspects of neural function in simulation models based on the PDP framework laid out in Chapter 2.

The relevant neurophysiology. There is, of course, a vast body of data about the details of human brain function, and it would be hopeless to try to summarize all of this information within the confines of our book. However, among these data are some emerging principles and observations about the characteristics of the mammalian brain that seem particularly relevant to parallel distributed processing.

In Chapter 20, Crick and Asanuma describe the principles and observations they see as most salient, based on their ongoing survey of the neurophysiology of the mammalian brain. The focus is on the details of neuronal interaction and on the developing state of knoweldge about the regional architecture of the visual system, as revealed through physiological and anatomical experiments, primarily in primates. The material in this chapter is useful background for any PDP modeler, and will be of particular interest to anyone interested in modeling the detailed properties of real neural circuitry.

In Chapter 21, Sejnowski takes a somewhat different approach, and considers several general questions about brain function. By and large, neurophysiological data do not answer such questions in a definitive way, but they do suggest several principles of brain function that should continue to help steer our efforts to capture the essential features of the computational processes that take place in the human brain.

The relation between PDP models and the brain. There are several different ways in which PDP models relate to the brain. Different pieces of work represented in this book have different relationships to the details of brain structure. All share in common that they are "neurally inspired" models and are explorations of "brain-style" processing. Beyond this, however, there are some important differences. Here, we briefly outline some of these different approaches, and say how they relate to the chapters in Part V.

Some PDP models are intended to explore the computational properties of "brain-like" networks. The models described in Parts II and III are generally of this type. They consider sets of units with specified characteristics, and study what can be computed with them, how many units are required to do certain computations, etc. These models focus on parallel processing mechanisms more or less in their own right, quite apart from facts about the details of brain structure or the details of human behavior and cognition.

There are two principle kinds of motivations for this computational type of work. One is that the mechanisms under study are sufficiently brain-like that they may shed light on the actual way in which computation is done in the brain. The fact that these models generally idealize various properties of the brain may in many cases be a virtue for understanding brain function, since idealization, as Sejnowski points out in Chapter 21, can often facilitate an analysis that leads to deeper understanding of the emergent properties of complex systems. Indeed, some of this work is driven by the goal of exploring the implications of specific properties of brain function, such as the stochastic nature of neural firing, as explored in harmony theory (Chapter 6) and in Boltzmann machines (Chapter 7). The other motivation is to explore the computational capacities of networks that appear on their surface to be well suited to certain kinds of information-processing tasks, such as search or representation building.

Other PDP models—and here we have in mind primarily the models in Part IV—attempt more directly to build accounts of human information processing capabilities, at a level of abstraction somewhat higher than the level of individual neurons and connections. In such models, the relationship between particular brain structures and particular elements of the models is not generally specified, but since the models attempt to capture the behavioral products of the activity of human brains, it is assumed that there is some relationship to real activity in the brain. The basic idea is that there is a mapping between elements of the model and the brain, but it is unknown and probably only approximate. A single unit may correspond to a neuron, a cluster of neurons, or a conceptual entity related in a complex way to actual neurons.

It might be thought that by adopting this more abstract approach, these models lose all contact with the underlying neurophysiological mechanisms. This is not the case. While models of cognitive processes may be developed without detailed regard for the underlying physiology, some of the characteristics of the brain clearly place constraints on the cognitive mechanisms. Some examples are the speed and precision of the basic computational elements, the general characteristics of their patterns of interconnection, the nature of the operations they can perform, the number of elements available in the brain, etc. (Chapter 4 provides a fuller discussion of these points). It becomes important, then, to develop some way of relating the more abstract, cognitive-level theory to the underlying neurophysiology. More fundamentally, this relation is central to conceptions of the relation between mind an brain. It is therefore of considerable importance to have an explicit theoretical framework for conceptualizing the exact nature of this relation. This point is addressed by Smolensky in Chapter 22.

In still other cases, the goal is to do neural modeling—to account for the facts of neuroscience rather than (or in addition to) the facts of cognition. This use of PDP models involves less idealization of the neural elements and more attention to the details of brain structure. Crick favors these applications for he feels that building up from the facts about real neural function is the best way to find out the way things really work—as opposed to the way things might work—in the brain. The last three chapters of Part V take this approach.

We hasten to add that we do not think any one of these uses is the "right" or "only" way to employ PDP models. Rather, we believe that work at each of these levels complements and reinforces work at the other levels, and that work at all of these levels will eventually allow us to converge on an understanding of the nature of the cognitive processes that the brain supports and the neural mechanisms underlying these processes. We believe that the use of the common PDP framework for all of these applications will facilitate this process.

PDP models of neural mechanisms. Chapters 23 through 25 each take their own approach to modeling information-processing activity in the brain. Chapters 23 and 24 are explicit attempts to develop models that capture aspects of what is known about the behavior of neurons, while Chapter 25 focuses on neuropsychological data.

In Chapter 23, Zipser takes as his goal the development of biologically plausible models of place learning and goal localization in the rat. The goal of the models is to account for localization and place learning behavior, and at the same time, incorporate knowledge gained from single-unit recording experiments. The first model described in the chapter accounts for the behavior of so-called "place-field" cells in the hippocampus of the rat, and is closely tied to the physiological data. Two other, more speculative models work toward an account of goal localization, about which less of the physiology is known.

In Chapter 24, Munro considers the plasticity of neural mechanisms, as revealed through studies of single units in visual cortex after various schedules of visual deprivation and other forms of intervention. He shows how a very simple neural model that focuses on the plasticity of individual units can account for much of the data on the critical period. The account is based on the simple observation that changes in the connections of a neuron will make more difference if its prior connections are weak than if they are strong. The critical period is seen, on this account, as a simple manifestation of the natural consequences of the strengthening of connections through experience, and not as a manifestation of some sort of preordained maturational process that turns off plasticity.

Chapter 25 is not concerned with modeling data on the behavior of individual neurons; rather, it is concerned with reconciling neuropsychological evidence about amnesia with distributed models of memory. In distributed models, such as the one described in Chapter 17, information of different ages is stored in superimposed form, in the same set of connections. This fact provides a natural way of accounting for one aspect of amnesia: the fact that amnesics exhibit the residual ability to learn, gradually, from repeated experiences, even though their memory for individual episodes is extremely weak. Distributed memory, however, seems incompatible with another aspect of amnesia: namely, the temporally graded nature of the retrograde amnesia—the loss of prior information—that accompanies the reduction in the capacity to learn new material. If all memories are stored in the same set of connections, why should more recent ones be more susceptible to loss or disruption than older ones? Chapter 25 reports simulations of various aspects of anterograde and retrograde amnesia, based on one possible answer to this question.

CHAPTER 20

Certain Aspects of the
Anatomy and Physiology of the Cerebral Cortex

F. CRICK and C. ASANUMA

Our aim in this chapter is to describe some aspects of our present knowledge of the anatomy and physiology of the cerebral cortex of higher animals which may be of interest to theorists. We shall assume that readers have at least an elementary knowledge of this subject, so that they know, for instance, about the structure of neurons and the basis of neuronal excitation and synaptic transmission. The text by Kandel and Schwartz (1981) could be used to cover this background knowledge.

It is clearly impossible to describe most of what is known, even though this represents a tiny fraction of what one would like to know. We shall select examples to illustrate the general points we want to make. It will soon emerge that while some things are known with reasonable certainty, much is unknown or, even worse, surmised only on rather incomplete evidence. For this reason alone, the object of this chapter is not to dictate to theorists what "units" they must use in their modeling. It might turn out that theory will show that a particular process, or implementation of a process, gives a very advantageous performance, even though the experimentalists can, as yet, see no sign of it. The wise thing at that point would be to look for it experimentally, since it may have been overlooked for one or another technical reason. This aside, theorists should at least try to learn whether the features they wish to use for their implementation do actually occur in the relevant part of the brain, and they should be duly cautious if the experimentalist can see no trace of them. Whether a theorist's unit can be a group of neurons is discussed later.

One other general point should perhaps be stated at the outset. Different parts of the brain are "wired" in radically different ways. It is thus not sensible to take one feature from, say, the olfactory bulb, another from the thalamus, and a third from the cerebellum, and combine them all together to account for a task that the cortex is expected to perform. Wherever possible, therefore, we shall choose examples from the mammalian cerebral cortex, both because so much work has been done on it and also because the problems theorists choose are often taken from aspects of human behavior that are mediated by the cerebral cortex. Excursions to other parts of the nervous system, such as the retina, cerebellum, and the olfactory bulb, will be made only when necessary to clarify certain points. Figure 1 illustrates a human brain and demonstrates the general location of some of its internal structures in relation to the cerebral cortex. However our aim in this chapter is not to describe the cerebral cortex as fully as possible, as one would need to do if one were concerned with its detailed workings, but

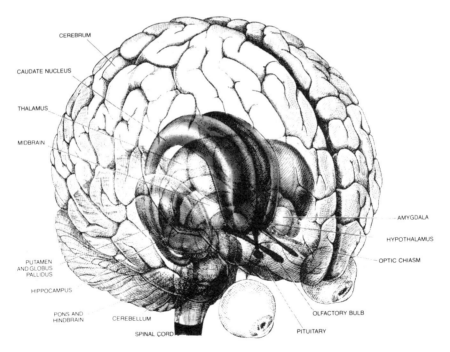

FIGURE 1. The human brain. The cerebral cortex is depicted transparently in this drawing so that some of the internal brain structures are visible. (From "The Organization of the Brain" by W. J. H. Nauta and M. Feirtag, 1979, *Scientific American*, *241*, p. 102. Copyright 1979 by W. H. Freeman & Co. Reprinted by permission.)

merely to point out certain features of the cortex which should not be overlooked in theoretical modeling.

Another general point that should be made is that in many cases theorists choose problems associated with either language or the human visual system without recognizing that there is at least one important difference between them. Put briefly, there is no animal model for language, nor is it possible to carry out many types of experiments on the language centers of the human brain for obvious ethical reasons. Most of the really useful new methods used in neuroanatomy, such as tritiated amino acid autoradiography, horseradish peroxidase histochemistry, and metabolic mapping with [^{14}C] deoxyglucose can only be used effectively on animals. We are in the embarrassing position of knowing a lot about the neuroanatomy of the macaque monkey while having only a very limited amount of similar information about the human brain. Similarly, the most powerful neurophysiological technique—the use of microelectrodes for isolating the electrical activity of single neurons (or small groups of neurons)—is not suited for extensive use on humans. This disadvantage is partly offset by the greater ease with which human psychophysical experiments can be done. There are also a number of techniques which can be used to study aspects of the neural activity from the outside. These include position emission tomography (PET scanning), magnetic field detectors, electroencephalography, (EEG) and scalp recordings of evoked potentials. Unfortunately either the spatial or the temporal resolution of these methods is usually inadequate, and, as a result, the interpretation of the results is often not clear cut.

In the long run, a theoretical model in biology can only be validated by a *detailed* comparison with experiment. All psychophysical tests show that the performance of the visual system of the macaque monkey is roughly comparable to our own. From this point of view, therefore, the solutions of visual problems should be easier to bring down to earth than linguistic ones. This does not mean that linguistic problems may not suggest valuable ideas about the working of the brain. It does mean that they may be more difficult to test at the level of neuronal organization and function.

The Neuron

The "classical" neuron has several dendrites, usually branched, which receive information from other neurons and a single axon which outputs the processed information usually by the propagation of a "spike" or an "action potential." The axon ramifies into various branches that make synapses onto the dendrites and cell bodies of other neurons.

This simple picture (Figure 2A) has become complicated in several ways: (For a more thorough, yet general account, see the book by Shepherd, 1979.)

- A neuron may have no obvious axon but only "processes" that seem to both receive and transmit information (Figure 2B). An example of such neurons is the various amacrine cells found in the retina (Cajal, 1892). Although neurons without axons also occur in the olfactory bulb (Cajal, 1911), they have not been convincingly demonstrated in other parts of the nervous system.

- Axons may form synapses on other axons. In the cerebral cortex these synapses have been found only upon the *initial* segments of the axons of certain cells (Figure 2C) (Peters, Proskauer, & Kaiserman-Abramof, 1968; Westrum, 1966).

- Dendrites may form synapses onto other dendrites (Figure 2D). Examples of this are known in the retina (Dowling & Boycott, 1966), the olfactory bulb (Rall, Shepherd, Reese, & Brightman, 1966), the thalamus (Famiglietti, 1970), the superior colliculus (R. D. Lund, 1972), and the spinal cord (Ralston, 1968), but such contacts appear to be rare or absent in the cerebral cortex.

- An axon may not propagate a spike but instead produce a graded potential. Because of attenuation, we should expect this form of information signaling not to occur over long distances, and indeed it is found largely in such places as the retina, where the distances between connected neurons are shorter than in many other neural tissues; possibly because the time requirements are different (Figure 2E) (Werblin & Dowling, 1969). It is also conceivable that graded potentials occur at more local levels (Figure 2F). For example, an axon terminal forming a synapse on a given cell may itself receive a synapse. The presynaptic synapse may exert only a local potential change which is therefore restricted to that axon terminal. (The existence of this sort of a mechanism has been suggested for the spinal cord [Kuno, 1964] and the thalamus [Andersen, Brooks, Eccles, & Sears, 1964], but to date, no examples of this arrangement have been reported in the cerebral cortex.)

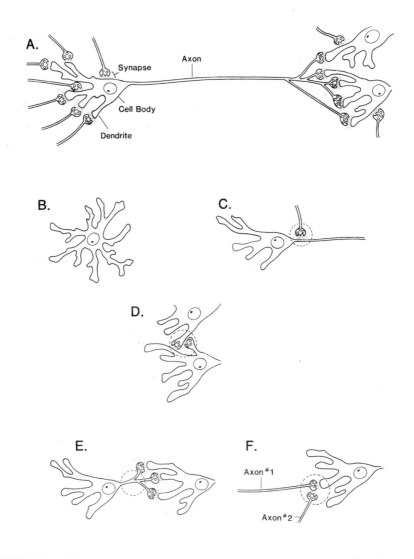

FIGURE 2. Highly schematized diagrams of the "classical" neuronal profile (*A*) and
some of its variants (*B-F*). *A*: The "classical" neuron receives synapses on its dendrites
and generates action potentials which travel down the axon. The axon subsequently
branches and forms synapses on the dendrites and cell bodies of other neurons. *B*: There
are neurons with no obvious axons. *C*: The initial segments of axons of neurons in the
cerebral cortex may receive synapses. Note the highly strategic position of this kind of
synapse. *D*: Dendrites forming synapses directly onto the dendrites of other neurons
occur in the olfactory bulb and the thalamus. *E*: Graded potentials (instead of action
potentials) can be effective if the axon is short. *F*: Graded potentials can also be effective
at local levels. Here, Axon #2 can modulate the efficacy of the synapse formed by Axon
#1 by producing a local potential change in the terminal of Axon #1.

Synapses

The great majority of synapses in the cerebral cortex are chemical, not electrical. A small star-shaped neuron (stellate cell) may receive a few hundred synapses, a small pyramid-shaped neuron (pyramidal cell) some thousands, and a large pyramidal cell some tens of thousands of synapses. Despite the large and variable number of synaptic contacts present upon neurons in the cerebral cortex, most synaptic contacts can be classified morphologically into two basic types (see Figure 3) (Peters, Palay, & Webster, 1976):

- *Type I.* These synapses have asymmetrical membrane specializations (the membrane thickening is greater on the postsynaptic side), and the presynaptic process contains fairly large (ca. 50 nm), round synaptic vesicles—believed to contain quanta, or packets of neurotransmitter. The synaptic cleft is usually about 30 nm across.

- *Type II.* These have symmetrical membrane specializations. The synaptic vesicles are smaller and, with the usual fixatives used for electron microscopy, are often ellipsoidal or flattened. (The shape of the vesicles depends on the details of the fixation and is not always a completely reliable criterion when comparing results reported by different investigators.) The synaptic cleft is usually 20 nm across and the zone of apposition is usually smaller than that of the Type I synapse.

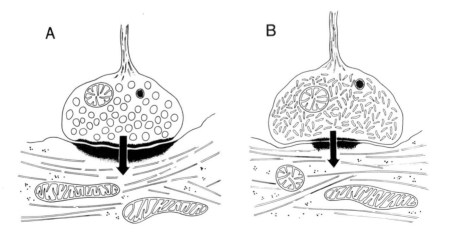

FIGURE 3. Idealized diagrams of a Type I (*A*) and a Type II (*B*) synapse. See text for explanation.

The importance of the classification into the two morphological types (originally recognized by Gray, 1959) is that *Type I synapses seem to be excitatory, whereas Type II synapses seem to be inhibitory.* We should add that the foregoing statement, though generally accepted by experimentalists, has not been systematically tested. In systems where the excitatory or inhibitory nature of a given synapse is well established, this correlation of morphology with physiology appears to be absolute (Uchizono, 1965).

There is another possible criterion for determining the character of synapses: This is the transmitter they use. In general, one is apt to assume that a given transmitter (or apparent transmitter) will usually do the same thing in different places, though there are well-established exceptions (depending on the nature of the postsynaptic receptors). Glutamate and aspartate always seem to excite, GABA (gamma-amino butyric acid) always seems to inhibit (Krnjevíc & Phillis, 1963). (It may come as a surprise to the reader to learn that for most cells in the brain we do not yet know what neurotransmitter they use.) The identity of the transmitters is usually determined immunocytochemically. Thus, an antibody staining for the enzyme glutamic acid decarboxylase (GAD), which is necessary for the production of GABA, can be used to identify some of the inhibitory synapses in that tissue.

Various other methods have been used to identify possible neurotransmitters, for example: injecting the putative transmitters on to neurons while recording from them, microassays to determine their level in the tissue, labeling of high affinity uptake systems, etc. Each technique has limitations on what it can show. At the moment it is difficult to identify the transmitters involved and their postsynaptic effects at most synapses in the central nervous system. That said, we can make a tentative list of possible generalizations about synapses, although most of them are only supported by our ignorance:

- No axon makes Type I synapses at some sites while making Type II at others.

- No axon in the mammalian brain has been shown to release two different *nonpeptide* neurotransmitters. (But it seems likely that many neurons, including cortical neurons, may release a "conventional" transmitter and a neuropeptide, or in some cases two or more neuropeptides.)

- There is no evidence *so far* in the mammalian brain that the same axon can cause excitation and inhibition at different synapses, but this is certainly possible since the effect of a given transmitter ultimately depends on the kinds of receptors present and their associated ion channels.

Peptides

A remarkable discovery over the last ten years or so has been the existence of many distinct peptides, of various sorts and sizes, which can act as neurotransmitters (see Iverson, 1984, for review). There are, however, reasons to suspect that peptides are different from more conventional transmitters such as acetylcholine or norepinephrine:

- Peptides appear to "modulate" synaptic function rather than to activate it by themselves.

- The action of peptides, in the few cases studied, usually appears to come on slowly and to persist for some time. That is, for times up to seconds or even minutes rather than for a few milliseconds or less as is the case for conventional transmitters.

- In some cases it has been shown that peptides act not at their place of release but at some distance away. This distance may be perhaps some tens of micra or further if carried by a vascular system (as in the path from the hypothalamus to the pituitary). Diffusion takes time. The slow time of onset would be compatible with the possible time delays produced by diffusion.

- There are many examples now known of a single neuron producing (and presumably releasing) more than one neuropeptide.

It has been argued that peptides form a second, slower means of communication between neurons that is more economical than using extra neurons for this purpose. Different peptides are used in the same tissue to enable this communication to have some degree of specificity. (We should remark that so far very little is known about either the *receptors* for peptides or the physiological role of most neuropeptides.)

THE CEREBRAL CORTEX

We shall assume that the reader has some familiarity with the structure of the cerebral cortex and with the behavior of the neurons it contains. For an excellent review of the functional architecture of the primary visual cortex, see Hubel and Wiesel, 1977. This section aims to expand that knowledge. We shall not deal here with the non-neuronal cells in the cortex (the glial cells, which may outnumber the neurons by 10-50 times) nor with its blood supply, though both these topics are of considerable practical and clinical importance.

The cerebral cortex is conventionally divided into the allocortex (comprising olfactory and limbic cortical areas) and the phylogenetically more recent neocortex, which is all the rest. We shall be concerned almost exclusively with the neocortex, the extensive development of which is characteristic of the mammalian brain, especially the behaviorally more interesting primates.

General Organization

The neocortex consists of two distinct sheets of neurons, one on each side of the head. Each sheet is relatively thin (typical thicknesses run from 1.5 to about 5 mm) and continuous. This is illustrated in Figure 4. Although it is highly convoluted in most of the larger mammals, the neocortex has no slits in it and, as far as we know, no insulating barriers within it. Since it is a continuous finite sheet, it must have an edge. This edge is surrounded by allocortical areas and by various non-cortical structures. The sheets on either side of the head are connected by a massive fiber bundle, the corpus callosum. In humans, each sheet has an area of roughly 1000 cm^2 (Henneberg, 1910). In the macaque monkey the figure is nearer 100 cm^2.

Each sheet of the neocortex is highly stratified. An example of the stratification in a typical cortical area is shown in Figure 5A. Historically and didactically, the neocortex has been subdivided into six layers (Lewis, 1878), although a more convenient parcellation can be made into four main layers, which can then be divided further. These four layers are listed below, along with their most prominent features.

- *A superficial layer* (usually referred to as layer I). This layer has rather few cell bodies and consists mainly of axons and apical dendrites. (The presence of this superficial cell-poor layer seems to be characteristic of a "cortical" arrangement of neurons, be it the neocortex, the allocortex, or the cerebellar cortex.)

- *An upper layer* (layers II and III). This layer contains the smaller pyramidal neurons which send their main axons to other cortical areas, either in the same hemisphere or on the opposite side.

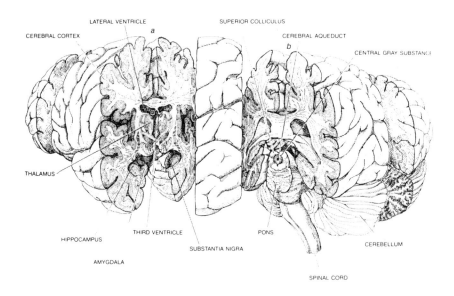

FIGURE 4. A human brain is sliced and opened like a book to demonstrate the continuity of each cortical sheet and its relation to some internal structures. (From "The Organization of the Brain" by W. J. H. Nauta and M. Feirtag, 1979, *Scientific American*, *241*, p. 92. Copyright 1979 by W. H. Freeman & Co. Reprinted by permission.)

- *A middle layer* (layer IV). In this layer are found the densely packed small stellate neurons whose axons commonly ascend vertically to terminate in the upper layers.

- *A deep layer* (layers V and VI). This layer contains the larger pyramidal neurons whose axons leave the cortex to terminate in subcortical structures such as the striatum, the claustrum, the thalamus, the brain stem, and the spinal cord. (Occasional pyramidal neurons are present in this layer which project to other cortical areas rather than projecting subcortically.)

This broad division covers all parts of the neocortex, but there is considerable regional variation in the relative amount of each layer. The middle layer in the primary sensory areas is usually rather thick, e.g., in the striate cortex of primates the middle layer is so pronounced and differentiated that it can be divided into four sublayers (Figure 5B).

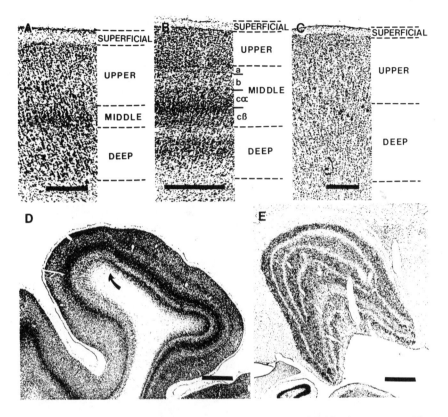

FIGURE 5. *Top:* Some examples of variations in cortical stratification patterns. The stain used is selective for cell bodies. The surface of the brain is at the top in each of these photomicrographs. *A:* Parietal cortex. In most cortical areas, the four main layers are easily recognized. *B:* Striate cortex. A marked differentiation of the middle layer is evident in primary sensory areas. *C:* Motor cortex. The middle layer is virtually absent in the primary motor cortex. *Bottom:* Stains selective for cell bodies are often used to differentiate cortical areas and thalamic nuclei. *D:* Cross-section of the junction between the striate cortex and its immediately adjacent area (area 18). The border is clearly evident (indicated by the arrow) due to the marked differentiation of the middle layer in the striate cortex (right of arrow), and the lack of such a differentiation in the middle layer of area 18 (left of arrow). *E:* The lateral geniculate nucleus is a laminated nucleus, which can easily be identified in cross-sections of the thalamus. Six distinct sheets of neurons can be recognized in the macaque and human lateral geniculate nucleus. All photomicrographs are taken from macaque monkey brains. Bars represent ½ millimeter in A-C, and 1 millimeter in D and E.

In contrast, the middle layer is virtually nonexistent in the primary motor area (Figure 5C).

In addition to the horizontal stratification of neuronal cell bodies, there is a pronounced vertical arrangement of dendritic and axonal arborizations in the neocortex (Figure 6). Not only do most of the incoming and outgoing axons travel vertically across the layers to enter or exit from the deep aspect of the cortical sheet, but many of the dendritic and axonal processes of neurons in the neocortex are vertically oriented (the ascending dendrites of pyramidal cells are particularly good examples of this—see Figure 14A,B).

The number of neurons per unit *volume* of the neocortex varies somewhat, but the total number of neurons underlying a given unit of surface *area* is remarkably constant from one area of the cortex to another and from species to species (Rockel, Hiorns, & Powell, 1980). In the unshrunken state, this figure is about 80,000 per mm^2 (Powell & Hendrickson, 1981). An exception is the striate cortex of primates, where the figure is about 2½ times as large (Rockel et al., 1980). The reasons for this regularity (and the exception) are not known.

FIGURE 6. The pronounced vertical orientation of many of the dendritic and axonal processes in the neocortex is evident in this diagram of the parietal cortex of an adult mouse. At the left is a diagrammatic representation of all neuronal cell bodies within one very thin section; at the center are the cell bodies and dendrites of some pyramidal neurons, and at the right are some different types of cortical input axons. The surface of the brain is at the top. (From "Cerebral Cortex: Architecture, Intracortical Connections, Motor Projections" by R. Lorente de Nó. In *Physiology of the Nervous System*, p. 282, edited by J. F. Fulton, 1943, New York: Oxford University Press. Copyright 1943 by Oxford University Press. Reprinted by permission.)

Cortical Areas

The neocortex, as already implied, appears to consist of several distinct areas, or "fields" (Rose, 1949). These differ somewhat in their histological appearance[1] (the striate cortex, for example, can be easily recognized in cross section by the presence of a distinct stripe through the middle layer [Figure 5D], although most areas are not so easily recognized.), anatomical connections, and the functions they perform. The hope is that in time it will be possible to parcel out unambiguously the entire neocortex into a number of distinct functional areas. Within each such area we may expect there to be considerable homogeneity of cell types, connections, and functions, all of which are likely to change rather abruptly when one crosses the border of each area and passes into another area. The number of distinct areas in the neocortex of humans (on one side) is likely to be of the order of 100. Presently, the most commonly accepted cortical parcellation scheme is the one that was established by Brodmann (1909) and is illustrated in Figure 7. Although extremely accurate in certain places, this map will undoubtedly be refined in future years.

It has yet to be shown that this simple concept of cortical area may not break down in parts of the neocortex. *If* it holds up, we should be able to count their exact number, so that we could say that in humans there are, say, 137 and not 136 distinct cortical areas. Eventually it should be possible to distinguish each area and thus construct a four-color map of the cortex.

This concept of cortical area appears to hold up fairly well in those cortical areas concerned with early visual processing. In primates there appear to be at least ten of them, covering the region occupied by Brodmann's areas 17, 18, and 19 (Figure 8) It applies very well to the striate cortex (area 17), sometimes called the first visual area (or VI), and to the area known as MT (or the middle temporal area). In the macaque, VI is an exceptionally large area, whereas MT is rather small, being less than 10% the size of VI (Van Essen, Maunsell, & Bixby, 1981; Weller & Kaas, 1983). The size of the other early visual areas will probably fall between these limits. It is important to not lose sight of the basic definition of a cortical area: ridiculously small subdivisions that do not reflect real functional differences can obscure the utility of this concept.

---

1 The common terms used for these differences are cytoarchitectonics and myeloarchitectonics. The former refers to the differences in neuronal density and to the relative development of individual cortical layers; the latter refers to differences in the distributions of axons (especially myelinated axons) within the cortex which varies from area to area.

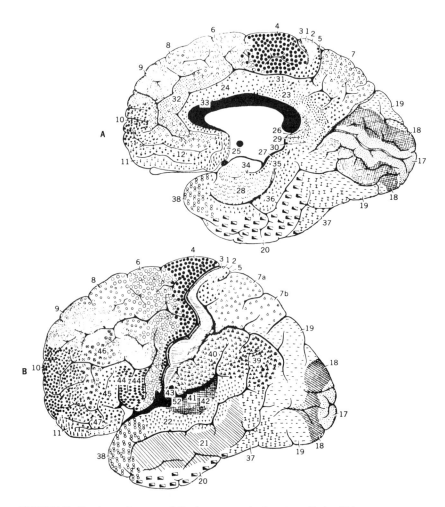

FIGURE 7. Brodmann's areas of the human cerebral cortex. Each of his areas are numbered and indicated by different symbols. *A*: Medial surface of the cerebral cortex (the black areas are occupied by fiber bundles crossing the midline to connect the two hemispheres). *B*: Lateral surface of the cerebral cortex. (From *Vergleichende Localisationslehre der Grosshirnrinde in Ihren Prinzipien Dargestellt auf Grund des Zellenbaues [Principles of comparative localization in the cerebral cortex presented on the basis of cytoarchitecture]*, by K. Brodmann, 1909, Leipzig: Barth. Copyright 1909 by Barth Publishing. Reprinted by permission.)

Cortical Inputs

An important feature of the neocortex is that almost all the outside information it receives (either from the sensory periphery or from other subcortical centers), with the exception of some olfactory

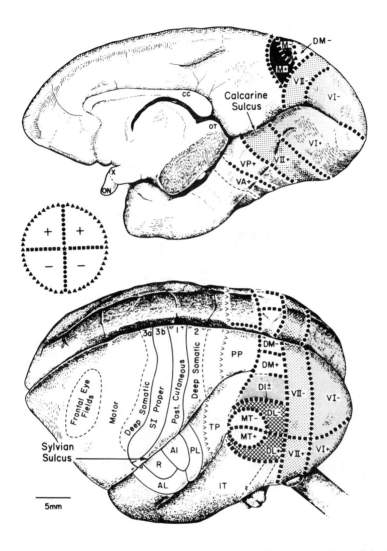

FIGURE 8. Visual processing areas of the owl monkey cerebral cortex. Each of the different types of shading represent different areas. +'s indicate the regions representing the dorsal half of the visual field. −'s indicate the ventral half of the visual field. DI, dorsointermediate visual area; DL, dorsolateral crescent visual area; DM, dorsomedial visual area; IT, inferotemporal cortex; M, medial visual area; MT, middle temporal visual area; PP, posterior parietal cortex; VA, ventral anterior visual area; VP, ventral posterior visual area; V1, first visual area; V2, second visual area. (From "Visual Response Properties of Neurons in Four Extrastriate Visual Areas of the Owl Monkey (*Aotus trivirgatus*): A Quantitative Comparison of the Medial, Dorsomedial, Dorsolateral, and Middle Temporal Areas" by J. F. Baker, S. E. Petersen, W. T. Newsome, and J. Allman, 1981, *Journal of Neurophysiology*, 45, p. 400. Copyright 1981 by The American Physiological Society. Reprinted by permission.)

information, passes through the thalamus. Input systems terminate on neurons in the thalamus, and these thalamic neurons, in turn, project to the cerebral cortex (Figures 9A and 10). Though the terminations of thalamic axons may account for only a small proportion of the total synapses in any given cortical area,[2] the thalamus is clearly the major

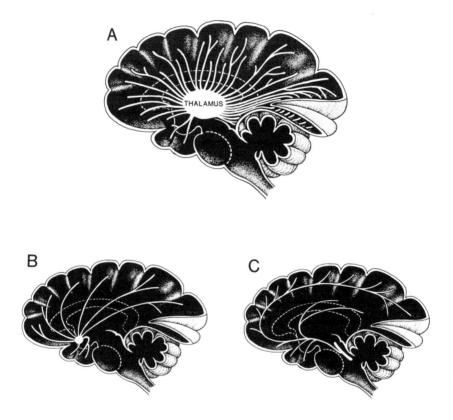

FIGURE 9. Some of the inputs to the neocortex. *A*: Most of the information entering the neocortex gets there through the thalamus. *B*: A diffuse cholinergic input arises in the basal forebrain. *C*: Diffuse noradrenergic and serotonergic inputs arise in the brain stem.

---

[2] Recent synapse counts indicate that in the monkey striate cortex, approximately 35% of the total synaptic population comprises middle layer synapses (O'Kusky & Colonnier, 1982). Reported percentages of thalamocortical synapses within the middle layer of the striate cortex range from 5% (Garey & Powell, 1971) to 29% (Tigges & Tigges, 1979). These data suggest that thalamocortical synapses account for 2-10% of the total synaptic population in the striate cortex, but this calculation does not take into account the thalamocortical synapses which terminate outside the middle layer (e.g., in layer I and in layer VI).

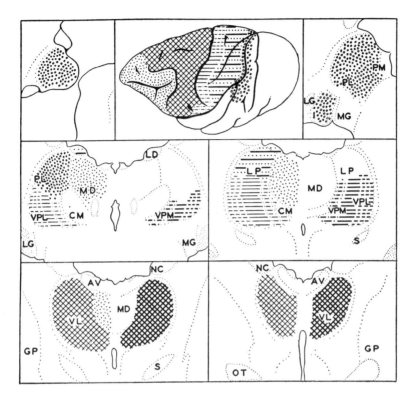

FIGURE 10. A schematic diagram demonstrating part of the systematic relationship of the thalamus with the cerebral cortex and the thalamic termination of several input systems in macaque monkeys. The cerebral cortex (top center) is viewed from the left side, and several frontal cross-sections of the thalamus are illustrated, with the caudalmost section at top left and the rostralmost section at bottom right. Similarly marked parts on the left sides of the thalamic sections and the cerebral cortex have thalamocortical connections. The terminations of afferent systems are represented on the right sides of the thalamic sections as follows: inputs from the cerebellum by heavy cross-hatching, somatic sensory inputs from the leg area by short heavy dashes, somatic sensory inputs from the arm area by heavy lines, and somatic sensory inputs from the face by heavy dots and dashes. Abbreviations in the thalamic sections indicate histologically identifiable nuclei. (From *The Primate Thalamus*, p. 189, by A. E. Walker, 1938, Chicago: University of Chicago Press. Copyright 1938 by the University of Chicago Press. Reprinted by permission.)

center through which the cerebral cortex has access to outside information. The thalamus is, therefore, often referred to as the "gateway" to the cerebral cortex.

There are a few exceptions to this general rule. The following pathways do not relay through the thalamus:

- Diffuse innervations to the cortex arise in a number of brain stem and basal forebrain areas (Figures 9B,C). Among them are a noradrenergic system that arises principally in the locus coeruleus (Andén et al., 1966), a serotonergic system that arises mainly in the dorsal raphé nucleus (Bobillier et al., 1976), and a cholinergic system that arises in the basal forebrain (Mesulam & Van Hoesen, 1976). These systems spread throughout the cortex both horizontally and vertically and do not appear to be organized with any topographic finesse. In addition to these, it has been suggested recently that there may be a diffuse GABAergic system innervating the neocortex, which arises in the hypothalamus (Vincent, Hökfelt, Skirboll, & Wu, 1983). The exact function of these diffuse inputs is not known, but it is important for theorists to be aware of their existence.

- A structure called the claustrum—situated deep to the insular region of the cortex, receives inputs from (Carman, Cowan, & Powell, 1964) and projects to (Macchi, Bentivoglio, Minciacchi, & Molinari, 1981) almost all areas of the cortex. Since, apart from its diffuse innervation from the brain stem, it receives no other input, it could well be described as a satellite of the cortex. Only the visually responsive part of the claustrum has been intensively studied, and it has been shown to be systematically connected with the striate cortex and the adjacent visual area (LeVay & Sherk, 1981a).

- Restricted neocortical projections arise in the hippocampus (Swanson, 1981) and amygdala (Porrino, Crane, & Goldman-Rakic, 1981). These tend to terminate most heavily in cortical areas that are removed from the primary sensory and primary motor areas.

The thalamus comprises a number of *specific nuclei*; these have well-defined inputs and project to restricted portions of the cerebral cortex. The thalamic nuclei, like the cortical fields, can be differentiated in terms of histological appearance, connections, and function. The lateral geniculate nucleus, for example, is the relay center for inputs from the retina that pass on to the striate cortex. It is a distinct, laminated nucleus which can be easily differentiated histologically from the surrounding structures (Figure 5E). Neurons in the lateral geniculate nucleus receive visual signals from the axons of ganglion cells in the retina and project, in turn, to the striate cortex.

The inputs to the cortex from the thalamus terminate primarily in the middle layer (Cajal, 1911). Where the middle layer is sparse or absent (as in the motor cortex), they usually terminate in the lower part of the upper layer (Jones, 1975a). In the electron microscope, the terminal boutons of thalamocortical axons are all Type I (Jones & Powell, 1970a) and are known to be excitatory. The axons of the thalamic neurons that project to the cortex rarely have collaterals within the main body of the thalamus (Friedlander, Lin, Stanford, & Sherman, 1981).

In addition to the specific nuclei, there are a number of less specific (sometimes called nonspecific) thalamic nuclei. The most prominent group of the less-specific nuclei is known as the intralaminar nuclei and occupies a thin, vertical, sheet-like zone extending anteroposteriorly through the center of the thalamus. While most of these are rather small, one—the centromédian nucleus—is very prominent in the human brain. The neurons of the intralaminar nuclei project both to the striatum and to the cerebral cortex (Jones & Leavitt, 1974; Powell & Cowan, 1967). Their cortical projection, instead of terminating in the middle layer of the cortex, terminates mainly in the superficial layer (layer I). Moreover, the cortical projections of the intralaminar nuclei are not confined to a single cortical field, but tend to be rather widespread. Our present knowledge of the less-specific thalamic nuclei and their role in cortical function is quite vague.

Each of the specific thalamic projections to the cortex is accompanied by a reverse projection from the cortex to the thalamus. The spatial organization of the reverse projection reciprocates, fairly precisely, the forward one. This reverse projection arises from cells at the bottom of the deep layer and terminates directly on the peripheral dendrites of thalamocortical relay cells. Their terminations are also Type I (Guillery, 1969; Szentágothai, Hámori, & Tömböl, 1966). Although they are very numerous, the function of these reverse projections is not known.

A very remarkable nucleus, called the thalamic reticular nucleus forms a thin shell around the main body of the thalamus. It is only a few cells thick. Its neurons are very large, with protrusions on their dendrites called spines (M. E. Scheibel & A. B. Scheibel, 1966). This nucleus does *not* project to the cortex but projects back into the thalamus (Jones, 1975b). It appears to receive small collaterals from most of the axons that pass between the thalamus and the cortex. It also gets some input from the traffic between the thalamus and the striatum. Its axons are inhibitory (Houser, Vaughn, Barber, & Roberts, 1980; Montero & Scott, 1981), and they have extensive axon collaterals *within* the reticular nucleus (Scheibel & Scheibel, 1966). Obviously this nucleus occupies a very strategic place in the brain. It deserves more attention, both from experimentalists and theorists (Crick, 1984).

In addition to their thalamic inputs, most cortical neurons receive inputs from other cortical areas either in the same hemisphere (in which case they are called associational inputs) or in the opposite hemisphere (where they are known as commissural inputs). It is important to note that a typical cortical area is not connected directly to all or even most other cortical areas. Usually it projects to a handful of other areas, although the areas that are removed from the primary sensory or motor areas tend to project more widely (Jones & Powell, 1970b; Pandya & Kuypers, 1969). But if a cortical area projects to another cortical area, its projections are usually topographically organized, at least on a coarse scale. That is, as far as position in the sheet is concerned, connections between areas are not random; neighboring neurons in a field tend to project to neighboring regions in other fields in some systematic way. Moreover, as a general rule, projections from one field are usually matched by a reciprocal projection from that field which is also topographically organized. To a considerable extent the forward and backward mappings coincide, at least on a coarse scale, but are not symmetrical in all details. Rockland and Pandya (1979) and Maunsell and Van Essen (1983) suggest that for the early visual areas, a forward projection (forward with respect to the retina) is likely to project predominantly into the middle layer, whereas the reverse projection is likely to avoid the middle layer and instead terminate largely in the superficial and deep layers (Figure 11).

Topographically organized maps occur in many cortical areas. Anatomical and electrophysiological studies show that in most cortical areas there is a more or less topographic representation of the periphery upon the cortical surface. Detailed maps of cortical representation patterns are available for the areas concerned with sensory input (and motor output). As might be expected, such maps vary somewhat from individual to individual. Their broad topography is usually not linear; the extent of cortex representing a given region of the periphery is roughly proportional to the peripheral innervation density. Thus, the fovea is heavily over-represented in the striate cortex (Figure 12A), and the hand occupies a bigger region than the trunk in the somatic sensory (Nelson, Sur, Felleman, & Kass, 1980; C. N. Woolsey, Marshall, & Bard, 1942) and motor cortical areas (C. N. Woolsey et al., 1952; see Figure 12B).

All this makes good sense. Because neurons in the cortex cannot act directly on more distant neurons in the same area (see later), such "maps" bring into proximity neurons which are likely to have related inputs. They also allow more space for the more densely innervated regions.

This systematic mapping of the periphery is likely to be a feature of many cortical areas, not merely those near the sensory input. In general terms, as one proceeds further from the sensory input, the

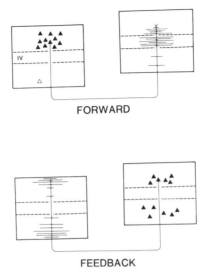

FORWARD

FEEDBACK

FIGURE 11. A schematic diagram of the characteristic laminar distributions of cell bodies and terminals in forward and feedback cortico-cortical pathways. Forward pathways arise mainly from the upper layer and terminate mainly in the middle layer. Feedback pathways arise from both upper and lower layers and terminate mainly outside the middle layer. Triangles represent cell bodies, and axons are represented by thin lines. (From "The Connections of the Middle Temporal Visual Area and Their Relation to a Cortical Hierarchy in the Macaque Monkey" by J. H. R. Maunsell and D. C. Van Essen, 1983, *Journal of Neuroscience, 3*, p. 2579. Copyright 1983 by The Society for Neuroscience. Reprinted by permission.)

mapping of the periphery becomes more diffuse. At the same time the neurons respond to more elaborate "features" of the input.

There are irregularities within a cortical area. It should not be inferred from the above description that cortical areas are relatively uniform. On the contrary, on the scale of 1 mm to 2 mm, patchiness is almost always the rule. The patchiness can, in many instances, be attributed to disjunctive distributions of inputs. Dense foci of terminations, separated by zones that are relatively free of terminations, have been described for the inputs from the thalamus and from other cortical areas.

This is especially clear in cases where the inputs are not continuous but discrete. For example, the thalamic input from either of the two eyes form nonoverlapping stripe patterns (not unlike fingerprints when viewed from the surface) in the middle layer of the striate cortex (Figure 13) (Hubel, Wiesel, & LeVay, 1977; LeVay, Connolly, Houde, & Van Essen, 1985). In the somatic sensory system of rodents, inputs

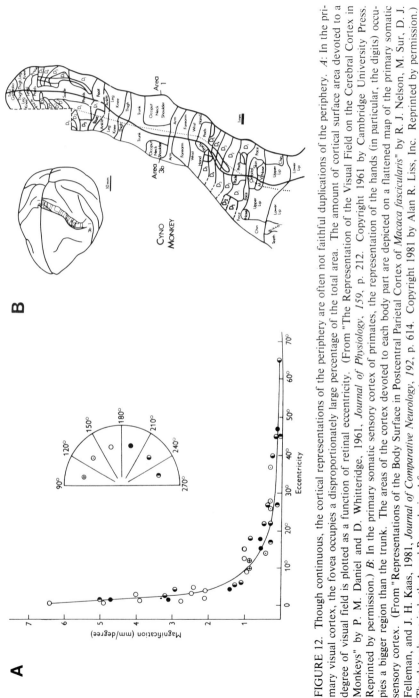

FIGURE 12. Though continuous, the cortical representations of the periphery are often not faithful duplications of the periphery. *A*: In the primary visual cortex, the fovea occupies a disproportionately large percentage of the total area. The amount of cortical surface area devoted to a degree of visual field is plotted as a function of retinal eccentricity. (From "The Representation of the Visual Field on the Cerebral Cortex in Monkeys" by P. M. Daniel and D. Whitteridge, 1961, *Journal of Physiology, 159*, p. 212. Copyright 1961 by Cambridge University Press. Reprinted by permission.) *B*: In the primary somatic sensory cortex of primates, the representation of the hands (in particular, the digits) occupies a bigger region than the trunk. The areas of the cortex devoted to each body part are depicted on a flattened map of the primary somatic sensory cortex. (From "Representations of the Body Surface in Postcentral Parietal Cortex of *Macaca fascicularis*" by R. J. Nelson, M. Sur, D. J. Felleman, and J. H. Kaas, 1981, *Journal of Comparative Neurology, 192*, p. 614. Copyright 1981 by Alan R. Liss, Inc. Reprinted by permission.) The data shown in both A and B are derived from macaque monkeys.

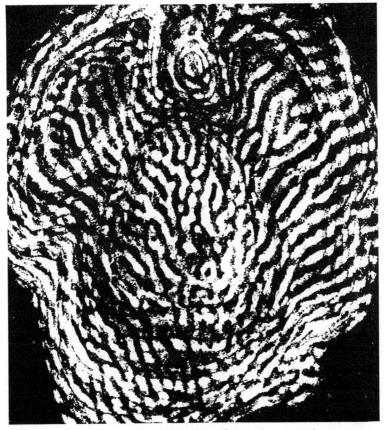

FIGURE 13. A surface view of the middle layer of the striate cortex in a macaque monkey in which the input from one of the eyes is labeled autoradiographically with tritium. In this figure, the label, in the form of exposed photographic silver grains, appears light against a dark background. The disjunctive, stripe-like patterns of input from each eye is clearly evident. Due to the curvature of the cortex, this figure is a montage of many photomicrographs. (From "The Complete Pattern of Ocular Dominance Stripes in the Striate Cortex and Visual Field of the Macaque Monkey" by S. LeVay, M. Connolly, J. Houde, and D. C. Van Essen, 1985, *Journal of Neuroscience, 5.* Copyright 1985 by The Society for Neuroscience. Reprinted by permission.)

from each mystacial vibrissa project, through independent channels in the thalamus, to separate foci in the middle layer of the somatic sensory cortex called "barrels" (T. A. Woolsey, 1978). Neural connections *between* barrels appear less strong than those *within* each barrel. Similar focal arrangements have been proposed for the thalamic input to other cortical areas.

Disjunctive, stripe-like patterns have also been demonstrated in certain terminations coming from other cortical areas. For example, the projection from the striate cortex to other visual areas is usually irregular. Stripe-like patterns of terminations have been described for their projections to area 18 (Gilbert & Wiesel, 1981; Maunsell, Newsome & Van Essen, 1980). In the primary auditory cortex, there are irregular rows of commissural terminations which in cross-section appear as multiple column-like clusters. Such clusters contain neurons that display certain kinds of binaural interaction (Imig & Brugge, 1978).

The various kinds of patches outlined above appear to underlie a tendency of neurons aligned vertically to display certain similarities in functional properties. Other response properties, not necessarily correlating with input signals, are also aligned vertically. For example, in the visual system of monkeys, the cells that project to the cortex have center-surround receptive fields, and the neurons in the upper and lower layers of the striate cortex respond specifically to lines oriented in a particular orientation (Hubel & Wiesel, 1968). Furthermore, neurons with similar orientation preferences are often aligned vertically, though this may not always be true. However, in a general sense it has been found that when recordings are made with a microelectrode inserted perpendicular to the surface of the cortex, the response properties of all the neurons encountered throughout the depths of the cortex tend to be similar.

These various types of patchinesses have suggested to some authors that the cortex is organized on a "modular" basis, either involving columns about ½ mm or so in diameter or into minicolumns, some 50 microns or so across (Mountcastle, 1978). In the cortex there is really no evidence for true modules of the type found, for example, in the fly's eye and rather little for more irregular modules. The exact basis on which the cortex is organized remains to be discovered.

Cortical Outputs

All the axons that leave a cortical area make Type I (excitatory) synapse in their projection fields. The subcortical projections of the cortex, in contrast to the subcortical inputs, are not subject to the constraint of always relaying through the thalamus. Thus, the cortex can project directly to the spinal cord or superior colliculus, but the spinal and collicular inputs to the cortex always relay in the thalamus.

As mentioned earlier, there is a close relationship between populations of cortical output neurons and their laminar positions. This is particularly evident for the populations of neurons that project subcortically. For example, the cortical neurons which project to the thalamus are situated at the bottom of the deep layer, whereas those that project to the striatum and brain stem tend to be concentrated in more superficial portions of the deep layer (Jones & Wise, 1977; J. S. Lund, R. D. Lund, Hendrickson, Bunt & Fuchs, 1975).

As with the input systems, most of the output systems of the neocortex are patchy. Patches of neurons have been demonstrated to project to other cortical areas and others to project subcortically (Jones & Wise, 1977; Murray & Coulter, 1981). Although the groups of cells giving rise to the various projections are segregated into layers, physiological studies indicate that neurons producing *similar* outputs are vertically aligned. This has been most convincingly demonstrated in the motor cortex, where microstimulation points producing isolated contractions of given muscles have been shown to be organized in cylindrical zones which extend throughout the depths of the cortex (Asanuma & Rosén, 1972).

THE NATURE OF NEOCORTICAL NEURONS

Experimental Methods

The shape of neurons in many areas of the brain has traditionally been studied in Golgi-stained material. The Golgi method stains, in their entireties, a few cells here and there, apparently "at random." The haphazard selectivity of this method allows investigators to visualize clearly most of the dendritic processes of individual neurons. Unfortunately, though, the technique is capricious and does not always stain the axon and axon collaterals completely. A more recent method, using intracellular injections of an appropriate marker such as the fluorescent dyes procion yellow or lucifer yellow or the enzyme horseradish peroxidase (HRP), is capable of producing a similar picture with better reliability and control, but is not suited for sampling large populations of neurons since each neuron must be separately injected.

Very fine details, such as the morphology of the synapse, cannot be seen using the light microscope. The electron microscope is necessary for this. The extremely high magnifications needed to differentiate such details, however, make it difficult to reconstruct (using a very large number of serial sections) even a very small neuron in its entirety. (This problem can be partially alleviated by combining standard electron microscopic procedures with the Golgi method or a variety of labeling methods.)

It is important to realize that each of the techniques presently available is limited in its sampling capacities. Many of the following conclusions, therefore, are rather general and apply to broad groupings of cortical neurons. Except in rare cases, there is little quantitative information on specific cell types.

Cell Types in the Neocortex

The most common and most characteristic neurons of the cerebral cortex are the pyramidal cells (Figure 14A). Such neurons, as their name implies, have a pyramid-shaped cell body. A thick apical dendrite ascends vertically from the apex of the cell body. It often has tufts of branches that arborize in specific laminae. A number of basal dendrites also extend out from the bottom of the cell; these branches spread horizontally for some distance. All dendrites of pyramidal cells possess large numbers of dendritic spines. The axons of pyramidal cells emerge either from the bottom of the cell body or from one of the basal dendrites and generally leave the area.

Though these basic features are common to all pyramidal cells, there are a variety of different types of pyramidal cells (Figure 14B). On the basis of cell body size alone, they range from about 10 microns up to 100 microns in diameter (Cajal, 1911; Conel, 1939). The branching patterns of the apical and the basal dendrites also differ tremendously, as do the target sites of the projection axons and the arborizations of axon collaterals which participate in local circuits. These variations do not appear to be random, and a number of classes of pyramidal cells have been described (Cajal, 1911; Lorente de Nó, 1943; J. S. Lund, 1973). Unfortunately, the information presently available for differentiating the various types of pyramidal cells is far from complete. It is unlikely that there are as few as ten distinct types: There could be as many as a thousand.

All axons that leave the cortex belong to pyramidal cells. The converse, however, may not be true. Though all pyramidal cells were once considered to be projection neurons (Cajal, 1911), recent data suggest otherwise. Using intracellular HRP injections, Gilbert and Wiesel (1983) have found a few pyramidal cells with extensive local axonal arborizations that do not project out of the striate cortex.

The remainder of the neurons in the cerebral cortex can be broadly categorized as "nonpyramidal cells" (Figure 14C). By definition, they comprise all neurons whose cell bodies are not pyramid-shaped. They constitute a large number of cell types which can be differentiated on the basis of the shape of the cell body, size, dendritic arborizations, and axonal ramifications. For example, some have relatively spherical cell bodies with dendrites that radiate in all directions and therefore are star-shaped, or "stellate." Others are spindle-shaped and have dendritic

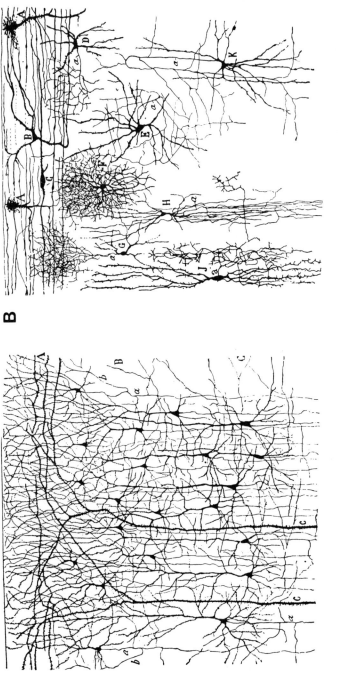

FIGURE 14. Drawings of Golgi-stained neurons of the cerebral cortex. *A:* The most characteristic neurons of the cerebral cortex are the pyramidal cells. They are characterized by a pyramid shaped cell body, a thick apical dendrite, and a number of basal dendrites. There are a variety of different types of pyramidal cells. *B:* The remainder of neurons in the cerebral cortex can be broadly referred to as nonpyramidal cells. There are a variety of different types of nonpyramidal cells. See text for further details. (From *Histologie du Systeme Nerveux de L'homme et des Vertébrés, T. 2.* [*Histology of the Nervous System of Man and of Vertebrates, Vol. 2*], by S. Ramón y Cajal, 1911, Paris: Maloine. Copyright 1911 by Librairie Maloine. Reprinted by permission.)

arborizations which are aligned vertically. Again, the number of distinct types is not known. It is unlikely to be as few as ten; it could run to several hundreds.

Instead of discussing in detail the varieties of nonpyramidal cells that are known, we should like to draw the reader's attention to two characteristics of nonpyramidal cells that are of significance in the context of this chapter. One is that the nonpyramidal cells are local circuit neurons, viz., they do not project out of a given cortical area. The other is that there are two basic types of nonpyramidal cells according to their dendritic morphology. One type has numerous spines on its dendrites, and the other has few or no spines.

A convenient categorization combines the entire population of pyramidal cells and also the spiny nonpyramidal cells into a group called "neurons with spines," and the remaining nonpyramidal cells into a group called "neurons without many spines." The following generalizations are applicable when cortical neurons are categorized in this way (see Table 1).

Neurons With Spines

The neurons with spines are the pyramidal cells, the star pyramids, and the spiny stellate cells. Such cells receive Type I synapses mainly on their spines. Usually, each spine has only a single Type I synapse;

TABLE 1

	Neurons With Spines	Neurons Without Spines
Input to Spines	Usually one Type I Occasionally one Type I plus one Type II	—
Input to Dendrites	Both types	Both types
Input to Soma	Type II only	Both types
Input to Axon Hillock	Multiple Type II on many pyramidals in layers II and III (see text)	—
Output	Type I	Usually Type II

Working Assumptions: Type I = excitation Type II = inhibition

a minority of spines has an additional Type II synapse. No spine seems to have only a Type II synapse by itself (Colonnier, 1968).

The Type II synapses on spiny cells tend to be fewer in number than Type I synapses. They do, however, occupy sites that are suited for effectively influencing impulse generation by the neuron (which generally takes place at the axon initial segment). They usually occur on the proximal shafts of the dendrites or on the cell bodies (Gray, 1959). A special type of Type II synapse is found on the initial segment of the axon of many pyramidal cells (Peters et al., 1968).

The axon of a spiny cell forms only Type I synapses (LeVay, 1973). Pyramidal cells almost always send the main branch of their axon out into the white matter below the cortex. Spiny stellates almost always have a local axon which does not leave the cortex but ramifies instead in close proximity to its cell of origin.

Neurons Without Many Spines

Although neurons without spines may have transient spines on their dendrites early in development, in the mature state they have relatively few, if any, spines.

Such cells, of which there are several obviously different morphological types, receive both Type I and Type II synapses on their dendritic shafts and on their cell bodies (Gray, 1959). The axons of nonspiny neurons do not descend to the white matter but ramify locally. Their axons are believed, in most cases, to form only Type II synapses (LeVay, 1973), but there may well be exceptions.

It remains to say that spiny cells are in the majority (80%). Of these, a fair number, perhaps as many 25% can be nonpyramidal in some cortical areas. The nonspiny cells are in a minority (20%). Unfortunately these percentages are only very approximate.[3]

---

[3] These approximations are derived from the results of Tömböl (1974), who found 59% of the neurons in the visual cortex of monkeys to be pyramidal neurons, 7.4% to be large stellate neurons, and 33.6% to be small stellate neurons. Of the population of small stellate neurons, 63% were found to occur in the middle layer. These results did not take into account the spiny stellate neurons which also need to be included in the spiny category. Since spiny stellate neurons are small (Cajal, 1911), restricted to the middle layer (J. S. Lund, 1973), and constitute the major component of the middle layer (J. S. Lund, 1973), we have added 21% to the overall percentage of pyramidal neurons reported by Tömböl, to obtain a very rough figure of 80% for our estimate of spiny neurons.

Excitation Versus Inhibition

We shall now discuss these generalizations (summarized in Table 1) on the assumption that most Type I synapses excite and most Type II synapses inhibit. The main point to grasp is the asymmetry between excitation and inhibition.

Excitatory neurons and excitatory synapses are clearly in the majority. Only excitation is sent out of any area of the cortex to other places. On the other hand, the inhibitory synapses are more strategically placed on the cells, being nearer the impulse initiating site at the axon hillock, at least for spiny cells. This is almost a necessity if the system is not to get out of hand, since any area of the cortex feeds back excitation to itself in large amounts. The synapses formed by axons entering the cortex are in a minority, sometimes a very small minority. The great majority come from the axons of other cortical cells: mostly, but not entirely, from those within a few hundred microns.

Inhibition seems to have priority in two ways. Inhibitory neurons have excitatory inputs on their cell bodies, so that they can be brought into action quickly. Excitatory neurons, on the other hand, receive inhibition at strategic sites. This preference for inhibition must be set against the fact that (excluding some of the diffuse innervation) the inputs entering the cortex are all excitatory. Thus any inhibition needed must be generated from this excitation. This requires an extra step and therefore will take time. It seems as if the cortex is arranged so that this time delay is minimized by the regularities discussed above.

In special conditions, such as epilepsy and under the influence of hallucinogenic drugs, the cortical system may go into full-scale oscillation, presumably because the balance between excitation and inhibition has been upset.

Special Cell Types

Nonspiny neurons are of various types. Many of them are stellate. Here we describe three unusual types of nonspiny neurons whose properties may be of special interest.

Chandelier cells. So-called because their axons end in a set of vertically oriented beaded terminal segments which make them look somewhat like chandeliers (Szentágothai & Arbib, 1974). An alternative name for them is "axo-axonic" cells (Somogyi, 1977).

These beaded terminal segments turn out to be the sites of multiple Type II synapses upon the axon initial segments of nearby pyramidal cells (Somogyi, 1977). Each chandelier cell axon has at least 50 to 200 such terminal segments (Somogyi, Freund, & Cowey, 1982), each of which may form 5 to 10 synapses (Fairén & Valverde, 1980; Somogyi et al., 1982). (There does not appear to be any corresponding cell which makes Type I synapses on axon hillocks.) Each pyramidal cell may receive up to six such terminal segments, presumably from different chandelier cells (Fairén & Valverde, 1980; Somogyi et al., 1982).

It is difficult to resist the impression that a chandelier cell can veto the output of a whole set of neighboring pyramidal cells. Whether this is really true is not known, but the fact that chandelier cells are probably GABAergic (Peters, Proskauer, & Ribak, 1982) and form Type II synapses strongly implies that they are inhibitory.

·The number of chandelier cells is not known. A not unreasonable guess would be that they form about 1% of the total neuronal population.[4] It is not known how many types of chandelier cells exist. So far no stain has been found which selectively stains them.

Basket cells. These cells have vertically oriented stem axons which give rise to several horizontally disposed collaterals. The collaterals subsequently give off obliquely or vertically directed fine terminal branches, which are commonly clustered and resemble baskets (Cajal, 1911; Marin-Padilla, 1969), though not all terminal branches resemble baskets (Jones, 1975c). These terminal branches often form loops of boutons around the cell bodies of pyramidal cells (10 to 20 terminal branches may converge upon certain pyramidal cells; Martin, Somogyi, & Whitteridge, 1983). The characteristic fine axonal sprays of these cells are rather difficult to visualize, so it is not certain that basket cells occur in all cortical areas. The terminal boutons of basket cells are Type II in morphology and are thought to contain GABA (Martin et al., 1983; Somogyi, Kisvárday, Martin, & Whitteridge, 1983).

Again, one cannot resist the impression that the basket cell is likely to exert a veto (or at least a partial veto) upon the output of the cortical projection neurons.

Bipolar cells. These are cells whose dendrites and axons form a very long thin bundle in the vertical direction (M. L. Feldman & Peters, 1978). Some of them have been shown to contain peptides such as somatostatin, cholecystokinin (CCK), and vasoactive polypeptide (VIP)

---

[4] This guess is based on three arbitrary assumptions: (a) all pyramidal neurons receive initial segment synapses from chandelier cells, (b) every chandelier cell innervates about 200 pyramidal neurons, and (c) every pyramidal neuron is innervated by about five different chandelier cells. Together, these suggest that there are about 40 pyramidal neurons for each chandelier cell. Since pyramidal neurons account for about 60% of the neurons in the visual cortex of monkeys (Tömböl, 1974), the above assumptions would suggest that chandelier cells form about 1.5% of the neuronal population of the monkey visual cortex. This is probably an overestimate.

(Emson & Hunt, 1981; Fuxe, Hökfelt, Said, & Mutt, 1977). These cells can be stained immunohistochemically. About 1350 VIP cells are estimated to occur under 1 mm^2 of surface in the rat visual cortex (this figure is uncorrected for shrinkage) (Morrison, Magistretti, Benoit, & Bloom, 1984), thereby making up about 1% of the total neurons. Whether bipolar VIP cells also have other nonpeptide transmitters is not yet known. The type of synapse made by each type of bipolar cell is not clear. Their relationship, if any, to the "columns" reported in the cortex is also unclear. Nevertheless, their narrow shape is intriguing.

The Behavior of Single Neurons in the Cerebral Cortex

The behavior of single neurons in the cerebral cortex will depend to some extent on the exact nature of the neuron concerned. Here we will treat only the general problem.

It is widely believed that the dendrites of most neurons behave in a largely passive manner. Their cable constants can be estimated from passive membrane transients that can be recorded intracellularly. In the cortex the total soma-dendritic input resistance ranges from 6.7 to 78 megohms with a mean of 24 megohms, while the membrane time constant tends to be relatively constant at around 8.2 milliseconds (Connors, Gutnick, & Prince, 1982). If the specific membrane capacitance is assumed to be about 1 μF/cm^2 (K. S. Cole, 1968), then the mean specific membrane resistance is relatively high at about 8,200 ohm·cm^2. This implies a relatively large length constant for most neurons in the cerebral cortex.

However, there are disturbing reports that the dendrites of some neurons (for example, the Purkinje cells of the cerebellum and the pyramidal cells of the hippocampus) may have spike generating patches (Llinás & Nicholson, 1969; Spencer & Kandel, 1961). Unfortunately, the experimental techniques to look for this are not easy and the interpretation of the results is not straightforward.

Clearly, it would make a tremendous difference if the dendrites of a neuron were not purely passive. To give just one example, if its dendrites are passive it might be argued that the apical dendrites of a pyramidal cell may mainly serve to "modulate" the inputs of the basal dendrites, because the apical dendritic shaft would attenuate any change of potential produced by synapses in the apical tuft so that, by itself, it might not be able to fire the cell. If the apical dendrite had spike generating capabilities, this argument would not be valid. It is clearly an urgent matter to decide just how potential changes are propagated in dendrites of different types of cells.

Another important theoretical parameter is the "weight" of individual synapses; that is, the size (and time course) of the potential change that synapse produces in the postsynaptic cell. Since in many theories these weights not only determine the behavior of the neuron but are thought to be important for memory, they are obviously of considerable significance. Such weights could be influenced by many factors, both presynaptic and postsynaptic. Moreover, the weights could be subject to transient changes due to many different biochemical processes. Parenthetically we may point out that there is a problem concerning long-term memory and the synaptic weights that may be associated with it. How do synapses manage to remember anything over a period of years in the face of relentless molecular turnover? This makes one wonder whether some single structural feature, either at the molecular level or at a higher level, might embody the "long-term weight." A naive guess might be that it is simply the *area* of the synapse, since in the neocortex this varies, from synapse to synapse, by a factor of 10 (Peters & Kaiserman-Abramof, 1969).

Type II synapses tend to be nearer the axon hillock than do Type I synapses. Thus, it can be argued that such inhibitory synapses on a given dendrite can exercise a partial veto on more distal excitation and thus the exact arrangement of the various synapses could be significant. On this view, a single neuron, rather than being a single integrating device for distributed excitation and inhibition, may be a more complex processing unit, with each dendritic branch acting, in a loose sense, as an integrating unit for its own inputs (Koch, Poggio, & Torre, 1982). It remains to be seen whether this new concept is really valid. Whether it is or not, it should be noted that it is a lot easier to implement addition, subtraction, and division than it is to implement multiplication in a single neuron. In logical terms, AND-NOT seems easier to implement than AND. However, because of the uncertainties in our knowledge, such generalizations are precarious.

In any single neuron a synapse nearer the cell body is likely to have a larger effect than one near the ends of the dendrites, but even this rather obvious deduction has been questioned for neurons whose dendrites generate action potentials. It is obvious that a lot needs to be learned about dendritic behavior before theorists have a solid body of facts to build on.

The Behavior of Groups of Neurons in the Cerebral Cortex

If little is known for certain about single neurons, even less is known about neuronal groups and their behavior in the cerebral cortex.

As we have already seen, in the cortex there are extensive axon collaterals. Many of these only extend laterally for relatively small distances—less than mm—but a significant fraction spread for several millimeters (Gilbert & Wiesel, 1983; Rockland & J. S. Lund, 1983). However, they never seem to spread as much as a centimeter, though this is not true for the diffuse inputs such as those from the brain stem. Thus, in mathematical terms, for one cortical area the connections seem to be "near-diagonal," assuming that we have a two-dimensional arrangement of cells and a four-dimensional connection matrix. Whether excitatory axons spread more or less than inhibitory axons is not clear. The data on this point are confusing.

A favorite theoretical model is one in which all cells of one type connect directly to all other cells of the same type. It seems very rare for a cell to connect to itself (the diagonal term), although occasional axon collaterals that terminate upon their parent cell have been described (Van der Loos & Glaser, 1972). A more disturbing criticism is that, among the sea of axon collaterals, we really have no evidence that cells of one type connect to other cells of the same type. A better guess would be that the collaterals usually contact cells of *other* types in that area of cortex, often to cells in a different layer. Our ignorance springs partly from the fact that we lack convenient experimental methods for studying which cell types actually connect to which.

Are there any general rules about the connection in the neocortex? An outline model, based on recent studies of the visual cortex (Gilbert, 1983; Wiesel & Gilbert, 1983), might be that the main extrinsic input to a cortical area (from the thalamus) is to the middle layer. From there the excitation spreads largely to the upper layer, and from there to the deep layer—first to layer V and then to layer VI. The flow of information for other inputs is less clear. Inputs from other cortical areas tend to end in the upper layer and presumably spread from there to the deep layer. This description is certainly grossly oversimplified in almost every respect but it may turn out to have an element of truth in it.

Rates of Firing

The average discharge rate of neocortical neurons is relatively slow—perhaps only 50 to 100 spikes per second, though the rate for a brief time can rise as high as several hundred spikes per second (see e.g., Lynch, Mountcastle, Talbot, & Yin, 1977). This means that the time between spikes is often 10 to 20 milliseconds. Bearing in mind how much "computation" the brain can do in 100 milliseconds, this seems to suggest that the *average* rate of firing of an individual neuron can only be transmitted rather approximately and that computations involving many reiterations are unlikely, at least for the initial processing of an input. The precise nature of this limitation deserves further study.

Most, but not all, cortical neurons have very low resting discharges: only a few spikes a second (see, e.g., Abeles, 1982). This is likely to cause problems when one needs to signal both positive and negative values of a function. One obvious way is for one set of neurons to signal the positive values and another to signal the negative ones. In the visual system this appears to be initiated by the retinal ganglion cells (which have separate ON-center and OFF-center receptive fields, Kuffler, 1953) which signal information from the retina to the lateral geniculate body (Cajal, 1892) and from there are relayed to the primary visual cortex and elsewhere.

A large "soma-dendritic" spike generally follows an action potential which is initiated at the axon hillock. This large spike is believed to take place at the cell body and possibly invade the proximal parts of the dendrites (Brock, Coombs, & Eccles, 1952a). This might conceivably wipe clean the slate each time a neuron fires, as well as tell each synapse that the cell had fired. There is evidence suggesting that such a mechanism might exist in spinal motoneurons. Virtually complete destruction of pre-existent postsynaptic potentials have been seen following soma-dendritic spikes in spinal motoneurons (Brock, Coombs, & Eccles, 1952b). Further data on this important point would be welcome.

Feature Detection

In certain parts of the neocortex, especially those near the sensory inputs, it has been shown that a particular neuron will respond best to a certain set of "features" in the input. Thus a so-called "simple cell" in the first visual area responds best to a line or edge of a particular orientation in a particular place in the visual field (Hubel & Wiesel, 1968). It may prefer a certain small range of spatial frequencies (Schiller, Finlay, & Volman, 1976). It may respond better to movement in one direction rather than that in the opposite direction (Hubel & Wiesel, 1968). Some cells are sensitive to the horizontal "disparity" between the input to the two eyes (Hubel & Wiesel, 1970a). Other cells may respond better to some wavelengths of light than to others (Michael, 1978) and so on.

The description and classification of features are matters of the first importance, but even for the early parts of the visual system our knowledge is still very incomplete. These details are of the greatest interest to theorists actively studying a particular system. Here we can only mention them in passing.

As one proceeds further from the sensory input, the mapping of the periphery becomes more diffuse. At the same time most of the neurons tend to respond to one or another feature of the stimulus— movement, color, etc.—while still others respond to more elaborate features in the input. For example, Zeki has shown that in the visual area V4 of the macaque, a cell's response to the wavelength of the light depends somewhat on the wavelength coming from fairly distant regions in the visual field (Zeki, 1978). In the first visual area, on the other hand, a neuron's response to wavelength is much more local (Michael, 1978). This makes sense since although the axon collaterals spread a similar cortical distance in both areas, V4 is smaller than V1, so that in the former the collaterals can reach more distant parts of the visual field, especially as the "mapping" in V4 is more diffuse than it is in V1 (Van Essen & Zeki, 1978).

As one proceeds further into the system it becomes increasingly difficult to discover exactly what "feature" a cell likes best. A few neurons in the cortex, lining the superior temporal sulcus and in the inferior temporal cortex of monkeys, appear to respond to pictures of faces (Bruce, Desimone, & Gross, 1981; Desimone, Albright, Gross, & Bruce, 1980; Perrett, Rolls, & Caan, 1982). If the eyes are blocked out in the pictures, the neurons fire less. If the features of the face are jumbled, the neurons do not respond at all. The exact location and the orientation of the pictures do not appear to be critical to the response of these neurons. It is claimed that other neurons in these complex cortical areas respond specifically to hands (Desimone et al., 1980), but for most of the neurons in these cortical areas, the best type of stimulus has eluded discovery. This is a case where a good theory might give useful pointers to an experimentalist. Since a neuron's output is simply the spikes it sends down its axon, the same pattern of spikes can conceivably be produced by a whole variety of different but related inputs. Thus, in a very real sense the firing of a single neuron conveys somewhat ambiguous information. It is widely believed that it is the pattern of a set of neurons which is best thought of as conveying information. It thus becomes important to know whether the input from a *single* cortical neuron can, by itself, fire a particular cell in its projection field or whether several neurons are required to do this. There is little evidence in the neocortex that a single neuron can, by itself, fire a cell, but exactly how many are needed can only be guessed at. Even where we know that the axon of one neuron contacts the dendrites of another neuron, we usually do not know *how many* distinct synapses it makes on the second cell. Methods that could answer this question would be very valuable.

Conceptually the problem is not straightforward, since the background "noise" (the occasional random firing of many neurons) is likely to bring a neuron closer to threshold and thus reduce the number of more active neurons required to fire the cell. Spontaneous, transient fluctuations of the membrane potential have been seen in all intracellularly examined cortical neurons (Connors, Gutnick, & Prince, 1982).

There seems to be rather little *correlated* firing of neurons in the neocortex. (For details, see the monograph by Abeles, 1982.) That is, neighboring neurons seldom fire at precisely the same time. When neurons do fire with some degree of synchrony, as they appear to do to produce alpha-waves in the EEG, they do so mainly when the mind appears to be idle. This apparent lack of synchronized firing suggests that the brain is not organized, as a modern digital computer is, on a repeating time cycle. However, the thalamus does appear to impose a degree of rhythm on its output, so it is possible that more subtle time effects exist and have been missed.

An important unresolved question concerning feature detection is whether it is inborn or learned. This is a difficult topic and we can only touch on it. At the moment it seems likely that "feature detectors" are to a considerable extent inborn but can be tuned up by experience, especially during certain "critical periods" in development. The neocortex does not appear in its details to be a general-purpose computing machine. Each area (with its connections) seems designed to carry out a specific function, even though this can be modified somewhat by experience. Natural selection has had every opportunity to develop the necessary neuronal machinery to cope with issues which, because of the nature of the external world, have changed little in evolution. Theorists almost always assume that they are cleverer than natural selection. This is usually a mistake.

In spite of the above remarks, it is reasonable to assume that the neocortex has evolved because it is good at a particular sort of computation and that, with appropriate local variations, it may be broadly the same in all parts. We may also expect that these basic processes will be complicated by somewhat elaborate neural gadgetry designed to make for better performance. What these basic computational processes are remains to be discovered.

Other sections of this book concern models constructed out of "units." These units often have properties similar in some respects to neurons in that they have multiple inputs, some sort of summation rule, a threshold rule, and a single output which is usually distributed to several other units. However, their inventors are always careful to point out that they are not intended to represent real neurons. Indeed, at this stage in the game, it would be foolish to attempt to do this. Nor are most of the units used satisfactory idealizations of real neurons. If the properties of real neurons present useful gadgets to neural modelers, they should not be mixed together in combinations that never occur together in the brain.

Another explanation offered by modelers in defense of their units is that a single unit really stands for a *group* of neurons. This might be acceptable to neuroscientists if it were carefully stated how this group might be built out of more or less real neurons, but this is seldom if ever done. Consequently, it is difficult to know whether a given "unit" is plausible or not.

Another approach to the difficulty is to claim that "units are place-holders for informational states" (J. L. McClelland, personal communication) and that the relationship between the neurons and such states may be complex. This may indeed be plausible, but from the neuroscientist's point of view it makes it almost impossible to test the models unless the relationship is spelled out in detail.

Another difficulty is that neural modelers seldom state exactly what their models are supposed to demonstrate. This difficult question is addressed more fully in the last chapter of this book.

Meanwhile we list here briefly some of the devices loved by theorists which, if interpreted literally, are not justified by the available experimental evidence:

- Neurons that excite some cells and inhibit others.

- Neurons that merely "change sign." For example, a neuron that accepts excitation from one neuron only and whose onput produces inhibition on one neuron only.

- Neurons that connect to all other cells of the same type.

- Neurons with distinctive synapses that do elaborate computations. Apart from spines which sometimes have both a Type I and a Type II synapse, such complications are rare or absent in the neocortex, though they do occur in the thalamus. However, separate dendrites may perform such a role in the cortex.

- A neuron that, by itself, can fire another cell. This does occur in the cerebellum (the climbing fiber on a Purkinje cell) (Eccles, Llinás, & Sakaski, 1966). It is not certain that it does not occur in the neocortex but the available evidence suggests that it is not common. However, chandelier cells and basket cells may, by themselves, be able to veto the firing of another cell.

The following are features found in the neocortex but often not used in theoretical models:

- Veto cells, which appear to veto many other cells and which probably need the summated activity of several distinct inputs to fire them.

- The various diffuse inputs, from the brain stem and elsewhere, which may be important, not only for the general level of arousal of the cortex (as in sleep) but also for potentiating the synaptic modification involved in laying down a memory.

ACKNOWLEDGMENTS

We thank our colleagues at the Salk Institute for many helpful discussions. We are especially grateful to W. Maxwell Cowan and Jay McClelland for their detailed comments on the manuscript. We also thank Betty Lang for her careful typing of the various drafts.

This work was supported by the J. W. Kieckhefer Foundation, the Samuel Roberts Noble Foundation, the System Development Foundation, and NIH Fellowship NS-07061.

CHAPTER 21

Open Questions About
Computation in Cerebral Cortex

T. J. SEJNOWSKI

In Chapter 20, Crick and Asanuma have attempted the difficult task of summarizing what is presently known about the physiology and anatomy of cerebral cortex. Here I will attempt to summarize what is not known. The goal of this chapter is to provide a framework within which to ask computational questions about cerebral cortex.

QUESTIONS ABOUT CEREBRAL CORTEX

Different areas of cerebral cortex are specialized for processing information from different sensory modalities, such as visual cortex, auditory cortex, and somatosensory cortex, and other areas are specialized for motor functions; however, all of these cortical areas have a similar internal anatomical organization and are more similar to each other in cytoarchitecture than they are to any other part of the brain. The relatively uniform structure of cerebral cortex suggests that it is capable of applying a general-purpose style of computation to many processing domains, from sensory processing to the most abstract reasoning. Whether the similarity between different areas of cortex is merely superficial or extends to the computational level is an experimental question that depends on theoretical issues.

Information processing and memory share the same circuitry in cerebral cortex, in contrast with digital computers where the memory and central processing unit are physically separated. The style of computation and the style of memory must therefore be closely related. This requirement is a very powerful one and should help narrow the range of possible candidate computational styles because, in addition to showing that a class of algorithms has significant processing capabilities, it is necessary to also show that the performance of the algorithms can be seamlessly improved by experience. This intimate relationship between the hardware and the software may make it possible to use constraints from both the computational level and the implementational level to investigate experimentally the representations and algorithms used in each cortical area (Ballard, in press; Sejnowski, in press).

The key issue about which we know least is the style of computation in cerebral cortex: How are signals in neurons used to represent information? How do networks of neurons cooperate in transforming the information? How are the results of a computation stored for future use? These questions will be the focus of this chapter, which concludes with some remarks on the role of computational models in understanding complex systems like cerebral cortex.

REPRESENTING INFORMATION

Almost all information that must be transmitted by neurons over distances greater than 1 millimeter is coded into action potentials. These all-or-none spike discharges last for about 1 millisecond and carry information by their presence or absence. When the technique for reliably recording action potentials from single cortical neurons was introduced, it was a surprise to many that the response from some cortical neurons in somatosensory cortex and visual cortex could be correlated with simple features of sensory stimuli (Hubel & Wiesel, 1962; Mountcastle, 1957). This early success put a special emphasis on the cellular level rather than either the subcellular or network levels and led to the idea that single neurons coded simple sensory features and perhaps simple percepts as well (Barlow, 1972).

It should be emphasized, however, that rarely does a single neuron respond solely to a single feature dimension and that the tuning curves along feature dimensions are usually broad. Thus, single neurons in sensory cortex can be thought to represent volumes in a high-dimensional space of features: The firing rate of a single cortical neuron no longer represents the analog value of a variable directly, but rather the probability of a variable lying within some volume of the space of features (Ballard, Hinton, & Sejnowski, 1983).

The perceptual interpretation of a local feature depends on the context of the visual scene in which the feature is embedded. If the response of a single neuron were to represent not merely a conjunction of local features, but an interpretation of those features in the context of the image, then the response should be influenced by parts of the image outside the classical receptive field. Recently, evidence has been found for strong surround effects in visual cortex which are antagonistic to the response properties of the receptive fields (Allman, Miezin, & McGuinness, 1985).

What makes these new surround effects especially interesting is that they are selective. As shown in Figure 1, some neurons with directionally selective receptive fields in extrastriate cortex can have their best responses modulated 100% depending on the direction of movement of the surrounding visual field (Allman et al., 1985). The surround effects in the middle-temporal area (MT), where receptive fields are typically 5-10°, can extend 40-80°. Significant surround effects related to illusory contours have also been reported in area V-II (von der Heydt, Peterhans, & Baumgartner, 1984), as shown in Figure 2. In another region of visual cortex, the V4 complex, neurons have been found whose surrounds are selectively tuned for orientation, spatial frequency, and color (Desimone, Schein, Moran, & Ungerleider, 1985). Some of the neurons in this area appear to be selective for color over a wide range of illumination: The wavelength-dependent response in the receptive field is influenced by the color balance of the surround. (Zeki, 1983a, 1983b).

These surround effects may be important for perceptual phenomena like motion parallax and color constancy that require comparison of local features within a larger context of the visual field (Allman, et al., 1985). The basis of these long-range influences is not known, but several sources may contribute: First, stimulus-specific information could spread laterally within cortex through intrinsic horizontal axonal arborizations that extend 2-4 mm (Gilbert & Wiesel, 1983; Rockland & J. S. Lund, 1983); second, reciprocal connections between cortical maps, particularly the descending feedback projections, could have extensive spread; third, inputs from noncortical structures such as the claustrum (LeVay & Sherk, 1981b) could influence the surrounds; and fourth, transcollosal connections might carry surround information, particularly between regions across the vertical meridian (Desimone et al., 1985).

The discovery of large nonclassical surrounds provides an important opportunity to explore the collective properties of cortical processing. The response properties within the classical receptive field probably represent local, intrinsic processing, but the surround effects represent the long-range pathways and the spread of information within cortical areas. The spread is stimulus specific and should prove to be as important as the primary response properties of the receptive field itself. For

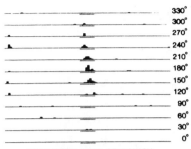

BACKGROUND DOTS STATIONARY

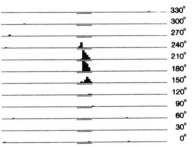

BACKGROUND DOTS MOVE AT 0°

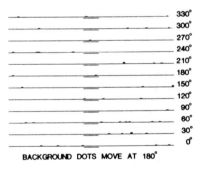

BACKGROUND DOTS MOVE AT 180°

FIGURE 1. Responses of a neuron in middle temporal (MT) cortex to a bar moving in different directions superimposed on a background of random dots. The bar was oriented orthogonally to the direction of movement. The results of each of the 12 directions (0° through 330°) are shown in histograms consisting of a before period, an underscored stimulus presentation period, and an after period. The largest histogram bin contains 26 spikes. When the background is moving in the same direction as the bar the response is entirely abolished, but when its movement is opposed to the direction of the bar the response is enhanced. (From "Stimulus Specific Responses From Beyond the Classical Receptive Field: Neurophysiological Mechanisms for Local-Global Comparisons in Visual Neurons" by J. Allman, J. Miezin, and E. McGuinness, 1985, *Annual Review of Neuroscience, 8*, p. 416. Copyright 1985 by Annual Reviews, Inc. Reprinted by permission.)

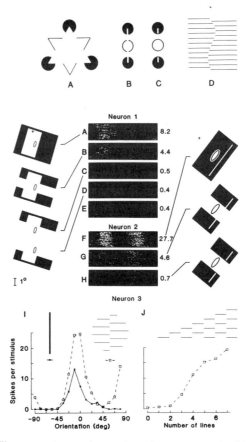

FIGURE 2. *Top:* Illusory contours. Apparent contours are perceived in *A*, *B*, and *D* at sites where the stimulus is consistent with an occluding contour. Small alterations in the stimulus can have dramatic effects on the appearance of these contours, as in *C* where thin lines have been added to *B*. *Bottom:* Responses of neurons in extrastriate visual cortex (area 18) of the monkey to edges, bars and stimuli producing illusory contours. The stimuli (insets) were moved back and forth across the receptive fields (ellipses). In each line of the raster scan a white dot indicates the occurrence of an action potential as a function of time. The mean number of spikes per stimulus cycle is indicated to the right. Neuron 1, which responded to the lower right edge of the light bar *(A)*, was activated also when only the illusory contour passed over its classical receptive field *(B)*. Either half of the stimulus alone failed to evoke a response *(C* and *D)*. Spontaneous activity is shown in *E*. Neuron 2 responded to a narrow bar *(F)* and, less strongly, to the illusory bar stimulus *(G)*. When the ends of the "bar" were intersected by thin lines, however, the response was nearly abolished *(H)*. In Neuron 3, the border between two abutting gratings elicited a strong response. The orientation tuning curves show corresponding peaks for bar and illusory contour *(I)*. When the lines inducing the contour were reduced in number to less than three, the response disappeared *(J)*. In contrast, neurons in primary visual cortex (area 17) did not respond to any of these stimuli. (From "Illusory Contours and Cortical Neuron Responses" by R. von der Heydt, E. Peterhans, and G. Baumgartner, 1984, *Science, 224*, p. 1261. Copyright 1984 by the American Association for the Advancement of Science. Reprinted by permission.)

example, different orientation-sensitive neurons could respond differently to a local edge depending on the meaning of the edge within the context of the surrounding image: whether it represents an occluding contour, surface texture, or a shadow boundary (Sejnowski & Hinton, in press).

The analysis of a single image requires the processing power of the entire visual system; in a sense, the interpretation of the image is the state of all the neurons in the visual system. In the language of features one would say that the object is internally represented by the particular combination of features currently activated. A problem arises when it is necessary to compare two objects, such as faces, seen at different times. One needs some means for binding together the most important combination of features at one moment and storing them for future use.

This binding problem is particularly difficult because most regions of cortex are only sparsely interconnected. If two groups of neurons have few direct connections then it is difficult to imagine how a conjunction of two facts represented in the two regions can somehow be stored. Several solutions to this binding problem have been suggested, but no one yet knows how it is actually solved in the nervous system. The binding problem is a touchstone for testing network models that claim to have psychological validity. For a discussion of binding in the context of shape recognition see Hinton (1981c).

One approach to the binding problem is to represent a complex object by the activity of only a few neurons, a so-called local representation similar to the "grandmother cell" hypothesis (J. A. Feldman, 1981). In this case the binding of two facts can be accomplished by dedicating one or more intermediate links between the neurons representing the features and the neurons representing the object. One problem with this approach is the combinatorial explosion of neurons and links that must be dedicated to the representation of even modestly complex objects. One consequence of the localist solution to binding is that some decisions to take an action may be based on the state of very few links between very few units. There is no convincing evidence for such "command fibers" anywhere in cerebral cortex. An alternative solution to the binding problem, based on a "searchlight of attention" is discussed in the section on temporal coincidence.

In summary, it appears from studies of single neurons in visual cortex that they generally respond to a conjunction of features on a number of different dimensions. The sharpness of the tuning for values on different dimensions varies, but in general, each neuron is rather coarsely tuned, and its receptive field overlaps with the receptive fields of other neurons. Coarse coding of features also holds for other sensory areas of cortex and motor cortex (Georgopoulos, Caminiti,

Kalaska, & Massey, 1983), although the evidence there is not nearly as good as in the visual system. These observations are consistent with the ideas about distributed representations described in Chapters 3, 7, and 18.

NEURONAL PROCESSING

The time available for processing puts strict constraints on the types of algorithms that could be implemented in cerebral cortex. Following a briefly presented image, the information coded as a pattern of hyperpolarizations in the photoreceptors is processed in the retina and coded into trains of action potentials by the retinal ganglion cells. Within about half a second following presentation of the image we can recognize an object in the image. Because photoreceptors are slow, a significant fraction of the response time, about 25–50 milliseconds, is required for the information to reach cortex and several hundred milliseconds are required for the motor system to produce a response, which leaves about 200–300 milliseconds for visual processing. This is a severe restriction on algorithms, such as cooperative ones, that require extensive exchange of information between local neurons. David Marr (1982) concluded that:

> cooperative methods take too long and demand too much neural hardware to be implemented in any direct way. The problem with iteration is that it demands the circulation of numbers around some kind of loop, which could be carried out by some system of recurrent collaterals or closed loops of neuronal connections. However, unless the numbers involved can be represented quite accurately as they are circulated, errors characteristically tend to build up rather quickly. To use a neuron to represent a quantity with an accuracy of even as low as 1 in 10, it is necessary to use a time interval that is sufficiently long to hold 1 to 10 spikes in comfort. This means at least 50 ms per iteration for a medium-sized cell, which means 200 ms for four iterations—the minimum time ever required for our cooperative algorithm to solve a stereogram. And this is too slow. (p. 107)

The timing problem is even more severe than Marr states because the responses of single neurons in cortex often vary from trial to trial and there are usually only a few spikes in a single response (Figure 3).

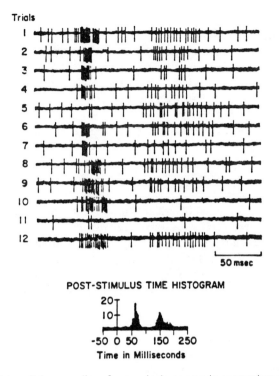

FIGURE 3. Extracellular recordings from a single neuron in extrastriate visual cortex of the cat. This neuron responded best to a slit of light obliquely oriented in a particular part of the visual field. The first 12 successive responses of the neuron to 50 exposures of the light are shown above, and the average response for 20 trials is shown below (Post-Stimulus Time Histogram). Even before the onset of the stimulus the neuron was spontaneously active. Although the pattern of firing varied from trial to trial, the ensemble average of the responses is repeatable. (From "Integrative Properties of Parastriate Neurons" by F. Morrell, in *Brain and Human Behavior*, p. 262, edited by A. G. Karczmar and J. Eccles, 1972, Berlin: Springer. Copyright 1972 by Springer-Verlag. Reprinted by permission.)

Therefore, in many experiments the spike train is averaged over a number of responses (typically 10) to obtain a post-stimulus time histogram. The histogram represents the probability of a spike occurring during a brief interval as a function of time following the stimulus. This suggests that for short intervals (5–10 milliseconds) and especially during nonstationary conditions, stochastic variables may be more appropriate than the average firing rate (Hinton & Sejnowski, 1983; Sejnowski, 1981).

A probabilistic code means that the probability of firing, rather than being represented by a number that must be accurately transmitted, can be represented directly as the probability for a neuron to fire during a short time interval. The use of a probabilistic code rather than one based on the average value of spike firing reopens the possibility of cooperative algorithms because in 200 milliseconds it is possible to perform 40 or more iterations with 3-5 milliseconds per iteration, the minimum interspike interval. This is enough time for some cooperative algorithms to converge to a steady state solution through a process of relaxation (Sejnowski & Hinton, in press). Interestingly, it was found that adding noise to the system during the relaxation often improved the convergence by helping the system overcome locally inappropriate configurations and achieve the best overall global state. Some of the details and consequences of these probabilistic algorithms are discussed in Chapter 7.

Nonetheless, "single-shot" algorithms that converge in one pass through a network, such as a linear transformation follo ed by thresholding (Duda & Hart, 1973; Kohonen, 1977), remain attractive, especially for the early stages of visual processing that are fairly automatic. Even for problems that require the global comparison of features, it would be desirable wherever possible to minimize the number of passes that information must make through a circuit without new information being added. However, contextual influences on processing might still require iterative computation, as illustrated in models like the interactive activation model of word perception.

The single-shot and relaxation strategies have been presented as alternatives, but a better way to view them is as extremes in a continuum of strategies, any of which may be adopted by cortex depending on the level in cortex and the nature of the problem. For the recognition of common objects, a network that can process the image in a single-shot may be learned through experience. For novel or more complex objects, a relaxation strategy may be more flexible and have a higher capacity. Since the relaxation strategy must start out with a guess anyway, it may as well be a good one and get better with practice. This is also a graceful way to use the finite resources available in visual memory. In Chapter 7, an example is given of a relaxation strategy which can improve its performance with practice.

Several nonlinear parallel models (J. A. Anderson, 1983; McClelland & Rumelhart, 1981; see also Chapters 14 through 17) make use of units that have continuous activation values. While the membrane potential of a neuron does have an approximately continuous value, the interaction between neurons with action potentials is clearly not continuous. Several ways have been proposed to relate the variables in these models more closely with neural properties. First, the continuous value may be considered an average firing rate; however, as explained earlier, the time average firing rate is ill-defined over short time intervals. Second, a single unit could correspond to a population of neurons, and the activation would then represent the fraction of the neurons in the ensemble that are firing during a short time interval (Wilson & Cowan, 1972). A third possibility is that the units are dendritic branches that interact nonlinearly.

Until recently, the dendrites of most neurons were thought to be governed mainly by the passive linear properties of membranes (Rall, 1970) and to serve mainly in the integration rather than processing of synaptic signals. If, however, there are voltage-dependent channels in dendrites, then the signals represented by membrane potentials are nonlinearly transformed and the dendrites must then be studied in smaller units, perhaps as small as patches of membranes (J. P. Miller, Rall, & Rinzel, 1985; D. H. Perkel & D. J. Perkel, 1985; Shepherd et al., 1985). The membrane patches and dendritic branches still have an integrative function, but to analyze that function requires a finer-grain analysis (Figure 4).

At an even finer level individual synapses may themselves interact nonlinearly because the conductance change at one synapse may serve as a current shunt and alter the driving force for other nearby synapses (Koch, Poggio, & Torre, 1982; Rall, 1970). This is particularly true for synapses that occur on dendritic shafts but is less important for synapses that are electrically isolated on spines. With nonlinear interactions between synapses and between patches of dendritic membrane many more processing units are available but this advantage is partially offset by the limited topological connectivity of dendritic trees. We need to explore the range of transformations that can be performed inside neurons and build up a vocabulary of elementary computational operations (Shepherd, 1978, 1985).

Whether nonlinear computations are performed at a subneuronal level is of fundamental importance to modeling parallel networks. If, at one extreme, each neuron acts like a classical linear integrator, then the sigma-pi units discussed in Chapters 2, 10, and 16 would have to be implemented in the brain using a single neuron for each multiplicative connection. If, on the other hand, nonlinear interactions can be implemented at a subneuronal level, our estimate of the computational power of a fixed number of neurons would be greatly increased, and it would be possible to directly implement several proposals that require multiplicative connections, such as Ballard's (1984) implementation of Hough transforms; Hinton's (1981c) mapping from a retinocentric to a viewer-centered frame of reference; or McClelland's (1985) method for "programming" the weights between one set of units using signals set up in another set of units. It is worth noting, though, that these schemes would generally require very precise anatomical connectivity and very accurate timing of signals to get the most out of the nonlinear processing capabilities of dendrites.

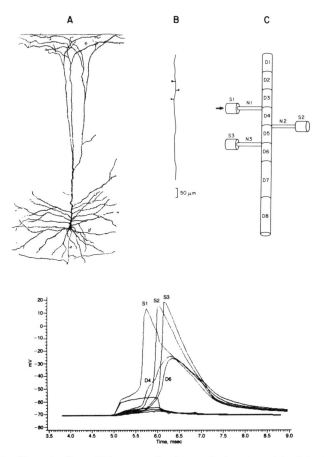

FIGURE 4. *Top:* *A*: Pyramidal neuron in the cerebral cortex stained by the Golgi method of impregnation. *B*: Schematic diagram of the terminal segment of an apical dendritic branch in the most superficial cortical layer in *A*. Only three spines out of the total array are indicated, spaced 50 μm apart. *C*: Diagrammatic representation of a compartment model of *B*. Dendritic compartments are symbolized by D, the necks of spines by N, and the spine heads by S. In addition to passive membrane resistance and capacitance, active sodium and potassium currents, following the Hodgkin-Huxley model, were incorporated into the spine heads. This is a simplified model of only a small portion of the apical dendrite. *Bottom:* Current pulses were injected into spine head S1. The traces show the simulated membrane potential in various compartments of the dendrite and spines following either a subthreshold current pulse or a suprathreshold current pulse. Note that when a spike occurs in S1 it propagates by saltatory conduction through the dendritic compartments down the chain of spines. Spines may also interact if they simultaneously receive input currents: The combination may reach threshold even though the inputs individually produce only subthreshold membrane potentials. (From "Signal Enhancement in Distal Cortical Dendrites by Means of Interactions Between Active Dendritic Spines" by G. M. Shepherd, R. K. Brayton, J. P. Miller, I. Segev, J. Rinzel, W. Rall, 1985, *Proceedings of the National Academy of Sciences USA, 82,* p. 2193. Reprinted by permission.)

TEMPORAL COINCIDENCE

Digital computers have a central clock that synchronizes the processing of signals throughout the central processing unit. No such clock has been found in the central nervous system on the millisecond time scale, but this does not rule out the importance of small time differences in neural processing. For example, the information about visual motion in primates at the level of the retina is represented as time differences in the firing pattern of axons in the optic nerve. In visual cortex, the relative timing information is used to drive cells that respond best to edges that are moving in particular directions (Koch & Poggio, 1985). In the realm of learning, the timing of sensory stimulation in the 10–50 millisecond range is known to be critically important for classical conditioning (Gluck & Thompson, in press; Sutton & Barto, 1981). Unfortunately, very little is known about the coding and processing of information as spatio-temporal patterns in populations of neurons. A few of the possibilities will be mentioned here.

The arrival time of impulses is extremely important in the auditory system where slight temporal differences between spike trains in the two cochlear nerves can be used to localize sound sources. It also appears that information in the relative timing of impulses in the same nerve is essential in carrying information in speech at normal hearing levels (Sachs, Voigt, & Young, 1983). Although it is difficult for neurons with millisecond time constants to make accurate absolute timing measurements, differences between arrival times down to 10 microseconds can be detected (Loeb, 1985) and therefore submillisecond timing information could also be important in cortex. This raises the possibility that the timing of arriving impulses might also be important in the cerebral cortex as well.

The transformation between the input current and the firing rate of the neuron has a range between threshold and saturation over which the relationship is fairly linear. However, at any given time only a small fraction of all cortical neurons is operating in the linear region. Therefore only a subpopulation of neurons is sensitive to the timing of synaptic events, namely, those neurons that are near the threshold region. This leads to the idea of a skeleton filter—the temporary network of neurons near threshold that can linearly transform correlations between spikes on input lines (Sejnowski, 1981). This is a way for temporarily changing the effectiveness with which some synapses can transmit information on the timing of synaptic events without actually altering their strength. It is not as flexible as a general modification scheme, such as the one suggested by McClelland (1985), because all the synapses originating from one neuron are modified by the same factor.

Recently, von der Malsburg (in press) has suggested a scheme for binding together distributed circuits that represent a set of facts by their simultaneous activation. He speculates that this temporal binding is implemented by the rapid changes in the strengths of synapses between neurons that are co-active within a few milliseconds. Crick (1984) has modified this proposal by suggesting that the binding occurs during longer intervals of about 50 milliseconds during which bursts of impulses, produced by "searchlights" of attention in the thalamus, provide the signal for rapid synaptic changes (Figure 5). The advantage of this approach is that the representations are distributed over a population of neurons and that simultaneous co-activation of a group of neurons in one area will impose simultaneous co-activation in another group that receives a projection from the first.

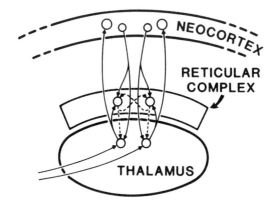

FIGURE 5. The main connections of the reticular complex, highly diagrammatic and not drawn to scale. Solid lines represent excitatory axons, dashed lines show inhibitory axons, and arrows represent synapses. The principal neurons in the thalamus that project to the neocortex have two modes of response depending on the membrane potential. If the membrane potential is initially depolarized, then ascending input to the thalamus (from the retina, for example) causes the principal neuron to fire at a moderate rate roughly proportional to the input. If, however, the neuron is initially hyperpolarized, for example by inhibitory inputs from the reticular complex, then the output from the principal cell is a rapid burst of spikes. According to the searchlight hypothesis, focal inhibition arising from the reticular nucleus produces sequentially occurring bursts in subsets of active thalamic neurons. The bursts are thought to last about 50 milliseconds and to produce short-term transient alterations in the synaptic strengths in cerebral cortex. (From "Function of the Thalamic Reticular Complex: The Searchlight Hypothesis" by F. H. C. Crick, 1984, *Proceedings of the National Academy of Sciences USA, 81,* p. 4587. Reprinted by permission.)

If a temporal code is used in cortex, then spatio-temporal correlations should show up in recordings from groups of neurons. In recordings from nearby cortical neurons, spatio-temporal correlations have been observed between spike trains, but the significance of these correlations is unclear (Abeles, 1982; Abeles, de Ribaupierre, & de Ribaupierre, 1983; Gerstein, Bloom, Espinosa, Evanczuk, & Turner, 1983; Shaw, Silverman, & Pearson, 1985; Ts'o, Gilbert, & Wiesel, 1985). However, it is already clear from these pioneering observations that the complexity of spatio-temporal signals from two or more neurons will require new techniques for data analysis and presentation. Several groups now have the technical ability to record simultaneously from many isolated neurons in cerebral cortex (Gerstein et al., 1983; Kuperstein & Eichenbaum, 1985; V. B. Mountcastle, personal communication, 1985; Reitboek, 1983). It will be especially interesting to record from alert animals attending to sensory stimuli and to search for correlated bursts of spikes and temporal coherence between spike trains.

NEURONAL PLASTICITY

Not only does cortex provide the capability of fast processing but it can be partially reconfigured with experience. There are experimental hints that the functional connectivity within cerebral cortex is far more fluid than previously thought. In a series of careful studies, Merzenich, Kaas and their colleagues have demonstrated that the spatial map of the body surface on the surface of primary somatosensory cortex of monkeys can be significantly altered by changes in activity.

In one series of experiments the map of the hand in somatosensory cortex was determined by multiple electrode penetrations before and after one of the three nerves that innervate the hand was sectioned (Merzenich et al., 1983), as illustrated in Figure 6. Immediately following nerve section most of the cortical territory which previously could be activated by the region of the hand innervated by the afferent nerves became unresponsive to somatic stimulation. In most monkeys, small islands within the unresponsive cortex immediately became responsive to somatic stimulation from neighboring regions. Over several weeks following the operation, the previously silent regions became responsive and topographically reorganized. In another set of experiments a hand region in somatosensory cortex was mapped before and after prolonged sensory stimulation of a finger; the area represented by the finger on the surface of cortex expanded and the average size of the receptive fields within the finger region diminished (Jenkins, Merzenich, & Ochs, 1984).

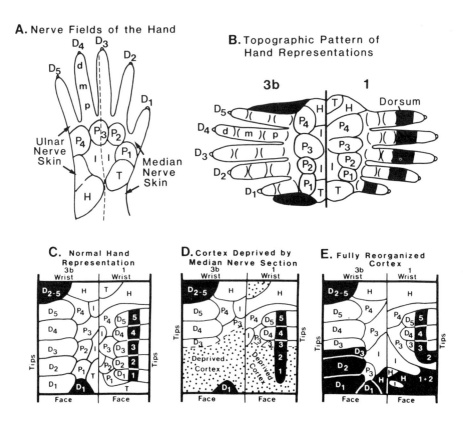

FIGURE 6. The effect of median-nerve section and ligation on the representations of the hand in somatosensory Areas 3b and 1 of the owl monkey. *A*: The radial side of the glaborous hand is innervated by the median nerve, the ulnar side by the ulnar nerve, and the dorsal surface by the ulnar and radial nerves. Digits (D) and palmar pads (P) are numbered in order. Insular (I), hypothenar (H), and thenar (T) pads, and distal (d), middle (m), and proximal (p) phalanges are also indicated. *B*: Pattern of topographic organization of the two hand representations, indicated without accurately reflecting cortical surface areas devoted to the representation of various surfaces of the hand. Cortex devoted to representation of dorsal surfaces of the digits is shown in black. *C*: Typical organization of the hand representations in a normal owl monkey. *D*: The same area of cortex following section of the median nerve. The part of cortex indicated by dots is deprived of normal activation by sensory stimulation. *E*: The organization of the two hand representations several months after median-nerve section and ligation. Much of the deprived cortex is activated by stimulation of the dorsal digit surfaces and the dorsum of the hand (black). In addition, palmar pads innervated by the ulnar nerve have an increased cortical representation. All "new" inputs are topographically ordered. Peripherally, the ulnar and radial nerves do not grow into the anesthetic median-nerve skin field. (From "Reorganization of Mammalian Somatosensory Cortex Following Peripheral Nerve Injury" by M. M. Merzenich and J. H. Kaas, 1982, *Trends in Neuroscience*, 5, p. 435. Copyright 1982 by Elsevier Biomedical Press. Reprinted by permission.)

The uncovering of previously "silent" synapses is the favored explanation for these experiments because the maximum shift observed, a few millimeters, is about the size of the largest axonal arborizations within cortex. The apparently new receptive fields that were "uncovered" immediately after nerve section could represent a rapid shift in the dynamical balance of inputs from existing synapses, and the slower reorganization could be caused by changes in synaptic strengths at the cortical and subcortical levels. This raises several crucial questions: First, what fraction of morphologically identified synaptic structures are functionally active in cerebral cortex? It is not known, for example, how many quanta of transmitter are released on average at any central synapse. Second, how quickly if at all can a synapse be transformed from a "silent" state to an active state, and what are the conditions for this transformation? Finally, how is this plasticity related to the representation and processing of sensory information? Perhaps the most serious deficiency in our knowledge of cerebral cortex concerns the physiology of individual central synapses, which are inaccessible by conventional techniques owing in part to the complexity of the cortical neuropil. New optical techniques for noninvasively recording membrane potentials and ionic concentrations may someday make it possible to study dynamic changes at central synapses (Grinvald, 1985).

The evidence for a rearrangement of the body map on the surface of cerebral cortex during experimental manipulations raises interesting questions about perceptual stability because this reorganization is not accompanied by confused or mistaken percepts of the body surface (Merzenich & Kaas, 1982). This suggests that as a neuron shifts its input preference, other neurons that receive information from it must reinterpret the meaning of the signal. If the connectivity of cerebral cortex is as dynamic under normal conditions as these experiments suggest, then many of the issues that have been raised in this chapter must be re-examined from a new perspective (Changeux, Heidmann, & Patte, 1984; Crick, 1984; Edelman, 1981).

ROLE OF COMPUTATIONAL MODELS

A complete description of every neuron, every synapse, and every molecule in the brain is not synonymous with a complete understanding of the brain. At each level of description the components must be related to the phenomena which those components produce at the next highest level, and models are a succinct way to summarize the relationships. A classic example of a successful model in neuroscience is the Hodgkin-Huxley model of the action potential in the squid axon. Here the bridge was between microscopic membrane channels (hypothetical at the time) and macroscopic membrane currents. The first step in making a model is to identify the important variables at both the lower and upper levels; next, a well-defined procedure must be specified for how these variables interact (an algorithm); and finally, the conditions under which the model is valid must be stated so that it can be compared with experiments.

The models that have been explored in this book do not attempt to reconstruct molecular and cellular detail. Rather, these connectionist models are simplified, stripped-down versions of real neural networks similar to models in physics such as models of ferromagnetism that replace iron with a lattice of spins interacting with their nearest neighbors. This type of model is successful if it falls into the same equivalence class as the physical system; that is, if some qualitative phenomena (such as phase transitions) are the same for both the real system and the model system (Ma, 1976). When they are successful these simple models demonstrate the sufficiency of the microscopic variables included in the model to account for the macroscopic measurements.

The emergence of simple parallel models exhibiting nontrivial computational capabilities may be of great importance for future research in neuroscience because they offer one of the few ways that neuroscientists can test qualitative ideas about the representation and processing of information in populations of neurons. Suppose that single neurons in an area responded to features of the visual input that could be important for computing, say, optical flow. Knowing the goal of the computation, one could design a parallel algorithm for implementing the computation of optical flow and then test it with a wide range of inputs. The process of specifying and testing an algorithm often reveals unexamined assumptions and refines the original motivation for the model. If one successful algorithm is found then the computational feasibility of the original proposal is strengthened; to test whether some form of the algorithm is actually implemented in cortex would be much more difficult; ultimately, the performance of the algorithm has to be compared with psychophysical testing.

Some neuroscientists may feel uncomfortable because connectionist models do not take into account much of the known cellular properties of neurons, such as the variety of membrane channels that have been found. What if the processing capabilities of cerebral cortex were to depend crucially on some of these properties? In this case it may not be possible to get networks of oversimplified model neurons to solve difficult computational problems, and it may be necessary to add new properties to the model neuron. The added capabilities would yield a better understanding of the roles played by these neural properties in processing information, and suggestions could emerge for useful properties which have not yet been observed. The present models are guideposts for thinking about the computational capabilities of neural networks and benchmarks that set the standards for future models.

One of the key insights that has already emerged from studying one class of simple nonlinear networks with recurrent collaterals is that amongst the large number of possible states of a network, only relatively few of these states, called attractors, are stable (J. A. Anderson, 1983; M. A. Cohen & Grossberg, 1983; Hinton & Sejnowski, 1983; Hogg & Huberman, 1984; Hopfield, 1982; Hopfield & Tank, 1985; Sejnowski, 1976; Wilson & Cowan, 1972). The existence of stable attractors is a feature that is likely to generalize to more complex networks. Objects and their relationships can be internally represented by these attractors, and the search for the best match between the world and the internal representation of the world by the dynamics of the network is much more powerful than previous template-matching procedures. This opens a large number of research problems, such as the issue of local vs. distributed representations and the binding problem, both of which have been discussed in this chapter. The identification and study of these issues in simple network models will greatly help us in understanding the principles that went into the design of the cerebral cortex.

ACKNOWLEDGMENTS

I am grateful to Neal Cohen, Francis Crick, Mark Gluck, Geoffrey Hinton, Chip Levy, and Gordon Shepherd who provided valuable suggestions on earlier versions of this chapter, and I especially wish to thank Jay McClelland for helping to shape the chapter in many ways, large and small.

CHAPTER 22

Neural and Conceptual Interpretation of PDP Models

P. SMOLENSKY

Mind and brain provide two quite different perspectives for viewing cognition. Yet both perspectives are informed by the study of parallel distributed processing. This duality creates a certain ambiguity about the interpretation of a particular PDP model of a cognitive process: Is each processing unit to be interpreted as a neuron? Is the model supposed to relate to the neural implementation of the process in some less direct way?

A closely related set of questions arises when it is observed that PDP models of cognitive processing divide broadly into two classes. In *local models*, the activity of a single unit represents the degree of participation in the processing of a known conceptual entity—a word, a word sense, a phoneme, a motor program. In *distributed models*, the strength of *patterns of activity over many units* determine the degree of participation of these conceptual entities. In some models, these patterns are chosen in a deliberately arbitrary way, so that the activity of a single unit has no apparent "meaning" whatever—no discernible relation to the conceptual entities involved in the cognitive process. On the surface, at least, these two types of models seem quite different. Are they as different as they seem? How are they related?

This chapter begins with a brief consideration of the neural interpretation of PDP models of cognition. These considerations serve mostly to lay out a certain perspective on the PDP modeling world, to make some distinctions I have found to be valuable, to introduce some terminology, and to lead into the main question of this chapter: How are distributed and local PDP models related? The chapter ends with a discussion of how, using the framework of PDP models, we might forge a mathematical relationship between the principles of mind and brain.

The following technique will be used to relate distributed to local models. We take a distributed model of some cognitive process, and mathematically formulate a *conceptual description* of that model, a description in terms of the conceptual entities themselves rather than the activity patterns that represent them. From some perspectives, this amounts to taking an account of cognition in terms of neural processing and transforming it mathematically into an account of cognition in terms of conceptual processing. The conceptual account has a direct relation to a local model of the cognitive process, so a distributed model has been mapped onto a local model.

The mathematical formulation of the conceptual description of a distributed model is straightforward, and the mathematical results reported in this chapter are all quite elementary, *once the appropriate mathematical perspective is adopted.* The major portion of this chapter is therefore devoted to an exposition of this abstract perspective on PDP modeling, and to bringing the consequent mathematical observations to bear on the cognitive issues under consideration.

The abstract viewpoint presented in this chapter treats PDP models as *dynamical systems* like those studied in mathematical physics. The mathematical concepts and techniques that will be used are those of linear algebra, the study of vector spaces; these techniques are discussed in some detail in Chapter 9. The formal parts of the discussion will be confined to footnotes and italicized passages; these may be skipped or skimmed as all results are discussed *conceptually* in the main portion of the text, which is self-contained.

NEURAL AND CONCEPTUAL INTERPRETATIONS

The interpretation of any mathematical model involves the mapping of a mathematical world into some observable part of the real world. The ambiguity in the interpretation of PDP models arises because the mathematical world of the model can be mapped into *two* observable worlds: the neural world and the world of cognitive behavior.

The neural world relevant here is discussed in Chapter 20—a world of receptor cells, neurons, synaptic contacts, membrane depolarizations, neural firings, and other features of the nervous system viewed at this level of description. The mathematical world is that described in Chapter 2—a world of processing units, weighted connections, threshold and sigmoidal functions, and activation.

Least precisely prescribed is the world of cognitive behavior. This world is mapped by experiments that probe perceptual behavior, reasoning and problem solving behavior, skilled motor behavior, linguistic behavior, memory task behavior, and the like. PDP models have been interpreted in terms of all these aspects of behavior, and to understand the common basis for these interpretations we must adopt a fairly general—and rather imprecise—way of speaking about cognition.

The connection between the formal structure of PDP models and cognitive behavior rests on theoretical knowledge constructs hypothesized to underlie this behavior. Consider perception first. Interpreting sensory input can be thought of as consideration of many *hypotheses* about possible interpretations and assigning degrees of confidence in these hypotheses. Perceptual hypotheses like "a word is being displayed the first letter of which is *A*," "the word *ABLE* is being displayed," and "the word *MOVE* is being displayed" are tightly interconnected; confidence in the first supports confidence in the second and undercuts confidence in the third. Thus assignment of confidence—*inference*—is supported by knowledge about the positive and negative evidential relations among hypotheses. This same kind of knowledge underlies other cognitive abilities; this kind of inference can support problem solving, the interpretation of speech and stories, and also motor control. The act of typing *ABLE* can be achieved by letting "the word *ABLE* is to be typed" support "the first letter to be typed is *A*" and inhibit "the first letter to be typed is *M*."

This way of thinking about cognition can be summarized by saying that behavior rests on a set of internal entities called hypotheses that are positively and negatively related in a knowledge base that is used for inference, the propagation of confidence. The hypotheses relate directly to our way of thinking about the given cognitive process; e.g., for language processing the hypotheses relate to words, syntactic categories, phonemes, meanings. To emphasize that these hypotheses are defined in terms of our concepts about the cognitive domain, I will call them *conceptual hypotheses* or the *conceptual entities*, or simply *concepts*; they are to be distinguished from the mathematical and neural entities—units and neurons—of the other two worlds.

The internal structure of the neural, mathematical, and conceptual worlds we have described are quite similar. Table 1 displays mappings that directly relate the features of the three worlds. Included are all the central defining features of the PDP models—the mathematical world—and the corresponding features of the portion of the neural and conceptual worlds that are directly idealized in PDP models.

TABLE 1

THE MAPPINGS FROM THE MATHEMATICAL WORLD TO
THE NEURAL AND CONCEPTUAL WORLDS

Neural	Mathematical	Conceptual
neurons	units	hypotheses
spiking frequency	activation	degree of confidence
spread of depolarization	spread of activation	propagation of confidence: inference
synaptic contact	connection	conceptual - inferential - interrelations
excitation/inhibition	positive/negative weight	positive/negative inferential relations
approximate additivity of depolarizations	summation of inputs	approximate additivity of evidence
spiking thresholds	activation spread threshold G	independence from irrelevant information
limited dynamic range	sigmoidal function F	limited range of processing strength

In Table 1, individual units in the mathematical world are mapped on the one hand into individual neurons and on the other into individual conceptual hypotheses.[1] These two mappings will be taken to define the *local neural interpretation* and the *local conceptual interpretation* of PDP models, respectively. These are *two separate* mappings, and a particular PDP model may in fact be intended to be interpreted with only one of these mappings. Using both mappings for a single model would imply that individual concepts, being identified with individual units, would also be identified with individual neurons.

In addition to the local mappings there are also *distributed* mappings of the mathematical world into each of the neural and conceptual worlds. In a distributed *conceptual* interpretation, the confidence in a

---

[1] What is relevant to this chapter are the general features, not the details, of Table 1. The precision with which the mappings are described is sufficient for present purposes. For a more precise account of the relation between the mathematics of certain PDP models and the mathematics of inference, see Hinton, Sejnowski, and Ackley (1984).

conceptual hypothesis is represented by the strength of a pattern of activation over a set of mathematical units. In a distributed *neural* interpretation, the activation of a unit is represented by the strength of a pattern of neural activity of a set of neurons.

It must be emphasized that the choices of neural and conceptual interpretations are truly independent. Some neural models (e.g., Hopfield, 1982) may have no direct conceptual interpretation at all; they are intended as abstract models of information processing, with no cognitive domain implied and therefore no direct connection with a conceptual world. The reading model of McClelland and Rumelhart (1981; Rumelhart & McClelland, 1982; see Chapter 1) has an explicit local conceptual interpretation; we can choose to give it no neural interpretation, a local neural interpretation (implying individual neurons for individual words), or a distributed neural interpretation. The Hinton (1981a) and J. A. Anderson (1983) models of semantic networks have explicitly distributed conceptual interpretations; they *can* be given a local neural interpretation, so that the patterns over units used in the models are directly interpreted as patterns over neurons. They can also be given a distributed neural interpretation, in which the units in the model are represented by activity patterns over neurons so that the concepts—patterns over units—correspond to new patterns over neurons.

The nature of the patterns chosen for a distributed interpretation— either neural or conceptual—can be important (although it is not always; this is one of the results discussed later). A distributed interpretation will be called *quasi-local* if none of the patterns overlap, that is, if every pattern is defined over its own set of units. Quasi-local distributed interpretation, as the name implies, forms a bridge between local and distributed interpretation: a quasi-local neural interpretation associates several neurons with a single mathematical unit, but only a single unit with each neuron.

Since quasi-local interpretations are special cases of distributed representations, the methods applied in this chapter to the general case of distributed representations could also be applied to quasi-local representations. Certain similarities to local representations can be expected to emerge, but the general results to be discussed suggest that it is a mistake to assume that the mathematical properties of the local and quasi-local cases will be essentially the same.

The primary reason for displaying Table 1 is to emphasize that the neural and conceptual bases for interest in PDP models are *completely* independent. Even if all neural interpretations are eliminated empirically, or if no neural interpretation is given a model at all, a conceptual interpretation remains a strong independent source of cognitive relevance for a PDP model.

At the same time, the independence of the neural and conceptual mappings makes it quite striking that *both* contact exactly the same class of mathematical models. Why should this be? Is it that we have ignored crucial features of the two worlds, features which would lead to quite different mathematical abstractions? A more encouraging possibility is this: Perhaps we *have* captured the essence of neural processing in PDP models. Perhaps implicit in the processing of neural firing patterns is another mathematical description, a description in terms of the *concepts* represented in those patterns. Perhaps when we analyze the mathematics of this conceptual description, we will find that it has the mathematical structure of a PDP model—that because of special properties of PDP models, at both the neural *and* conceptual levels of description, *the mathematical structure is the same.*

This wildly optimistic scenario (depicted schematically in Figure 1) will be called *the hypothesis of the isomorphism of the conceptual and neural levels*—the *isomorphism hypothesis* for short. We will find that it is in fact exactly true of the simplest—one might say the purest—PDP models, those without the nonlinear threshold and sigmoidal functions. For models with these nonlinearities, we shall see how the isomorphism

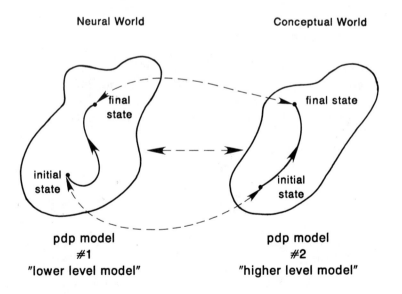

FIGURE 1. The dashed line indicates a mapping from a PDP model representing neural states to a PDP model representing conceptual states. The mapping is an isomorphism if, when the two models start in corresponding states and run for the same length of time, they always end up in corresponding states.

hypothesis fails, and explore a phenomenon within these models that distinguishes between the neural and conceptual levels.[2]

It is an open empirical question whether the conceptual entities with which we understand most cognitive processes are represented by the firing of a single neuron, by the firing of a group of neurons dedicated to that concept, by a pattern of firing over a group of neurons involved in representing many other concepts, by neural features other than firing, or by no neural features at all. The purpose of this chapter is to show how to use PDP models as a mathematical framework in which to *compare the implications* of assumptions like those of local and distributed models.

The plan of attack in this chapter is to compare two related mathematical models, each of which can be given either neural or conceptual interpretations. They can be thought of as describing a single neural net with both a local and a distributed model, or as implementing inference over a single set of conceptual hypotheses using both a local and a distributed model. The comparison of these two models will thus tell us two things. It will show how a description of a neural net in terms of its patterns compares with a description in terms of its individual neurons. It will also provide information about how behavioral predictions change when a local model of a set of concepts is converted to a distributed model.

The comparison between the two mathematical models will constitute an investigation of the isomorphism of levels hypothesis. Model 2 will be a description of the *dynamics of the patterns of activation* of Model 1; the hypothesis is that the description at the higher level of patterns (Model 2) obeys the same laws as—is isomorphic to—the description at the lower level of individual units (Model 1). To permit all the relevant interpretations to apply, I shall call Model 1 simply the *lower-level model* and Model 2 the *higher-level model*.

---

[2] Strictly speaking, an *isomorphism* insists not only that there be a mapping between the neural and conceptual world that preserves the dynamics, as indicated by Figure 1, but also that the map establish a one-to-one correspondence between states in the two worlds. Actually, it seems reasonable to assume that the neural world is a larger space than the conceptual world: that a conceptual state lumps together many neural states, or that the set of possible neural states includes many that have no conceptual counterpart. This would render the "conceptual level" a "higher" level of description. These considerations will manifest themselves formally in the section "Pattern Coordinates," but for now the relation of isomorphism, despite its symmetry, serves to emphasize the true strength of the hypothesis under consideration.

FROM NETWORK OF PROCESSORS TO DYNAMICAL SYSTEM

This section introduces the perspective of PDP models as *dynamical systems*. Traditionally, a PDP model is viewed as a network of processors communicating across links; I will call this the *computational viewpoint*. To illustrate the difference between the computational and dynamical system perspectives, it is useful to consider a prototypical dynamical system: a collection of billiard balls bouncing around on a table.

A common exercise in object-oriented computer programming is to fill a video screen with billiard balls bouncing off each other. Such a program creates a conceptual processor for each billiard ball. Each "ball" processor contains variables for its "position" and "velocity," which it updates once for each tick of a conceptual clock. These processors must exchange messages about the current values of their variables to inform each other when "bounces" are necessary.

Billiard balls *can* be seen from a computational viewpoint as processors changing their local variables through communication with other processors. Physics, however, treats the position and velocity values simply as real variables that are mutually constrained *mathematically* through the appropriate "laws of physics." This is characteristic of the dynamical system viewpoint.

To view a PDP model as a dynamical system, we separate the data and process features of the units. The activation values of the units are seen merely as variables that assume various values at various times, like the positions and velocities of billiard balls. The changes in these variables over time are not conceptualized as the result of prescribed computational processes localized in the units. In fact the processes by which such changes occur are unanalyzed; instead, mathematical equations that constrain these changes are analyzed.

The equations that determine activation value changes are the analogs of the laws of physics that apply to billiard balls; they are the "laws of parallel distributed processing" that have been described in Chapter 2. A version of these equations can be written as follows. Let $u_\nu(t)$ denote the activation of unit ν at time t. Then its new value one unit of time later is given by

$$u_\nu(t+1) = F\left[\sum_\mu W_{\nu\mu} G(u_\mu(t))\right]. \qquad (1A)$$

Here F is a particular nonlinear sigmoid function, an increasing S-shaped function that takes a real number as input (the net activation flowing into a unit) and gives as output a number in the range $[-m, M]$

(the new activation of the unit). G is a nonlinear threshold function: $G(x) = x$ unless x is less than a threshold, in which case $G(x) = 0$. $W_{\nu\mu}$ is the strength of the connection *from* unit μ *to* unit ν.

The "knowledge" contained in a PDP model lies in the connection strengths $\{W_{\nu\mu}\}$ or *weight matrix*, **W**. The nonlinear functions F and G encode no knowledge about the cognitive domain of the model, and serve to control the activation spread in non-domain-specific ways.

Thus the heart of a PDP model is its weight matrix; the rest of the machinery can be viewed as bells and whistles added to get the weight matrix to be "used properly" during inference. The purest PDP models, from this point of view, consist only of the weight matrix; from the preceding equation, F and G are removed:

$$u_\nu(t+1) = \sum_\mu W_{\nu\mu} u_\mu(t).$$

The absence of the controlling nonlinearities make these models difficult to use for real modeling, but for our analytic purposes, they are extremely convenient. The main point is that even for nonlinear models with F and G present, it remains true that at *the heart of the model is the linear core,* **W**. For this reason, I will call the dynamical systems governed by the preceding equation *quasi-linear dynamical systems*. In this chapter, the analysis of PDP models becomes the analysis of quasi-linear dynamical systems. These systems will be viewed as elaborations of linear dynamical systems.

KINEMATICS AND DYNAMICS

Investigation of the hypothesis of the isomorphism of levels is a purely mathematical enterprise that bears on the questions of interpreting PDP models. This section provides an introduction to the mathematical structure of dynamical systems.

There are two essential components of any dynamical system. First, there is the *state space S*, the set of all possible states of the system. In our case, each state **s** in S is a pattern of activation, i.e., a vector of activation values for all of the units. The second component of a dynamical system is a set of *trajectories* s_t, the paths through S that obey the *evolution equations* of the system. These trajectories can start at any point s_0 in S. For activation models, s_0 is the initial activation pattern determined by the input given to the model (or its history prior to our observation of it), and the corresponding trajectory s_t is the ensuing sequence of activation values for all later times, viewed as a path in the state space S. (This "path" is a discrete set of points because the values of t are; this fact is not significant in our considerations.) The evolution equations for our quasi-linear dynamical systems are Equations 1A of the previous section.

Corresponding to these two components of dynamical systems are two sets of questions. What is the state space S? Is it finite or infinite? Bounded or unbounded? Continuous or discrete? How can the points in S be uniquely labeled? What structures relate points to one another? These properties define the *geometry* of state space, what I will call the *kinematics* of the dynamical system. In kinematics, the evolution equations and trajectories are ignored; time plays no role. Only the properties of the points in state space themselves are considered.

The questions in the second set pertain to the trajectories. Are they repetitive (periodic)? Do they tend to approach certain special states (limit points)? Can we define quantities over S that differ from trajectory to trajectory but are constant along a trajectory (conserved quantities)? These are the questions about the system's *dynamics*, and their answers depend strongly on the details of the evolution equations.

It may seem that the questions of dynamics are the real ones of interest. However, it is useful to consider kinematics separately from dynamics for two reasons. First, the link between kinematics and dynamics is strong: The kinds of evolutionary equations that can sensibly be assumed to operate in a dynamical system are limited by the geometry of its state space. For example, geometrical structures expressing the *symmetries* of spacetime or elementary-particle state spaces restrict *severely* the possible evolutionary equations: This is the basis of the theory of relativity and gauge field theories of elementary particles. In our case, imposing boundedness on the state space will eventually lead to the breakdown of the isomorphism of levels. The second reason to emphasize kinematics in its own right is that the questions of *interpreting* the dynamical system have mostly to do with interpreting the states, i.e., with kinematics alone.

In this chapter we are concerned primarily with interpretation, and the discussion will therefore center on kinematics; only those aspects of dynamics that are related to kinematics will be considered. These aspects involve mainly the general features of the evolution equations—the linearity of one component (**W**) and the nonlinearity of the remaining components (**F,G**). More detailed questions about the trajectories (such as those mentioned above) address the *behavior* of the system rather than its interpretation, and will not be considered in this chapter.[3]

---

[3] J. A. Anderson, Silverstein, Ritz, and Jones (1977) show the power of concepts from linear algebra for studying the dynamics of PDP models. Some of their results are closely related to the general observations about nonlinear models I will make in the last portion of this chapter.

KINEMATICS

The first question to ask about a state space is: How can the points—states—be labeled? That is, we must specify a *coordinate system* for S, an assignment to each activation state s of a unique set of numbers.

Each such s denotes a pattern of activation over the units. Let us denote the activation value of the νth unit in state s by $u_\nu(\mathbf{s})$; this was formerly just denoted u_ν. These functions $\{u_\nu\}$ form the *unit coordinates* for S. Each function u_ν takes values in the set of allowed activation values for unit ν. In the standard PDP model, this is the interval $[-m, M]$. One could also consider binary units, in which case the functions u_ν would take values in the set $\{0,1\}$ (or $\{-1,1\}$). In any event, if all units are of the same type, or, more specifically, if all have the same set of allowed values, then all unit coordinates take values in a single set.

It is sometimes helpful to draw a very simplified example of a state space S. Using unit coordinates, we can plot the points of S with respect to some Cartesian axes. We need one such axis for each u_ν, i.e., each unit. Since three axes are all we can easily represent, we imagine a very simple network of only three units. The state space for such a network is shown in Figure 2. In Figure 2A, the case of activation values in $[-m, M]$ is depicted. In this case, S is a solid cube (with side of length $m+M$). Figure 2B depicts the case of binary activation values $\{0,1\}$; in this case, S is the eight vertices (corner points) of a cube. Except where specified otherwise, in the remainder of this chapter the standard case of continuous activation values will be assumed.

Thus if the network contains N units, S is an N-dimensional space that can be thought of as a solid hypercube. Any other point of N-space outside this hypercube is excluded from S because it corresponds to activation for at least one unit outside the allowed range. Thus, states with "too much" activation have been excluded from S.

"Too much" has so far been defined according to each unit individually; it is interesting to consider whether states should be excluded if they correspond to "too much" activation among the units collectively. This would amount to excluding from S even some of the points in the hypercube.

Here are two ways one might eliminate states with too much activation in total. The first way is to require that of the complete set of N units, only $N_{\max}$ can be active (i.e., have nonzero activation) at any one time. As Figure 3 shows, this removes much of the hypercube and leaves S with a rather bizarre shape. This S is topologically quite different from the original hypercube.

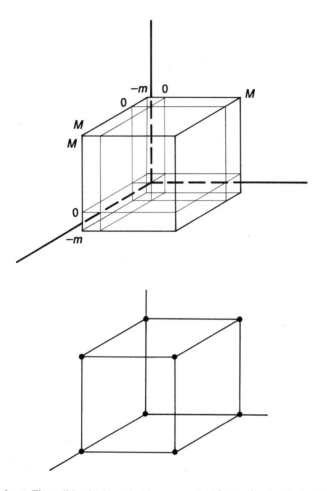

FIGURE 2. *A*: The solid cube bounded by $-m$ and M forms the standard state space for PDP models containing continuous units. *B*: The eight corners of the cube bounded by 0 and 1 forms a modified state space, corresponding to PDP models containing binary units.

A less unusual approach is to define "too much activation in total" by the condition

$$\sum_{\nu} |u_{\nu}| > a_{\max}.$$

This results in an S that is depicted in Figure 4. Unlike the bizarre S of Figure 3, this new S is not topologically distinct from the original hypercube.

Imposing this kind of limitation on total activation would turn out to have much less of an effect on the dynamics of the model than would the limitation on the number of active units.

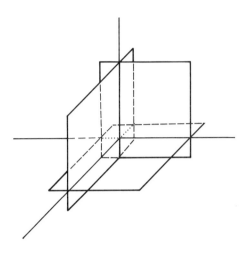

FIGURE 3. If only two of three units are allowed to have nonzero activation, the state space is formed from three intersecting planes.

Redefining the "total activation" to be the Euclidean distance of the plotted point from the origin would not change the conclusion that S is not topologically distinct from the hypercube. In fact, limiting the Euclidean distance or the sum of activations, or using the original hypercube, are all special cases of defining S to be a "ball" with respect to some metric in N-space. It is a fact that all such balls are topologically equivalent (e.g., Loomis & Sternberg, 1968, p. 132).

In the remainder of this chapter, S will denote the standard hypercube as depicted in Figure 2A, the state space of the general nonlinear activation model. The state space of the simplified, linear, activation model will be denoted S_L. This space, as we shall see in the next section, is simply all of N-dimensional Euclidean space, where N is again the number of units in the network. (For example, there is no need to draw S_L because, for $N = 2$, it is the entire plane of the paper!) S is clearly a subset of S_L; in S, the unit coordinates of any state fall within the restricted range $[-m, M]$, while in S_L the unit coordinates can be any real numbers.

The unit coordinates provide a convenient description of S for many purposes. However, it is important to realize that points of S can also be described with an infinitude of other coordinate systems. In a distributed interpretation, new coordinates that give a simple description of these patterns will turn out to be better than unit coordinates for interpreting states of the system. We shall construct these *pattern*

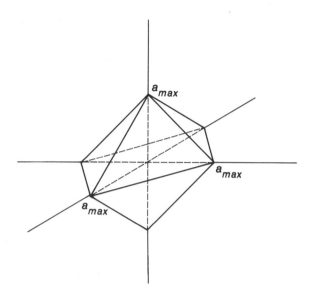

FIGURE 4. This solid octahedron is the state space obtained by restricting the sum of the magnitudes of the units' activations to be less than a_{max}.

coordinates shortly, but this construction uses a property of S we must now discuss: its vector space structure.

VECTOR SPACE STRUCTURE

A central feature of parallel distributed processing is its exploitation of *superposition* of knowledge during computation. Each unit that becomes active exerts its influence in parallel with the others, super-imposing its effects on those of the rest with a weight determined by its level of activation. As we shall see, in linear models, this is a mathematically precise and complete account of the processing. As emphasized in the section "From Network of Processors to Dynamical System," even nonlinear PDP models are quasi-linear systems, and knowledge is used in the same fashion as in linear models. Thus super-position plays a crucial role in all PDP models.

Superposition is naturally represented mathematically by *addition*. In the simplest case, addition of individual numbers represents super-position of unidimensional quantities. Adding multidimensional quanti-ties (like states of activation models) is mathematically represented by *vector addition.* The concepts of vector and vector addition are best viewed together, as data and process: "Vector" formalizes the notion of "multidimensional quantity" specifically for the purpose of "vector addition." [4]

Actually, the notion of superposition corresponds to somewhat more than the operation of addition: Superposition entails the capability to form *weighted* sums. This is important for parallel distributed processing, where the complete state of the system typically corresponds to a *blend* of concepts, each *partially* activated. Such a state is mathematically constructed by *summing* the states corresponding to the partially activated concepts, each one *weighted* by its particular degree of activation.

Using unit coordinates, the operation of weighted summation is simply described. The activation of a unit in a state that is the weighted sum of two other states is simply the weighted sum of the activations of that unit in those two states. In other words, the unit coordinates of the weighted sum of states are the weighted sum of the unit coordinates of the states. Given two states s_1 and s_2, and two weights w_1 and w_2, the weighted sum s is written

$$s = w_1 s_1 + w_2 s_2.$$

What I have already said about the unit coordinates is then written

$$u_\nu(s) = w_1 u_\nu(s_1) + w_2 u_\nu(s_2).$$

Using unit coordinates, the evolution equation of quasi-linear systems, Equation 1A, can be written

$$u_\nu(s_{t+1}) = F[\sum_\mu W_{\nu\mu} G(u_\mu(s_t))]. \tag{1B}$$

The reason all the unit coordinates u_ν of the state s_{t+1} are guaranteed to lie in the allowed range $[-m, M]$ is that the function F takes all its values in that range. This nonlinear function is what ensures that the trajectories do not leave the bounded cube S of states. If F were absent, then the coordinate u_ν would just be the weighted sum (with weights $W_{\nu\mu}$ for all values of μ) of the quantities $G(u_\mu(t))$; this need

---

[4] It is common to use the term "vector" for any multidimensional quantity, that is, a quantity requiring more than one real number to characterize completely. This is, however, not faithful to the mathematical concept of vector unless the quantity is subject to superposition. The precise meaning of "superposition" is captured in the axioms for the operation of "addition" that defines it (see Chapter 9).

not lie in the allowed range.[5] So in the simplified linear theory in which F and G are absent, the evolution equation

$$u_\nu(\mathbf{s}_{t+1}) = \sum_\mu W_{\nu\mu} u_\mu(\mathbf{s}_t) \qquad (2A)$$

imposes no restriction on the range of values of the coordinates; trajectories may wander anywhere in N-space.

Thus we see that the kinematical restriction of state space to the bounded region S has lead to the insertion of the bounded (and therefore nonlinear) function F into the dynamical equation. If state space is extended to all of N-space, i.e. S_L, then the linear dynamical equation above is permissible.

The linear evolution equation can be written more transparently in terms of state vectors rather than unit coordinates. Define N vectors $\mathbf{w}_\mu$ by

$$u_\nu(\mathbf{w}_\mu) = W_{\nu\mu}.$$

$\mathbf{w}_\mu$ *is the vector of weights on connections from unit μ. It is also the activation vector that would exist at time $t+1$ if at time t, unit μ had activation 1 and all other units had activation 0. Now because the evolution is linear, the state at time $t+1$ produced by a general activation pattern at time t is just the weighted sum of the patterns that would be set up by individual unit activations at the units, with the weights equal to the actual activation of the units. That is,*

$$\mathbf{s}_{t+1} = \sum_\mu u_\mu(\mathbf{s}_t) \mathbf{w}_\mu.$$

(This vector can be seen to obey the linear evolution equation given above by evaluating its νth unit coordinate, using the rule for coordinates of weighted sums, and the defining coordinates of the vectors $\mathbf{w}_\mu$.)

This equation explicitly shows the blending of knowledge that characterizes parallel distributed processing. The vector $\mathbf{w}_\mu$ is the output of unit μ; it is the "knowledge" contained in that unit, encoded as a string of numbers. The state of the system at time $t+1$, $\mathbf{s}_{t+1}$, is created by forming a weighted superposition of all the pieces of knowledge stored in all the units. The weight for unit μ in this superposition determines how much influence is exerted by the knowledge encoded by that unit. This weight, according to the previous equation, is $u_\mu(\mathbf{s}_t)$. That is just the degree to which unit μ is active at time t.

Another useful form for the linear evolution equation uses matrix notation:

$$\mathbf{u}(t+1) = \mathbf{W}\,\mathbf{u}(t). \qquad (2B)$$

---

[5] By imposing special restrictions on the weights, it is possible to ensure that the weighted sum lies in S, and then the nonlinearity F can be eliminated. But like F, these restrictions would also have strong implications for the dynamics.

u is the $N \times 1$ column matrix of unit coordinates u_ν, and **W** is the $N \times N$ matrix of values $W_{\nu\mu}$. This matrix is relative to the unit coordinates; shortly we shall switch to other coordinates, changing the numbers in both the square matrix **W** and the column matrix **u**.

PATTERN COORDINATES

Now we consider activity patterns used in a distributed neural or conceptual interpretation. For concreteness, consider the pattern $<+1,-1,+1,-1>$ over the first four units. To denote this pattern, we define a vector **p** the first four unit coordinates of which are $<+1,-1,+1,-1>$; the remaining unit coordinates are zero. Now consider the state **s** with unit coordinates $<.3,-.3,.3,-.3,0,0, \ldots, 0>$. This state can be viewed in two ways. The first is as the superposition of four states: $\mathbf{u}_1$ with unit coordinates $<1,0,0, \ldots, 0>$, $\mathbf{u}_2$ with unit coordinates $<0,1,0, \ldots, 0>$, etc., with weights respectively $+.3, -.3, +.3,$ and $-.3$. This is the *unit view* of **s**. The second view is simpler: **s** is simply .3 times **p**:

$$\mathbf{s} = .3\mathbf{p}.$$

This is the *pattern view* of **s**.

The general situation is comparable. If there is a whole set of distributed patterns, each can be represented by a vector $\mathbf{p}_i$. Any given state **s** can be represented in two ways: as the superposition of the vectors $\mathbf{u}_\nu$, with weights given by the unit coordinates of **s**, or as a superposition of pattern vectors $\mathbf{p}_i$. If the patterns comprise a distributed conceptual interpretation, the weights in this latter superposition indicate the system's confidence in the corresponding conceptual hypotheses.

Let's consider a slightly less simplified example. Let $\mathbf{p}_1$ be the vector **p** above, and let $\mathbf{p}_2$ correspond to the activation pattern $<+1,+1,+1,+1>$ over the first four units. Then the state **s** with unit coordinates $<.9,.6,.9,.6,0,0, \ldots, 0>$ can be viewed either as

$$\mathbf{s} = .9\mathbf{u}_1 + .6\mathbf{u}_2 + .9\mathbf{u}_3 + .6\mathbf{u}_4$$

or as

$$\mathbf{s} = .15\mathbf{p}_1 + .75\mathbf{p}_2.$$

The first representation shows the activation pattern of units, while the second shows **s** to be a weighted sum of the two patterns with respective strengths .15 and .75.

It is useful to consider this example geometrically as well as algebraically. In Figure 5, **s** is drawn (only the first two unit dimensions are depicted). The projections of this vector onto the unit axes (defined by the vectors $\mathbf{u}_1$ and $\mathbf{u}_2$) are .9 and .6, while the projections onto the vectors $\mathbf{p}_1$ and $\mathbf{p}_2$ are .15 and .75. These conceptual vectors define the axes of the *pattern coordinate system* for state space.

In a PDP model with a distributed interpretation, the interpretation of the state of a system requires the use of the pattern coordinate system. The mathematics of linear algebra (discussed in Chapter 9) tells how to convert state descriptions from unit coordinates to pattern coordinates, and vice versa. All that is required is the specification of the patterns.

Before considering the conversion between unit and pattern coordinates, one observation needs to be made. Consider a distributed conceptual interpretation. (Exactly the same considerations apply to a distributed neural representation.) If confidence in a group of conceptual hypotheses are to be separable, then the pattern

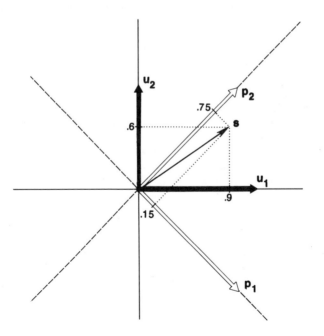

FIGURE 5. The unit basis vectors $\mathbf{u}_1$ and $\mathbf{u}_2$ define the axes for the unit coordinate system. The state **s** has unit coordinates $<.9,.6>$. The pattern vectors $\mathbf{p}_1$ and $\mathbf{p}_2$ define the axes for the pattern coordinate system. The state **s** has pattern coordinates $<.15,.75>$.

vectors representing them must be linearly independent. Suppose, for example, that $\mathbf{p}_3$ *is not independent of* $\mathbf{p}_1$ *and* $\mathbf{p}_2$ *; say,*

$$\mathbf{p}_3 = .4\mathbf{p}_1 + .3\mathbf{p}_2$$

Then the state $\mathbf{s}$ *representing confidence .2 in Hypothesis 1, .15 in Hypothesis 2, and 0 confidence in Hypothesis 3,*

$$\mathbf{s} = .2\mathbf{p}_1 + .15\mathbf{p}_2 + 0\mathbf{p}_3 \ ,$$

will be identical to the state representing confidence .5 in Hypothesis 3, and 0 confidence in Hypotheses 1 and 2:

$$\mathbf{s} = 0\mathbf{p}_1 + 0\mathbf{p}_2 + .5\mathbf{p}_3 \ .$$

Thus Hypothesis 3 is not separable from Hypotheses 1 and 2; in fact any of the three hypotheses can be written as a superposition of the other two, so it is better to say that the Hypotheses 1, 2, and 3 are not *independently represented by the activation patterns described by the vectors* $\mathbf{p}_1$ *,* $\mathbf{p}_2$ *,* $\mathbf{p}_3$*.*

Thus we must assume that the distributed representation involves a set of separable hypotheses that are represented by a linearly independent *set of vectors* $\{\mathbf{p}_i\}$*. Therefore if there are* N *units, so that state space is* N*-dimensional, there may be at most* N *conceptual vectors. If there are exactly* N *such vectors, then* $\{\mathbf{p}_i\}$ *forms a* basis *for* S_L*: Every state in* S_L *is uniquely expressible as a superposition of the patterns, and therefore interpretable in terms of the conceptual hypotheses. If there are fewer than* N *conceptual vectors, then there will be states of the model that are not conceptually interpretable, since no superposition of the vectors* $\{\mathbf{p}_i\}$ *will equal the state. This may be no problem if the dynamics of the model (i.e., W) tends to keep trajectories away from such states.*

When a distributed interpretation involves fewer than N *patterns, only the vectors in a* subspace *of the state space* S_L *are interpretable, and the pattern coordinates allow description only of this subspace. This reduction in expressivity is to be expected in a passage from a lower- to higher-level description. In any event, when there are fewer than* N *patterns, extra pattern vectors can be freely created to expand* $\{\mathbf{p}_i\}$ *to a complete basis. To simplify the analysis, we shall assume this to be done, noting that states involving these extra vectors are not completely interpretable.*

The unit and pattern coordinates of a state $\mathbf{s}$ are the components of the vector $\mathbf{s}$ with respect to the unit basis $\{\mathbf{u}_\nu\}$ and the pattern basis $\{\mathbf{p}_i\}$, respectively. To translate between these bases, we need the change-of-basis matrix $\mathbf{P}$ defined by the entries

$$\mathrm{P}_{\nu i} = u_\nu(\mathbf{p}_i).$$

The ith column of this matrix is simply the pattern of activation over the units defining the ith pattern. Then the unit coordinates of a state $\mathbf{s}$, the components of $\mathbf{s}$ with respect to the unit basis $\{\mathbf{u}_\nu\}$, can be computed from the conceptual components of **s**, the components of **s** with respect to the pattern basis $\{\mathbf{p}_i\}$, by

$$\mathbf{u} = \mathbf{P}\,\mathbf{p}. \tag{3A}$$

In this matrix equation, **u** and **p** are the $N \times 1$ column matrices of coordinates of **s** in the unit and pattern systems, respectively.

To compute the pattern coordinates from the unit ones, we need the *inverse* matrix of **P**:

$$\mathbf{p} = \mathbf{P}^{-1}\mathbf{u}. \tag{3B}$$

The existence of an inverse of **P** *is guaranteed by the linear independence of the* $\{\mathbf{p}_i\}$.

Let's consider a simple case with two units supporting the two patterns $<1,2>$ *and* $<3,1>$. *Here the pattern matrix is*

$$\mathbf{P} = \begin{bmatrix} 1 & 3 \\ 2 & 1 \end{bmatrix}.$$

Thus the state $\mathbf{s}_1$ *representing .5 of pattern one and .3 of pattern two has pattern coordinates* $<.5,.3>$ *and unit coordinates*

$$\mathbf{u}_1 = \begin{bmatrix} 1 & 3 \\ 2 & 1 \end{bmatrix} \begin{bmatrix} .5 \\ .3 \end{bmatrix} = \begin{bmatrix} 1.4 \\ 1.3 \end{bmatrix}.$$

The two coordinate systems and $\mathbf{s}_1$ *are shown in Figure 6.*

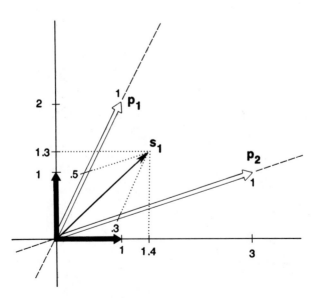

FIGURE 6. The state $\mathbf{s}_1$ in the pattern and unit coordinates.

To go from unit coordinates to pattern coordinates, we use the inverse of the pattern matrix:

$$\mathbf{P}^{-1} = \begin{bmatrix} -.2 & .6 \\ .4 & -.2 \end{bmatrix}.$$

Thus the state $\mathbf{s}_2$ *with unit coordinates* $<4,3>$ *has pattern coordinates*

$$\mathbf{p}_2 = \begin{bmatrix} -.2 & .6 \\ .4 & -.2 \end{bmatrix} \begin{bmatrix} 4 \\ 3 \end{bmatrix} = \begin{bmatrix} 1 \\ 1 \end{bmatrix}.$$

(It is easily verified that $<4,3>$ *is the sum of* $<1,2>$ *and* $<3,1>$ *!) The state* $\mathbf{s}_2$ *is shown in Figure 7.*

The evolution Equation 2B for the linear model can now be transformed, by multiplication by $\mathbf{P}^{-1}$, from unit to pattern coordinates:

$$\mathbf{p}\,(t+1) = \mathbf{I}\,\mathbf{p}\,(t). \qquad (4A)$$

Here the matrix $\mathbf{W}$ of interunit connection weights has become the matrix

$$\mathbf{I} = \mathbf{P}^{-1}\,\mathbf{W}\,\mathbf{P}. \qquad (5)$$

The meaning of this matrix is clear upon writing the evolution equation out in component form:

$$\mathbf{p}_i\,(t+1) = \sum_j \mathbf{I}_{ij}\,\mathbf{p}_j\,(t). \qquad (4B)$$

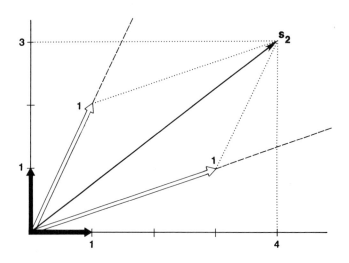

FIGURE 7. The state $\mathbf{s}_2$ in the unit and pattern coordinates.

Thus, I_{ij} is the interconnection strength from *pattern j* to *pattern i*. In a distributed conceptual interpretation, this number expresses the sign and strength of the evidential relation from the *j*th conceptual hypothesis to the *i*th conceptual hypothesis. Thus **I** governs the propagation of *inferences*.

The important result is that *the evolution equation for the linear model has exactly the same form in the pattern coordinate system as it had in the unit coordinates.*

THE ISOMORPHISM HYPOTHESIS HOLDS IN LINEAR SYSTEMS

The analysis at the end of the preceding section shows that for a *linear model* the evolution equation has exactly the same form in pattern coordinates as in unit coordinates. In other words, *there is an exact isomorphism between the lower and higher levels of description in linear models.*

This isomorphism has been viewed so far as a mapping between two descriptions of a given model. It can also be viewed as a mapping between the behavior of two different PDP models. One is the original model with which we began, a model supporting a distributed interpretation, having N units with interconnection weight matrix **W**. Let's call this the *lower-level model*, M_l. The *higher-level model M_h* has a unit for every *pattern* of the distributed interpretation and has interconnection matrix **I**. The law governing its processing (Equation 4) is exactly the same as that of the lower-level model (Equation 2).

The isomorphism maps states of the lower-level model into states of the higher-level model. Take any state s_l of the lower-level model. To find the corresponding state s_h of the higher-level model, express s_l in pattern coordinates. The coordinate for the *i*th pattern—its strength in state s_l—gives the activation in state s_h of the *i*th unit of M_h. (In other words, the *i*th *pattern* coordinate of s_l equals the *i*th *unit* coordinate of s_h.)

So for example if the state **s** *of the lower-level model is a superposition of the first two patterns, say,*

$$s = .6p_1 + .3p_2$$

then the corresponding state in the higher-level model would have activation .6 in Unit 1, .3 in Unit 2, and 0 elsewhere.

The mapping between the lower- and higher-level models is an isomorphism because the evolution equation for one model is mapped into the evolution equation of the other, so that the *behavior* of one model is exactly mapped onto that of the other. That is, if the two models start off in corresponding states, and are subject to corresponding inputs, then for all time they will be in corresponding states.[6]

The significance of the behavioral equivalence of the two models can be brought out in an example. Consider a linear version of the letter perception model (that is, remove the nonlinearities from the equations of the model, maintaining the original nodes and interconnections). As originally proposed, this model, M_h, has a local conceptual interpretation. Suppose we wanted to rebuild the model with a distributed conceptual interpretation, with hypotheses about words and letters represented as activation patterns over the units of a model M_l.

First, each conceptual hypothesis (i.e., unit in the original model) would be associated with some specific activation pattern over units in M_l. The inference matrix of M_h, which gives the positive and negative strengths of connections between conceptual hypotheses in the original model, would be algebraically transformed (following Equation 5), using the activation patterns defining the distributed interpretation. This new matrix defines the correct weights between units in M_l. This sets up the lower-level model.

To run the model, inputs are chosen and also an initial state. Both are originally specified in conceptual terms, and must be algebraically transformed to M_l (following Equation 3A). This defines the inputs and initial state of the lower-level model. Then the lower-level model is run for a length of time. The model's response to the input is determined by taking the final state, representing it in pattern coordinates, and reading off the activations of the corresponding conceptual hypotheses.

What the isomorphism tells us is that after all this effort, the result will be *exactly* the same as if we had simply run the higher-level model.

The higher-level model can be viewed as a conceptual description of the lower-level model in which details of the "implementation patterns" have been ignored. The behavioral isomorphism implies that these details have no effect at all on the behavior of the model. However, the two models do implement the same processing differently, and they will differ in how they respond to modifications of the processing mechanism itself. Thus the behavioral effect of destroying a unit will

---

[6] If the lower-level model has external inputs to the units, exactly the same transformation that maps *states* of M_l to states of M_h also maps *inputs* of M_l to inputs of M_h. This can be verified by taking the matrix form of the new linear evolution equation,

$$\mathbf{u}(t + 1) = \mathbf{W}\,\mathbf{u}(t) + \mathbf{i},$$

where $\mathbf{i}$ is the $N \times 1$ column matrix of external inputs, and performing the same operations as in the preceding section to transform it from the unit to pattern basis.

be different for the two models, and the demands on connection adaptation, i.e., learning will differ. Understanding these phenomena in terms of the lower- and higher-level models provides an opportunity to exploit more of the power of the techniques of linear algebra. They will be discussed in the following two sections, which can be skipped without diminishing the comprehensibility of the subsequent discussion.

LOCALIZED DAMAGE AND THE ISOMORPHISM OF LEVELS

Suppose we remove one unit from the lower-level model. What will be the corresponding modification in the isomorphic higher-level model? That is, what will be the effect on the model's ability to process the patterns that are meaningful in the distributed interpretation?

Removing unit ν can be viewed as insisting that it have activation zero and no incoming connections. This amounts to allowing only states $\mathbf{s}$ that have

$$u_\nu(\mathbf{s}) = 0.$$

This change in the kinematics of the system brings with it a change in the dynamics. We need to follow this change through, transforming it to the higher-level model.

The evolution equation must be changed so that only allowed states are reached; this amounts to saying that the activation of unit ν will never change from zero. This can be done as follows. The column matrix $\mathbf{W}\,\mathbf{u}(t)$ has for its νth element the inputs coming in to unit ν; rather than this for the νth element in $\mathbf{u}(t+1)$ we want zero. So we apply a "damage matrix" that "projects out" the component along $\mathbf{u}_\nu$:

$$\mathbf{D}_{\mathbf{u}_\nu} = 1 - \mathbf{u}_\nu\,\mathbf{u}_\nu^{\mathrm{T}}.$$

(Here 1 is the unit or identity matrix, and $\mathbf{T}$ is the matrix transpose operation.) The new evolution equation becomes

$$\mathbf{u}(t+1) = \mathbf{D}_{\mathbf{u}_\nu}\,\mathbf{W}\,\mathbf{u}(t).$$

Introducing $\mathbf{D}$ is equivalent to the more obvious step of replacing the νth row of $\mathbf{W}$ by zeroes, or of simply removing it altogether, along with the νth element of $\mathbf{u}$. However, it is difficult to map these surgeries onto the higher-level model to see what they correspond to. By instead performing the "damage" by introducing a new linear operation (multiplication by $\mathbf{D}$), we can again use simple linear algebra and transparently transform the "damage" to the pattern basis.

We can view D *as a projection onto the states orthogonal to* u_ν *if we introduce the inner product in which the unit basis vectors are orthonormal. Then the inner product of two states* s *and* s', (s, s'), *can be easily computed using the column matrices* u *and* u' *of the unit cooordinates of these states:*

$$(s, s') = u^T u'.$$

While the unit basis is orthonormal, the pattern basis may not be. The relevant matrix is M *with elements*

$$M_{ij} = (p_i, p_j) = \sum_\nu u_\nu(p_i) u_\nu(p_j) = \sum_\nu P_{\nu i} P_{\nu j},$$

i.e.,

$$M = P^T P.$$

If M *is the unit matrix* 1, *then the pattern basis is orthonormal, also, and the inner product of two states* s *and* s' *can be computed using the column matrices* p *and* p' *of their pattern coordinates using the same simple formula as in the unit coordinates:*

$$(s, s') = p^T p' \qquad \text{[orthonormal patterns only]}.$$

Otherwise one must use the general formula of which this is a special case:

$$(s, s') = p^T M p'.$$

(Since $u = Pp$, *this is just* $u^T u'$.*)*

   Now we apply the standard procedure for changing the dynamical equation to the pattern basis. This gives

$$p(t+1) = D_d I p(t).$$

Here, the "damage vector" d *is the column matrix of pattern coordinates of* u_ν:

$$d = P^{-1} u_\nu$$

and D_d *is again the matrix that orthogonally projects out* d:

$$D_d = 1 - d d^T M.$$

(If the pattern basis is not orthonormal, the inner product matrix M *must appear in the orthogonal projection matrix* D; *if the pattern basis is orthogonal, then* $M = 1$ *so it has no effect.)*

   So the corresponding damage in the higher-level model is removal of the pattern represented by d: *All allowed states will be orthogonal to this pattern. Introducing* D *here is equivalent to making a change in the inference matrix* I *that is more complicated than just converting a row to zeroes. All connections coming into the patterns that employ the deleted unit will be altered in a rather complicated way;* I *is replaced by* $D_d I$.

Thus corresponding to the damage produced in the lower-level model by removing unit ν is the removal of all states in the higher-level model that are not orthogonal to a particular state of the higher-level units. This state corresponds under the isomorphism to $\mathbf{u}_\nu$; it represents exactly the superposition of lower-level patterns in which the activations on all lower-level units exactly cancel out to 0, except for the removed unit, which ends up with an activation of 1.

Let's return to the two-unit example from the section "Pattern Coordinates." Suppose the first unit is removed. This eliminates the horizontal axis from the state space; that is, now only states that have zero x-coordinate are allowed. This is shown in Figure 8. How does this look in pattern coordinates? The damage vector **d** *has unit coordinates* $< 1,0>$. *Its conceptual coordinates are therefore*

$$\mathbf{p} = \begin{bmatrix} -.2 & .6 \\ .4 & -.2 \end{bmatrix} \begin{bmatrix} 1 \\ 0 \end{bmatrix} = \begin{bmatrix} -.2 \\ .4 \end{bmatrix}.$$

(The weighted sum of the two patterns, with weights $<-.2,.4>$, *is seen to be 0 on the second unit, 1 on the first.) Thus in the pattern coordinates, the damage has been to remove the state* $<-.2,.4>$, *leaving only states orthogonal to it. This of course is just a different description of exactly the same change in the state space: see Figure 9.*

Thus removing a lower-level unit corresponds to removing a pattern of activation over the higher-level units; under a distributed conceptual interpretation, this amounts to removing "pieces of concepts"—the pieces relying on that unit.

The same analysis can be run in reverse, to show that "removing a higher-level unit" (e.g., a concept) amounts to performing a rather complicated change in the weight matrix that has the effect of eliminating from the state space those states not orthogonal to the pattern

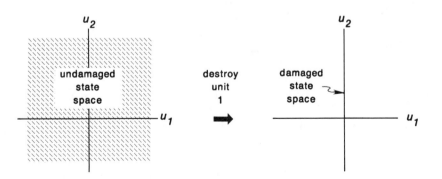

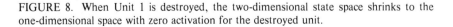

FIGURE 8. When Unit 1 is destroyed, the two-dimensional state space shrinks to the one-dimensional space with zero activation for the destroyed unit.

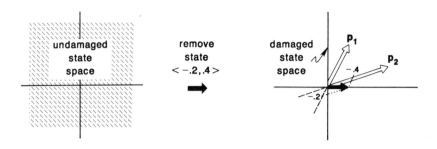

FIGURE 9. In pattern coordinates, the destruction of Unit 1 is described as the removal of the state $<-.2,.4>$, leaving only the states orthogonal to it.

corresponding to the deleted higher-level unit. In short, localized damage in one model is distributed damage in the other. Geometrically, the picture is similar in the two cases; localized damage removes the states orthogonal to a coordinate axis, while distributed damage removes the states orthogonal to a vector that is not a coordinate axis. Since the higher-level coordinates simply employ different axes than the lower-level coordinates, the picture makes good sense.

LEARNING OF CONNECTIONS AND THE ISOMORPHISM OF LEVELS

One major conceptual distinction between local and distributed interpretations is that in the former case the individual elements of the interconnection matrix can be readily interpreted while in the latter case they cannot. In a version of the reading model with a local conceptual interpretation, the positive connection from the unit representing "the first letter is *A*" and that representing "the word is *ABLE*" has a clear intuitive meaning. Thus the connection matrix can be set up intuitively, up to a few parameters that can be adjusted but which have clear individual interpretations. By contrast, the interconnection matrix of the modified reading model defined in the section "The Isomorphism Hypothesis Holds in Linear Systems," with a distributed conceptual interpretation, had to be obtained by algebraically transforming the original interconnection matrix: This matrix cannot be obtained by intuition, and its individual elements (the connection strengths between individual units in the distributed model) have no conceptual interpretation.

This way of generating the interconnection matrix for the distributed model seems to give a primal status to the local model. There is, however, another way to produce the interconnections in the distributed model: through a learning procedure. The two procedures I will discuss are the original Hebb learning procedure and its error-correcting modification, the delta rule.

I will only briefly and imprecisely discuss these two learning procedures; a fuller discussion is presented in Chapter 2, Chapter 8, and Chapter 11. In the Hebb procedure, the interconnection strength between two units is increased whenever those units are simultaneously active. This can be implemented by adding to the strength the product of the units' activation. In the delta rule, this is modified so that what is added to the strength is the product of the input unit's activation and the *difference* between the output unit's activation and the value it should have according to some outside "teacher."

The relevant results about these procedures, for present purposes, are these: The delta rule will eventually produce the interconnection matrix that minimizes the error, measured relative to the teacher. The Hebb procedure will give essentially the same result *in the special case* that the activation patterns the model must respond to are mutually *orthogonal*. In this case, the error-correction feature of the delta rule is superfluous.

There is an intuitive explanation of this result. Suppose a new input activation pattern must be associated with some output pattern. In general, that input pattern will, by virtue of previously learned associations, produce some output pattern. The delta rule adds into the interconnections just what is needed to *modify* that output pattern, to make it the correct one. The Hebb procedure adds connections that will themselves produce the output pattern *completely* from the input pattern, ignoring connections that have already been stored. If the new input pattern is orthogonal to all the already learned patterns, then a *zero* output pattern will be produced by the previously stored associations. [7] That is why, with orthogonal inputs, the simple Hebb procedure works.

The point is that in general, the delta rule will produce the correct interconnection matrix for a distributed model, as it will for a local model. This represents a degree of parity in the status of the two models. However, this parity has its limits. The Hebb rule will in general *not* produce the correct matrix for a distributed model, unless the patterns have been carefully chosen to be orthogonal. (A simple example of such orthogonality is a strictly local model in which each input is uniquely represented by a single unit.)

Thus the lower- and higher-level models may differ in whether Hebb learning works; this is because what is local to the connection between

---

[7] Orthogonality means that the inputs to each output unit cancel each other out completely.

two units in one model will in general not be local in the other, and some special arrangement—orthogonality of patterns—is needed to ensure that nonlocalities do not cause any problems.

This issue of the orthogonality of the patterns relates back to kinematics. To analyze what the individual components of the interconnection matrix need to be, it is necessary to add to the vector space of states the additional geometric structure of an inner product; it is this that tells whether states are orthogonal. The inner product is defined so that the unit basis is orthonormal. When the pattern basis is orthonormal as well, the transformation from unit to pattern coordinates is a rotation, and it preserves the constraints on the interconnection matrix elements. In particular, the adequacy of the Hebb procedure is invariant under the transformation from the lower- to higher-level model. In the general case, when the pattern basis is not orthonormal, this invariance is broken.

NONLINEARITY AND RESTRICTED STATE SPACE

Having established the validity of the isomorphism of levels hypothesis for linear PDP models, it is time to consider quasi-linear systems with nonlinearities. To understand the effects of these non-linearities, it is helpful to go back to kinematics.

The state space S_L of linear PDP models is all of N-space, where N is the number of units in the model. By contrast, the standard state space S of general PDP models is the solid hypercube in N-space described in the section "Kinematics." This represents states in which each unit has activation within the limited range $[-m, M]$. Such a restriction is motivated within the neural interpretation by the limited range of activity of individual neurons; it is motivated within the conceptual interpretation by the desirability of limiting the possible influence of a single conceptual hypothesis. (PDP models are feedback systems that tend to be difficult to control unless activation values have a ceiling and floor. Nonlinearities also allow multilayer PDP networks to possess greater computational power than single-layer networks: see Chapter 2.)

Whatever the motivation, the restriction of states to the cube S, instead of allowing all states in N-space S_L, means that the general linear evolution equation (with unrestricted weight matrix **W**) is unacceptable. Introducing the nonlinear sigmoidal function F of Equation 1 with range $[-m, M]$ solves this problem by brute force.

Unlike S_L, the hypercube S looks quite different in different coordinate systems. In unit coordinates, all coordinates are limited to $[-m, M]$; with a different set of axes, like those of pattern coordinates, the allowed values of a given coordinate vary, depending on the values of the other coordinates. (See Figure 10.) No longer is it possible for the description of the system to be the same in the two coordinate systems. As we shall see explicitly, the exact ismorphism of levels that holds for linear models breaks down for nonlinear models.

Another characterization of the kinematic effect of restricting S to a hypercube is that now states can be distinguished according to their position relative to edges, faces, corners, and so on. Choosing to represent a concept, for example, by a point in the corner of S will produce different behavior than choosing a point in the middle of a face. These distinctions simply cannot be made in the linear state space S_L.

A crucial effect of limiting the state space to S is that now *some* superpositions of states will be in S while others will not; it will depend on the position of the states relative to the boundary of the hypercube. Because some patterns cannot superimpose, i.e., coexist, they must instead *compete* for the opportunity to exist. Other patterns can coexist, so the choice of patterns in an interpretation will matter. In particular,

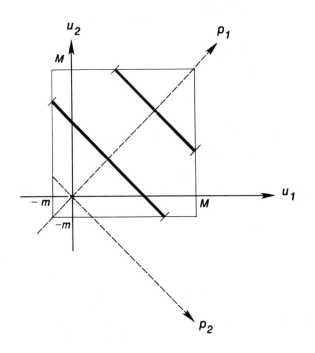

FIGURE 10. In unit coordinates, the allowed values of u_2 are $[-m,M]$ regardless of the value of u_1. In the pattern coordinates shown, the allowed values of p_2 depend on the value of p_1, as shown by the two diagonal bars. These bars correspond to two different values of p_1, and their length indicates the allowed values of p_2.

the choice of nonoverlapping patterns (e.g., over single units)—a local or quasi-local interpretation—will produce different behavior than the choice of overlapping patterns—a genuinely distributed interpretation.

Take for concreteness the allowed unit coordinate range $[-m, M]$ to be $[-1, +1]$.

Consider for example a local conceptual interpretation in which Concepts 1 and 2 are represented by Units 1 and 2, respectively. In unit coordinates, then, the representation of the hypotheses are $<1,0,0,0, \ldots ,0>$ and $<0,1,0,0, \ldots ,0>$. Then the superposition of these two states is simply $<1,1,0,0, \ldots ,0>$. This falls within S, and is therefore kinematically allowed.

By contrast, consider the simple distributed representation in which the two hypotheses are respectively represented by $<1,1,0,0, \ldots ,0>$ and $<1,-1,0,0, \ldots ,0>$. Now the superposition of the two states is $<2,0,0,0, \ldots ,0>$ which is kinematically forbidden. The situation is graphically depicted in Figure 11.

The basic observation is that superimposing states representing maximal confidence in two conceptual hypotheses is kinematically allowed when the corresponding patterns do not overlap—always true in a local interpretation—but kinematically forbidden when they do overlap—sometimes true in a genuinely distributed interpretation.

The kinematic restriction that states stay inside the hypercube has a dynamical consequence in the sigmoidal function F. It will now be verified that F does indeed lead to a greater difficulty in superposing $<1,1>$ and $<1,-1>$ than in superposing $<1,0>$ and $<0,1>$. (All coordinates in this note will be unit coordinates.)

Let $\mathbf{F}$ be the function that takes a vector, passes all its unit coordinates through F, and uses the resulting numbers as the unit coordinates of the output. I will show that F retards the growth in length of a vector along the edge (the $<\alpha, 0>$ direction or $\mathbf{e}$) more than along the diagonal (the $<\beta, \beta>$ direction or $\mathbf{d}$). These retardation factors are

$$R_{\mathbf{e}} = \frac{|\mathbf{e}|}{|\mathbf{F}(\mathbf{e})|} = \frac{\alpha}{F(\alpha)}$$

and

$$R_{\mathbf{d}} = \frac{|\mathbf{d}|}{|\mathbf{F}(\mathbf{d})|} = \frac{\sqrt{2}\beta}{\sqrt{2}F(\beta)} = \frac{\beta}{F(\beta)}.$$

As shown in Figure 12, these retardation factors are the reciprocals of the average slope of the F curve between the origin and the x values α and β, respectively. Since the F curve is concave downward, as x increases, this average slope diminishes so its reciprocal increases. That is, the retardation will be greater as α and β grow; F squashes vectors more and more strongly the closer they get to the edge of the state space. A fair comparison between the retardation along $\mathbf{e}$ and $\mathbf{d}$ requires that these vectors be of equal length. In that case, α is greater than β (by a factor of $\sqrt{2}$); this means the retardation is greater for $\mathbf{e}$, i.e., along the edge.

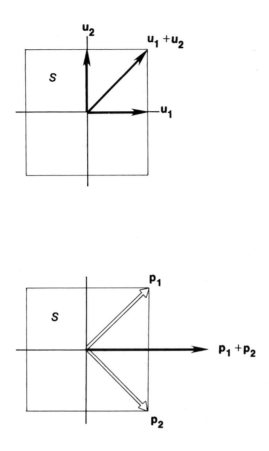

FIGURE 11. *A*: In a local interpretation in which Concepts 1 and 2 are represented by vectors $\mathbf{u}_1$ and $\mathbf{u}_2$, the superposition $\mathbf{u}_1 + \mathbf{u}_2$ lies in S (along the diagonal). *B*: In a distributed interpretation in which Concepts 1 and 2 are represented by vectors $\mathbf{p}_1$ and $\mathbf{p}_2$, the superposition $\mathbf{p}_1 + \mathbf{p}_2$ lies outside of S (in the direction of an edge).

Competition of patterns that are forbidden to superimpose—what I will call *natural competition*—does not exist in linear models. The subsequent discussion of nonlinear models will focus on natural competition and how it distinguishes between local and distributed models.

Natural competition resembles inhibition, which of course can be present in linear models. There is a crucial difference, however. Inhibition is *explicitly inserted into the interconnection matrix*. Competition arises independently of the interconnection matrix; it depends on the overlap of patterns and on the nonlinearity. Loosely, we can say that whether two states naturally compete depends only on *kinematics* (their position in state space relative to the edges of the space), and not on

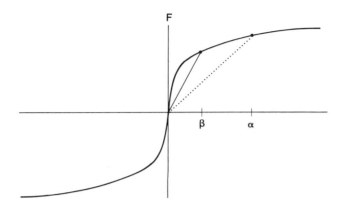

FIGURE 12. The slope of the solid line, $F(\beta)/\beta$, exceeds the slope of the dotted line, $F(\alpha)/\alpha$, because $\beta < \alpha$ and F is concave downward (for positive inputs).

the *dynamics* (interconnection matrix); whether two states mutually inhibit depends on the dynamics but not on the kinematics. If natural competition is to be exploited computationally, distributed interpretations are needed.

For expository convenience, for the remainder of this chapter, except where otherwise indicated, we take $M = 1$ and $m = 1$, so the allowed activation range is $[-1,1]$.

The plan for the remainder of this chapter is first to explicitly show the breakdown of the isomorphism hypothesis for nonlinear models, next to analyze natural competition and its effects, and finally to summarize the results and discuss the value of a mathematical approach to the mind/brain problem.

THE ISOMORPHISM HYPOTHESIS FAILS IN NONLINEAR MODELS

To investigate the isomorphism hypothesis in nonlinear PDP models, as before we take a model with a distributed interpretation and transform the description from unit to pattern coordinates. We then compare the new form of the evolution equation with the original form; if the form has changed, the hypothesis fails. To start with, we restore the nonlinear sigmoidal function F to the evolution equation; we will discuss the thresholding function G later.

Our evolution equation in unit coordinates is Equation 1A with G removed:[8]

$$u_\nu(t+1) = F[\sum_\mu W_{\nu\mu} u_\mu(t)].$$ (6A)

Using Equations 3 and 5, this can be transformed to the pattern coordinates:

$$p_i(t+1) = \sum_\nu P^{-1}{}_{i\nu} F\left[\sum_k P_{\nu k} \sum_j I_{kj} p_j(t)\right].$$ (6B)

Now in the linear case, when F is absent, the pattern matrices P^{-1} and P cancel each other out: the patterns don't matter. By definition of nonlinearity, here F prevents the cancellation, and the equation in pattern coordinates is more complicated than it is in unit coordinates. *There is no isomorphism of levels.* The choice of patterns can matter behaviorally.

As in the linear case, we can construct a higher-level model, with one unit for each pattern. The evolution equation for the higher-level model is Equation 6B; it is not the equation for a PDP model, so the higher-level model, while behaviorally equivalent to the lower-level model, *is not a PDP model.* Here is how the higher-level model works. The units add up the weighted sum of their inputs from other units, using the inference matrix I, just as in the linear case. However, what happens next is complicated. To determine its new value, a unit must find out what the weighted sum of inputs is *for all the other units*, form a weighted sum of these (with weights determined by the patterns), and then pass the resulting value through the nonlinear function F. But this is not the end. Now the unit must find out what number all the other units have gotten out of F, then form a weighted sum of these (again, with weights determined by the patterns). This finally is the new value for the unit.

Thus, what distinguishes the higher- and lower-level models is that the nonlinearity in the lower-level model is applied *locally in each unit* to its weight sum of inputs, while in the higher-level model the nonlinearity is applied *globally* through consultation among all the units. It is this nonlocal nonlinearity that makes the higher-level model not a PDP model.

---

[8] Using the vector-valued function F defined in the previous section, the evolution equation can be written in a concise, coordinate-free form:

$$s(t+1) = F[U s(t)]$$

where U is the linear transformation on S which is represented by matrix W in unit coordinates and by matrix I in pattern coordinates. This equation is the nonlinear generalization of Equation 2B.

The higher-level model just considered has the same behavior as and different processing rules from the lower-level model. We can also consider another higher-level model that has the same processing rules as and different behavior from the lower-level model. This model also has one unit for each pattern of the lower-level model, and also uses the interconnection matrix I. However, each unit computes its new value simply by passing its weighted sum of inputs through F, without the added complications of consulting other units as in the other higher-level model. Because these complications are removed, the behavior of this model will be different from the lower-level model; that is, if the two models were started in corresponding states and given corresponding inputs, they would *not* continue to stay in corresponding states. However, the linear core of this higher-level model uses the correct matrix I, so its behavior should be related to that of the lower-level model in meaningful ways. A more precise comparison has not been carried out.[9]

NATURAL COMPETITION IN DISTRIBUTED NONLINEAR MODELS

One aspect of the level isomorphism for linear models is that there is no competition between patterns that share units; they simply superimpose without conflict. As discussed in the section "Nonlinearity and Restricted State Space," however, in nonlinear models, patterns that fall on the boundary of the state space space—involving saturated units— often cannot superimpose and remain in the state space. This gives rise to natural competition between strong patterns that require common units.

In this section I will analyze a simple example of natural competition, contrasting the cases of distributed and local models. I will work through the effect for a *particular* nonlinear function F, and then show it obtains for all suitable functions F. I will then show explicitly how

---

[9] A possibility that deserves investigation is that the pattern coordinates determined by specific activation patterns should be defined some way other than through superposition, i.e., other than as a change of basis in a vector space. If this approach succeeded in saving the isomorphism hypothesis for the nonlinear case, it could destroy it for the linear case, for which change of basis is guaranteed to leave the evolution equation invariant. To save the isomorphism hypothesis in the nonlinear case, however, the pattern coordinates would probably have to be determined *jointly* by the distributed patterns *and* the nonlinearities; a successful procedure might well reduce to change of basis in the limit as the nonlinearity disappears.

the effect can simulate inhibitory connections between units representing incompatible hypotheses in the higher-level model.

*Consider a **PDP** model with a distributed interpretation involving the patterns $<1,1>$ and $<1,-1>$. A superposition of these two states with weights .3 and .2, respectively, would produce, in a linear model, the state with unit coordinates $<.5,.1>$ (this state has pattern coordinates $<.3,.2>$ of course). In the nonlinear model, the numbers .5 and .1 are each passed through the function F to get the unit coordinates of the new state. Consulting the F function of Figure 13, we see that $F(.5) = .9$ and $F(.1) = .6$, so the new state has unit coordinates $<.9,.6>$. The pattern coordinates of this state, as shown in the section called "Pattern Coordinates," are $<.75,.15>$. Thus the strength of pattern 1 relative to pattern 2 has been amplified by the factor*

$$\frac{.75/.15}{.3/.2} = 3.33.$$

The dominance of the stronger pattern has been enhanced by the nonlinearity, just as the dominance of stronger nodes is increased by mutual inhibition. This is natural competition.

This competitive effect is not present if the patterns do not overlap. In fact the nonlinearity diminishes dominance in this case. If the "patterns" are $<1,0>$ and $<0,1>$—i.e., if the model is local—and we again consider combining .3 of the first with .2 of the second, then the weighted sum is $<.3,.2>$ and the new state, after passing the unit coordinates through F, is $<.78, .7>$ The "amplification" factor is therefore

$$\frac{.78/.7}{.3/.2} = .74.$$

In other words, the nonlinearity is actually countercompetitive for local models.

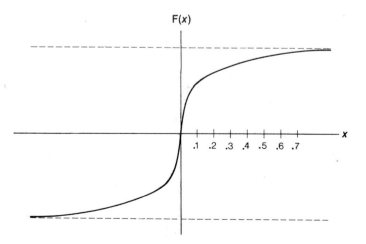

FIGURE 13. An example of a sigmoidal function F.

For these patterns, the conclusions that the amplification factor is greater than one for the distributed model and less than one for the local model is not dependent on the particular F values used above; they hold for any negatively accelerated F function. The distributed amplification is (letting .3 be replaced by x and .2 by y):

$$\frac{\frac{1}{2}[F(x+y) + F(x-y)] / \frac{1}{2}[F(x+y) - F(x-y)]}{x / y}.$$

Letting x + y = w and x − y = z, this becomes

$$\frac{\frac{1}{2}[F(w) + F(z)] / \frac{1}{2}[F(w) - F(z)]}{\frac{1}{2}[w+z] / \frac{1}{2}[w-z]} = \frac{\frac{1}{2}[F(w) + F(z)] / \frac{1}{2}[w+z]}{[F(w) - F(z)] / [w-z]}.$$

As shown in Figure 14, this is the ratio of the slopes of two lines that can be drawn on the graph of F, and because F is negatively accelerated, this ratio must be greater than one.

The local amplification is simply

$$\frac{F(x) / F(y)}{x / y} = \frac{F(x) / x}{F(y) / y}.$$

This is the ratio of the average slope of F between the origin and x and between the origin and y; as already pointed out in section 10, the negative acceleration of F implies that this slope will be less for the greater interval (x), so the local amplification is always less than one.

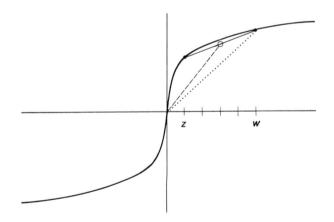

FIGURE 14. The slope of the solid line is $[F(w) - F(z)] / (w - z)$. The point marked × is the midpoint of this line—the vector average of the endpoints—with coordinates $<\frac{1}{2}[w+z], \frac{1}{2}[F(w)+F(z)]>$. Thus the slope of the dashed line is

$$\frac{\frac{1}{2}[F(w) + F(z)]}{\frac{1}{2}[w + z]}.$$

Since F is concave downward, the slope of the dashed line is greater (slope of dashed line > slope of dotted line > slope of solid line).

It is useful to explicitly relate natural competition to mutual inhibition; we shall do this for the above case. The pattern coordinates of the state created in the nonlinear model by superposing $<1,1>$ with strength x and $<1,-1>$ with strength y are

$$p_1 = \tfrac{1}{2}[F(x + y) + F(x - y)]$$
$$p_2 = \tfrac{1}{2}[F(x + y) - F(x - y)].$$

Now let us substitute

$$F(x + y) = s_+ [x + y]$$
$$F(x - y) = s_- [x - y].$$

s_+ *is the average slope of F between the origin and $x + y$; this is less than s_-, the average slope of F between the origin and $x - y$ (see Figure 15). This substitution gives, on regrouping,*

$$p_1 = \alpha x - \gamma y$$
$$p_2 = \alpha y - \gamma x$$

where

$$\alpha = \tfrac{1}{2}(s_+ + s_-)$$
$$\gamma = \tfrac{1}{2}(s_- - s_+).$$

These equations, displayed graphically in Figure 16, show that the effect of the non-linearity on the $<x, y>$ combination is the same as having an inhibition of strength γ between units representing the initial and final strengths of the different patterns, along with an excitation of strength α between units for the initial and final values of the same patterns. The magnitude of the inhibition γ depends on the difference in average slope of F between the origin and $x - y$ on the one hand and $x + y$ on

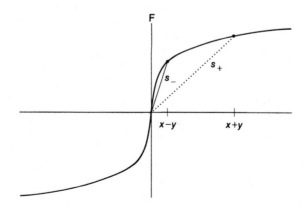

FIGURE 15. Because F is negatively accelerated, the slope of the solid line, s_-, is greater than the slope of the dotted line, s_+.

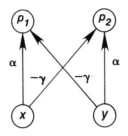

FIGURE 16. A network displaying the effective inhibition produced by natural competition between overlapping patterns $\mathbf{p}_1$ and $\mathbf{p}_2$.

the other. In the region sufficiently close to the origin that the graph of F is nearly linear, γ is zero. Thus for sufficiently weak patterns there is no competition; how weak is "sufficient" will depend on how large the linear range of F is.

The conclusion, then, is that with overlapping patterns like those occurring in truly distributed models, F *amplifies* differences in the strengths of patterns; with nonoverlapping patterns like those of local or quasi-local models, F has the opposite effect. This is natural competition at work.

When viewed at the higher level of patterns, natural competition acts like inhibition between overlapping patterns. The nonlinearities *automatically* create significant inhibition between *strong* patterns but insignificant inhibition between *weak* patterns. So for a distributed conceptual interpretation, it is impossible to have a *high* degree of confidence in more than one conceptual hypothesis represented on a given set of units, but the system can simultaneously entertain without difficulty a low degree of confidence in several such hypotheses. If we think of the units on which conceptual hypotheses are represented as some kind of semantic features, then two hypotheses that call for different values of the same features—two hypotheses represented by overlapping patterns—are hypotheses that are semantically incompatible. Thus distributed representation offers an attractive form of automatic inhibition between mutually incompatible hypotheses.

These considerations suggest that we could try to make a higher-level PDP model approximately behaviorally equivalent to a distributed nonlinear lower-level PDP model by inserting additional inhibitory weights between units representing competing patterns. These extra weights might approximately serve the function of the complicated nonlocal nonlinearity in the truly isomorphic non-PDP higher-level model. Of course, assigning fixed magnitudes to those weights would only approximate the real situation in the lower-level model in which the degree of competition varies with the strength of the patterns. With nonlinear models, it does seem that higher-level models must incorporate some form of complexity that goes beyond the standard PDP framework in order to really capture the subtlety of the interaction of patterns in the lower-level model; the degree of failure of the isomorphism hypothesis seems to be significant.

The failure in the isomorphism hypothesis induced by F is compounded by the nonlinear thresholding function, G. In the previous section we saw that the nonlinearity introduced by F becomes nonlocal in the higher-level model. Exactly the same analysis applies to the nonlinearity introduced by G.

It is interesting to consider whether a *local* thresholding of the activation of patterns in the higher-level model, since it can't be created by putting a local G in the lower-level model, can be affected some other way. Natural competition functions like a threshold on inhibition. When a concept is weakly present, it creates essentially no inhibition; when it is strongly present, it generates considerable inhibition. Thus, it is as though there were a threshold of activation below which a concept is incapable of influencing other concepts inhibitively. The threshold is not sharp, however, as there is a gradual transition in the amount of inhibition as the concept strengthens.

CONCLUSION

If mind is to be viewed as a higher-level description of brain, some definite procedure is needed to build the higher-level description from the lower. One possibility is that the higher level describes *collective* activity of the lower level, for example, patterns of activation over multiple lower-level units. In the analysis of dynamical systems, it is customary to create such higher-level descriptions, using higher-level units that are collective aspects of many lower-level units. The higher-level description employs new mathematical variables, each of which is defined by combining many variables in the lower-level description. This requires that the lower-level description be expressed in terms of mathematical variables.

PDP models constitute an account of cognitive processing that does rest on mathematical variables: the activation of units. Thus these models can be described at a higher level by using the preceding method. The lower-level description of the system amounts to a set of variables and an evolution equation; the higher-level description is created by defining new variables out of many old ones and substituting the new variables for the old in the evolution equation.

Analyzing PDP models in this way leads to several observations. The core of the evolution equation, where the knowledge of the system is employed, is linear. The dynamical systems corresponding to PDP models are quasi-linear. A central kinematic feature of these systems is that the state spaces lie in vector spaces, and linear operations play a very special role. Superposition is fundamental to the way parallel distributed processing works. Linear algebra gives us both concepts for understanding PDP models and techniques for analyzing them. It tells us that there are many equally good coordinate systems for describing the states, and how to transform descriptions from one coordinate system to another. In particular, it suggests we consider a coordinate system based on the patterns, and use this to frame our higher-level description. It tells us how to transform the knowledge (interconnection matrix) from one level of description to the other. Linear algebra tells us that linear operations are invariant under this kind of transformation, and therefore if the evolution equation is linear, its form will be the same in the two descriptions. This is the isomorphism of levels for linear PDP models.

Linear algebra also alerts us to the fact that edges to the state space interfere with superposition. This leads to an evolution equation that modifies pure linear superposition by adding nonlinear operations. The breakdown of superposition at the boundaries of state space leads to competition between states that cannot superpose. The lack of invariance of the bounded state space under linear operations leads to a breakdown of the isomorphism of levels in the corresponding nonlinear PDP models.[10]

The concepts of linear algebra add considerably to our understanding of PDP models, supplementing the insight that comes from simulating PDP models on computers. We can analyze with precision, for example, the effects of localized damage or of synaptic modification upon the higher-level description of the processing. We can understand why the choice of patterns doesn't matter in linear models and why it does in nonlinear models. We can even get some handle on what kind of effects these choices have for nonlinear models, although the picture needs much more clarification.

---

[10] Note that the nonlinearity in quasi-linear systems is minor compared to that in highly nonlinear systems. In PDP models, the knowledge is contained in the linear part—it is this part that is learned, for example, in adaptive models—while the nonlinearity is uniform across the units, and fixed from the outset. In a highly nonlinear model, the activation value of a unit would be a nonlinear function of the other activations, with parameters encoding the unit's knowledge that are used in arbitrary ways, rather than merely as coefficients in a weighted sum. In such a model, superposition and linear algebra might have no relevance whatever.

The isomorphism of levels hypothesis constitutes a mathematically analyzable question about the mind/brain duality, within the framework of PDP models of cognition. These models serve well because of their independent interpretations as models of neural and conceptual processing.

While the isomorphism of levels hypothesis speaks to a fairly philosophical issue, the analyses of this chapter show that no strong argument about the hypothesis can have validity unless it makes sharp enough distinctions among models to differentiate between linear and nonlinear PDP models: The matter demands a certain degree of mathematical care. From an appropriate formal viewpoint, however, we have seen that conceptual accounts of mind and physiological accounts of brain *can* be two descriptions of a single cognitive system.

ACKNOWLEDGMENTS

While I carry full responsibility for the views expressed in this chapter, the critical role played by insights of Jay McClelland and Dave Rumelhart is fully and gratefully acknowledged. This chapter is actually in major part the result of a joint effort, and I thank Dave and Jay for the ideas and understanding they have shared with me. Many thanks, too, to Eileen Conway and Mark Wallen for excellent graphics and computer support. The work was supported by the Systems Development Foundation, the Alfred P. Sloan Foundation, National Institute of Mental Health Grant PHS MH 14268 to the Center for Human Information Processing, and Personnel and Training Research Programs of the Office of Naval Research Contract N00014-79-C-0323, NR 667-437.

CHAPTER 23

Biologically Plausible Models of
Place Recognition and Goal Location

D. ZIPSER

In this chapter I describe three parallel network models that deal with some elementary aspects of spatial cognition—namely, recognizing places and locating goals. The performance of these models was examined using computer simulations in which the networks guide the movements of a simulated observer. These models are constrained by both the requirements of accomplishing the tasks and by a desire to achieve some degree of biological plausibility. The requirement of biological plausibility has the advantage of generating models that can have immediate relevance to psychology and neurobiology but it has the disadvantage of limiting the range of problems that can be approached. This limitation is not one of principle, but grows out of our currently limited understanding of how computation is organized in the brain. More general problems can often be attacked when the constraint of biological plausibility is not invoked and unrestricted symbol manipulation is allowed. Before tackling biologically plausible models, it is useful to look at what has been done in this unrestricted computing context because it will give some insight into the major issues in spatial cognition.

From the realm of "pure" artificial intelligence research, for example, we have the *fuzzy map* theory (Davis, 1981; McDermott, 1980). The

The first third of this chapter is a modification of the article "A Computational Model of Hippocampal Place Fields" by D. Zipser, which appeared in *Behavioral Neuroscience*, 1985, 99, 1006-1018. Copyright 1985 by the American Psychological Association, Inc. Adapted with permission.

goal here was to develop a system that could deal with the spatial relations between objects so that queries about them could be answered in the context of a language understanding system. Among the questions the fuzzy map system is designed to answer are: "Am I headed towards home? How do I get from here to Carnegie Hall? What buildings are between Woolsie and Sterling? What is the nearest Chinese restaurant to here? What street is the best route from Yale to the hospital?" These questions are answered in a linguistic framework and new information is assimilated through language so no attempt was made to guide actual navigation behavior or to learn from the environment.

Clearly, the range of possible facts about the spatial relationships between objects necessary to answer questions of this kind is far too great to be explicitly stored in memory. What is needed is a representation that stores basic information so that inference can be used to determine the required facts when the system is questioned. In the fuzzy map system, each landmark or object has its own reference frame fixed rigidly to it. The combination of object and associated reference frame is called a *frob*. Frobs are connected to each other to form maps by storing *mquants*, each of which contains information about the translation, rotation, or scale factors relating a pair of frobs. Queries about spatial relations are answered by searching a data structure consisting of mquants. An important feature of the fuzzy map formulation is that it separates the components of the transformation relating frobs so that rotation, translation, and scale factors are in separate mquants. This has a number of advantages, one of the most important of which is that partial information about the spatial relations between frobs can be used. For example, two frobs connected only by a scale factor can be compared with respect to size even if their other spatial relations are unknown. Another important feature of the fuzzy map concept is its "fuzziness." The components of the transformation between frobs are not specified exactly, but are specified as ranges. This allows inexact, semiqualitative information to be incorporated into the map data structure.

Another approach to spatial processing, in large scale environments, is found in the work of Kuipers (1978, 1983) who is concerned with developing psychological models. A particularly interesting feature of Kuipers' approach is the concept of *view,* which he defines as "the sensory image received by the observer at a particular point" (Kuipers, 1983, p. 217). The critical importance of views is that they are used to recognize places. Places are important because their representation can serve as a node to which related information can be attached. The critical operation on views is comparison to determine equality. For example, by comparing the current view with a data base of stored views, it is possible to gain access to all the information associated with the current place. The concept of view is very relevant biologically because there are neurons in the brain that fire only within the context of a particular view. One of the important things that can be associated with a place is information about getting from that place to another place. Kuipers represents the paths between places as a sequence of actions and views. The idea is that if you are at a particular place, and take a certain action, it will generate a new view. This view, in turn, can be used to index a new action, which will produce a new view, and so forth, tracing out a path to the goal. This strategy almost certainly corresponds to one of those used by humans and animals.

In addition to path information, it is also possible to associate actual knowledge about the direction and distance to a goal with a place. The existence of absolute knowledge of goal location, relative to external landmarks, has been clearly demonstrated in the case of rats that are forced to swim in a tank of milky water (Morris, Garrud, Rawlins, & O'Keefe, 1982). A small platform is placed below the surface of the cloudy water so that it is invisible to the rat. The rat is then put into the tank and swims around randomly until it locates the platform. The rat climbs up on the platform, rises up, and observes the landmarks in its environment. On subsequent immersions, the rat no longer searches randomly but swims directly to the remembered position of the platform. Since the rat will swim directly to the platform from any initial position in the tank, it must be computing the location of the platform, using remembered information about the location of external landmarks.

The models described in this paper were designed to emulate the kind of goal seeking behavior seen with the rat in cloudy water. Such models require that a place, or more generally, an environmental context, can be recognized. In addition, they need a mechanism to bind goal location to place representations and a way to use this information to guide behavior. Finally, the models need a mechanism to assimilate new information so that goals can be located in unfamiliar environments. The first model discussed is concerned with place recognition and it can be quite biologically plausible because a lot is known about the neurobiology of place recognition. The subsequent models of goal location must be more speculative since much less is known about the mechanism of this activity.

Place Recognition

One of the most remarkable correlations between single neuron activity and events in the world is found in the place-field cells of the hippocampus (Becker, Walker, & Olton, 1980; O'Keefe, 1976; O'Keefe & Dostrovsky, 1971; Olton, Branch, & Best, 1978). In the rat, for example, there are single units that fire at their maximum rate only when the animal is at a particular location relative to a set of distal landmarks. These locations are called place fields. Removing too many landmarks or radically altering their spatial relations abolishes the place-field response of these units (O'Keefe, 1979; O'Keefe & Nadel, 1979). Recently, it has become possible to make accurate measurements of the location, size, and shape of place fields, and to determine how they change when the environmental landmark cues are manipulated (Muller, Kubie, & Ranck, 1983). Quantitative experiments of this kind show that the response of place-field units is determined in a systematic, rule-governed manner by the configuration of the surrounding scene. Muller, Kubie, and Ranck have demonstrated this in a particularly elegant fashion. They showed that when every feature of the environment is scaled to a different size, the relative location of place fields and their areas scale with the environment. This, as we will see later, places a nontrivial constraint on the computational mechanisms that can be used to model location in rats.

To account for these experimental observations, I have developed a computational model that relates the configuration of distal landmarks to the location, size, and shape of place fields. The input to this model is the configuration of landmarks in the environment together with the location of the observer. The output of the model represents the activity of a place-field unit in the observer's brain. The model can be used to simulate experiments on a computer, the results of which can be compared with real experiments to test the correctness of the model. For example, suppose it is known from experiment that an animal can use a set of simple objects—such as rectangular cards—as place cues, and it is also known that the size of the retinal images of these cards is used as an indicator of location. Then the model can use this information to provide quantitative predictions of how the shape and location of place fields will change when the size, tilt, or location of these cards is changed. Conversely, suppose that the way in which place fields change when the environmental cues are manipulated has been determined experimentally. Then the model can help to determine what features of the environmental cues are being measured to locate the animal.

It might seem strange to try to construct a model to do this since so much information processing takes place between the input of sensory information and the excitation of place-field units. Obviously, no theory currently feasible can account for all this processing in detail, and I make no attempt to do so. Rather, my model employs a biologically plausible architecture to compute the activity of place-field units from the configuration of distal landmarks using only the geometrical constraints inherent in the problem itself and as few specific hypotheses about brain structure and function as possible. This approach implies that even if the model proves to be a valid predictor, it will not directly provide new physiological knowledge. What it will do is give a precise description of the computations being carried out by the place-field system. Once these computations are understood, the task becomes to explain how neural networks in the brain actually implement them. I feel that trying to explain how neurons might accomplish a clearly defined computation is a more tractable task than trying to relate an animal's behavior directly to the functioning of its neurons.

The Model

The basic question that the model must address is: How do the place-field neurons know that the observer, in whose brain they reside, is at the location where they are supposed to fire? I consider this to be a pattern-recognition problem in which a stored representation of a scene is compared to a representation of the current scene. Roughly speaking, the better the match between these two representations, the closer the viewer is to the viewpoint of the stored scene. The activity of place-field units can be thought of as reflecting the closeness of this match. A matching task of this kind could be very complicated if details from the entire scene had to be considered. However, the experiments show that only a few discrete objects need to be used. This greatly simplifies the modeling problem and also makes sense since location information about only a small set of objects is sufficient to locate a viewer in the environment. An example of how several measurements combine to specify location is shown in Figure 1 where the point P is uniquely located by its distance from the three landmarks a, b, and c.

In order to see how a place-field unit might make use of this kind of information, imagine that at some time when an observer was at location P, in Figure 1, representations of the objects a, b, and c were recorded in some way in memory. Further assume, that the descriptions of each of these objects contained the value of the distance to P. Suppose that at some later time the observer returns to the vicinity of P. The observer's sensory system can now generate a set of current descriptions representing the objects, together with their current distances. If these representations, other than their distance components, are not affected too much by the viewing position, then the current representation of each object will differ from its stored representation

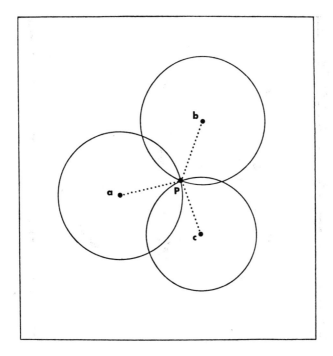

FIGURE 1. The location of point *P* is uniquely determined by its distance from the three landmarks located at *a*, *b*, and *c*. The circles are parametric curves along which the distance of an observer from a landmark remains the same. Other measures of location such as the size of the retinal image of a landmark also give parametric curves, but these are not generally circles centered at the landmark. (From "A Computational Model of Hippocampal Place Fields" by D. Zipser, 1985, *Behavioral Neuroscience, 99*, p. 1007. Copyright 1985 by the American Psychological Association. Reprinted by permission.)

only to the degree that the object's current distance from the observer differs from its distance to *P*. A neuron whose output response is a summated measure of the similarity between these current and stored representations for each landmark will have the properties of a place-field unit.

The model deals with the problem of implementing a place-field unit that can carry out the required matching by breaking the task into two stages, each of which is carried out by a "layer" of simple processors roughly analogous to neurons or small neural networks. This arrangement is shown in Figure 2. Each unit in the first layer is specific for one landmark and one place field, and each is able to compare the current value of a location parameter with a value of the same parameter stored in the unit. The output of these units remains 0 while their landmarks are not in the current scene. When a first-layer unit's

SCENE

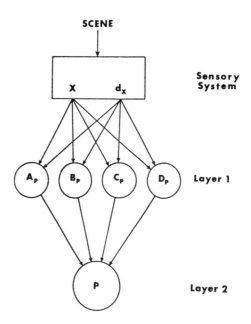

FIGURE 2. The model is shown here as implemented in a network of processors. The "sensory system" processes input from an environmental scene and produces two outputs. These are X, which is a description of a landmark in the current scene, coded so that it will be recognizable by units in Layer 1, and d_x, which is the value of a location parameter, such as distance, valid for the current position of the observer and landmark X. Each unit in Layer 1 is concerned with just one landmark and place field. When a description of this landmark appears at the output of the sensory system, the current value at the d_x output is compared with the value, d_a, stored in the unit. Thus, if the value at $X = A$, then the unit labeled A_p will generate an output by applying its response function to $d_x - d_a$. The procedure used to evaluate the model assumes that the whole scene is scanned while the observer remains still, and it also assumes that the outputs of Layer 1 units, once computed, remain fixed until the observer moves again. The Layer 2 unit integrates the outputs of the Layer 1 units to get an overall measure of how closely the current scene matches the scene as viewed from the place field centered at P. (From "A Computational Model of Hippocampal Place Fields" by D. Zipser, 1985, *Behavioral Neuroscience*, 99, p. 1008. Copyright 1985 by the American Psychological Association. Reprinted by permission.)

landmark is detected by the sensory system, its response increases as the current and stored location parameter values become closer. The response of these first-layer units reaches its maximum value when the current and stored location parameters are equal. The task of the sensory system is to detect the landmarks and generate descriptions of them that can be recognized by the processors in the first layer. The sensory system must also measure the current value of whatever parameters are used to determine location. The model, however, is not concerned with the details of how this is done. It is also not concerned with how this information is represented or transmitted to the processors in the first layer. The computations involved in the model do not require any detailed knowledge of these processes.

The output of the processors in the first layer serves as input to the unit in the second layer, which acts like a place-field neuron. Only one of these is shown in Figure 2 because the current model is only concerned with the output of a single place-field unit. The output of the second-layer unit must be a monotonically increasing function of the sum of its inputs from the first layer in order to represent the similarity between the current scene and the scene as recorded at the center of the place field. The output of this second-layer unit is used to define the location and shape of place fields in simulated experiments.

So far, the response functions for the units in each of the two layers have only been specified in general terms. In order to precisely define the model for computer simulation, some real functions must be chosen for these roles. The properties of the model are not very sensitive to the exact nature of these functions so long as they have the required general properties. This insensitivity to detail is illustrated in Figure 3 where place fields have been calculated using the two different first-layer tuning functions shown below:

$$R1 = \begin{cases} 0 & \text{if } |d_x - d_x^*| > \sigma \\ 1 & \text{otherwise} \end{cases}$$

$$R1 = \exp{-\left[(d_x - d_x^*)^2 / \sigma^2\right]} \tag{1}$$

where $R1$ is the first-layer unit response function, d_x and d_x^* are the stored and current values of a location parameter (distance from landmark X to P in the case of Figure 3), and σ is the matching criteria parameter. The second-layer unit response function is

$$R2 = \begin{cases} 0 & \text{if } \sum_x R1(x) - \Theta \leqslant 0 \qquad\qquad (2) \\ \sum_x R1(x) - \Theta & \text{otherwise} \end{cases}$$

where $R2$ is the second-layer unit response function, x is the set of all landmarks used for the match, and Θ is the threshold parameter. The outlines of the place fields in Figure 3 indicate the curves along which the function $R2 = \Theta$. While the two place fields are not identical, the outlines and locations generated by the two different first-layer matching functions are quite similar. Since the step function and the Gaussian function used are about as different as this kind of monomodal

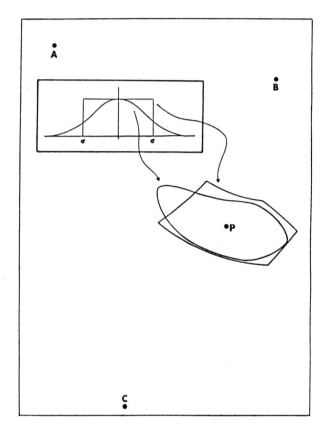

FIGURE 3. The boundaries of the place fields centered at *P* have been computed using two different Layer 1 response functions. The location parameters used were the distances to landmarks at *A*, *B*, and *C*. The two Layer 1 functions used are plotted in the insert and the arrows indicate which function generates which place-field boundary. (From "A Computational Model of Hippocampal Place Fields" by D. Zipser, 1985, *Behavioral Neuroscience*, 99, p. 1009. Copyright 1985 by the American Psychological Association. Reprinted by permission.)

tuning function can get, the similarity in place-field shape indicates the general insensitivity of the model to the nature of these functions. While the outlines of the place fields generated are similar, the internal structure of the fields are different. The Gaussian matching function (Equation 1) gives the kind of smooth peak observed experimentally in place fields.

It is clear that what has been described is not a single model but a class of computational schemata related by the way in which they use location parameters in a two-layer architecture. The individual models in the class differ in just what features are measured to serve as location parameters and in which response functions are used in each layer.

Location Parameters

In Figure 1, distance was used as a measure of landmark location. Any other measurable feature whose value is a function of observer location can also be used. Somewhat more formally, a location parameter is a function depending on the relative location of landmark and observer that has a single value for each point in the environment. The inverse is not true. That is, a single value of a location parameter may map to many different locations. In the examples used so far, distance has been the location parameter, but in fact, distance is probably not being used by the rats in the Kubie, Muller, and Ranck (1983) experiments. This is demonstrated in Figure 4 which shows what happens when a triangle of landmarks a, b, c, is expanded by dilation to a larger triangle a', b', c'. The experimental results on environmental dilation predict that the place field should increase in size and still be centered at P. However, if the remembered distances are used, there is no longer a triple intersection at P. Actually, there is no triple intersection anywhere, which implies that there will be either no place field at all or, at best, one with a greatly *reduced* size, just the opposite of what is actually observed. This suggests that remembered values of the absolute distance to landmarks are not being used by the rat to activate place-field units in the dilation experiments.

If distance is not being used as a measure of landmark location, what is? Something is required that, like distance, defines curves in the environment whose intersections can be used to localize the viewer, but, unlike distance, scales properly with dilation. Among the measurable features that have these properties are the visual angle subtended by two points on a landmark and the retinal area of the landmark image. Angle and area differ from distance in several ways. Unlike distance, they cannot refer to single points but require landmarks with some finite extent. Also, unlike distance, angle and area can be determined by a single measurement. Distance requires either two measurements or knowledge of the absolute size of the landmark. Because the notion of "landmark" is somewhat arbitrary, measurements can often just as easily be made that reflect the relationship between landmarks. For example, the gap between two trees could make just as good a landmark as each of the trees themselves. Retinal area, it should be noted, is a very simple thing for a visual system to compute since after a figure-ground separation, all that is required is that the number of

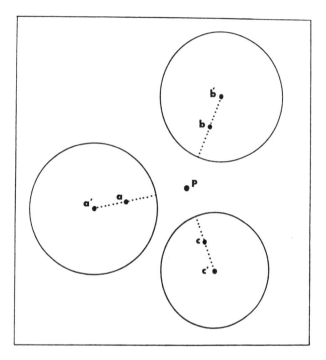

FIGURE 4. The scene from Figure 1 characterized by landmarks at *a*, *b*, and *c* has been dilated by a factor of 1.5 which has relocated the same landmarks at *a'*, *b'*, and *c'*. The circles indicate the curves along which the location parameters (the distances, indicated by the dotted lines in this case) have their undilated values. These circles no longer intersect at *P* making it unlikely that a place-field mechanism that dilates with the scene could depend on remembered distances. (From "A Computational Model of Hippocampal Place Fields" by D. Zipser, 1985, *Behavioral Neuroscience, 99*, p. 1010. Copyright 1985 by the American Psychological Association. Reprinted by permission.)

units in the figure be totaled. Of course, other factors must be included to explain conscious perception that corrects for apparent retinal size.

Since location parameters, by definition, have a single value at each point in space, the properties of a particular parameter can be visualized by plotting the locus of a single value of the parameter. Examples of the curves generated in this way for distance, angle, and area are given in Figure 5. These are the curves that arise assuming the landmark is a square card that looks the same from both sides. Actually, most natural landmarks look different from the back, so that generally only the curves on one side of the landmark will play a role in place-field location. The shape of these location parameter curves will depend on the shape of the landmark.

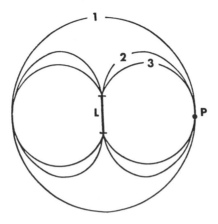

FIGURE 5. *L* is a landmark and *P* is a viewpoint. Curve 1 is a locus of all viewpoints a constant distance from the center of *L*. Curve 2 is a locus of all viewpoints from which a square standing vertically on *L* would have a constant area on a spherical retina. Curve 3 is a locus of all viewpoints where a line of length *L* would subtend a constant visual angle. The visual angle, *A*, subtended by a line of length *a* is given by:

$$A = 2 \arctan \sqrt{(s-b)(s-c)/s(s-a)}$$

where *b* and *c* are the distances from the endpoints of the line to the focal point and *s* = ½ *(a + b + c)*. Note that *A* is invariant to dilation by a factor *d* since replacing *a*, *b*, and *c* by *da*, *db*, and *dc* does not affect the value of the above equation. The area of any object can be approximated by the sum of the areas of a set of triangles which cover its visible surface. The retinal area of a triangle is:

$$Area = 4f^2 \arctan \sqrt{\tan(S/2)\tan((S-A)/2)\tan((S-B)/2)\tan((S-C)/2)}$$

where *f* is the focal length; *A*, *B*, and *C*, are the visual angles of the three sides of the triangle; and *S* = ½ *(A + B + C)*. Since *A*, *B*, and *C* are dilation invariant, so is the retinal area of the object. (From "A Computational Model of Hippocampal Place Fields" by D. Zipser, 1985, *Behavioral Neuroscience*, 99, p. 1011. Copyright 1985 by the American Psychological Association. Reprinted by permission.)

Simulated Experiments

In order to see how well the model works, experiments of the kind used to study place fields were simulated on a computer. The simulated experimental environment, as detailed above, is shown in the computer generated display in Figure 6 (details of the simulations are given in the legend to this figure). The outer rectangle, which represents the limit of the simulated environment, contains the landmarks (shown as straight lines). The simulation assumes that the animal has a place field centered at point *P*. The location of the simulated observer is systematically moved over a uniform lattice of points

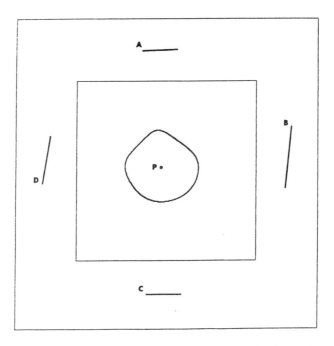

FIGURE 6. This figure shows the output from a typical simulated experiment on the screen of a Symbolics 3600 LISP machine. The experiment is initialized by first placing landmarks, (simulated square cards in this case) in the outer square. The size, position, and number of landmarks can be input explicitly as a set of numbers or determined interactively by the use of a "mouse." Once the scene is set, an initial viewpoint P, which will be the center of the place field, is chosen. Then values for σ and Θ are picked and the simulation is started. First, the simulation program calculates the areas of the landmarks from the viewpoint P. These areas are used as the stored values of the location parameter. Then the viewpoint is systematically scanned through a raster of 145,000 points in the inner rectangle. At each of these points the retinal areas of landmarks are computed and used as the current value of the location parameters in evaluating the response of the Layer 2 unit that represents the place-field unit. A mark is placed in the inner rectangle wherever the value of the Layer 2 unit just crosses threshold. The curve generated by these marks is the border of the place field. The landmarks in this figure are quite symmetrically placed around the scene that leads to a fairly symmetrically shaped place-field border. Asymmetrically placed landmarks would produce a less symmetrical field (see Figures 7 and 8). The size and, to some degree, the shape of the place field are determined by the choice of σ and Θ. See the text and Figure 7 for more about σ and Θ. (From "A Computational Model of Hippocampal Place Fields" by D. Zipser, 1985, *Behavioral Neuroscience*, 99, p. 1012. Copyright 1985 by the American Psychological Association. Reprinted by permission.)

covering the inner rectangle. The observer is limited to this inner rectangle so as to keep the landmarks distal because the model cannot deal with either proximal landmarks or the complexities of having the observer wander around in between landmarks. At each viewpoint, the activity of the place-field unit is computed using the model. In this, and all following simulations, the Gaussian matching function (Equation 1) and a threshold function (Equation 2) are used for the response functions of the first- and second-layer units respectively, and retinal area is the location parameter. The shape and location of place fields are visualized by plotting their boundaries. At each viewpoint scanned in the inner rectangle, information from all landmarks is used. That is, no account is taken of viewer orientation. The significance of this simplification will be discussed later.

More About Landmarks

In the experiments that originally demonstrated the dependence of place-field unit activity on distal landmarks, a variety of objects such as a fan, a white card, a buzzer, and a light, all arranged at 90° intervals around the experimental area, were used (O'Keefe & Nadel, 1979). The only manipulation carried out on these cues was either to remove them or to interchange their locations. While this served well to demonstrate the basic phenomenon in a qualitative fashion, a more precisely defined set of landmarks and closer control over their location is required to determine the quantitative relationship between landmark configuration and place-field unit activity. At the present time not enough information is available to determine what would constitute a good, simple set of landmarks. In the computer simulations described here, planar, square landmarks of different sizes, that stand vertically at any position in the outer rectangle, are used. As viewed from above, in the computer-generated display, these square landmarks appear as straight lines. Squares were chosen because the computation of their projected retinal areas is reasonably tractable. The computation of the projected retinal area of an arbitrarily shaped object from a large number of different positions is quite complex. In all the rest of the simulations described here, the retinal area of these squares will be the location parameter used. It seems reasonable that a set of such square cards of different sizes, textures, and reflectances might well serve as adequate cues in real place-field experiments. Indeed in the Kubie, Muller, and Ranck (1983) experiments, a rectangular card played an important role as an environmental cue.

Place-Field Shape and Size

The two parameters σ and Θ in Equations 1 and 2 affect the size and, to some degree, the shape of the simulated place fields. The way

σ and Θ affect place fields is shown in Figure 7, in which σ and Θ are systematically varied while the landmarks and place-field center are kept fixed. The results shown in Figure 7 indicate that although these two parameters do not have identical effects on the shape and size of place fields, their effects are quite similar, particularly in the region of moderately high Θ and moderately low σ, which give realistic looking place fields. Real place-field units are probably subject to a good deal of inhibitory interaction with other hippocampal units and many other modulating influences. For this reason, the model described here is unlikely to give a very accurate picture of the shape variation of place fields with landmark manipulation.

Place-Field Location

Extensive simulations indicate that both the location, shape, and size of place fields is affected by the number, size, and location of the landmarks. The effect of landmark manipulation on the *shape* of place fields is relatively small and might be difficult to observe experimentally. However, the much larger effects of changing landmark location, size, or orientation on the *location* and *size* of place fields can easily be detected experimentally and will serve as the main test of the model.

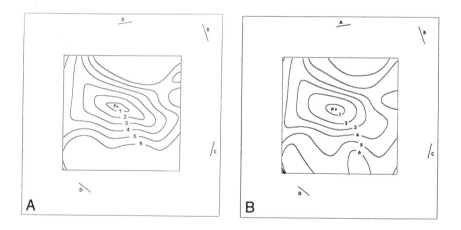

FIGURE 7. This figure was generated in the same way as Figure 6 except that rather than displaying the place-field boundary for just one set of values of σ and Θ, a range of values was used to create a contour map of the place fields. *A*: σ has been scanned while Θ is held at 2.4. The σ values for the different curves are: $\sigma_1 = 0.1$, $\sigma_2 = 0.25$, $\sigma_3 = 0.4$, $\sigma_4 = 0.55$, $\sigma_5 = 0.7$, $\sigma_6 = 0.85$. *B*: σ is fixed at 0.6 and Θ is scanned: $\Theta_1 = 3.9$, $\Theta_2 = 3.4$, $\Theta_3 = 2.9$, $\Theta_4 = 2.4$, $\Theta_5 = 1.9$, $\Theta_6 = 1.4$.

For example, the one hard quantitative experimental result that we have in this area comes from the dilation experiments (Kubie et al., 1983). Experiments similar to this were simulated using the model described here. The results of this simulation are shown in Figure 8. First, a place field centered at *P* was generated using landmarks *A*, *B*, and *C*. All the coordinates specifying the location of *A*, *B*, and *C* were then multiplied by 2 to give a dilation. The dilated scene was then translated back to a convenient location in the outer rectangle to

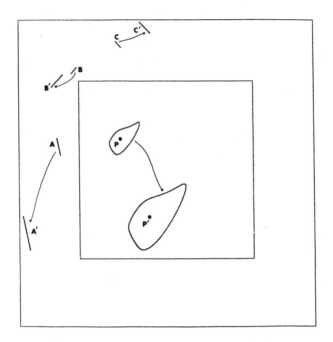

FIGURE 8. This figure shows the result of a simulated dilation experiment. First, the landmarks *A*, *B*, and *C* of the indicated size were set up and the viewpoint *P* chosen as the place-field center. The field centered at *P* was generated by the simulation. Then the dilation, which increased the size of each landmark and the distance between them by a factor of two, was computed using the known coordinates of the landmarks. This enlarged scene was then shifted so that it could be put in the outer rectangle. This resulted in landmarks *A′*, *B′*, and *C′*. A place-field boundary was then computed by the simulation using the same values of the stored location parameters originally determined at *P*. This led to the generation of a new larger place field at a different location. When the point *P* was dilated and translated by the same amount as the landmark *A′*, *B′*, and *C′*, it mapped to a location indicated by *P′*. The fact that *P′* occupies the same position in the new place field as *P* did in the original, together with the fact that the new place field is a two-fold magnification of the original, are consequences of the dilation properties of retinal area used as a location parameter. (From "A Computational Model of Hippocampal Place Fields" by D. Zipser, 1985, *Behavioral Neuroscience, 99*, p. 1015. Copyright 1985 by the American Psychological Association. Reprinted by permission.)

give landmarks A', B', and C'. The same dilation and translation were applied to the point P itself to give the point P'. The landmarks A', B', and C' were used to generate a new place field using the same recorded values of the location parameters employed to generate the original place field centered at P. The new place field is centered at P' and has the same shape but is twice the size of the original. This result demonstrates that when retinal area is used as a location parameter, the model gives the expected result for a dilation experiment. While the actual shape and size of a place field depends on model parameters, the way it dilates is the result of the inherent mathematical properties of the location parameter used—area in this case. (See the caption of Figure 5 for proof that the area has the expected dilation properties.) Other simulations in which the size or location of landmarks was changed in ways other than uniform dilation, produced large shifts in the location and size of the place fields. The finding of similar place-field position shifts accompanying corresponding landmark manipulations in real experiments serves as strong confirmation that the model described here embodies the correct rules relating landmark configuration to place-field properties.

Scope of the Place-Field Model

The model place-field described here is both oversimplified and incomplete, yet within its domain of applicability, it is testable by experiment. One way in which the model is incomplete is that it ignores the problem of left-right asymmetry. Early experiments that indicated that distal landmarks determined place fields also showed that interchanging the landmarks on the right and left side extinguished the location-specific activity of place-field units. This indicates that the animal has the ability to distinguish right from left, whereas the model as described is symmetrical. The ability to distinguish right from left is a relatively straightforward feature to incorporate into the model and one way of doing this is described later in this chapter in connection with goal-finding models.

Another oversimplification of the place-field model is that it ignores the problem of viewer orientation. It is known from experiment that many place-field units show sensitivity to the orientation of the observer (McNaughton, Barnes, & O'Keefe, 1983). For example, a rat running through a maze arm in one direction may excite a place-field unit at a particular location that is not excited when the rat runs back through the same location in the opposite direction. Orientation is important for goal location and a mechanism for incorporating it into a place-field system is described in the next section.

Goal Location

Knowing where you are is only worthwhile if this information can be used for some useful cognitive purpose. For example, to answer the question "How do I get home from here?", information about the location of a goal (home in this case) must be available. For the rest of this chapter, I will deal with the problem of developing mechanisms that can answer this question. The problem is trivial, of course, if the goal is perceivable from the current location. The problem becomes interesting only when the goal is too far away or too inconspicuous to detect. In all of what follows, it will be assumed that the observer must find the way home using a set of locally visible landmarks and that the observer has no access to absolute direction information such as "Which way is north?" and so forth. This does not significantly reduce the generality of the problem, since the determination of an absolute direction such as north would, in general, depend on knowledge about the configuration of the local landmarks anyway.

The unoriented place fields described above are insufficient for goal location. The reason for this is that in the absence of absolute direction information, the location of a goal must be given relative to both the location *and* orientation of the observer. One way to approach this issue is to incorporate orientation into the place field itself. I call an oriented place field a *view field.* Many of the place fields observed in the hippocampus are, in fact, oriented as pointed out above. That is, some hippocampus neurons fire at a particular place only when the rat is pointed in a specific direction. Orientation can be incorporated into place fields by adding some weak, topological constraint, such as classifying landmarks as being either on the right, in the center, or on the left. Distinguishing right, left, and center is a fairly straightforward task for the sensory system of any bilaterally symmetrical animal.

In the following sections I describe two models of goal-finding. One, the *distributed view-field model,* uses a large number of oriented place-fields, each with associated goal location information, scattered around the environment. The other model is based on a more extensive computational analysis and requires fewer units but more elaborate calculations.

The Distributed View-Field Model

The distributed view-field model for goal location is based on the network shown in Figure 9 which is used as the "cognitive map" to guide an observer to a goal. The first layer of units, called object units,

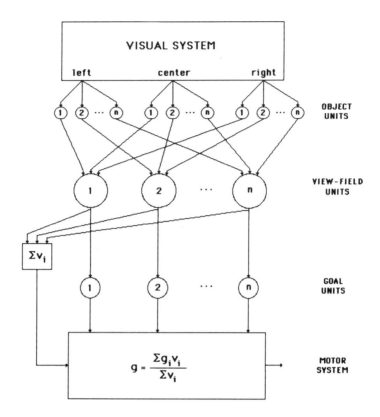

FIGURE 9. Network used in the distributed view-field model. The visual system output consists of three sets (left, center, and right) of three numbers representing an object's type, distance, and orientation relative to the observer. The output of an object unit is 0 unless its recorded type number matches its current input from the visual system. In this case, the output is the product of two Gaussian tuning functions (see Equation 1), one for distance and one for orientation. The output of the view-field units is the same kind of sum given in Equation 2. The output of the goal units is two components of a vector pointing to the goal each multiplied by the input from a view unit.

is analogous in function to the Layer 1 units of the preceding place-field model shown in Figure 2. The primary difference is that the object units in Figure 9 are divided into left, center, and right groups. Object units are gated on only when the landmarks they recognize appear in the appropriate region of the visual field, i.e., left, center, or right. The view-field units in Figure 9 are analogous to the Layer 2 units in Figure 2. Together, the object units and the view-field units recognize an oriented place field, i.e., a view field, in the same general way as the corresponding units in Figure 2. However, now the output of the view-field unit also depends on orientation as well as location. Once a view-field unit for a particular place is created, information about the location of a goal can be associated with it. In the network shown in Figure 9, goal location is encoded in an additional unit whose output is interpreted by the motor system as the direction of the goal. To actually head to a goal from any particular location, an observer with this kind of network would first examine the landmarks in front of it, which would lead to the activation of a set of object units and then to the activation of an associated view unit. This in turn would activate a goal unit, the output of which could direct the observer to the goal.

There are several problems to be solved before this general strategy can actually be applied to controlling the movements of an observer. First, it is impossible to have a unique view-field unit for every place and orientation in the environment. At best, one could hope to have a rich supply of view-field units whose place fields spanned the environment. Another problem is that even with a large number of view-field units, it is unlikely that the observer is ever at the exact center of a place field. This means that some form of interpolation is required to compute the location of the goal from the current position of the observer. Another problem is to specify how information about the environment gets assimilated by the network in Figure 9 in the first place. Aside from the technical issues of recruiting the appropriate units and connecting them together with the desired weights, there is the conceptual problem that arises from the need to know the location of a goal at the time that its association with a particular view field is made. There is no trivial solution to this problem. One approach is to remember and constantly update the location of a particular goal as the observer moves through the environment. This active memory of goal location can then be used to establish the place-field goal relationships when new units are recruited. Once enough units have been instantiated, the territory can be thought of as familiar and it will no longer be necessary to have an active memory of the location of a goal in order to find it. This is the strategy used in the distributed view-map model described here. This approach has the drawback that it demands both processing resources and memory accuracy. These requirements could possibly be met for modest periods of time and over short distances with no distractions, but they could cause problems in more extended situations.

In the distributed view-map model, the viewer has a large set of view fields scattered over the environment. Associated with each of these view fields is a vector pointed towards the goal. If the observer is in the center of one of these view fields and oriented in the optimal direction, the associated goal unit indicates the direction of the goal. However, the observer is seldom at the center of a view field so information from a set of nearby view fields must be interpreted. The first step in this process is recognition of the currently visible set of left, center, and right landmarks that are used to gate the appropriate ensemble of view fields. Associated with each of these view fields is a vector pointing home. This is illustrated in Figure 10, which shows a viewer, V, oriented towards the central landmark, C, and nine view fields, each labeled m. Of course, the oriented view fields are not actually written on the ground in the environment but are encoded in the brain of the observer. Although the whole ensemble of view fields that have the same set of three landmarks will be gated on, the activity of each individual unit will depend on how closely its parameters match the current situation. The closer the viewer is to the center and orientation of any particular view field, the more active the unit representing it becomes. In the distributed view-field model, the observer interpolates the location of the goal by using a weighted average of the direction vectors of all of the active goal units. The way this weighted average is formed is that each goal unit multiplies its output vector, indicating goal location by its input from a view-field unit. The motor system then sums all of these weighted vectors and normalizes them to the sum of the activity

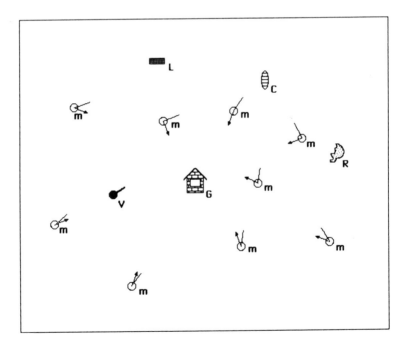

FIGURE 10. A diagrammatic representation of the view-field model. G is the goal, V is the observer, L, C, and R are landmarks. The positions at which view fields have been recorded by the observer are indicated by ms. Each view field consists of a line representing the orientation of the observer when the view field was recorded and an arrow representing the direction of the goal relative to this orientation.

of all the view-field units (see Figure 9). Thus, the goal vectors associated with close-matching view fields play a large part in the sum, since their associated units are highly active, whereas the vectors from less closely fitting view fields play a smaller role. This strategy requires the view fields to be reasonably well distributed around the environment.

In order to test the model, the network of Figure 9 was used to actually control the behavior of a simulated observer moving around in a model environment. Before describing this simulation in detail, it's necessary to deal with the problem of getting environmental information into view fields. Rather than building in spatial information, a procedure for assimilating it while the viewer moves around was incorporated into the model. The naive viewer was supplied with a network similar to that shown in Figure 9, but in which all object units and goal units had zero values for all their relevant parameters. At fixed time intervals, while the observer explored the environment at random, a signal is sent to recruit object and goal units. Upon receipt of this signal, the object units being recruited record the values necessary to recognize and locate the currently visible set of landmarks, while the goal unit records the location of the goal relative to the current position and orientation. The object units receive their information from the current output of the visual system, whereas the goal unit to be trained obtains its values from a working memory. Once all the available units are recruited, no more view fields can be learned and all subsequent goal-directed travel is dependent on the existing set of view fields.

To observe the properties of the distributed view-field model, a simulated observer was developed with two motivational states, "curiosity" and "guilt." In the curiosity state, the observer explores the environment by a random pattern of movement. In the guilt state, the observer attempts to return home using the information recorded in its network of view-field units. The assimilation of new information takes place at all times, independent of the motivational state of the observer. Thus, new information can be assimilated either during random exploration or during return to home.

The actual simulation program to carry out all this is quite complex, because in addition to simulating the cognitive mapping network of Figure 10, the environment, together with some information about the observer's location and visual system, had to be modeled. This project was expedited by the availability of a general purpose parallel simulating system called P3 (Parallel Process Programmer). The P3 system is described in more detail in a separate chapter of this book, but briefly, P3 is used to simulate networks consisting of an arbitrary number of unit processors connected together by signal lines. The complete network of interconnected units is called a P3 plan. Each unit in the plan must have a computer program, called a method, associated with it that updates its outputs according to some function of its current inputs and state variables. The input to units described in a P3 plan can come only from other units in the plan. That is, there is no outside world in the P3 description of a network. This means that the environment, as well as the nervous system network within the model, must be described by individual units. The simulated model can interact with the outside world in two ways. It can be initialized and observed as it functions, using facilities provided by the P3 simulation system. In addition, the method programs can communicate with the investigator through the keyboard and display using the available features of the method language. The methods can also use the file system. However, all communication between units must be via connections described in the P3 plan.

A simplified view of the P3 plan used to simulate the model described here is shown in Figure 11. There is one very complex unit called the *environ unit* which carries out a multitude of tasks. It keeps a record of and displays for observation the current position of the viewer, the location of the goal (home in this case), and the position of all relevant landmarks. The output of the environ unit provides the viewer-centered information necessary to activate the object units in the model. In this way, the environ unit serves as the sensory system of the viewer. Another output of the environ unit is working memory, which has the current goal location used by goal units when they are assimilating new information. The input to the environ unit comes from the *mover unit* which serves as the motor system for the viewer. The environ unit uses this information to update the location and the orientation of the observer. Thus, we see that the environ unit serves as the entire world outside the particular part of the brain being simulated by the mapping network under test, that is, it constitutes the musculature, the visual system, and the short-term memory of the observer together with the physical world outside the viewer. Obviously the method program for the environ unit is quite complex. The shaded units in Figure 11 represent the major parts of the cognitive map networks under test. These are the elements that are changed significantly when different models of the viewer's space representation system are studied. The rest of the P3 plan can remain essentially the same, constituting a major saving in effort.

The model and the associated methods were coded in the P3 system. An example of what the screen of the Symbolics 3600 LISP machine looks like during the progress of a simulation of the model is shown in Figure 12. Information about the state of various units, at one time cycle, is displayed in the large panel on the left. The "world display" window of the environ unit appears in the panel on the right. The behavior of the model is most easily analyzed by examination of the

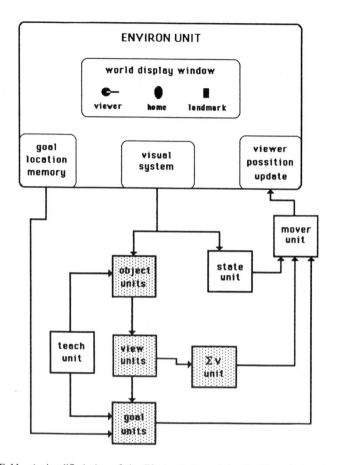

FIGURE 11. A simplified plan of the P3 simulation of the distributed view-field model. The shaded boxes represent the part of the model that contains the specific map information. The mover is actually comprised of several units that compensate for the delay between observation and resulting movement. This compensation is required because other motions may occur during this period. The symbols in the world display window of the environ unit are those used in the computer display that it drives (see Figure 12).

world display window. The observer, marked V in the figure, is just about to reach the goal at the center, having started towards the goal at a point marked S. The path of the observer during the "guilt" state, when it is trying to go home, is characterized by evenly spaced dots. The path of the observer during the "curiosity" state is seen as a set of much more arbitrarily spaced dots. One dot is added to these paths at each cycle of the simulation. At the outer extremes of the field, the landmarks appear as rectangles. In this particular simulation, eleven landmarks were used. Although they all look the same on the screen,

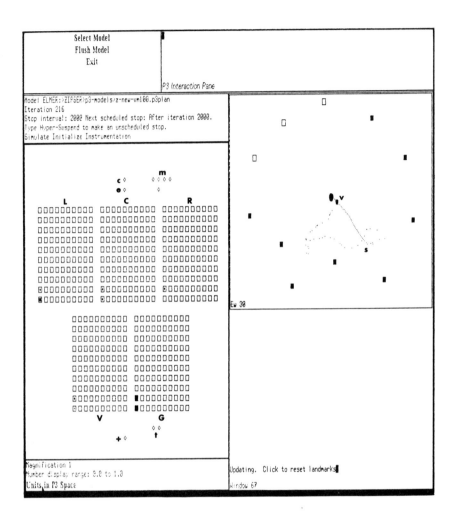

FIGURE 12. Symbolics 3600 LISP machine display of the distributed view-field model. See the text for explanation.

they are in fact distinguished by the viewer. The three landmarks that constitute the current view are highlighted by making them hollow rectangles slightly larger than the unviewed landmarks.

Figure 12 represents the state of the system after 216 iterations of the simulation. Only 10 out of the pool of 100 view units have been recruited, so this is relatively early in the exploration phase. Examining the path from S to V in the world display window, shows that the viewer is able to go directly home along a moderately straight path. In general, this is the case early in a simulation, before many view fields have been recruited. The panel on the left shows the output activity of the object units, the view-field units, and the goal units. No information is being displayed about the teach unit, the environ unit, the state unit, or the mover unit. Actually, the mover unit consists of a group of several units that work together to provide the appropriate motion information based on the output of the array of goal units. Some caution needs to be used in looking at a state display of this type since there is a delay of one cycle between each layer of the network. Thus, what's seen at the output of the goal units is a consequence of events occurring one cycle earlier in the view units, and two cycles earlier in the L, C, and R object units. The important point to notice here is that only two goal units show any output activity at all, and both of these have activity of the same sign. This means that at this point the movement of the viewer is being controlled by only a small set of view fields. This is consistent with the fact that the simulation is early in the assimilation phase.

After the assimilation phase is over and all units have been recruited, the pattern of view fields is as shown in Figure 13. This figure was generated from the world display window of the environ unit, which has the option of showing the actual positions corresponding to the observer's store of view fields. Figure 14 shows the path of the observer attempting to get home after simulation cycle 3081 when the assimilation phase is long over. The path starts at S, and the observer, after much thrashing about, has reached point V, and not yet gotten close enough to home for the home detector to have triggered. Note that the path starts at S and continues in a relatively straight fashion, but not directly towards home. Before passing too far beyond the goal, the observer has made a sharp turn to the left and moved towards the goal on a very uncertain and constantly changing path. Then, after approaching quite close to the goal, the observer again turned, this time in a path that took it somewhat away from the goal. As we look at Figure 14 at the current cycle, the observer has been thrashing around in the general vicinity of the goal but has not been able to actually reach it. Eventually the observer will get home, but it will take a great many cycles. This behavior is quite typical of what happens when an attempt is made to go home after all 100 view-field units have been recruited. To the right of the world display window, the current states of the array of the view-field units and goal units are shown. These are not both from the same cycle, rather the view-field units are shown from one cycle before the goal units so that the active view-field units are the ones that determined the activity of the goal units in the lower array. Notice that a large number of view-field units are active and that a correspondingly large number of goal field units are active. However,

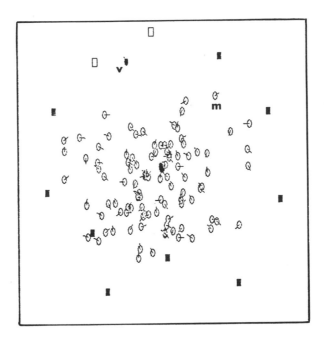

FIGURE 13. The distribution of view fields after all 100 available units have been recruited. A typical view field, labeled *m*, has a line indicating the orientation of the observer at the time it was recorded. The goal vector is left out for clarity but it would point to the dark oval at the center. The current position of the observer is indicated by *v*.

some goal units have positive output values as indicated by solid rectangles, while others have negative values as indicated by open rectangles. The reason for this is that some of the view fields are on opposite sides of the goal, which provides the observer with conflicting information about the sign of the direction of the goal. This conflicting information is the cause of the viewer's inability to get to the goal in a smooth fashion. This conflict situation only causes a serious problem when the observer is quite close to the goal because when it is further away, information from view fields on the opposite side of the goal contribute very little to the weighted sum of goal vectors. Although the problem is inherent in the model, by adjusting various parameters that control the density of view-field placement and the strength of their interaction, it is possible to reduce the size of the region of ambiguity to the point where it is small compared to the area in which the goal is directly perceivable.

The distributed view-field model simulated above has the virtue of simplicity, since the viewer is not required to carry out any complex

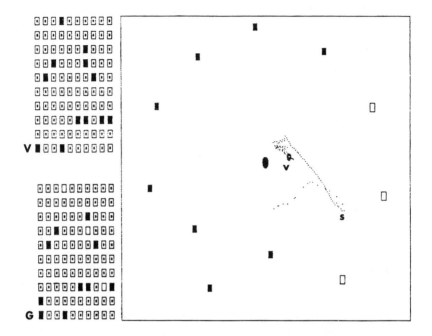

FIGURE 14. Behavior of observer, V, in the distributed view-field model with all units recruited. The state of the view-field units, V, on the 3080th cycle and of the goal units on the next cycle are shown on the left. The reason for the complex path followed by the observer is given in the text.

computations to navigate. Its algorithm for assimilating new information is extremely unsophisticated, operating only on the basis of time. One price paid for this simplicity is the need for a large number of units to store assimilated information. One way in which this algorithm might be improved is to assimilate new information only when the present information is inadequate. This inadequacy is easy to detect since the unit that sums the activity of all the view-field units can also be used to determine when there is insufficient information available. The way this is done is to take advantage of the fact that when the sum of the activities of the view-field units is high, the observer is in familiar territory; when this sum is low, the viewer is in unfamiliar territory where it would be wise to assimilate some new information. This would do away with redundant view fields that are generated by the clock method. On balance, the distributed view-field model meets the criterion of biological plausibility since it is based on the concept of place-field units that are known to exist. These units may well play a role in goal navigation since the hippocampus, where they are located, is required for navigation to unperceived goals. The magnitude of the ambiguity problem that this model has in the vicinity of the goal can probably be reduced to a point where it is not any more troublesome than the uncertainties about goal location that would arise from other sources of error in any model.

The β-Coefficient Model

In this section, a navigation scheme called the β-coefficient model (β model for short) is described that overcomes some of the drawbacks of the distributed view-field model described previously. Rather than requiring many separate view fields, the β model needs to record information about a set of landmarks and a goal only once. This recorded information can then be used to locate the goal anywhere the landmarks are visible. The β model also indicates goal location unambiguously, so the problem of sign that caused confusion near the goal in the view-field model is eliminated. Since only one recording is required, the actual number of units needed for the β model can be much less than for the view-field model. The β model can do away with the need for an updated memory of goal location whenever the set of landmarks to be used are visible from the goal itself since in these cases the recording can be made at the goal. If additional recordings of new sets of landmarks are made at locations where both a new and an old set are visible, the need for an updatable memory of goal location can be eliminated because the old set of landmarks can be used to establish the goal location for the new set. The price paid for these good features is an increase in computational complexity together with a loss of an immediately obvious link to current neurobiological data. Both of these drawbacks could be more apparent than real. More about this after the β model has been examined in detail.

To understand the β model, it's necessary to delve a little more deeply into the computational aspects of representing location. The problem we wish to solve is illustrated in Figure 15. An observer located at position V records the position of the three landmarks L, C, and R together with the location of the goal G. At some later time, the same observer is imagined to be at point V' at which the locations of the landmarks L, C, and R can be determined. The observer's problem is to locate the goal G, which cannot be seen from the position V', using the information recorded at point V. It can be shown that there exist three constants, β_L, β_C, and β_R, which depend only on the

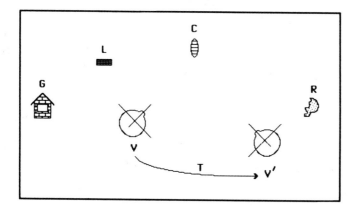

FIGURE 15. The observers at V and at V' can both locate the landmarks L, C, and R in their viewer-centered reference frames but only at V can the goal G be perceived. The problem is to solve for T so that the observer at V' can locate G.

relative locations of L, C, R and G, and have the following nice properties:

$$x'_G = \beta_L x'_L + \beta_C x'_C + \beta_R x'_R$$
$$y'_G = \beta_L y'_L + \beta_C y'_C + \beta_R y'_R$$
$$1 = \beta_R + \beta_C + \beta_C$$

where x'_G, y'_G, x'_L, y'_L, x'_C, y'_C, and x'_R, y'_R are the coordinates of G, L, C, and R in the reference frame of the observer at V'. (For a derivation of the βs see Appendix A.) Since the βs depend only on the relative locations of the landmarks and the goal and not on the location of the observer, they can be computed at any place where the location of the landmarks and the goal are known, that is, at V in our example. Because the βs are constant, it is only necessary to have a single record of their values to locate the goal from anywhere the landmarks can be viewed. The significance of this set of β equations relating the location of G to the location of the landmarks is that the coordinates of the goal can be obtained as a linear combination of the current values of the landmark coordinates and the remembered β. This means that neural-like elements can perform the computations if their synaptic weights correspond to the values of β.

A network using the β-coefficient concept and consisting of three units is shown in Figure 16. The job of the context gate unit is to determine that the current left, center, and right landmarks are the same as were used to originally program the β units. When the context unit recognizes the appropriate set of landmarks, it gates on the β units.

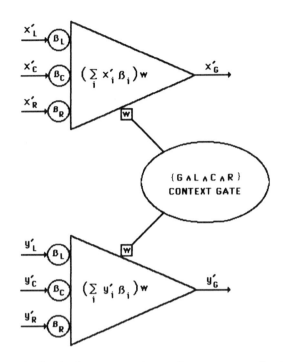

FIGURE 16. A pair of units that compute the x and y components of goal location using the method of β-coefficients. In the units, the βs are implemented as weights or lines coming from a sensory system that can measure the current values of the coordinates of three landmarks, L, C, and R. Each of the triangular β units is gated by a context gate unit (oval shape) which has an output of 0 unless the desired goal is G and the current landmarks are L, C, and R in the appropriate configuration, in which case the output of the gate unit goes to 1. If the value of $w = 1$ also, the correct value of the goal coordinates appears at the output of the β units.

These units perform the appropriate linear summation operations and their output represents the coordinates of the goal. One of the sets of units in Figure 16 is required for every combination of three landmarks with a goal. The output of the appropriately gated β units is an exact indication of the location of the goal from any point in the environment. This means that there will be no problem approaching the goal from any angle and there will be no confusion from information coming from the opposite side of the goal as there was in the previous example. Note that only one set of β units and a context unit is required for each goal-landmark combination as opposed to many units for each combination in the distributed view-field model described above. If information about the configuration of the environment is recorded at the goal itself, as may be the case for the rat on the platform in cloudy water, no working memory of goal location is required.

An important issue in the β model is how the βs are initially computed and recorded. Since we can get no hints from neurobiology, a wide range of possibilities are available. For example, we might imagine that the βs are computed locally by each β unit when it is recruited. Such a computation could take the form of a learning rule that adjusted the value of the βs in accordance with an error signal generated by comparing the current output of a β unit with the correct output, i.e., the coordinates of the goal that, of course, must be known when the βs are programmed. Another approach is to have a single, central β computer that calculates the appropriate values of β and transmits them to the newly recruited β units for recording. In the absence of hard data, it is a matter of taste which of these one considers to be more biologically plausible.

There is a completely different way to go about programming the β units that does not involve setting the values of β at all. In this approach it is assumed that there is a pre-existing pool of β units each with a set of fixed β values. It is then the values of the strengths of connections between a context unit and the β units in the pool that are set to produce the desired result. The idea is that the output of each β unit is weighted by the value of the weight, w, which connects it to a context unit. If these ws are chosen correctly, the sum of the weighted outputs of all the β units in the pool gives the appropriate goal coordinates. In this system a set of ws rather than a set of β units have to be learned. This could have some advantages, for example, the number of β units in the pool is fixed and can be much smaller than the number of context units. This means that fewer units are required to instantiate a w programming version of the model than a β programming version. Another advantage of the w version is that setting the w values is synonymous with recruiting the β units so that recruitment and programming are accomplished as a single operation rather than as two separate ones. As with the computation of β, the determination of the ws can be local to the β units or carried out by a central w computer.

In the simulation used here to test the β model, a version of w programming that had a pool consisting of only four β units with a central w computer was used. The way this system worked to program the β units with the correct connection strengths to the context units is explained in Appendix B.

A simulation analogous to that used to test the view-field model was employed to evaluate the β-coefficient scheme. Only those parts of the P3 plan that determine the representation of the spatial map were changed. The plan of this model is shown in Figure 17. The context units are very simple recognizers. Their output is 0 unless the particular set of left, center, and right landmarks to which they are sensitive is currently being viewed, in which case their output goes to 1. The teach

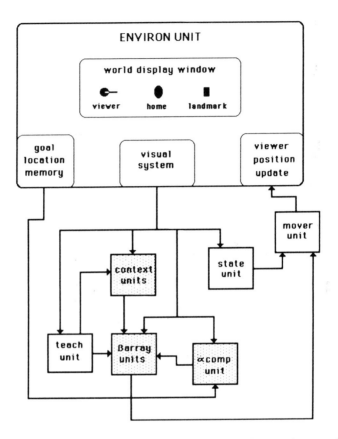

FIGURE 17. A simplified diagram of the P3 plan used to simulate an observer with a β array map of its environment. The shaded units constitute the map and they are the only significant difference between this plan and the one shown in Figure 11.

unit supplies a signal to the context units and the β units to recruit them when new contexts are assimilated. The teach unit requires modification since it no longer makes sense to teach at uniform time intervals. The teach unit produces a pulse only when no context units are active, that is, when the current view is unfamiliar.

The performance of this model is superb. The viewer always goes exactly home at the appropriate time. The results of a typical simulation episode are shown in Figure 18. Notice the straight paths heading directly towards the center of the goal. The number of context units required in this model is roughly equal to the number of landmarks, plus a few extra to take care of cases where a landmark can be the central object with different sets of flanking objects. The loops in the paths, found near the environment window borders in Figure 18, are

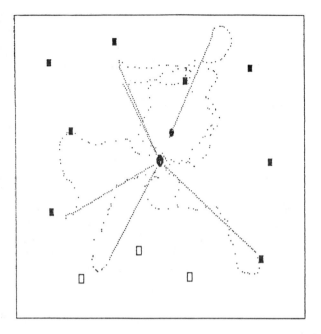

FIGURE 18. The performance of the β array model simulated with P3 and shown on the display of a Symbolics 3600 LISP machine. See text for details.

due to that fact that the state unit can detect when the observer leaves the area with landmarks. When this happens, the context unit instructs the move unit to turn the observer left until a set of three landmarks become visible. This feature is only included to keep the observer on the screen and has no significance for the network being tested. Even when the number of context units is insufficient to assimilate all possible views of the landmarks, the observer manages to find its way home. It simply searches until it finds an appropriate view for which it has assimilated information and then proceeds to use it to head home. Behavior degrades quite gracefully in this case and the number of landmarks has to exceed the number of context units by a great deal before serious difficulties are encountered.

Aside from being able to accomplish the task in an efficient and accurate fashion, the β-coefficient network has some other attractive features. It is, for example, able to explain the behavior of rats in the cloudy water experiment, since while on the platform, the rat can assimilate the appropriate β-related information for groups of room cues. The distributed view-field model described earlier has a much more difficult time with the cloudy water paradigm. It would require that the rat jump off the platform and swim around the tank to pick up a large set of view-field information. The problems that arose in the view-field scheme near the goal would be particularly annoying to the rat in this case since there are no visual cues to serve in goal detection, which become apparent as the rat approaches the submerged platform.

Conclusion

The three models that have been presented here differ from each other in the degree to which they encompass known experimental observations. The first model of place fields is based so strongly on neurophysiological data that its predictions are immediately amenable to testing. The second model, which adds a goal vector to place fields, is more speculative in the sense that it hypothesizes an explicit output role for the place-field neurons. It seems very likely that these neurons do play a role in navigation since removal of the hippocampus, which contains them, cripples navigational ability (Morris et al., 1982). Both of these models clearly meet the criterion of biological plausibility since they attempt to explain observable data in terms of neural-like models.

The β model is of a different sort. It attempts to explain the observed data by proposing a novel representational and computational scheme. This scheme is biologically plausible in the sense that it is implementable with neural-like elements operating in a highly parallel fashion. However, the β model can draw neither support or opposition from current experimental data. Indeed, this is just what we would expect from a theory that proposes a mechanism that is not completely intuitive but derives from a formal analysis of the problem domain. This is to be expected since it is very unlikely that experimentalists would find data relevant to the β model, even if it is correct, until they are specifically looking for it. This, of course, is exactly what theory is supposed to do, ask questions which would not be asked in the normal course of experimentation.

APPENDIX A

To see where the βs come from, we solve for the transformation T that maps all points in the V system to the V' system and then divide the expression for T into a constant part that depends on landmark and goal configuration and a variable part that depends on observer location. T must be found using only information about the recorded and current location of the landmarks L, C, and R. Once T has been obtained, it can be applied to the coordinates of the goal G in the V system to map them into the coordinates of G in the V' system. If information about three landmarks is used in two dimensions, homogeneous coordinates can be employed and T can be a linear transformation (Foley & VanDam, 1983). Having T linear will greatly simplify the nature of the computations necessary to use the transformation and will make the networks that carry it out much more biologically plausible. The derivation of T is shown below:

Let $x_i = (x_i, y_i, z_i)$ be the homogeneous coordinates of the ith object. (In the case of Figure 15, $i = L$, C, R, or G.) Then T is defined so that $x'_i = x_i T$. To solve for T let

$$X = \begin{pmatrix} x_L \\ x_C \\ x_R \end{pmatrix} \qquad X' = \begin{pmatrix} x'_L \\ x'_C \\ x'_R \end{pmatrix}$$

Then, $T = X^{-1}X'$ where X^{-1} is the inverse of X that exists as long as L, C, and R are distinct and not collinear.

Now we have T defined in terms of the coordinates of the three landmarks recorded at V and the coordinates of the same landmarks observed from V'. Since we are using homogeneous coordinates, the third coordinate of all objects can be taken as 1.

The observer at V' is now able to compute T. Applying T to the recorded coordinates of the goal G will give the coordinates of G in the observer's current reference frame. That is,

$$x'_G = x_G T$$

or substituting for T,

$$x'_G = x_G X^{-1} X'. \tag{1}$$

Equation 1 points out an interesting fact. Namely, the first two terms on the right-hand side are independent of viewer location and can be computed once—at the initial siting of the landmarks. This computation, in fact, yields three constants.   Once determined, these three constants are sufficient to locate the goal G in any reference frame in which the locations of the landmarks L, C, and R can be determined. I call these constants β-coefficients and their definition is given below:

Define:

$$\beta = (\beta_L, \beta_C, \beta_R) = x_G X^{-1},$$

then:

$$x'_G = \beta_L x'_L + \beta_C x'_C + \beta_R x'_R$$
$$y'_G = \beta_L y'_L + \beta_C y'_C + \beta_R y'_R$$
$$1 = \beta_L + \beta_C + \beta_R.$$

APPENDIX B

Four β units with fixed, predetermined values of β were used in the simulation described here. They were arranged in a 2 × 2 array as shown in Figure 19. When instructed by the teach unit to establish a new weight for connection to a newly instantiated context unit, each β unit calculated the appropriate weight by multiplying together the values of its two inputs from the α-computer. The problem is what should the outputs of the α-computer be to assure that each β unit generates the correct weights? Assuming that the contribution of each

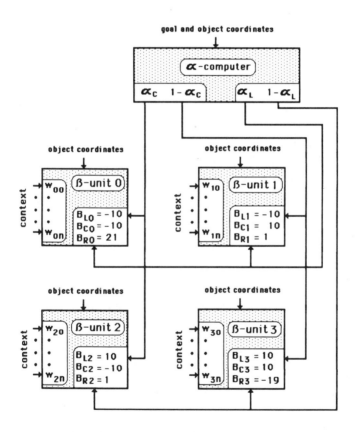

FIGURE 19. An array of four predetermined β units and the α-computer unit used to train them. Each β unit shows the actual values of the predetermined βs used in the simulation illustrated in Figure 18. The x and y components of the goal location vector output by each β unit are not shown in the figure for clarity. The calculations performed by the α-computer and β units when they are determining goal location and when they are being trained are given in the text.

of four predetermined β units to the x'_G component of the goal vector is given by

$$x'_{Gi} = w_{ij}(\beta_{Li}x'_L + \beta_{Ci}x'_C + \beta_{Ri}x'_R)$$

where $i = 0, 1, 2, 3$ are the four predetermined β units, w_{ij} is the weight for the ith β unit in the jth context, and β_{Li}, β_{Ci}, and β_{Ri} are the predetermined, constant β values for each β unit. The condition we want satisfied is $\sum_i x'_{Gi} = x'_G$. It was found that this is satisfied when[1]

$$w_{0j} = \alpha_{Lj}\alpha_{Cj}$$
$$w_{1j} = \alpha_{Lj}(1 - \alpha_{Cj})$$
$$w_{2j} = \alpha_{Cj}(1 - \alpha_{Lj})$$
$$w_{3j} = (1 - \alpha_{Lj})(1 - \alpha_{Cj})$$

where

$$\alpha_{Lj} = (\beta_{Lj} - \beta_{L,0})/(\beta_{L,0} - \beta_{L,2})$$
$$\alpha_{Cj} = (\beta_{Cj} - \beta_{C,0})/(\beta_{C,0} - \beta_{C,1})$$

and $\beta_{L,0} = \beta_{L,1}, \beta_{L,2} = \beta_{L,3}, \beta_{C,0} = \beta_{C,2}, \beta_{C,1} = \beta_{C,3}$.

---

[1] Note that the terms on the right of the equations are the outputs of the α-computer.

CHAPTER **24**

State-Dependent Factors
Influencing Neural Plasticity:
A Partial Account of the Critical Period

P. W. MUNRO

PLASTICITY AND LEARNING

In this chapter I discuss the state dependence of learning rate using the development of ocular dominance in visual cortex (Area 17) as a model system. Several qualities make this system an excellent example of neural plasticity. Foremost and most obvious of these qualities is the clear and abundant evidence of plasticity in Area 17. Secondly, the susceptibility of ocular dominance properties to a very broad range of manipulations has been extensively analyzed and the circuitry connecting the retina to visual cortex is better understood than any other neural system of comparable complexity. Furthermore, for the purposes of this analysis, the system can be characterized in terms of two-dimensional "ocularity states," which serve to illustrate many of the mathematical concepts in this book with a relatively simple representation.

An account of early learning is presented in this chapter that can be described at several different levels. While the underlying concepts are actually quite simple, they are also in their essence mathematical and are thus expressed in mathematical language. To help make the presentation more understandable to all readers, I have tried to elaborate most of the points in more intuitive terms.

The aim of this chapter is to suggest that the high plasticity frequently observed in the early stages of a learning process may reflect, at least in part, an expected reduction in the modifiability of neuronal response characteristics as their synaptic connections are strengthened as a result of experience. This contrasts with previously proposed alternative possibilities that this phenomenon is either maturational or that it is centrally gated by global modulatory signals.

The organization of the chapter is as follows: The present section introduces the reader to the subject and the approach of the chapter. The following section describes a general developmental phenomenon called the *critical period* during which neural networks are observed to be relatively plastic, then focuses on how plasticity is manifested during the development of ocular dominance in visual cortex, both during and after the critical period. In the third section, the reader is introduced to the notion of state-dependent plasticity, a concept that is then applied to give an account of the time course of ocular dominance plasticity. In the fourth section, these ideas are compared with experimental data. In the final section, the ramifications of state-dependent plasticity are discussed in relation to experimental data, existing theories, and their predictive power.

Focus on single-unit learning. Unlike most of the chapters in this book, which explore the behavior of whole networks, the present chapter is focused at the level of the single neuron and its afferent synapses. This does not reflect a difference in belief about the appropriate level of analysis, but rather it is grounded in the simple pragmatic observation that at the single-unit level, a fairly good account of the relevant phenomena can be given. Put another way, the critical period is not seen as an emergent property of an interactive network in this approach, but as a phenomenon that might be traceable to the effects of prior experience on the behavior of individual units.

Learning rate as a factor in PDP models. Notions of learning rate (in general) and the critical period (in particular) bear on models for learning by parallel distributed processing systems and have not been extensively studied by theorists apart from specific applications. A popular approach to selecting a learning rate is the pragmatic strategy in which one chooses a step size that yields a good solution in an acceptable time frame; however, rate properties of adaptive systems deserve more attention, as they may provide an important link between theoretical models and empirical data. Several existing theories of plasticity in cortex are concerned with the sequence of learning and ignore the *rate* of learning, either ignoring the critical period or making ad hoc assumptions to account for it. Conversely, in and of themselves, the scope of the ideas to be presented in this chapter is limited to rate and do not include an explicit rule for synaptic modification. Since it lacks such an explicit learning rule (that could be used to predict final states), it must be emphasized that the present *partial account* is not sufficiently complete to constitute either a theory or a model of cortical plasticity.

Two cautionary notes regarding terminology. First, it should be noted that although the terms *plasticity* and *learning* are used interchangeably in this chapter, there is a difference in nuance between them. Plasticity carries with it a sense of physical modifiability but not necessarily toward adaptation, whereas learning has a flavor of storing information from the environment with less emphasis on the neural (synaptic) substrate. Both, however, refer to *adaptation*, an environment-driven metamorphosis that enables a system to generate responses appropriate to that environment. Second, one should be particularly careful in interpreting the term *state* (especially as in "state of the system") here. In many of the network models, the state describes the instantaneous activation levels of all the units. Here, the same term refers to the connectivity values of the afferents to a unit. In the context of a network model, the *activation state* is generally changing very rapidly as the assembly of units settles in response to a particular stimulus according to information implicit in the *connectivity state*, which modifies more slowly under the influence of some self-organizing adaptive principle.

THE CRITICAL PERIOD

It is well established that plasticity varies over the course of ontogenetic development. During so-called *critical* or *sensitive* periods of plasticity, certain perceptual[1] mechanisms have been found to be more susceptible to adaptive modification by the stimulus environment, whereas the susceptibility is low or nonexistent at other times. Sensitive periods have been demonstrated in many systems, such as song learning in birds (Konishi, 1978), socialization in dogs (Scott, 1962, 1968), and phoneme discrimination in humans (Eimas, 1978; Miyuwaki et al., 1975). However, the phenomenon has been most extensively analyzed with respect to the development of binocular interaction in mammalian (particularly cat and monkey) visual cortex.

It is certainly not expected that all of these examples will necessarily have the same physiological substrate, nor would it be correct to

---

[1] Weak evidence exists for sensitive periods at "higher" (i.e., more cognitive) levels of mental function. The most compelling results are based on findings that children recover from trauma-induced aphasia more rapidly and completely than adults (who rarely, if ever, fully recover). For an overview, see Lenneberg (1967).

assume that any behavioral observation of sensitive periods in learning should be reflected by sensitive periods in neurophysiological measures of single-unit plasticity. However the converse inference, namely, that plasticity in single units is sufficient (but not necessary) for macroscopic learning phenomena, seems valid. Therefore the reader should keep in mind that the framework to be presented below for neuronal plasticity is not restricted to visual cortex, but might be applicable much more generally.

Ocular Dominance and its Susceptibility to the Environment

Interest in visual system plasticity and the generation of relevant data enjoy a symbiotic relationship that has led to the production of a great deal of both. A brief overview is given in this section of a part of this data that is based on measuring changes in the *ocular dominance* (OD) statistics of a large sample of neurons (from about 50 to over 1,000, depending on the experiment). This method, devised by Wiesel and Hubel (1963), consists of assigning each unit to one of seven ocular dominance classes based on the relative values of its optimal responses to monocular inputs from both eyes. Since there is an innate statistical bias for a neuron to respond preferentially to the eye on the opposite side of the head, the ocular dominance classes are ordered according to the influence on the neuron of the ipsilateral eye relative to the contralateral eye. A Class 1 neuron responds exclusively to input from the eye contralateral to it, a Class 7 neuron responds only to the ipsilateral eye, a Class 4 neuron gives roughly equal responses to each eye, and neurons in the other classes are responsive to both eyes with corresponding degrees of preference.

The stimulus environment can be crudely controlled by suturing shut neither, either, or both of the eyelids. The resulting paradigms are respectively called normal rearing (NR), monocular deprivation (MD), and binocular deprivation (BD). These can be combined into rearing schedules such as reverse suture (RS) in which a period of MD is followed by opening the closed eye and suturing shut the previously open eye, and alternating monocular occlusion (AM) which consists of several such reversals.

Several results can be interpreted in terms of an age-defined critical period, though the boundaries of the critical period seem to vary somewhat with the particular details of the experimental procedure. Hubel and Wiesel (1970b) found that MD drives all neurons to strongly prefer the open eye, provided the animal is still sufficiently young—neuronal sensitivity to MD seems to peak at about 4 weeks and disappears by about 3 months (see Figure 1). More recently, other researchers

FIGURE 1. The ocular dominance histogram (ODH) shift associated with monocular deprivation (MD). The classic result of Hubel and Wiesel is shown in this figure adapted from their 1970 paper. The ODH is shown for a kitten deprived of vision in one eye from Day 10 to Day 37. Units that cannot be driven through either eye are classified as visually unresponsive (VU) and are represented here by an unfilled bar. (The ODH of a normally reared [NR] kitten is nearly symmetric.)

(Cynader, Timney, & Mitchell, 1980) report an ocular dominance shift in kittens subjected to monocular deprivation for 3 months after rearing under normal conditions from birth to 5 months of age. Because it is such a striking effect, requiring only 24 hours or less during the height of the critical period (Olson & Freeman, 1975; Peck & Blakemore, 1975), the ocular dominance shift under monocular deprivation is often used as a test for assessing cortical plasticity.

With the exception of BD, other rearing schedules distort the relatively uniform *ocular dominance histogram* (ODH) of normal kittens and cats. Binocular deprivation generally leads to a histogram with all OD classes represented, as in newborn and normally reared cats. However, unlike normally reared cats, newborn and BD cats have neuronal responses described as "sluggish" and "unpredictable" with ill-defined receptive fields (Cynader, Berman, & Hein, 1976; Wiesel & Hubel, 1965). In another rearing paradigm, artificial strabismus (AS), the eyes receive noncorrelated inputs. Both surgical (severing some of the extraocular muscles) and optical (attaching eyeglasses with prisms) techniques have been used to simulate strabismus. Strabismic kittens exhibit U-shaped ocular dominance histograms, i.e., the neurons become highly monocular, but do not show an overall preference for one eye over the other. While AS yields a sensitivity time course that closely matches MD (Hubel & Wiesel, 1965), the same cannot be said of all procedures.

The results reviewed thus far appear consistent with a fairly strict maturational hypothesis for cortical sensitivity. However, evidence against this view comes from a very interesting result that has been obtained by Cynader et al. (1976) who have raised binocularly deprived kittens throughout and beyond the period associated with plasticity (up to 10 months) and then opened just one eye. They observed a strong shift in ocular dominance after a 3-month period of monocular experience. Cynader (1983) extended this result by raising the kittens in the dark for up to 2 years, followed by 9 to 18 months of monocular experience. This phenomenon, described at the end of this section, provides the most compelling evidence for a link between a neuron's connectivity state and the plasticity of the state.

Plasticity-Modulating Processes

The hypothesis that plasticity is strictly age-related falls down in the face of another kind of evidence as well; it has been found that plasticity can be modulated by a variety of chemical and/or physical processes (see Table 1). One might suppose that the critical period is explicitly controlled by one or more of these processes (i.e., it is part of the "design" of the system). However, as will be seen, this is not the only alternative hypothesis. For now, consider the following evidence concerning the roles of a variety of modulators of plasticity, focusing on norepinephrine in particular.

Norepinephrine. Catecholamines attracted attention as potential modulators of neuronal plasticity during the 1960s (see, for example, Crow, 1968). The catecholamine norepinephrine (NE) is pervasively distributed throughout the central nervous system from a localized region in the brain stem, the *locus coeruleus* (LC); thus the anatomy of the norepinephrine system makes it an attractive candidate for a global

TABLE 1

PROCESSES THAT MODULATE CORTICAL PLASTICITY

Natural Site of Modulation	Experimental Stimulus Modality	References
Locus Coeruleus	pharmacological	Kasamatsu & Pettigrew (1979)
Proprioception (Extraocular)	mechanical	Freeman & Bonds (1979) Buisseret et al. (1978)
Reticular Formation	electrical	Singer (1979)
Medial thalamic nuclei	electrical	Singer (1982) Singer & Rauschecker (1982)

modulator of plasticity. The LC/NE system has been implicated in maze-learning behavior of rats by Anlezark, Crow, and Greenway (1973), who showed that bilateral LC lesions impaired learning and resulted in decreased cortical concentrations of NE. Neurophysiological evidence that NE modulates neuronal plasticity has been found recently in rat hippocampus by Hopkins and Johnston (1984) and earlier in cat visual cortex by Kasamatsu, Pettigrew, and Ary (1979).

In a series of experiments (see Kasamatsu, 1983, for a comprehensive review), Kasamatsu and his colleagues showed that when applied to visual cortex, 6-hydroxydopamine (6-OHDA), a neurotoxin that locally destroys LC projections, prevents the ocular dominance shift in MD kittens. They went on to show that plasticity could be restored with direct injections of norepinephrine. Kasamatsu et al. (1979) also attempted to induce plasticity in normally reared adult cats via NE injections. This experiment yielded an asymmetric U-shaped histogram biased toward the open eye. That is, NE was shown to produce observable plasticity in a system that otherwise would not demonstrate environmentally driven modification.

Attempts to replicate and extend these results have had mixed outcomes. Consequently, the role played by NE in modulating plasticity is very controversial at the present time. While NE certainly has *some* effect on the susceptibility of neurons to their stimulus environments, the situation is not as simple as one might think on the basis of Kasamatsu's initial reports. A study by Bear et al. (1983) exemplifies the mixed nature of the results and provides a partial explanation. They obtained catecholamine depletion using two different techniques to administer 6-OHDA. One group of kittens received 6-OHDA using the osmotic minipump procedure of Kasamatsu et al. (1979) and the other group received injections intraperitoneally (into the abdomen) before the third postnatal day. Plasticity was disrupted only in the former case; however, subsequent biochemical analysis (HPLC) revealed that the depletion was more effective in the latter group. The results were quite surprising: While the former group replicated the earlier result, the latter group showed no observable disruption of plasticity even though pharmacological analysis indicated that the NE depletion was more severe.[2]

Other examples of induced plasticity. Stimulation of at least three different sites has been shown to enhance plasticity in visual cortex of adult cats. Mechanical stimulation of the extraocular muscles generates

---

[2] By way of explanation, Bear et al. suggest that if the LC mechanism is disrupted sufficiently early, a compensatory mechanism may be activated. Such a mechanism, they point out, may have been discovered by Harik et al. (1981) in the rat.

a proprioceptive (feedback) signal which has been shown to be sufficient for plasticity with respect to both orientation selectivity (Buisseret,
Gary-Bobo, & Imbert, 1978) and ocular dominance (Freeman & Bonds,
1979). Plasticity-gating capabilities as revealed by electrical stimulation
techniques have also been reported for the reticular formation (Singer,
1979; Singer & Rauschecker, 1982) and the medial nuclei of the
thalamus (Singer, 1982; Singer & Rauschecker, 1982).

The Effect of the Environment on the Critical Period

The conjecture that NE is a global modulator of neural plasticity does
not necessarily imply that it governs the sensitive period. In this
chapter an alternative conjecture is entertained; namely, that the potential for change possessed by a neuronal state is a function *of that state.*
(It may perhaps be influenced by global factors as well.) The following
result, mentioned earlier, of Cynader et al. (1976) provides supporting
evidence. Kittens were raised in total darkness for periods as long as a
year, after which rearing continued under the MD paradigm. Complete
shifts to the open eye were found in each case, suggesting that visual
input influences not only the response properties of the visual system
by way of some plasticity mechanism, but the functioning of the
mechanism itself. Ramachandran and Kupperman (in press) have
taken these results further to show that the ocular dominance shift
under MD is not the only plasticity phenomenon to be delayed (or prolonged) by dark rearing. They have not only obtained ocular dominance shifts in dark-reared animals, but they have observed dominance
reversal upon reverse suture in animals at ages up to 6 and 6.5 months.
The details of how much and what kind of experience are necessary
and/or sufficient to delay the critical period are far from clear at this
time. Mower, Christen, and Caplan (1983) report that 6 hours of binocular experience at 6 weeks eliminates the plasticity-preserving effect of
dark rearing; but Ramachandran and Kupperman (in press) have shown
that kittens raised normally for a full month and subsequently reared in
darkness show plasticity in Area 17 for at least 6 months.

THE OCULARITY STATE AND ITS EFFECT ON PLASTICITY

The theoretical formulation in this chapter is intended to serve primarily as a partial explanation of the sensitive period for the development of ocular dominance in visual cortex. The central concept is that

retino-cortical connectivities are weaker in the newborn and are hence more susceptible to change than the stronger synaptic connectivities that develop as a result of experience. Assuming that the ocular dominance class depends on the relative synaptic weights (the *direction* of the synapse vector), this means that the effect on the ODH of a given change in connectivity is greater when the absolute values of the weights (the *magnitude* of the synapse vector) are relatively small. This notion can be supported by a very diverse variety of learning rules (i.e., synaptic modification rules), and so the specific formulae used here are secondary to the major theme. They are included for illustrative purposes.

The Model Neuron

Let the response r of an *ideal neuron* be given by a continuous nondecreasing function σ of its net depolarization x, and let x in turn be a weighted sum of the neuron's stimulus components s_i; i.e., the units have a *semilinear* (see Chapter 2) activation function:

$$r(t) = \sigma(x(t)) = \sigma\left(\sum_{i=1}^{N} w_i(t)s_i(t)\right). \tag{1}$$

The weights w_i are collectively designated by the weight vector $\mathbf{w}$, and the stimulus elements s_i are collectively designated by the stimulus vector $\mathbf{s}$. Since the weights are associated with synaptic connectivities, $\mathbf{w}$ is called the *synaptic state*. One can clearly see that under this assumption the neuron's *response characteristics*, as measured by a set of test stimuli, are determined by $\mathbf{w}$. Adaptive learning (i.e., plasticity) is a dynamic process in which the synaptic state is driven by the stimulus environment such that the response characteristics with respect to the same environment approach some optimizing criterion. Several theoreticians (see Chapter 2) have analyzed the adaptive capabilities of various networks (idealized anatomical configurations) that self-organize according to rules (idealized physiological processes, usually at the scale of the synapse) that express the time rate of change of synapse weight w_i as a function v of local variables:

$$\frac{dw_i}{dt} = v(r(t), w_i(t), s_i(t), \cdots). \tag{2}$$

By hypothesizing an explicitly time-dependent factor into the above modification rule, the notion of a fixed critical period could be easily incorporated. Global modulators such as NE can also be factored into the rule by appropriate independent functions.

However, as will be seen, no such ad hoc description is necessary. Rather, for certain initial conditions on $\mathbf{w}$, there exist modification rules from which plasticity decreases naturally under the influence of patterned input.

The Ocularity Plane

Figure 2 is a schematic view of the retino-geniculo-cortical pathway, the most direct (just two synapses) from the ganglion cells of the retina to the visual cortex. Note that the individual components of the stimulus vectors incident to a cortical cell are actually the activities of neurons in the lateral geniculate nucleus responding to retinal signals.[3] Despite findings that some LGN cells respond binocularly (Sanderson, Bishop, & Darien-Smith, 1971),[4] the components s_i are idealized as being strictly monocular. Let $C(I)$ be the set of all afferents to a given neuron from LGN neurons responsive to the contralateral (ipsilateral) eye. Let each neuron's *ocularity state* $\mathbf{W} \equiv (W_C, W_I)$ be given by its net input connectivities from each eye, and let the *ocularity plane* be the associated state space:[5]

$$W_C \equiv \sum_{i \in C} w_i \qquad\qquad W_I \equiv \sum_{i \in I} w_i. \tag{3}$$

---

[3] The retina is not the only neural structure that projects to the LGN. Other sources of LGN input include the reticular complex and feedback from Layer 6 of visual cortex (Gilbert & Kelly, 1975).

[4] Sanderson and his colleagues found that only about 9% of geniculate neurons in the cat could be excited from the nonpreferred eye, but that only 18% are truly monocular, the remaining majority having inhibitory fields in the nondominant eye. In monkey the LGN seems to maintain a higher degree of segregation; Marrocco and McClurkin (1979) report that 87% of neurons in macaque LGN are strictly monocularly driven.

[5] This is actually a special case of a more general formulation in which the ocularity state $\mathbf{W}$ is given by the net depolarizations of the neuron by arbitrary (fixed) monocular stimuli to each eye. Namely, this is the special case in which the test stimuli are uniform activations of all afferents. The typical laboratory procedure is quite different. In a given neuron the optimal stimulus is found for each eye. Then the two corresponding responses are compared in order to determine the ocular dominance class. Hence, the test stimuli are state dependent. A more detailed theory (such as Bienenstock, Cooper, & Munro, 1982) would be required to simulate this experimental method. However, the simple approach used here is adequate for the topic at hand.

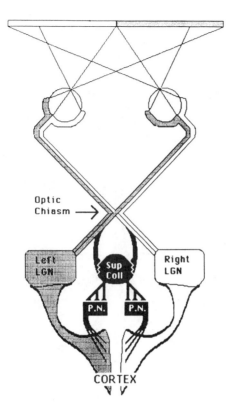

Optic Chiasm →

Shading Key

Right visual field information
Left visual field information
Tectocortical pathway

FIGURE 2. Circuitry of the visual system. Of the information pathways from retina to visual cortex, the retino-geniculo-cortical pathway, shown here in gray and white, is the most direct and the most well understood. The optic nerves approach the optic chiasm in two bundles, one associated with each eye. Here, they are separated and recombined into two groups according to whether they originate from the left or right side of the retina—i.e., they are grouped as a function of which "side" of the world they look at, such that information from the left side of the visual field is generally directed to the right side of the brain and vice versa. Other pathways from retina to cortex exist, for example, the tectocortical pathway: retina → superior colliculus → pulvinar nucleus → cortex (shown here in black).

Figure 3 depicts these assumptions with respect to a neuron in visual cortex. Macy, Ohzawa, and Freeman (1982) define a continuous measure of neuronal ocular preference, the ocular dominance index (ODI)

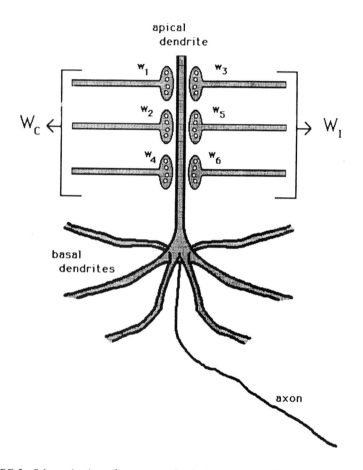

FIGURE 3. Schematic view of a neuron. Cortical neurons actually receive several types of inputs from several sources. The highlighted inputs represent monocular signals from LGN, though these do not necessarily project only to the apical dendrite as illustrated. These are segregated in the figure to illustrate the relationship between **W** and the synaptic weights w_i. In this example of just six synapses, $C=1,2,4$ and $I=3,5,6$. (In general these two sets need not have the same number of elements.) These synapses are thought to be purely excitatory. Other inputs, both excitatory and inhibitory, come from proximal cortical units, other areas of cortex, and other structures in the brain.

to be the quotient between the optimal ipsilateral response and the sums of the individual monocular optimal responses:

$$\text{ODI} \equiv \frac{r_{ipsi}}{r_{ipsi} + r_{contra}}. \tag{4}$$

Given *fixed* monocular test stimuli, the ODI can be roughly mapped onto the ipsilateral connectivity fraction z, a function of $\mathbf{W}$:

$$z(\mathbf{W}) \equiv \frac{\sigma(W_I)}{\sigma(W_I) + \sigma(W_C)}. \tag{5}$$

Using this index, the ocular dominance categories can be mapped onto regions of the ocularity plane (Figure 4). Two prominent aspects of the diagram are determined by the particular function σ. These are: (a) a threshold connectivity W_θ for both components of $\mathbf{W}$ below which no action potentials are elicited by the corresponding test stimuli and (b) the curvature of the iso-ocularity contours. (These are straight radial lines if σ is a linear function.)

The power of the ocularity plane as a tool in the theoretical analysis of ocular dominance plasticity will become clear presently. Scatterplots of experimental data on the ocularity plane may prove more valuable for presenting results than the ocular dominance histogram if accurate techniques for measuring $\mathbf{W}$ become available.

A hypothesis for the critical period. The introduction of the ocularity plane makes the task of describing the relationship between a neuron's state and its plasticity much easier than it would be otherwise.

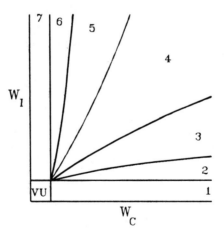

FIGURE 4. The ocularity plane. The plane is defined by axes corresponding to the net connectivities from both eyes to a cortical neuron. Loci of constant ODI are mapped onto contours that radiate from the point $\mathbf{W}_\theta \equiv (W_\theta, W_\theta)$, where the value W_θ corresponds to the firing threshold for some test stimulus. These have been chosen here to mark arbitrary boundaries that might be used to delimit the seven ocular dominance classes defined by Wiesel and Hubel (1963). Running orthogonal to these contours is, by definition, the gradient field for ocular dominance.

This description is repeated in the next section using more formal mathematical language for those who are interested, but the gist can be understood quite well without wading through the mathematics. The arguments in this chapter are founded on the following assumptions:

A1. Initially the synaptic weights are small.

A2. As it is molded under the influence of a structured stimulus environment, the pattern of synaptic weights is refined and reinforced with experience, and thus tends to grow.

A3. The rate of change of synaptic weights may decrease or remain invariant, with changes in the magnitude of the weight vector; in fact, the rate of change of synaptic weight may even *grow* as the magnitude of the weight vector grows, as long as it grows less rapidly than the weight vector itself.

These assumptions are readily understood in the context of the ocularity plane. Since the initial weights are small (A1), the ocularity state is initially near the point where the ocular dominance classes converge $(\mathbf{W}_\theta)$. In this region, plasticity is high because it requires only a small change in the synaptic weights to move the state from one OD class to another. As its weights increase (A2), the neuron's ocularity state is driven away from the origin where increasingly larger weight changes are required to produce shifts in ocular dominance class. Assuming the ability to change weights does not also grow in proportion to the growth of the weight vector (A3), this results in an observed loss of plasticity.

State Dependence of Plasticity

Given z, a scalar measure of the neuronal state, the plasticity p associated with a particular state W can be defined for any stimulus environment E. The plasticity is the time rate of change in z under the influence of E:

$$p \equiv \left\langle \frac{dz}{dt} \right\rangle_E \equiv \frac{\partial z}{\partial W_C} \overline{V}_C + \frac{\partial z}{\partial W_I} \overline{V}_I \equiv \nabla z \cdot \overline{\mathbf{V}} \qquad (6)$$

where $\overline{\mathbf{V}} \equiv \langle d\mathbf{W}/dt \rangle_E$ is the average "velocity" of the state in the W-plane and ∇z is a vector pointing in the direction of steepest change in z with magnitude equal to that change (the components of ∇z are simply the partial derivatives of z with respect to W_C and W_I). The plasticity can thus be represented as the inner (dot) product between two vectors: the velocity of the state and the gradient of the ODI. The inner product of two vectors can be expressed as the product of their magnitudes and the angle between them (see Chapter 9). These are considered separately below.

Dependence on the magnitudes of ∇z and $\overline{V}$. Whether the loss of plasticity associated with the critical period is a function of neuronal state ultimately depends on the behavior of the product between the magnitudes of these two vectors for large magnitudes of $\mathbf{W}$. Since $|\nabla z|$ decreases with $|\mathbf{W}|$ ($|\nabla z|$ goes as $|\mathbf{W}|^{-1}$ on average), the issue reduces to the relationship between $|\overline{V}|$ and $|\mathbf{W}|$. This is exactly the issue addressed in Assumption A3 above, since $|\mathbf{W}|$ is the magnitude of the weight vector and $|\overline{V}|$ is its speed in the ocularity plane. In keeping with the conventional Hebb-like approach (i.e., the change in each synaptic weight depends only on the activities of the corresponding presynaptic and postsynaptic units), consider the indirect effect of $\mathbf{W}$ on V mediated by the neuronal activity rather than any explicit dependence of V on the weights. It now becomes very important to consider the form of the function σ relating the response r to the quantity x (see Equation 1). This function is sometimes called a *squashing* function since it tends to saturate for high values of x (as shown, for example, by Chapman, 1966) because of the upper limit on the neuronal firing rate (i.e., the absolute refractory period). Hence, if $\overline{V}$ depends on r (not x), it will (like r) level off as x increases.

It would be sensible for $\overline{V}$ to *decrease* with increasing x. The argument supporting this possibility is based *teleologically* on the assumption that *each unit optimizes some function of its stimulus environment according to a gradient descent process.* In Chapter 8, it is shown that optimization of least square error in a multilayered system leads to an explicit rule for learning in which each synapse modifies according to a function separable into three factors: the presynaptic (afferent) activity, an error signal propagated from higher levels, and *the derivative of the squashing function of the postsynaptic unit.* More generally, one can show that the derivative $\sigma'(x)$ is a factor regardless of the function being optimized, whether it be a least square error (as in Chapter 8); selectivity for orientation, direction, or spatial frequency (as in many cells in visual cortex); or some other quantity. This is demonstrated quite easily. Let Q be the optimized function, a quantity measured over the set of response values r^i evoked from a given unit by a set of test stimuli $\mathbf{s}^i$. If $r^i \equiv \sigma(x^i)$ (where $x^i \equiv \mathbf{w} \cdot \mathbf{s}^i$), then the gradient ∇Q breaks down as follows:

$$\frac{\partial Q}{\partial w_j} = \sum_i \frac{\partial Q}{\partial r^i} \frac{dr^i}{dx^i} \frac{\partial x^i}{\partial w_j} = \sum_i \frac{\partial Q}{\partial r^i} \sigma'(x^i) s_j^i. \tag{7}$$

With increasing exposure to the world, units may become highly selective so that they come to respond at intermediate levels where $\sigma'(x)$ is high (see Figure 5) less frequently, and may hence become less plastic with respect to that environment.

The first derivatives of certain squashing functions can be expressed as functions of the squashing functions themselves—i.e., they satisfy $\sigma'(x) = F(\sigma(x))$. A simple (i.e., easily implemented) function F gives a technical advantage that may be exploited at two levels: (a) by the biochemical mechanisms that underlie plasticity and (b) by the

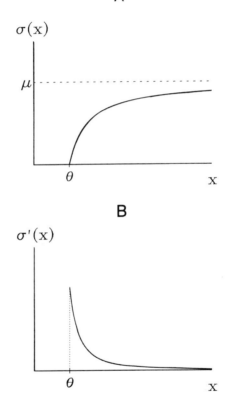

FIGURE 5. A squashing function and its first derivative. A: An easily computed function that incorporates the salient features (namely, a threshold, θ, and monotonic approach to an asymptote, μ) of neuronal response characteristics is given by the function that is defined by $\sigma(x) \equiv \mu(x-\theta)/(x_{1/2}+x-2\theta)$ for $x > \theta$ and vanishes for $x < \theta$, where $x_{1/2}$ is the value of x at which the function attains half its maximum value (i.e., $\sigma(x_{1/2}) = \mu/2$). B: The derivative $\sigma'(x)$ of this particular function is shown for $x > \theta$. The decay in the derivative as x increases can be implicated in a decay in learning rate for systems that follow gradient descent. This effect is potentially quite substantial as indicated by the very small values of the derivative.

computer algorithms used to simulate them. A few such pairs of functions and derivatives are given in Table 2 along with the function F that expresses the function relating each pair.

Dependence on the angle between ∇z *and* $\overline{V}$. Analysis of the time course of the OD shift under MD (Blasdel & Pettigrew, 1978) has shown that the binocular OD classes are depleted before those associated with the closed eye, resulting in a U-shaped OD histogram. Since the inner product between the vector fields (and hence the observed plasticity p) tends to increase as the fields are more *parallel*, the elevated plasticity in the binocular regions suggests that the fields may be more parallel in these regions than in the monocular regions of the ocularity plane. Since the shape of the ∇z curves is known to be somewhat circular about the origin, or *concave inward*,[6] the approximate shape of the $\overline{V}$-field can be deduced. It follows the $\overline{V}$-field trajectories are directed radially near the axes and oriented tangentially in the binocular regions; i.e., they are *concave outward* (Figure 6A). The resulting elevated plasticity level for the binocular states can be seen in Figure 6B which illustrates the plasticity function in the context of a stimulus environment restricted by monocular deprivation (i.e., $E = MD$) using a simplified version of the Bienenstock et al. (1982) formula for $\overline{V}$. Note that the states furthest from equilibrium (those at the "beginnings" of the trajectories) are initially driven *toward* the origin, but are "deflected." A useful analogy (suggested by a remark of Paul Smolensky's) is that the origin "repels" states under the influence of a patterned environment. It is thus seen that while the states *tend* to grow, sudden changes in the statistics of the visual environment (e.g.,

TABLE 2

SAMPLE SQUASHING FUNCTIONS AND FIRST DERIVATIVES

$\sigma(x)$	$\sigma'(x)$	σ' in terms of σ
$\dfrac{x}{1+x}$	$\dfrac{1}{(1+x)^2}$	$\sigma'(x) = (1-\sigma)^2$
$\dfrac{1}{1+e^{-x}}$	$\dfrac{-e^{-x}}{(1+e^{-x})^2}$	$\sigma'(x) = \sigma(1-\sigma)$
$1-e^{-x}$	e^{-x}	$\sigma'(x) = 1-\sigma$

---

[6] Concaveness can be assigned a direction corresponding to the second derivative (acceleration) of a trajectory.

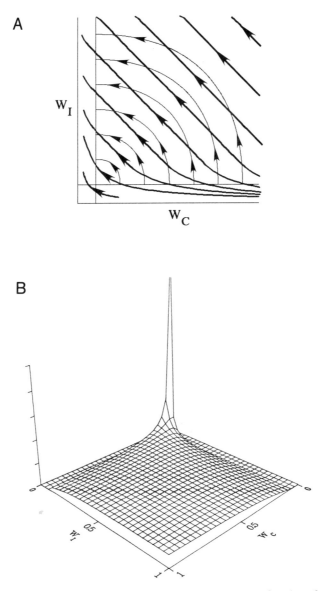

FIGURE 6. Plasticity in a monocularly deprived environment as a function of the ocular-ity state. *A*: The velocity field **V** (bold) is shown together with the gradient field ∇z. *B*: Plasticity is plotted as a function of the state **W**. Note that the origin is located in the *far* corner of the base plane so that the region obstructed by the large peak is minimal.

RS) may deliver an "impulse" in the direction of the origin which is eventually overcome by the system dynamics to drive the state out-ward, most likely in a direction different from its approach.

These last two paragraphs have formally expressed a notion that is actually simpler than the mathematical nature of the presentation might indicate. The fundamental idea is that the *observed* plasticity depends on the objective change in the neuronal state (i.e., $\mathbf{V}$) *relative to the plasticity measure* (in this case, the ocular dominance index z). Aspects of this relative change that have been considered are the absolute magnitude $|\overline{\mathbf{V}}|$ of the change in neuronal state, the sensitivity $|\nabla z|$ of the plasticity measure, and the angle between the change in the state and the contours of constant z.

COMPARISON WITH EXPERIMENTAL DATA

Developmental models of visual cortex have for the most part focused on the final (equilibrium) states inherent in the dynamics of hypothetical systems (e.g., Amari & Takeuchi, 1978; Bienenstock et al., 1982; Cooper, Lieberman, & Oja, 1979; von der Malsburg, 1973), ignoring the variability in the convergence rates of these systems that results from abrupt changes in the visual environment such as those manipulations used to test plasticity. Ocularity plane analysis is an attractive technique in the design of a detailed rate-descriptive model for learning.

The time course of the reverse suture paradigm. Qualitative analysis of the *shape* of the trajectories (as in Figure 6A), however, leads to a prediction for the RS paradigm. It appears that early reversal of monocular deprivation should bring the states closer to the origin as they migrate from the neighborhood of one axis toward the other axis (see Figure 7). Hence they might easily go from one monocular class to the other via the VU region—i.e., pass between the origin and the plasticity singularity $\mathbf{W}_\theta$. Consider states which begin further from the origin on the axis associated with the newly closed eye. After reversal, their paths are more likely to traverse the binocular regions of the ocularity plane. Therefore the intermediate (U-shape) stage of RS is expected to be less sharp, i.e., there should be more binocular units observed if the reversal is performed relatively late (but before plasticity vanishes).

The time course of the ocular dominance shift under RS has been analyzed by Blakemore and Van Sluyters (1974). As in the early stages of MD rearing, the OD histogram initially becomes somewhat U-shaped. Casual consideration of this in terms of the $\mathbf{W}$-plane leads to the conclusion that the trajectories must be strongly concave outward, matching the result derived in the previous section for simple MD. One explanation for these results rests on the assumption that many

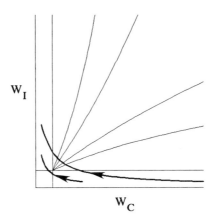

FIGURE 7. Trajectories corresponding to reverse suture. Two examples are shown. One path (that closer to the origin) starts from a weaker, and hence presumably less mature state. This state travels from Class 1 to 7 without entering the intermediate classes.

units maintain subthreshold synaptic input from the closed eye during monocular deprivation for some period of time. [7] The possibility that these inputs eventually disappear presents another candidate for the loss of plasticity by monocular deprivation. Computer simulation results of the Blakemore and Van Sluyters (1974) experiment are shown together with the experimental data in Figure 8.

Alternating monocular occlusion. Consider a stimulus distribution *E* corresponding to the experimental paradigm for alternating monocular occlusion. If the alternation is sufficiently frequent, states in the neighborhood of the diagonal line $W_C = W_I$ should display low plasticity since they are driven along that line, if at all; hence they move in a direction orthogonal to the ODI gradient (i.e., along a contour of constant ODI). This is illustrated by the plasticity function in Figure 9 and accounts for the W-shaped OD histograms (i.e., Columns 1, 4, and 7) obtained by Blasdel and Pettigrew (1979) in some of their alternating deprivation studies.

How is connectivity related to responsivity? A scatter-plot of a sample of cortical neurons in the ocularity plane adds a dimension to the

---

[7] Evidence supporting the existence of such input has been found in cat visual cortex by sudden removal of the open eye after a period of monocular deprivation (Blakemore, Hawken, & Mark, 1982; Kratz, Spear, & Smith, 1976), but Dräger (1978) reports the opposite finding for mice.

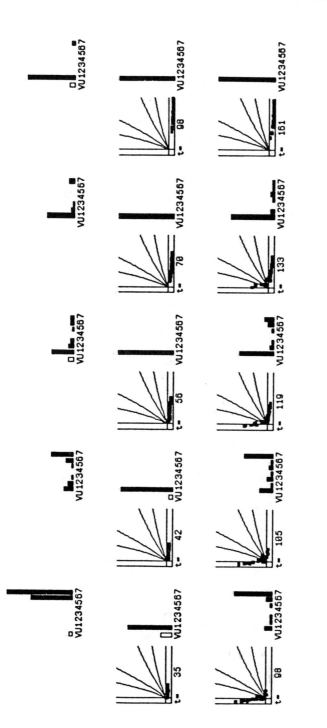

FIGURE 8. Simulation of reverse suture experiments run by Blakemore and Van Sluyters (1974). In each case, animals were deprived in one eye for some variable period (35 days, 42 days, 56 days, 70 days, 98 days) and then subjected to a suture reversal which was left in place for 63 days before recording. *Top row:* Experimental results. *Middle row:* Simulation state at time of suture reversal. Scatter plots in the ocularity plane are shown with the corresponding ocular dominance histograms. *Bottom row:* Simulation state at time of recording (compare with top row).

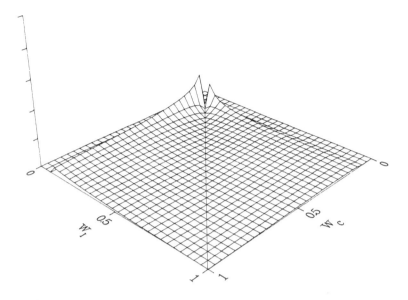

FIGURE 9. Plasticity under alternating monocular occlusion. Note the lack of plasticity along the diagonal. In an AM environment, highly binocular units are not as plastic as units with a bias for one eye. They are effectively "torn" between committing themselves to one eye over the other.

information in the ocular dominance histogram. In their study, Freeman, Sclar, and Ohzawa (1982) produce a similar representation which maps units onto a ODI vs. responsivity plot. These can be thought of as the basis of another (polar-like) coordinate system in the ocularity plane (see Figure 10). The ocularity plane would be much more viable as an analytic tool if a precise relationship could be established between synaptic connectivity and responsivity. A finer measure of response strength might be accomplished using intracellular electrodes to count individual EPSPs (and perhaps IPSPs as well). Lacking this, an objective and more sensitive measure of the degree of response should be adequate to the extent that it ought to correlate well with plasticity. Not only should the *degree* of response be included in such a measure, but such an index might incorporate several aspects of the response including not only the firing frequency, but also (for example) the latency of firing.

The role of norepinephrine: A conjecture. The prolongation of the critical period induced by BD observed by Cynader and the norepinephrine effect reported by Kasamatsu and Pettigrew (1979) extend

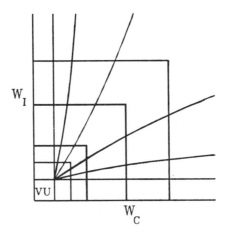

FIGURE 10. Ocular dominance and responsivity as polar-like coordinates. Accurate measures of **W** may be difficult to come by. One can only calculate rough approximations on the basis of the corresponding responses. Some estimates might be made on the basis of existing data by using ocular dominance as an indication of the "angle" and responsivity as a rough measure of the "magnitude" of **W**.

the concept of critical period far beyond the original work by Hubel and Wiesel. Of the many differences between these two approaches, there is one which is fundamental to the nature of learning by biological systems—namely, that while the rate of learning in biological information-processing networks can be made subject to *extrinsic* factors, it may well be influenced by *intrinsic* properties of the learning mechanism, the information patterns as they are learned, and interactions between them.

It has been established in this chapter that a theory of plasticity need not presuppose the existence of a global modulator to account for the critical period. What role, then, does the norepinephrine system play? After all, it has been well demonstrated, both experimentally and on teleological grounds, that it can modulate plasticity.

The conjecture that the LC releases NE as part of an attentional mechanism[8] seems to resolve the issue. Rather than acting as a clock controlling the developmental sequence of plasticity, the activity of NE probably fluctuates on a finer time scale. Under this assumption, non-recognition of a stimulus might trigger a "novelty alarm" in the form of an LC/NE plasticity signal. Thus, another factor in the critical period

---

[8] The notion that NE is a medium for externally gated plasticity originated in the early 1960s based on studies of the reward system (Poschel & Ninteman, 1963). Whether and how catecholamines are implicated in reward and reinforcement remains controversial. (For a review, see Wise, 1978.)

may be the loss of novelty brought about by experience; this suggests an explicit experience-dependent mechanism for plasticity—one that would *amplify* the effect described in this chapter and would therefore be difficult to analyze independently.

Grossberg (1980) offers a comprehensive description of the critical period that incorporates all of the mechanisms in Table 1. This is quite different from the approach taken in this chapter, in which I have attempted to separate the critical period from these other factors. Given the broad disparity between these two points of view, an experiment to discriminate between them should not be difficult to design. For example, what are the states of these mechanisms in cats reared in darkness for several months from birth?

DISCUSSION

State Dependence and Existing Modification Rules

Assumptions A1, A2, and A3 can be incorporated into many dynamical models of plasticity in cortex, as can hypotheses based on global modulation (see next subsection). Consider three varieties of mathematical formulations, characterized by the factors determining the magnitudes of the vectors representing the final equilibrium states (see Figure 11):

- *Case 1: Throughout modification, the magnitudes of the neural states are explicitly constrained; for example, the sum of the components is held constant* (e.g., Perez, Glass, & Shlaer, 1975; von der Malsburg, 1973; Chapter 5). Unfortunately, explicit constraints on magnitude are not consistent with the assumption that the state vectors grow during learning. Certain models of this kind employ such constraints because of a tendency for the weights to grow without limit; for these models, only an *upper bound* is really needed and the inherent tendency for growth may provide the necessary reinforcement. However, some models do require that strict magnitude constraints be imposed (e.g., those with even less stable dynamical equations). While it may not be impossible to adapt these models to Assumptions A1, A2, and A3, the solution is not likely an elegant one; for example, a requirement that the magnitude be conserved at a fixed value M might be altered such that M increases according to some complex ad hoc rule up to some maximum value M_{max}.

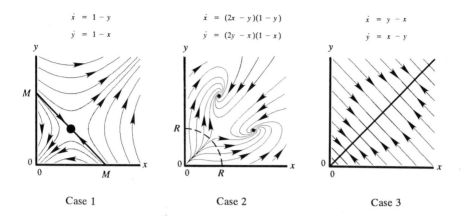

FIGURE 11. Three kinds of dynamical systems. These three examples (instances of Cases 1, 2, 3 in the text) illustrate certain important characteristics of models relevant to the assumptions stated earlier for state-dependent plasticity. The dynamical systems used are given above each graph in terms of two generic variables, x and y. *Case 1*: In this case, the trajectories are driven to either the origin or to infinity, unless the constraint $x + y = M$ is imposed. *Case 2*: Here, there are two global equilibrium points. Thus, under the assumption that the initial states are close to the origin (e.g., within the dotted circle $x^2 + y^2 < R^2$), the states will tend to grow. *Case 3*: All states are driven to the nearest point satisfying $x = y$ and hence all trajectories tend to preserve the sum $x + y$ of the initial state. Thus, there is no tendency for the states to grow.

- *Case 2: While no explicit constraints are placed on the weights, the system dynamics drive all states outside some nonzero radius R about the origin* (e.g., Amari & Takeuchi, 1978; Bienenstock et al., 1982). Models of this class are most readily adapted to Assumptions A1–A3 since they have a "natural" tendency to drive states with sufficiently small magnitudes away from the origin. The value R (like M in Case 1) gives a scale for selection of the initial weights—i.e., "small" in A1 is interpreted here as small relative to R. Three kinds of behavior are possible in the limit $t \rightarrow \infty$; these are (a) convergence to an equilibrium state with magnitude greater than R, (b) oscillations (e.g., limit cycles) that come no closer to the origin than R, or (c) divergence. It must be emphasized that such models must be carefully tuned if they are to give an accurate account of learning rate. This tuning is not necessarily limited to selecting parameters for the learning rules; good matching to the data may require alterations of the form of the rules themselves such as the derivative factor in Equation 7.

- *Case 3:* There is no minimum radius such as that in the previous case; equilibrium states exist arbitrarily close to the origin (e.g., Munro, 1984). A dynamical system of this variety presents a particularly hostile context for Assumptions A1–A3. There is no reason to expect the weights to grow no matter how small the initial weights are. This highlights the fundamental dependence of the assumptions on *scale.*

Global Modulators as Dynamical Variables

Global modulation of cortical plasticity, whether it is mediated by the locus coeruleus, proprioception, the brain stem, or the thalamus, can be integrated into a synaptic modification formula in any of several ways. Let us assume a general modification rule that is separable into a presynaptically dependent function ψ and a postsynaptically dependent function ϕ:

$$\frac{dw_i}{dt} = \psi(s_i)\,\phi(r). \tag{8}$$

Some simple conjectures as to the role of a modulatory signal (α) are that it may influence plasticity in a direct multiplicative fashion or by exerting an effect on the presynaptic or postsynaptic activities:

$$\frac{dw_i}{dt} = \begin{cases} \alpha\,\psi(s_i)\,\phi(r) \\ \psi(s_i(\alpha))\,\phi(r). \\ \psi(s_i)\,\phi(r(\alpha)) \end{cases} \tag{9}$$

The first of these possibilities is consistent with the data from both Cynader's and Kasamatsu's labs. That is, a global modulatory signal may control the speed (magnitude of the velocity vector) of the neuronal state as it is driven by the environment, but the distance between ocular dominance classes depends on the state. Thus the observed plasticity is a function of *both* the synaptic state and global factors—e.g., norepinephrine. Destruction of the catecholamine terminals in some region by 6-OHDA might then bring synaptic modification to a complete halt.

Secondary global modulatory action via influence of the presynaptic activity is not a very promising approach since it would distort the pattern of afferent activity. This effect would be disastrous except in the trivial case, namely, if it were equivalent to a uniform scale factor. Of course, the action would then be formally identical to the third conjecture for the role of α, in which it influences the postsynaptic activity (assuming that r depends monotonically on a variable that is linear in s_i, like x in Equation 1).

Norepinephrine is known to influence responsivity in neurons of the visual cortex. Kasamatsu and Heggelund (1982) found that this influence was not easily characterized; for example, overall responsiveness would increase, decrease, or not change with roughly equal probability. However, they did notice certain correlations with respect to cell type. Videen, Daw, and Rader (1984) also found that the relationship between NE and neuronal response is quite complicated. Thus the third of the above functions may have some support, though the exact form of the dependence of α on NE concentration has not been specified.

Predictions

Theoretical neuroscientists are often criticized (often with good reason) for neglecting their duty to suggest experiments that test their theories. They counter with accusations, which also are valid, to the effect that the bulk of theoretical work passes through the literature unnoticed by the experimental community and that many, if not most, predictions remain untested. Arguments assigning blame aside, the interaction of theory and experiment in neuroscience is not in a healthy state. In an effort to improve the situation, this section outlines some crude designs that may lead to laboratory verification of the ideas covered in this chapter.

A very informative experiment would be to analyze the time course of the loss of plasticity along the path of information flow (Layer 4c vs. other layers in the monkey) after a period of dark rearing. By comparing this result with the observation that Layer 4c loses plasticity for ocular dominance earlier than other layers in monkey Area 17 (Blakemore, Garey, & Vital-Durand, 1978; LeVay, Wiesel, & Hubel, 1980; Vital-Durand, Garey, & Blakemore, 1978), it should be possible to separate the maturational from the environmental influences on the critical period. A strict interpretation of the analysis in this chapter predicts that the entire process should be postponed and hence the results after dark rearing should not substantially differ from the established results. The outcome of the experiment could conceivably be mixed indicating that both environmental/informational and maturational factors exert influence. It seems reasonable, indeed quite likely, that at lower levels (such as the thalamus), maturational effects on the critical period outweigh the effect described in this chapter.

Reinstating plasticity after patterned rearing. Cynader's experiments have demonstrated that the visual system is capable of "waiting" in a plastic state for patterned input. This result may be indicative of a more general relationship between responsivity and plasticity.[9] One idea that grows out of the present approach is that plasticity may in fact be reduced in part because cells in the mature animal rarely respond in the intermediate range, where the derivative of the cell's response with respect to its net input is relatively large (see Figure 5B). If, as suggested earlier, synaptic modification depends on this derivative, then cells that are responding vigorously will not modify their connections. Thus these ideas predict that if the cells could be made to respond less vigorously, they could thereby become more plastic, provided the new response level were in the appropriate range. Of course, this is not the only factor that determines plasticity; the size of the weight vector plays a very important role in the extent to which a change of a fixed size will alter the ODI. That is, of the two quantities ∇z and $\sigma'(x)$ that are candidates for the loss of plasticity, only the latter can be manipulated in the adult. Hence, the observed plasticity resulting from a reduction of response magnitude in adult animals should be weaker than the natural plasticity of younger animals.

The following outline sketches an experimental paradigm for testing the relationship between responsivity and plasticity: One method to reduce responsivity without introducing the potential pitfalls of pharmacological techniques would be to apply diffusive contact lenses. Degrading the contrast of the stimuli has the further property of reducing orientation selectivity and hence simulates this aspect of kittenhood as well. This has the advantage of further simulating properties of the young animal and the disadvantage of confounding orientation selectivity and neuronal responsivity as factors in neural plasticity. Once the lenses have been calibrated with respect to neuronal responsivity, many different rearing paradigms should be tried since various factors may induce differential effects across different paradigms. Differences in morphological development (at both geniculate and cortical levels) are, for example, typically observed for different kinds of rearing; it may therefore prove advantageous to study the effects of manipulating the environments of normally reared animals since they have not suffered

---

[9] Cynader plays down the generality of this relationship, citing his observation that neuronal responsivity reaches adult levels before the peak of the critical period for animals raised in patterned environments. However, it should be noted that synaptic weights can continue to grow even after adult levels of responsivity are observed. This will naturally be the case in units whose activation saturates according to the general types of activation functions considered here and throughout this book.

the potentially irreversible and possibly plasticity-limiting damage that is associated with deprived rearing.

Regardless of the experimental technique, a clear demonstration that links weak neuronal responsivity to increased plasticity promises to both provide evidence that neuronal response characteristics evolve such that they follow a gradient descent with respect to some environment-related variable and lend strong support to the notion that plasticity is state dependent.

BEYOND VISUAL CORTEX

The ideas which have been described with respect to ocular-dominance plasticity in visual cortex are abstracted in this section and considered as they might apply to higher (more cognitive) levels of cortex. In general, it seems plausible to assume that the important informational content of a stimulus pattern in cortex depends upon the *relative* values of its components, which define radially oriented regions of the state space, since nearly every level of neural processing shows evidence of lateral inhibition, a known mechanism for pattern normalization. The absolute magnitude of the stimulus loses its relevance in this case and its informational content is reduced by one degree of freedom to the *direction* alone. Gati and Tversky (1982) refer to the magnitude and direction of a vector stimulus as its *qualitative* and *quantitative* aspects, respectively. One can clearly see that in this case the assumptions A1, A2, and A3 may apply under these circumstances. According to this distinction, the plasticity of a self-organizing process is expected to exhibit state dependence as a function of the *quantitative* aspect of the state across a *qualitative* measure on the stimulus environment. Small assemblies or modules of such units can be made to evolve to arbitrary regions of the input space (as in Reilly, Cooper, & Elbaum, 1982), which can be thought of as categories.

Stages of Processing

Environmentally driven sequential development. The state dependence of plasticity has interesting implications for a hierarchically organized system. Consider an organizational scheme in which information passes through a sequence of several neural assemblies, each responsible for some stage of processing.[10] Assume that synaptic connectivities are initially weak throughout the entire system. Each stage lies in a state of weak connectivity and high plasticity until it receives sufficient exposure to some pattern set. At the onset of exposure to patterned input (e.g., eye-opening in kittens), only the input assemblies receive strong signals. The earliest stages in the system are therefore the first to lose plasticity, and so increasingly longer sensitive periods may be expected for the development of higher cognitive functions.

This consequence of the ocularity plane framework is consistent with data from visual cortex of both cat and monkey. Daw and Wyatt (1976) found evidence that the sensitive period for direction-selective cells is earlier, both in its onset and offset, than the sensitive period for ocular dominance. While these may not represent different stages of processing, the experiment clearly supports the notion of a link between plasticity and the nature of the stimulus features.

In monkey (and in cat as well), the LGN projection to Area 17 terminates most predominantly in Layer 4c (Hubel & Wiesel, 1972); this layer therefore receives the earliest, most "raw" input to cortex. Considerable evidence, both physiological (Blakemore et al., 1978) and morphological (Vital-Durand et al., 1978) indicates that in monkey visual cortex, plasticity is diminished in this layer earlier than in the other layers. These findings have been replicated by LeVay et al. (1980), who attribute the effect to maturational factors rather than implicit aspects of plasticity.

Based on developmental data from the rat somatosensory system, Durham and T. A. Woolsey (1984) have formulated a "sequential instruction hypothesis for sensory pathway pattern development" (p. 445). In spite of their conjecture that "a given station might 'instruct' the next more central one" (p. 425), they stick to a more traditional interpretation of their data: ". . . critical periods are linked to developmental periods since they coincide with times during which axons are normally innervating their targets and are segregating" (p. 446). It is not clear therefore whether the instruction to which they refer is exogenous (i.e., sensory information) or endogenous (e.g., genetically precoded chemical transmission).

Another factor in considering the developmental aspects of different levels of the system is the structure of the pattern environments presented to each stage. As successive levels self-organize, the representation of the world becomes more symbolic. During its ascent

---

[10] The visual system seems to be organized in such a fashion. However, the information flow is not strictly serial, rather it seems to be directed along "functional streams" (Van Essen & Maunsell, 1983), which have mixed hierarchical/parallel attributes.

through the system, information is continually recoded in a manner such that the stimulus sets become more discrete. Shepard and Podgorny (1975) make this point with the following example:

> Whereas we can continuously shift a color (for example, blue) in brightness, hue, and saturation until it becomes as similar as we wish to any other color (for example, green), we cannot continuously deform a word 'blue' to another word 'green' without passing through intermediate configurations that are not words at all. (pp. 189-190)

This observation pertains to plasticity in that as stimulus sets become more discrete, and perhaps more orthogonal, their constituents are better separated; hence learning by feature-detecting units is much more rapid. So it is conceivable that for such systems, the equilibrium points may not be so far from the origin; thus, these neurons may enjoy extended or indefinite sensitive periods! This, of course is consistent with our experience; people learn to recognize patterns quite rapidly (relative to the hours or days required for neurons to become orientation selective, for example) and do not generally lose that ability.

SUMMARY

In this chapter I have proposed a factor that may account for the high rate of learning observed in the early stages of many learning processes. This factor of state dependence may in fact be the principal component responsible for the critical period; it emerges from three basic assumptions regarding the mathematical relationships between synaptic connectivity values and their time derivatives. The notion of state-dependent learning rate has been discussed here in the context of the observed susceptibility of ocular dominance characteristics to environmental manipulations. This approach suggests an account for several specific observations in visual cortex and also suggests further experimental tests of the relationship between ocularity state and plasticity.

On a broader level, it is hoped that this chapter will contribute to greater interaction between theory and experiment in this field. Toward this end, analytical tools such as the ocularity plane may prove useful both for making experimental data more accessible to theoretical analysis and for examining and describing aspects of the mathematical models.

ACKNOWLEDGMENTS

Over the past three years, several individuals were very generous with their time and energy to apply their knowledge and insight to the issues raised in this chapter. These include Leon Cooper, Michael Paradiso, and Alan Saul of Brown University; Paul Heggelund of the University of Oslo; V. S. Ramachandran of UCSD; and several members of the PDP research group at the UCSD Institute for Cognitive Science, in particular, David Rumelhart, Paul Smolensky, and especially Jay McClelland. This work has been supported by United States Office of Naval Research Contract N00014-81-K-0136, the Norwegian Council for Science and the Humanities (NAVF), the System Development Foundation, and grant PHS MH-14268-11 from the National Institute of Mental Health to the Center for Human Information Processing at the University of California, San Diego.

CHAPTER 25

Amnesia and
Distributed Memory

J. L. McCLELLAND and D. E. RUMELHART

In several chapters in this book, we have argued for distributed models of learning and memory. In most of these cases, we have considered primarily *psychological* and *computational* reasons to prefer distributed models. In this chapter, we ask, can distributed models shed any light on the *biological* basis of memory? One possible answer would be "no"—we could fall back on the claim that distributed models are abstract descriptions, not concrete descriptions of the physiology of memory. Indeed, many of the specific distributed models we have considered in this book are somewhat "unphysiological" in several of their details. But the general idea of distributed memory (at least, within localized regions of the brain, as discussed in Chapter 3) does seem sufficiently consistent with what we know about the brain that the hypothesis that memory is physiologically distributed seems worth considering.

In this chapter, we consider this hypothesis in light of the phenomenon of *bitemporal amnesia*—the deficit in memory that is produced by a bilateral insult to the medial temporal lobes of the brain. Bitemporal amnesia is interesting from the point of view of distributed models because two distinct aspects of the phenomenon seem to suggest very different things about the biological plausibility of distributed models.

One prominent aspect of bitemporal amnesia is that it produces a *retrograde amnesia* that is *temporally graded.* After the precipitating insult, the individual is unable to remember recent events, but memory for remote information appears to be intact. If there is recovery, as there is in many cases of bitemporal amnesia, much of the recent information that had been lost will return.

These aspects of amnesia seem to contradict the most basic assumptions of a distributed, superpositional model of memory. These models hold that all memories, old and new, are stored in the same set of connections. If this is so, why is it that an amnesic insult selectively disturbs the newer memories? And why is it that the memories that at first seemed to be lost can later be retrieved? The phenomenon seems to beg for an interpretation in which what is lost is access to that part of the memory store in which recent memories are held, rather than one in which all memories are superimposed in the same set of connections.

On the other hand, another prominent aspect of bitemporal amnesia seems to be highly consistent with a distributed model. Bitemporal amnesia produces a profound *anterograde* amnesia, as well as a *retrograde* deficit. That is, after the amnesic insult there may be a profound deficit in the ability to acquire new information. This is particularly true when amnesics are tested for their ability to recall or recognize specific individual events to which they have been exposed since onset of the amnesia. However, the amnesic deficit is not so profound, even in the severest cases, that the patient is unable to learn from repeated experience. For example, H. M., an extremely profound amnesic, is quite aware of his deficit, presumably as a result of repeatedly having been confronted with it. Milner (1966) reports that he often greets people by apologizing for not recognizing them, giving his memory deficit as his excuse. He remembers that he cannot remember, even though he cannot remember any particular occasion when he failed to remember.

This aspect of amnesia is quite naturally and directly accounted for by distributed models. We need only assume that the amnesic insult has resulted in a reduction in the size of the changes that can be made to connection strengths in response to any given event. Smaller changes will result in very weak traces of each individual episode or event, but, over repeated trials, what is common to a number of experiences will be gradually learned.

In summary, we appear to be faced by a paradoxical situation. One prominent aspect of bitemporal amnesia appears to argue against distributed models, while another appears to argue in favor of them.

In this chapter, we confront this paradox. First, we consider in more detail many of the basic aspects of retrograde amnesia. Then, we propose a model that appears to be capable of accounting for these facts within the context of a distributed model of memory. Several simulations are presented illustrating how the model accounts for various aspects of the empirical data on bitemporal amnesia, including the temporally graded nature of retrograde amnesia and the ability to extract what is common from a set of related experiences. In a final section of the chapter, we consider some recent evidence suggesting that for certain kinds of tasks, amnesics show absolutely no deficits.

Basic Aspects of Amnesia

The term bitemporal amnesia was introduced by Squire (1982) to refer to the syndrome that is produced by a number of different kinds of insults that affect the medial portions of the temporal lobes in both hemispheres of the brain. The syndrome may be produced by bilateral electroconvulsive therapy (still widely in use as a treatment for severe depression), bilateral removal of the medial portions of the temporal lobes (as in patient H. M.), head trauma, or in several other ways. The syndrome is marked by the following characteristics (see Squire, 1982, for a more detailed discussion):

- The anterograde and retrograde amnesias produced by the insult appear to be correlated in extent. While there are some reports of dissociation of these two aspects of amnesia, it is well established in cases of amnesia due to electroconvulsive therapy that anterograde and retrograde amnesia are correlated in severity; both develop gradually through repeated bouts of electroconvulsive therapy.

- The anterograde amnesia consists of a deficit in the acquisition of new knowledge accessible to verbal report or other explicit indications that the subject is aware of any particular prior experience; somewhat more controversial, it also consists of a more rapid loss of information once it has been acquired to a level equal to normal levels of acquisition through repeated exposure.

- The retrograde amnesia consists of an inability to give evidence of access to previous experiences within a graded temporal window extending back over an extended period of time prior to the amnesic insult. The size of the window varies with the severity of the amnesia, and good evidence places it at up to three year's duration based on careful experimental tests.

- Most strikingly, memories that appear to be lost after an amnesic insult are often later recovered. As the ability to acquire new memories returns, so does the ability to remember old ones that had previously been lost. The recovery is gradual, and it is as if the temporal window of retrograde amnesia shrinks. There is generally a residual, permanent amnesia for events surrounding the insult that caused the amnesia, extending variously from minutes to days from the event.

A Resolution to the Paradox

As we have already noted, the temporally graded nature of the retrograde aspect of bitemporal amnesia appears to suggest that recent memories are stored separately from older ones. However, it is possible to account for this aspect of the phenomenon in the context of a distributed model if we make the following assumptions. First, we assume that each processing experience results in chemical/structural change in a large number of connections in which many other traces are also stored, but that each new change undergoes a gradual consolidation process, as well as a natural decay or return to the prechange state. Thus, the changes resulting from a particular experience are widely distributed at one level of analysis, but at a very fine grain, within each individual connection, each change in its efficacy has a separate consolidation history.[1] Second, we assume that consolidation has two effects on the residual part of the change: (a) It makes it less susceptible to decay; and (b) it makes it less susceptible to disruption. These assumptions can explain not only the findings on the temporally graded nature of retrograde amnesia, but also the fact that memory appears to decay more rapidly at first and later decays more slowly.

So far this explanation simply takes existing consolidation accounts of the amnesic syndrome (e.g., Milner, 1966) and stipulates that the changes are occurring in synapses that they share with other changes occurring at other points in time. However, we need to go beyond this account to explain two of the important characteristics of the bitemporal amnesic syndrome. First, the hypothesis as laid out so far does

---

1 When we speak of connections between units, even if we think of those units as *neurons*, we still prefer to use the term *connection* somewhat abstractly; in particular, we do not wish to identify the connection between two units as a *single* synapse. Two neurons may have a number of different physical synapses. The total strength of these synapses determines the strength of the connection between them.

not explain recovery; second, it does not explain the coupling of anterograde and retrograde amnesia.

To capture these two important aspects of the syndrome, we propose that there exists a factor we call γ (*gamma*) that is depleted by insult to the medial temporal lobes. Gamma serves two functions in our model: (a) it is necessary for consolidation; without γ, new memory traces do not consolidate; and (b) it is necessary for expression; without γ, recent changes in the connection between two units do not alter the efficacy of the connection; they are just ineffectual addenda, rather than effective pieces of new machinery. Implicit in these assumptions is a third key point that γ is only necessary during consolidation. Fully consolidated memories no longer need it for expression.

Some Hypothetical Neurochemistry

To make these ideas concrete, we have formulated the following hypothetical account of the neurochemistry of synaptic change. While the account is somewhat oversimplified, it is basically consistent with present knowledge of the neurochemistry of synaptic transmission, though it should be said that there are a number of other ways in which connection strengths could be modulated besides the one we suggest here (for an introductory discussion of current understanding of synaptic function and synaptic modification, see Kandel & Schwartz, 1981).

The account goes as follows. The change to the connection from one unit to another involves adding new receptors to the postsynaptic membrane (the one on the input unit) (see Figure 1). We assume that both

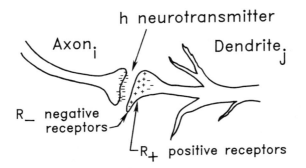

FIGURE 1. A connection between two units, as we conceptualize it in the amnesia model. Note that both positive and negative changes involve addition of new receptors. See text for discussion.

positive and negative changes involve the addition of receptors; in both cases, there must be new structure to consolidate for the model to work properly. In the figure, we have drawn the connection between two units as though it occurred at a single synapse and was not mediated by interneurons, though neither of these assumptions is excluded by the quantitative structure of the model.[2]

A cartoon of one of the receptors is shown in Figure 2. Receptors are, of course, known to be the physical structures whereby neurotransmitters released by the presynaptic neuron influence the potential of the postsynaptic neuron. To be functional, though, our hypothetical receptors must be clamped in place at each of several γ-binding sites by molecules of γ—this is the aspect of the model that is the most speculative. The probability that a site is bound depends, in turn, on the

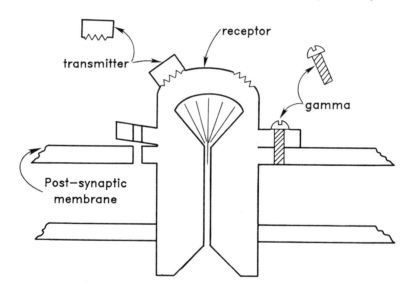

FIGURE 2. A cartoon of a receptor, showing its location in the postsynaptic membrane and illustrating the role of the transmitter substance, and of the hypothetical substance γ, which acts to bind the receptor into the membrane.

---

[2] As pointed out in Chapter 20, real neurons are generally thought to have either excitatory or inhibitory connections but not both. Our model could be brought into line with this idea if we assumed that negative (inhibitory) connections between two units actually occurred at excitatory synapses onto inhibitory interneurons, rather than on direct connections between two neurons. Connections onto these inhibitory interneurons would have to be trained, of course, using something like the generalized delta rule (Chapter 8). This revision of our assumptions would increase the complexity of the model but would not change its basic properties; therefore we have retained the less realistic assumption that positive and negative increments can be stored in the same connections.

concentration of γ in the environment of the synapse. In this model, consolidation amounts to the "hardening" or fixation of the γ-binding sites, while they are occupied by a molecule of γ. Thus, consolidation can only occur at bound sites. Consolidation is a process like the setting of glue, but it is thought to be probabilistic and all-or-none at each site rather than continuous.

As we have already seen, γ is essential for consolidation. In addition, we assume that it is necessary for the receptor to function. Once a site is consolidated, however, γ is irrelevant to it, just as a clamp is irrelevant once a glue-joint is set. Thus, unconsolidated sites depend on γ, but consolidated ones do not.

On this view, bitemporal amnesia simply amounts to taking away the clamps. Old, fully consolidated synaptic changes no longer require them, and new ones cannot function without them and will decay without becoming consolidated. But what of memories in an intermediate stage of consolidation? Here, we assume the consolidation process has gone far enough so that the structures will not break up rapidly without γ, but that it has not gone so far that they actually function effectively without it. When γ returns, after a period for recovery, they may still be there, so they will be able to function again and even continue to consolidate.

A Quantitative Formulation of the Model

Let us now formalize these assumptions in a quantitative model. We assume that time is broken up into a number of discrete ticks. In the simulations each tick represents about an hour of real time. On each tick, an unconsolidated site is bound by γ with a probability ρ, given by the *law of mass action* (this law governs a large number of chemical and biochemical processes):

$$\rho = \frac{\gamma}{1 - \gamma}.$$

This equation has the property that at high concentrations of γ (much greater than 1), all unconsolidated sites will be bound, but at low concentrations (less than about .2), the probability of being bound is roughly linear with γ.

Now, in each tick, an unconsolidated site may become consolidated or "fixed" with some probability f, but only if it is bound. Thus, the probability of consolidation of unbound site i per tick is just

$$p_c(site_i) = f\rho.$$

For a receptor to be *functional* at a particular tick, *all* its sites must be either consolidated or bound with γ. Each unconsolidated site is assumed to be independent of the others, so the probability that receptor i will be active, $p_a\,(receptor_i)$, is just

$$p_a\,(receptor_i) = \rho^u$$

where u is simply the number of unconsolidated sites.

Finally, receptors may be lost from the postsynaptic membrane. Each site contributes multiplicatively to the probability that the receptor will be lost. That is, the probability that receptor i will be lost is simply the *product* of the susceptibilities for each site. The susceptibility of consolidated sites, θ_c, is assumed to be small enough so that for completely consolidated receptors the probability of loss is very very small per tick; though over the course of years these small probabilities eventually add up. The susceptibility of unconsolidated sites, θ_u, is relatively large. For any given receptor, some number c of its sites are consolidated at any given time and u sites are not. The probability of receptor loss per tick, $p_l\,(receptor_i)$ simply becomes

$$p_l\,(receptor_i) = (\theta_c)^c\,(\theta_u)^u.$$

Relation to Other Accounts of Amnesia

Most attempts to account for temporally graded retrograde amnesia quite naturally involve some form of consolidation hypothesis, and our model is no exception to this. However, other accounts either leave the nature of the consolidation process unspecified (e.g., Milner, 1966) or give it some special status. For Wickelgren (1979), who has the most concretely specified account of retrograde amnesia, memory trace formation involves a "chunking" or unitization process whereby each memory trace is organized under its own superordinate or "chunk" unit. A number of other authors have proposed accounts with a similar flavor (e.g., Squire, N. J. Cohen, & Nadel, 1984).

In keeping with the view that our model can be implemented in a distributed memory system, our model of consolidation does not involve anything like chunking of a memory trace under a single superordinate unit. Instead, it simply involves the fixation of memory traces in a time-dependent fashion, dependent only on a single, global factor: the concentration of γ.

This difference means that our model gives the hippocampus a rather different role than it is taken to have in other theories. Theorists generally have not imagined that the hippocampus is the actual site of memory storage, for on that view, it would be difficult to explain why retrograde amnesia is temporally graded, unless only recent memories are thought to be stored there. But the hippocampus is often thought to play a very important role in memory trace formation. To Wickel-gren, for example, the hippocampus is the organ of unitization—it is the units in the hippocampus that bind the pieces of a memory trace together into chunks. In our model, we imagine that the primary role of the hippocampus in memory formation is to produce and distribute γ to the actual memory storage sites. This does not mean that we believe that this is the *only* function of the hippocampus. An organ as complex as the hippocampus may well play important information processing roles. However, as we shall see, this role is sufficient to provide quite a close account of a number of aspects of the amnesic syndrome.

Simulations

The primary goal of the simulations was to demonstrate that, with the simple assumptions given above, we could account for the main aspects of the coupled phenomena of anterograde and retrograde amnesia, using a single set of values for all of the parameters of the model, only allowing γ to vary with the assumed amnesic state of the subject. Since the phenomena range over a wide range of time scales (hours or even minutes to years), this is by no means a trivial matter.

Rather than embedding the assumptions about amnesia in a full-scale distributed model, we have simply computed, from the above assumptions, what the residual fraction (or residual *functional* fraction) of the memory trace would be at various times after the learning event, under various conditions. The assumption here, basically, is that each memory trace is made up of a large number of receptors distributed widely over a large number of connections. The fraction of the total that remains and is functional at any particular time gives the "strength" of the memory trace. To relate the results of these simulations to data, it is sufficient to assume that the size of the residual functional fraction of a memory trace is monotonically related to accuracy of memory task performance.

Of course, memory task performance, and indeed the effective residual strength of a memory trace, does not depend only on the hypothetical biochemical processes we are discussing here. For one thing, there is interference: New memory traces acquired between a learning event and test can change connection strengths in such as way as to actually reverse some or all of the changes that were made at the time of the original encoding event, producing what were classically known as retroactive interference effects. There will be proactive interference effects as well in a distributed model (see Chapter 3). Additionally, as time goes by, there will be changes in the mental context in which retrieval takes place; all of these factors will contribute to the apparent strength of a memory trace as observed in experiments. The point of our model of amnesia is not to deny the importance of such factors; we simply assume that performance in a memory task varies with the residual functional fraction of the original trace, all else being equal.

The values of the parameters are shown in Table 1. The significance of the particular values chosen will become clear as we proceed.

Anterograde amnesia: Smaller functional fractions at all delays. As we have said, our model assumes that amnesia amounts to a reduction in the size of γ. Reducing the size of γ does not reduce the size of the memory trace—the number of receptors added—but it does greatly reduce their effectiveness: For a receptor to be functional, *all* of its sites must be either bound with gamma or consolidated. Initially, before any consolidation has occurred, the probability that a receptor will be functional, or active, is

$$p_a(receptor_i) = \rho^n$$

where n is the number of sites on the receptor. For normals, we take ρ (that is, $\gamma/(1 - \gamma)$) to be .5. Since each receptor has three sites, p_a will be $.5^3$ or .125; if amnesia reduces the concentration of γ by a factor of 9, the effect will be to reduce ρ to .1, and p_a to .001. In general, with n_{sites} equal to 3, reducing ρ by a particular fraction of the normal

TABLE 1

PARAMETERS USED IN SIMULATIONS
OF THE AMNESIA MODEL

Parameter Name	Value
n_{sites}	3
f	.00007
θ_c	.02
θ_u	.25
γ_{Normal}	1.0

Note: Rate parameters f, θ_c, and θ_u are given on a per-hour basis.

value will reduce the effective strength of the initial memory trace by the *cube* of that fraction.

Correlation of retrograde and anterograde amnesia. The model produces retrograde amnesia, as well as anterograde amnesia, for the unconsolidated portion of a memory trace. The reason for this is that the *expression* of unconsolidated memories depends on γ; thus, it applies both to posttraumatic memories and to memories formed before the trauma. Indeed, the severity of anterograde and retrograde amnesia are perforce correlated in the model since both depend on γ for the expression of traces that have not had time to consolidate.

Retrograde amnesia: Older traces are less dependent on gamma. One of the most interesting aspects of retrograde amnesia is the fact that it is temporally graded. Indeed, in the data collected by Squire, Slater, and Chace (1975), shown in Figure 3, it is not only graded, but bitemporal amnesics actually show worse memory for recent events than for those about three to five years old. This matches the clinical impression for patients such as H.M., of whom it is reported that his retrograde amnesia initially extended over a period of one or two years

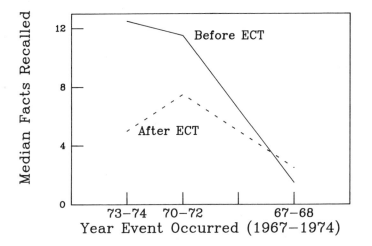

FIGURE 3. The striking temporally graded retrograde amnesia observed in patients whose amnesia was induced by electroconvulsive therapy. Patients served as their own controls, based on an alternate form of the test given prior to the beginning of treatment. (From "Retrograde Amnesia: Temporal Gradient in Very Long-Term Memory Following Electroconvulsive Therapy" by L. R. Squire, P. C. Slater, and P. Chace, 1975, *Science*, *187*, Copyright 1975 by the American Association for the Advancement of Science. Reprinted by permission.)

(Milner, 1966), and from victims of head trauma (Russell & Nathan, 1946). This pattern has been replicated many times, and the tests used by Squire et al. rule out artifacts that have plagued clinical assessments of the severity of retrograde amnesia. This inverted U-shaped curve for the relation between age of memory and memory test performance provides quite a challenge to theories of retrograde amnesia. However, this effect is a natural consequence of our model since old memories, though based on smaller residual traces than newer ones, are less dependent on γ for their expression. Indeed, when a receptor reaches a point where all of its sites are consolidated, it no longer depends on γ at all.

A simulation capturing the essential features of temporally graded retrograde amnesia as represented in Squire et al. is shown in Figure 4. The simulation produces a continual erosion in functional strength for normals which is almost linear against the log of time over the range of times covered by the simulation. In contrast, for amnesics, the function is decidedly (inverted) U-shaped: Functional trace strength reaches a peak at about 2 to 3 years with these parameters and then falls of gradually thereafter, following the same trajectory as the strength of the trace for normals. The location of the peak in the RA function depends on all of the parameters of the model, but the primary ones are the consolidation rate parameter f, which is .00007, and the rate of decay from consolidated memory, which is $(\theta_c)^n$, or

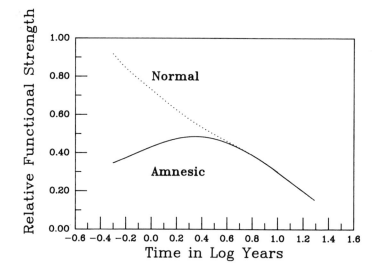

FIGURE 4. Simulation of temporally graded retrograde amnesia. Effective trace strength as a function of time is years preceding sudden onset of amnesia. Effective trace strength is normalized so that a value of 1.0 corresponds to the normal value at five months.

8×10^{-6}. It should be noted that these figures are *per hour*. The consolidation parameter translates into a consolidation rate of about 50% per year, per receptor site. The average time it takes for all of the sites on a receptor to become consolidated is longer than this, of course, but only by about a factor of 2 for the case of $n_{sites} = 3$; this is essentially the factor that determines where the curve for amnesics will catch up with the curve for the normals. The decay rate from fully consolidated memory, which translates into about 7% per year or 50% per decade, essentially determines the overall slope of the normal function and the tail of the amnesic function.

Recovery of lost memories: The return of the partially consolidated trace. Perhaps even more interesting than the fact that retrograde amnesia is temporally graded is the fact that it recovers as the ability to acquire new memories recovers. In the case of retrograde amnesia induced by electroconvulsive therapy, Squire, Slater, and Miller (1981) showed that the severe retrograde amnesia for pretreatment memories recovers over the course of several months, at the end of which test performance is back to pretreatment levels. In our model, since retrograde amnesia is due to the fact that loss of γ renders traces ineffective, it is not surprising that the return of γ will render them effective again. However, the phenomenon is somewhat more subtle than this, for recovery is not generally thought to be complete. There is usually some loss of memory for events in the time period preceding the onset of the amnesia, and the precipitating event is almost never recalled; this is particularly striking in head trauma patients, who often do not know what hit them, even if they had seen it at the time (Russell & Nathan, 1946).

To examine how well our model can do at reproducing these aspects of amnesia, we ran the model in the following simulated amnesia-recovery experiment. The model was made amnesic at some time t_a and was left in this state until some time t_r, at which point we assumed recovery occurred. Of course, real recovery is gradual, but for simplicity, we assumed that it was a discrete event. We then asked what fraction of a trace laid down before the onset of amnesia remained, relative to the fraction that would have remained had there been no insult. The results of the simulation are shown in Figure 5. Each curve shows the strength of the recovered trace, relative to the strength it would have had with no intervening amnesia, as a function of the duration in months of the amnesic episode. Clearly, for memories a year or more old, the trace recovers to nearly premorbid levels at the end of the amnesic episode, even if it lasts as long as a year. For memories laid down within the day of the the amnesic event, however, the bulk of the trace is gone by the end of the amnesic episode, even if it lasts only

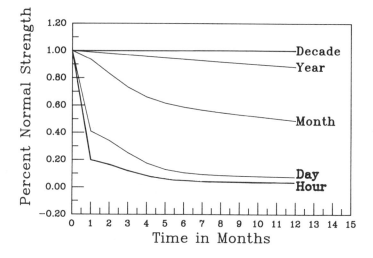

FIGURE 5. Simulated recovery of premorbid memories as a function of time in the amnesic state and age of memory at onset of the amnesia. The age ranges from an hour to a decade, as indicated by the labels on the curves.

a month. Memories that are a month old at the onset of the amnesia show an intermediate pattern. If the amnesia is relatively brief, they survive quite well; but if it lasts several months, they weaken considerably, relative to the strength they would have had in the absence of amnesia.

The loss of memory trace strength during amnesia is a result of the great reduction in the opportunity for consolidation during the amnesic interval. We now turn to a more direct consideration of this matter.

Do amnesics forget faster than normals? A number of studies (Huppert & Piercy, 1978; Squire, 1981) have reported that bitemporal amnesic subjects appear to forget more rapidly than normals, even if equated with normals for the amount of initial learning. The effect is generally rather small, and it is controversial because the equating of groups on initial performance requires giving amnesics considerably more training than normals, over a longer period of time. These differences could possibly change the basis of learning and other qualitative aspects of the task as experienced by amnesic and normal subjects. It is interesting, then, to consider whether our model would predict such a difference.

The model does predict a difference in rate of trace decay between amnesic and normal subjects. Though γ does not influence the rate of trace decay directly, it does influence the rate of consolidation, and consolidation drastically influences the rate of decay. Completely unconsolidated traces decay at a rate of $.25^3 = 1.5\%$ per hour, or about 30% per day, and are reduced to 1/100,000 of their initial strength in a month. Consolidated traces, on the other hand, decay at a rate of only $.02^3 = .0008\%$ per hour, or less than 1% per month. As each site becomes fixed, it retards decay by a factor of 12. Thus, to the extent that consolidation is occurring, memory traces are being protected against rapid loss; without any consolidation, trace strength falls precipitously. An illustration of this effect is shown in Figure 6. At higher values of γ, the drop of trace strength decelerates much earlier than at lower values of γ, leaving a much larger residual trace.

Unfortunately, the effect shown in Figure 6 does not happen in the right time scale to account for the differential decay of normal and amnesic memory over hours, as reported by Huppert and Piercy (1978) and Squire (1981). In fact, the effect does not really begin to show up until after about 10 days. The reason is clear: The consolidation rate is so slow (.0007/hour, or 5% per month) that very little consolidation happens in the first month. Thus, it appears that the value of the consolidation rate parameter required to account for temporally graded retrograde amnesia and U-shaped memory performance on a time scale of years is much too slow to account for differences in trace decay on a time scale of hours.

One possible reaction to this state of affairs would be to search for alternative interpretations of the apparent differences between normals

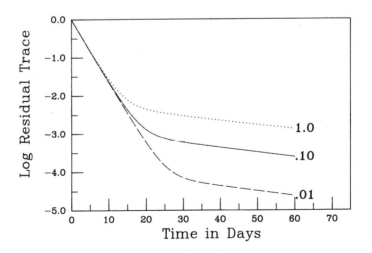

FIGURE 6. Log of the residual strength of a trace acquired at Day 0, as a function of the concentration of γ. The values of 1.0 and 0.1 correspond to the values described as normal and amnesic in the text.

and amnesics in day-scale decay rates. However, there are other reasons to believe that there is consolidation on a shorter time scale than we get with our model and the parameters in Table 1. For one thing, a single bout of electroconvulsive therapy that produces a brief and mild amnesia nevertheless appears to produce permanent loss of memory for the shock treatment itself. Such an effect seems to suggest that there is consolidation going on over a shorter time-scale.

The model might capture all of the data if we assumed that there are two separate phases to consolidation, both of them dependent on γ: one that occurs on a relatively short time scale and is responsible for the differences in day-scale decay rates, and one that occurs on a very long time scale and is responsible for extended temporally graded retrograde amnesia. As things stand now, traces decay rather slowly over hours, but, over the course of a month, they are reduced to about half of one percent of their original strength. Though we do not know exactly how to scale trace strength against response probability, it seems likely that we forget more quickly over hours but more slowly over days and months than in the present version of the model.

Summary. The model provides, we think, an appealing, unified account of most aspects of anterograde and retrograde amnesia, simply by assuming that the amnesic insult depletes γ and that recovery amounts to its gradual return to pretraumatic levels. By adding an additional stage of consolidation, the model could be made to span the very wide range of time scales, ranging from hours to years, of the coupled phenomena of anterograde and retrograde amnesia as they appear in the bitemporal amnesic syndrome.

Most importantly, the model shows clearly that there is no incompatibility between the phenomenon of temporally graded retrograde amnesia and distributed representation. So far, however, our account of amnesia has not really depended on the features of our distributed model. In the next section we will consider aspects of the amnesic syndrome which do seem to point toward distributed models.

RESIDUAL LEARNING AND SPARED LEARNING IN BITEMPORAL AMNESIA

As we noted briefly before, there are some domains in which amnesics exhibit what are generally described as *spared learning* effects: They show no noticeable deficits when compared to normal subjects. There is now a very large literature on these spared learning effects.

The following summary seems to capture the basic characteristics of what is spared and what is not.

While amnesics seem to be highly deficient in the ability to form accessible traces of particular individual episodic experiences, they seem to be completely spared in their ability to learn certain types of skills that require no explicit access to the previous processing episodes in which the skill was acquired (N. J. Cohen, 1981; N. J. Cohen, Eichenbaum, Deacedo, & Corkin, 1985). In addition, they show apparently normal repetition priming effects in experiments involving such tasks as perceptual identification, in which the subject must simply identify a briefly flashed word in a short exposure (see Schacter, 1985, for a review). These effects may be strongly and strikingly dissociated from the subjects' verbally expressed recollections. Thus, H. M. has acquired a skill that allows him to perform perfectly in solving the Tower of Hanoi problem, without becoming aware that he has actually ever performed the task before and without knowing (in a conscious, reportable sense) even what constitutes a legal move in the Tower Puzzle (N. J. Cohen et al., 1985). Also, amnesic subjects show normal effects of prior exposure to words in perceptual identification and related tasks, without necessarily having any awareness of having seen the words or even participating in the priming portion of the task. Between these two extremes lies a gray zone. Within the domains where learning is impaired, even the densest amnesics seem to learn, however gradually, from repeated experience (Schacter, 1985). First, we will consider these *residual learning* effects from the point of view of distributed memory. Then, we will examine the more striking *spared learning* effects.

Residual Learning in Bitemporal Amnesia

As we noted early in this chapter, distributed models provide a natural way of explaining why there should be residual ability to learn gradually from repeated experience within those domains where amnesics are grossly deficient in their memory for particular episodic experiences. For if we imagine that the effective size of the increments to the changes in synaptic connections is reduced in amnesics, then the basic properties of distributed models—the fact that they automatically extract the central tendency from a set of similar experiences and build up a trace of the prototype from a series of exemplars—automatically provides an account of the gradual accumulation of knowledge from repeated experience, even in the face of a profound deficit in remembering any specific episode in which that information was presented. Distributed models are naturally incremental learning models, and thus they provide a very nice account of how learning could occur through the gradual accumulation of small traces.

We call the hypothesis that anterograde amnesia amounts to reducing the effective size of the increments the *limited increment hypothesis*. For bitemporal amnesics, the effective size of the increments is limited by the depletion of γ; in other forms of amnesia (which also show similar kinds of residual learning) the size of the increment might be limited in other ways. According to the limited increment hypothesis, residual learning is simply a matter of the gradual accumulation of information through the superimposition of small increments to the connection strengths.

To illustrate this point, we have carried out a simulation analog of the following experiment by N. J. Cohen (1981). Amnesic subjects and normal controls were seated in front of an apparatus with a movable lever. On each trial of the experiment, the subject was asked to move the lever until it reached a stop set by the experimenter. The experimenter then moved the lever back to the start position and removed the stop. After a variable delay, the subjects were asked to reproduce the previous movement. Such trials are referred to as *reproduction trials*.

At the end of each group of three trials, subjects were asked to reproduce their impression of the average distance they had been asked to move the lever, based on all of the preceding trials in the experiment. Such trials will be called *averaging trials*.

The results of the reproduction task were as expected from the fact that amnesics have very poor memory for specific experiences; at very short delay intervals, amnesics did as well as normals, but at longer intervals, they were grossly impaired, as measured by the deviation of the reproduced movement from the training movement (Figure 7). However, amnesics did no worse than normals at reproducing the average movement. The experiment was divided into four parts: In the first and last parts, the movements were all relatively long; and in the two intermediate parts, the movements were all relatively short (some subjects had the long and short trials in the other order). At the end of each block of trials, both groups accurately reproduced the average movement for that block (For the long blocks, movements averaged 42.6 degrees for the normals and 41.3 for the amnesics; for the short blocks, movements averaged 30.8 for the normals and 30.6 for the amnesics).

We simulated this phenomenon using the distributed memory model described in Chapter 17. Briefly, that model consists of a single module, or set of units, with each unit having a modifiable connection to each other unit. The units in the module receive inputs from other

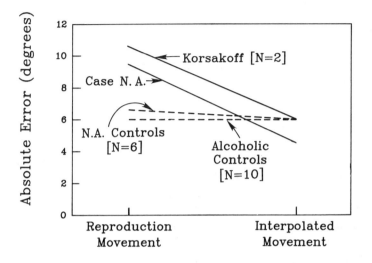

FIGURE 7. Accuracy of reproduction movements by amnesics and normal controls in the lever placement experiment described in text. (From "Neuropsychological Evidence for a Distinction Between Procedural and Declarative Knowledge in Human Memory and Amnesia" by N. J. Cohen, 1981, doctoral dissertation, University of California, San Diego. Copyright 1981 by N. J. Cohen. Reprinted by permission.)

units via the modifiable connections, as well as external inputs from stimulus patterns. Processing in the module begins with all units at a resting activation of 0 and the presentation of an external input pattern. In this case, the module consisted of 16 units, and each input pattern was a vector of 16 excitatory or inhibitory inputs. When a pattern is presented, it begins to drive the activations of the units up or down as a result of its direct effects; the units then begin to send excitatory and inhibitory signals to the other units via the modifiable connections. For patterns that have previously been stored in the connections among the units, the internal connections produce an enhancement of the pattern of activation over and above what would be produced by the external input alone; if, however, the external input is very dissimilar (orthogonal) to the patterns that have been stored in the module on previous trials, there will be little or no enhancement of the response.

On each trial of the experiment, a new distortion of the same 16-element prototype pattern was presented to the module, and connection strengths were adjusted after each trial according to the delta rule (see Chapter 17 for details). We then tested the module in two ways: First, to simulate Cohen's reproduction test, we looked at the magnitude of the model's response to the pattern it had just been shown. For the averaging test, we looked at the magnitude of the model's response to the prototype. Note that these test trials were run with connection strength modification turned off, so each test was completely uncontaminated by the previous tests.

In keeping with the limited increment hypothesis, we assumed that the difference between amnesics and normals in Cohen's experiment could be accounted for simply by assuming that amnesics make smaller changes to the strengths of the connections on every learning trial. To show that the model shows residual learning of the prototype under these conditions, we ran the simulation several times, with three different levels of the increment strength parameter η from the equation for the delta rule, which we reproduce here:

$$\Delta w_{ij} = \eta \delta_i a_j .$$

The results of the simulation are shown in Figure 8. As the figure indicates, the larger the size of η, the more strongly the model responds to the immediately preceding distortion of the prototype. But, after a few trials, the response to the *central tendency* or prototype underlying each distortion is as good for small values of η as for larger ones. In fact, response to the prototype is actually *better* when the model is "amnesic" (low η) than when it is "normal" (high η); in the latter state, the connections are continually buffeted about by the latest distortion, and the model has trouble seeing, as it were, the forest for the trees.

In the figure, there is a gradual improvement in the response to the immediately preceding stimulus for small increment sizes. This occurs only because the stimuli are all correlated with each other, being derived from the same prototype. For a sequence of unrelated stimuli, the response to each new input shows no improvement over trials.

This pattern of performance is very reminiscent of the pattern seen in several of the experiments performed by Olton and his colleagues (see Olton, 1984, for a review). They have trained rats to run in two different mazes, each having two choice points. At one of the choice points, the response was always the same for a given maze (we call this the maze-dependent choice); at the other choice point, the response that had to be made varied from trial to trial, based on the response the rat made on the preceding trial (we call this the trial-dependent choice). The principal finding of these experiments is that rats with hippocampal lesions show gross impairment in the ability to make the right trial-dependent choice, but show no impairment in the ability to make the right maze-dependent choice if they were trained up on the task before surgically induced amnesia. The acquisition of the maze-dependent choice is slowed in rats trained after surgery, but these animals eventually reach a point where they can perform as well as normals. Such animals show near chance performance in the trial-dependent choice

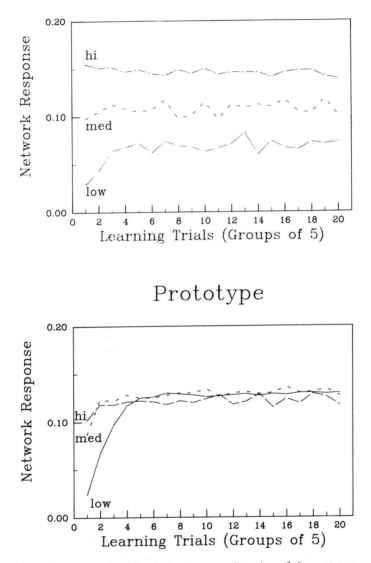

FIGURE 8. Network response for the most recent distortion of the prototype pattern and for the prototype pattern itself, as a function of test trials, at three different levels of the increment size parameter η. Network response is the dot product of the external input to each unit with the internal input generated by the network.

after surgery, even if they had already acquired the ability to do this part of the task before the surgery.

Such a pattern of results is completely consistent with the limited increment hypothesis: Performance in the trial-dependent choice requires, by the design of the task, that the subject rely on a memory trace of the preceding trial of the experiment, whereas performance on the maze-dependent choice can be based on a composite memory trace acquired gradually over repeated experience in the same maze. No separate mechanisms for retaining recent episodes, as opposed to more general memories, is required.

In summary, there are a variety of phenomena, both in human and animal amnesia, which fit in very well with the kind of gradual, residual learning we see in our distributed model. Distributed models naturally and automatically pull out what is common to a set of experiences, even if, or one might even say especially when, the traces of the individual experiences are weak.

It is worthwhile to note that this property of distributed models would not be shared by all learning models, especially those that rely on some mechanism that examines stored representations of specific events in order to formulate generalizations, as in the ACT* model (J. R. Anderson, 1983), or in Winston's (1975) approach to learning. For on such models, if the individual traces are impaired, we would expect the generalization process to be impaired as well. Of course, such models could account for these effects by assuming that each episode is stored in two different ways, once for the purpose of learning generalizations and once for the purpose of remembering the details of particular experiences. Thankfully, our distributed model does not require us to duplicate memory stores in this way; residual learning based on small increments drops out of the basic superpositional character of the model.

Spared Learning of Skills

More striking that these residual learning effects is the phenomenon of spared learning: The fact that the acquisition of a variety of general skills occurs at roughly the same rate in normal and amnesic subjects. This fact has been taken as evidence that the brain maintains a distinction between those structures underlying explicitly accessible episodic and semantic information on the one hand and those underlying general cognitive skills on the other (N. J. Cohen et al., 1985).

While this conclusion is certainly plausible, it is worth noting that there are other possibilities. One that we have considered is the possibility that limited increments to connection strengths make a difference for some kinds of tasks but not for others. The simulations reported in the previous section indicated that this can sometimes be the case; in fact, they indicated that, as far as extracting the prototype or central tendency of an ensemble of experiences is concerned, it can sometimes be better to make smaller changes in connection strengths.

The preserved *skill learning* observed in many tasks appears to be the kind of learning that may be relatively unaffected by the size of the changes made to connections. For we can view skill learning as the process of learning to respond to *new* stimuli in a domain, based on experience with previous examples. For example, consider the mirror-reading experiment of N. J. Cohen and Squire (1980). In this experiment, subjects were required to read words displayed reflected in a mirror so that they had to be read from right to left. In this task, both amnesic and normal subjects learn gradually. Though normals learn to read specific repeated displays much more quickly than amnesics, both groups show equal transfer to novel stimuli.

To assess transfer performance in the simple distributed model described in the previous section, we observed the response of the model to new input patterns, after each learning trial. The results of the simulation are shown, for three levels of η, in Figure 9. Though there are initial differences as a function of η, these differences are considerably smaller than the ones we observe on reproducing old associations. And, at a fairly early point, performance converges, independently of the level of η. As with learning the prototype, there is a slight advantage for smaller values of η, in terms of asymptotic transfer performance, though this is difficult to see in the noise of the curves.

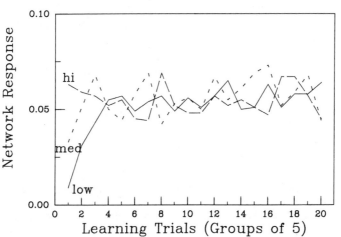

FIGURE 9. Network response to transfer patterns (new distortions of the prototype) for three levels of η, as a function of learning.

This simulation does not capture spared learning of skills perfectly, since our simulated normals approach asymptote more rapidly than our simulated amnesics. However, if such skills really consisted of many small skills, the differences might not be noticeable. We have not yet developed a version of our model in which there are no differences as a function of η. We still consider it an open question as to whether we will succeed. It may turn out that there are other distributed models (perhaps involving hidden units) in which rate of learning is quite insensitive to large differences in sizes of increments on certain kinds of measures for certain kinds of tasks. This is a matter we are continuing to pursue as our explorations of memory and learning continue.

Setting this possibility aside, let us suppose for the moment that episodic and semantic memory are learned in one memory system, dependent on medial temporal lobe structures, and general skills are learned in a different system. This view raises a question: Why should this be? Why should the brain make this distinction? We can actually provide one possible answer to this question based on our observations of the properties of the simple distributed model presented in the simulations. These observations suggest that large changes in connection strengths may be better for storing specific experiences but may do more harm than good for gradually homing in on a generalizable set of connection strengths.[3]

On the basis of these observations, we might propose that the temporal lobe structures responsible for bitemporal amnesia provide a mechanism that allows large changes to connections to be made in parts of the system in which memories for specific experiences are stored, but other parts of the cognitive system make use of a different mechanism for changing connections that results in the smaller changes that are at least as good as larger ones for learning what is common to a set of experiences. However, as we have already suggested, we remain unconvinced that such a distinction is necessary. It may turn out that learning of generalizable skills is simply insensitive to the size of the changes made in connection strengths.

---

[3] As noted previously, spared learning effects also show up in single-trial priming experiments. One view of these effects, consistent with the separate systems view, is that they reflect the subtle effects of single trials in those parts of the system where skills and procedures are learned. Again, an alternative would simply be that the priming task is less sensitive to the magnitude of the changes in connection strengths and more sensitive to the relative size of changes made by different stimuli.

CONCLUSION

In this chapter, we have considered the phenomenon of bitemporal amnesia in the light of models of distributed memory. We have described a hypothetical mechanism that can account for temporally graded retrograde amnesia without assuming that recent memories are stored separately from older ones. We have demonstrated how the ability to learn gradually from repeated experiences is an automatic consequence of assuming that amnesia simply amounts to the reduction of the effective size of the changes in the connections in a distributed memory. And we have indicated how a distributed approach can allow us to suggest reasons why large changes to connection strengths might make more of a difference in forming explicit representations of facts and episodes than in laying down the connections required for cognitive skills.

Obviously, there is considerable room for further work to test and to extend our views. If our hypotheses are correct, and if γ really is a chemical, then we might hope that someday someone may discover just what the chemical is, and will then go on to show that normal memory depends only on γ and not on some information processing activity that takes place in the hippocampus, as has frequently been suggested (Squire et al., 1984; Wickelgren, 1979). Considerably more theoretical work will be required to build a tight connection between those tasks in which spared learning is observed empirically and the situations in which large increments to the weights do not result in superior learning. In the meantime, we hope this chapter has demonstrated that what we know about amnesia is not only consistent with the idea of distributed, superpositional memory, but that certain aspects of the amnesic syndrome—in particular, residual learning in domains where amnesics show deficits—actually support the idea.

ACKNOWLEDGMENTS

The work reported here was supported in part by a grant from the System Development Foundation, in part by contracts from the Office of Naval Research (N00014-79-C-0323, NR667-437 and N00014-82-C-0374, NR 667-483), and in part by a NIMH Career Development Award (MH00385) to the first author. We would like to thank Neal Cohen, Morris Moscovitch, Daniel Schacter, and Larry Squire for several useful discussions of various aspects of amnesia on several different occasions.

PART **VI**

CONCLUSION

CHAPTER 26

Reflections on
Cognition and Parallel Distributed Processing

D. A. NORMAN

This book presents a novel approach to the study of cognitive processes. The overall proposal is for a major change that alters the fundamental mechanisms of human information processing. Still, despite the impressive results, it is not always obvious just what impact this work should have upon the study of cognitive mechanisms. Yes, the underlying components might change, but what is the effect on models of cognition? Why should cognitive scientists care about the class of models discussed in this book—parallel distributed processing models: PDP. What difference do they make? What really has been learned about cognition? Anything? And what has been left out: Where are the strengths and the weaknesses? Where do we go from here? This chapter addresses these issues.

I start with the issue of level, for I can already hear my colleagues muttering—those who are not part of the enterprise, of course:

> So what? So you have this work on low-level processing structures. Why should it make any difference to our work, or to cognitive science in general? After all, the study of cognition is a high-level pursuit and this PDP stuff is awfully low level—all this talk of weights and synapses and connections between low-level units. Oh sure, it is all very nice and even significant, but it has nothing to do with cognitive science: It has no effect on anything we are doing. We care about language and thought,

problem-solving and memory, not about neural mechanisms and Hebbian rules for adjusting synaptic weights.[1]

Well, in many ways those concerns are correct. Most of what is presented in this book differs radically from the kind of studies that are called "cognitive." Most are what would normally be called "low-level," concerned with "implementation details," with the hardware and the neurophysiology. Why should it be of concern to the normal, everyday cognitive scientist? And, equally important, how does this work fit with the substantive concerns of the cognitive scientist? What is useful, what is questionable, and what has been left out?

It is still too early to give firm answers to all these questions. But I can say where I stand. I am an in-between cognitive scientist. I have watched the developments reported in this book with great interest. Much of my work in the last ten years has been in the spirit of these models, some of it, of course, done with my long-time collaborator, Dave Rumelhart. I have long thought this philosophy to be powerful and important. But I have also disagreed with some aspects, for I have had some problems with the work. It is probably useful to discuss these here, at the end of the book. Some of these problems may reflect inherent, fundamental difficulties. Some may just be temporary problems that persist simply because they haven't been worked on. And some may be pseudoproblems, resulting either from my own misunderstandings or from the current point of view, but that will disappear when the "proper" approach is adopted. Let me start off with the good things: What does this approach—and this book—deliver? Then I will turn to the problems.

Before discussing "the PDP approach," I must remind you that there really isn't a single PDP approach. No simple listing of either strengths or weaknesses is really fair, for there are many different contributors, each with a somewhat different approach with its own set of virtues and vices. Although much of the work has a common thread, the details and even the operative philosophies are oftentimes quite different. This level of disagreement is healthy for a developing science, although it may cause confusion to the unwary onlooker.

---

[1] I wrote this paragraph without realizing that my prophecy had already been confirmed: See the essay by Broadbent (1985) and the partially responsive answer by Rumelhart and McClelland (1985), recapped in Chapter 4.

WHAT IS NEW: THE PDP APPROACH TO THE STUDY OF COGNITION

"This is a difficult book, difficult but important." With this sentence, Rumelhart and I opened our introduction to the book *Parallel Models of Associative Memory,* the earlier work on the study of parallel, distributed models of psychological processing edited by Hinton and J. A. Anderson (1981). The sentence still applies, perhaps with a bit less emphasis on the difficult, a bit more on the important.

This book starts to put it all together, to make significant advances in the mathematics and underlying bases of the approaches and to explore their implications for the study of cognition. The result is a new form of model for understanding the mechanisms of human (and animal) information processing, models that have several important aspects:

- A major rethinking of the levels at which we must think of cognitive processes, involving, among other things, an intimate relationship between neuroscience and psychology.

- A new form of computational model involving large numbers of highly interconnected parallel computational units, a form that takes into account constraints from studies of the nervous system, constraints from psychological data, and a deep understanding of computational requirements.

- New conceptualizations of cognition that change the emphasis from symbolic processing to states that reflect the quality of the match between the data to be described and the possible configurations of the underlying processing structures. These lead to new modes of thinking of cognition, modes that use metaphors such as "stability," "harmony," "local" and "global minima," and "temperature."

- An emphasis on learning—the mechanisms are built around continual adjustment and learning.

Everyone agrees that full understanding of cognition requires a combination of efforts that cut across disciplinary boundaries. Indeed, this is the basic premise underlying the establishment of *cognitive science.* Nonetheless, even relatively recently, there has been little combination of efforts. Scientists in one field tended to think work in other fields interesting, but irrelevant to their own. Thus, cognitive psychologists tended to think that studies of the brain were interesting but irrelevant;

neuroscientists acted as if models of cognition or psychological data were too speculative and unrelated to neural activity (amusing pastimes, but not science); those in artificial intelligence thought the emphasis should be on control structures and representation, and paid little attention to other disciplines. This has changed. There is now a spirit of genuine cooperation. The research reported in this book shows the results. Neurological data offer strong constraints and invaluable guides to the model building efforts: The work reported here both builds upon what is known of the nervous system and asks specific questions to be decided by future research. Psychological evidence is taken seriously: The models must be able to reproduce this form of evidence. And the models must reflect the real demands of computational plausibility, which means that the models must work, must produce answers in appropriate times and with appropriate degrees of accuracy. Each field pushes the other. Each field suggests the data that should be provided by the other, suggests questions to be examined, and sometimes even suggests in which directions future efforts should be directed. The overall effort reflects the appropriate interplay of theory and data, with genuine interaction among the disciplines: a good model for a cognitive science.

On the Computer Metaphor

It has always bothered me that models of psychological processing were thought to be inspired by our understanding of the computer. The statement has always been false. Indeed, the architecture of the modern digital computer—the so-called Von Neumann architecture—was heavily influenced by people's (naive) view of how the mind operated. [2] Now that whole debate can finally be put aside: The work presented here in no way can be interpreted as growing from our metaphor of the modern computer. Here, we are talking about a new form of computation, one clearly based upon principles that have heretofore not had any counterpart in computers. These models are highly parallel, with thousands or millions of elements interacting primarily

---

[2] Perhaps I had better document this. Simply read the work on cybernetics and thought in the 1940s and 1950s prior to the development of the digital computer. The group of workers included people from all disciplines: See the Macy Conferences on Cybernetics, or "Her Majesty's Conference on Thought Processes." Read the preface to Wiener's book on cybernetics. Everyone who was working together—engineers, physicists, mathematicians, psychologists, neuroscientists (not yet named)—consciously and deliberately claimed to be modeling brain processes.

through activation and inhibition of one another's activity. Each element is highly interconnected, with perhaps tens of thousands of interconnections. Although the processing speed of each element is slow—measured in milliseconds rather than the picoseconds common in today's computers—the resulting computations are fast, faster than is possible with even the largest and fastest of today's machines. Parallel computation means that a sequence that requires millions of cycles in a conventional, serial computer can be done in a few cycles when the mechanism has hundreds of thousands of highly interconnected processors. These neurologically inspired computational processes pose new requirements on our understanding of computation, suggest novel theoretical explanations of psychological phenomena, and suggest powerful new architectures for the machines of the future.

Although the models reported here are new, they carry on a tradition that has long existed, a tradition of building models of neurological processes. Work in neural networks, in cybernetics, and in brain theory has preceded the present efforts. The studies of perceptrons, a computational device invented by Frank Rosenblatt in the 1950s clearly anticipated many of the results in use today (see Rosenblatt, 1962). The critique of perceptrons by Minsky and Papert (1969) was widely misinterpreted as destroying their credibility, whereas the work simply (but elegantly) showed limitations on the power of the most limited class of perceptron-like mechanisms, and said nothing about more powerful, multiple layer models. Grossberg developed many of the ideas in popularity today with his mathematical models of brain structures, starting in the late 1960s through today (see Grossberg, 1982).

Strengths of the PDP Approach

For me, the work has been important for several reasons. In my personal opinion, many students of cognition have boxed themselves into corners, forcing themselves to develop ever more complex rule structures, and ever more complex algorithms for dealing with the behavior of people. Each new experimental finding seems to require a new theory. Clearly, some new approach has been needed. But what?

The PDP approach offers a very distinctive counterapproach. Basically, here we have an adaptive system, continually trying to configure itself so as to match the arriving data. It works automatically, prewired if you will, to adjust its own parameters so as to accommodate the input presented to it. It is a system that is flexible, yet rigid. That is, although it is always trying to mirror the arriving data, it does so by means of existing knowledge, existing configurations. It never expects to make a perfect match, but instead simply tries to get the best match possible at any time: The better the match, the more stable the system. The system works by storing particular events, but the results of its operations are to form generalizations of these particular instances, even though the generalizations are never stored directly. The result, as has been illustrated throughout the chapters of the book, is that although the system develops neither rules of classification nor generalizations, it *acts* as if it had these rules. Thus, the system really mirrors experience; the regularities of its operation result partially from the regularities of the world, partially from the interpretations of the beholder. It is a system that exhibits intelligence and logic, yet that nowhere has explicit rules of intelligence or logic.

These ideas have particular charm for many of us. The attempt always to find a best fit to existing data by maximizing harmony (or minimizing energy or discrepancy) gives mechanism to a common hope among many modern cognitive theorists (e.g., Shepard, 1984). The heavy dependence upon particular instances and the lack of explicit generalizations fit a reasonable body of evidence about the importance of specific experiences (e.g., see Kahneman & D. T. Miller, in press; Medin & Schaffer, 1978). The way by which these systems try to accommodate themselves to the data by minimizing energy or maximizing harmony results in preferred states or interpretations, where the preferences reflect the particular events the system has experienced. This leads to categorization and classification of the input signals, but in a flexible manner—categorization by distance from prototypes, a mode that is consistent with much recent evidence. Here is a system that incorporates learning as a fundamental, essential aspect of its behavior, that makes no attempt to form categories or rules, yet that acts as if it were a prototype-matching, categorization system that has rules and strategies.

The schema. What about the schema, that omnipresent, powerful tool of modern cognitive theory? Well, as Rumelhart, Smolensky, McClelland, and Hinton show in Chapter 14, the schema still exists, but in somewhat different form than that which schema theorists have thought of it (schema theorists such as Rumelhart and I, I might add). I like this new formulation. More than that, I think this an important step in the understanding of thought. Schemas are not fixed structures. Schemas are flexible configurations, mirroring the regularities of experience, providing automatic completion of missing components, automatically generalizing from the past, but also continually in modification, continually adapting to reflect the current state of affairs. Schemas are not fixed, immutable data structures. Schemas are flexible interpretive states that reflect the mixture of past experience and present circumstances.

Because the schema is in reality the theorist's interpretation of the system configuration, and because the system configures itself differently according to the sum of all the numerous influences upon it, each new invocation of a schema may differ from the previous invocations. Thus, the system behaves as if there were prototypical schemas, but where the prototype is constructed anew for each occasion by combining past experiences with the biases and activation levels resulting from the current experience and the context in which it occurs.

Essential properties of human information processing. Some years ago, Bobrow and I listed some properties we felt were essential components of the human cognitive system (Bobrow & Norman, 1975; Norman & Bobrow, 1975, 1976, 1979). We constructed our lists through observations of human behavior and reflection upon the sort of processing structures that would be required to yield that behavior. We concluded that the system must be robust, relatively insensitive to missing or erroneous data and to damage to its parts. Human cognition appears to work well in the face of ambiguity, incompleteness, and false information. On the one hand, the system is not only robust, it is also flexible and creative. On the other hand, the system continually makes errors. Speech is riddled with incomplete sentences, with erroneous words and false starts. Actions are riddled with "slips" and mistakes. People have learned to deal with these problems, in part by elaborate (although mostly subconscious) correcting mechanisms in language, in part through humor and tolerance for the slips that characterize everyday behavior. Nonetheless, these characteristics of human cognition seem to provide important clues about the nature of the processing mechanisms. We argued that the system had to work by descriptions rather than precise specifications, by partial information rather than complete information, and by competition among competing interpretations.

We argued for a set of essential properties: graceful degradation of performance, content-addressable storage, continually available output, and an iterative retrieval process that worked by description rather than by more traditional search. Although these requirements seemed to us to be both necessary and sensible, supported by the evidence, a common complaint was that these were hand-waving speculations, dreams, that there was no method for implementing these concepts. And if they couldn't be built, then they didn't exist—the ideas were wrong. Well, the PDP mechanisms described in this book have exactly the properties we required. Hurrah for our side.

New metaphors for thought. A very useful result of the PDP work is to provide us with new metaphors with which to discuss cognitive activity. These metaphors are more important than might be realized. They accommodate the transfer of knowledge from one application domain to another. The metaphors of PDP are of weights and goodness-of-fit, of stability, of local and global minima or maxima, and of the means to move from one stable state (a local minimum) to another (perhaps a global minimum) by "shaking up the system" or "raising its temperature." We see why systems have what psychologists call set or functional fixity: The systems are stable, even if in only a local minimum; it takes much less evidence to get into a stable state than it takes to get out of it. To change a state is not necessarily easy. You have to shake up the system, heat up the temperature. Don't let it freeze into position. New interpretations suddenly arise, with no necessary conscious experience of how they came about: a moment of nothing, and then clarity, as the system heats up, bounces out of one stable configuration and falls into a new configuration. The descriptions flow smoothly from the tongue. Good metaphors and comfortable descriptions are not scientific criteria. But they make a large difference in the psychological acceptance of the system and in the general utility of the ideas.

Levels of description. And what about the most common complaint of all: Who cares about this low-level stuff? I think that the answer is clear: all of us. Now, be careful about how this is interpreted. I, personally, am not particularly excited by much of the discussions of synaptic adjustment rules and the details of the operation of single neurons (and worse, single synapses). But careful attention to these low-level details is essential in understanding the appropriate primitive structures of cognition.

The whole point for the cognitive scientist is to understand cognition: perception, language, thought, social interaction, action, emotion. . . . To do so, we insist on explanation of the internal processing structures that give rise to cognitive activities. This is why we have spoken of representation, of the mechanisms of memory and attention, of the rule-structures of language, and of the plans, schemas, and beliefs of people. But in developing models of internal activity, we must adopt an appropriate set of primitive descriptors. So far, the primitives have been based upon symbols: symbol manipulating rules, production systems, control structures, and representational systems. How do we know these are the appropriate descriptors? We don't. The choice of primitive building blocks depends upon the choice of computational device. Selection of the PDP computational system changes things dramatically: New conceptualizations emerge, new primitive levels of description are required, new explanations of cognition are possible.

By attending to the seemingly uninteresting level of the adjustment of synaptic weights, we end up with a whole new conceptualization of how concepts are encoded. The point is that some low-level stuff makes large differences in the way that the entire system operates, which in turn makes huge differences in the types of high-level descriptions that are most important for the cognitive scientist. Who would have thought that a person interested in problem solving or thinking should be interested in energy states or measures of "harmony" among units at a "neural-like" level? But guess what? They turn out to be important. A focus upon these levels turns out to change the way we view cognition. The PDP approach forces a re-evaluation of the importance of what would otherwise have been thought to be low-level concepts.

Weaknesses of the PDP Approach

The approach has weaknesses. In part, it is hard to see exactly how to apply it to many of the difficult issues in the study of cognition. In general, the closer to perception or to motor output, the easier it is to see how the work applies. Thus, there seems to be no question about the significance for pattern recognition, for vision or speech understanding, or even for categorization. There have been a few studies that show promise for the control of motor movements (Hinton, 1984; Rumelhart & Norman, 1982). The question arises whether there are fundamental limitations in these models or whether the lack of examples simply represents the early stage of development and the initial concentration of effort upon the mathematics, the modeling techniques, and examples drawn from more restricted domains. The answers, of course, are not known.[3]

The type-token problem. Among the problems that have long bothered me about the PDP approach are some technical issues in

---

[3] Yes, it is possible to make a PDP network balance a pole (Barto, Sutton, & C. W. Anderson, 1983), but this is really an exercise in demonstrating learning, and the perceptual input and motor output required are, well, trivial. The study isn't trivial—it is a very nice demonstration—but the theme is the adjustment of weights, not perception, not motor control, and not decision or thought. In this section I examine several of these problems: the type-token problem, the problem with variables, the need for an extra, evaluative structure capable of overseeing and passing judgment on others, and then some general issues in learning and consciousness.

computation. There is a list of them, but the two most significant are the type-token distinction and the handling of variables. The type-token problem is to be able to handle different instances of the same concept, sometimes at the same time. Thus, if the system has the knowledge that "John eats a sandwich" and that "Helene eats a sandwich," the system has to treat these as different sandwiches. This capability is not easy for PDP systems. PDP networks are very good at representing general properties, properties that apply to classes of objects. This is where their power to generalize, to generate default values automatically, arises. But the complementary skill of keeping individual instances separate seems much harder.

The issue is addressed in the book; it is discussed in Chapter 3, "Distributed Representations," by Hinton, McClelland, and Rumelhart; in the McClelland and Rumelhart chapters on distributed memory and amnesia; and in McClelland's chapter on "The Programmable Blackboard Model of Reading" (and probably elsewhere that I missed). All this discussion is somewhat symptomatic of the difficulties: In traditional symbolic representations, for example, semantic network and frame-based approaches to representation, the issue is straightforward—the solution is natural. Here, however, considerable complications must be introduced to handle the issues, and even then it is not clear that the problem is entirely solved.

One question that should be raised, however, is whether we really wish the problem to be solved. It is interesting to note that in our typing model, Rumelhart and I concluded that the kinds of errors people make in typing require a system in which there is no type-token distinction at the level of control for the typing of individual letters (Rumelhart & Norman, 1982). The lack of distinction gave rise to some difficulties, but we argued that these are exactly the difficulties that humans face. McClelland makes a similar argument in his chapter on the blackboard model of reading. Thus, it may turn out that the problems faced by PDP systems in handling the type-token distinction are actually virtues with regard to their ability to model human cognition, especially if people have the same problems.

Variables. Once upon a time I worried about how the nervous system could have evolved to do all the things computational systems seemed required to do. Among them was the ability to call subroutines. The problem with calling subroutines was that it required variables: not only the values of the actions to be performed, but the name or symbol of the thing that had requested the subroutine—this is how the system would know where to resume operations once the subroutine was completed. Symbol processing systems are fundamentally variable-interpretation processes, and I couldn't see how they could have evolved. [4] Well, now along comes the PDP approach: In one sense, my problem is solved—there are no variables, save for activation values. All that is represented is particular instances. No need to represent variables: The variable is represented by the configuration of units that are active. But I still am not satisfied.

Now I worry about how PDP units can get away without representing variables. Aren't variables necessary? What about thought? If I solve a problem mentally don't I have to postulate hypothetical situations in my head, evaluate them, and make decisions? How do I plan my day, deciding which activities should be done first, which last, which require others to precede them, and which can be postponed? How do I compose music, or do science? Don't I need to have mental variables, symbols that I manipulate? My answer is "yes." I think this lack is a major deficiency in the PDP approach. I believe this problem can be solved by having several levels of systems, each specialized for different things. The PDP system is fine for perception and motor control, fine for categorization. It is possibly exactly the sort of system required for all of our automatic, subconscious reasoning. But I think that more is required—either more levels of PDP structures or other kinds of systems—to handle the problems of conscious, deliberate thought, planning, and problem solving.

The need for an extra, evaluative structure. A problem with the PDP models presented in this book is that they are too specialized, so concerned with solving the problem of the moment that they do not ask how the whole might fit together. The various chapters present us with different versions of a single, homogeneous structure, perfectly well-suited for doing its task, but not sufficient, in my opinion, at doing the whole task. One structure can't do the job: There do have to be several parts to the system that do different things, sometimes communicating with one another, sometimes not.

The argument seems especially relevant in considering learning. Though many of the learning rules are self-correcting, and therefore tend to converge in one way or another on optimal performance, they seem to be insufficiently sensitive to the global goals and evaluations made by the organism in which they are implemented. While it is now well accepted that the intention to learn, per se, is not a major determinant of learning, this intention mobilizes cognitive activities that result in better learning. Not much is said about these intentions, their

---

[4] Of course, I can't figure out how the bones and impedance-matching structures of the ears could have evolved, or the complex tendons that control the tens of bones in the hands could have evolved either: My lack of imagination is not restricted to information processing.

source, or the ways in which they influence the system's learning and performance.

When it comes to learning, it is frequently the case that something has to watch over the operations and act as a trainer. But this trainer is separate from the learning mechanisms. And it has to be able to evaluate the quality of performance. How does this take place? What is this second, overseeing mechanism? And how did it get started? How did the trainer know what task it was to train, and when? And how did it acquire the knowledge of what was good performance if it was a task that the person had never before performed? Even in the competitive learning environment (see Chapter 5 by Rumelhart & Zipser) in which learning can take place without an overseeing, evaluatory mechanism, it is often advantageous to "train" the system by careful presentation of the items that are to be classified. Thus, Rumelhart and Zipser "trained" their system on various stimulus situations—"teaching" is what they called it. But how does this come about? Who does the "teaching" or "training"? In some cases, it is another person, but not always.

It is often the case that complex behavior has to be monitored to see if things are proceeding well. This requires some sort of evaluation of the output of the system. But what mechanism does this evaluation? And what mechanism decides what the desired output is to be? I believe this requires an "overlooking" system, one that compares expectations with outcomes. This second, evaluative, system might very well also be a PDP structure, but it must be separate from the first, overlooking the first and operating upon it.

More than one system at a time is required. A second problem is that because any single PDP network can only settle into one state at a time, it is probably the case that multiple systems are required: modules of PDP systems, perhaps. Now, this is not really a criticism of the PDP model. In fact, the property that a PDP network reaches a single state at any moment is both interesting and important: It gives rise to some of the strong points of the model. The point is that although the system is highly parallel and very fast when viewed at the level of computational operations, it is highly serial and relatively slow when viewed at the "higher level" of interpreting and analyzing the resulting state changes. This dual perspective is a strong virtue for the modeling of human cognition, for this duality reflects current understanding. People interpret the world rapidly, effortlessly. But the development of new ideas, or evaluation of current thoughts proceeds slowly, serially, deliberately. People do seem to have at least two modes of operation, one rapid, efficient, subconscious, the other slow, serial, and conscious. The problem, however, is that people can do multiple activities at the same time, some of them quite unrelated to one another. So, in my opinion, a PDP model of the entire human information processing system is going to require multiple units. That is, the complete model requires that the brain consist of several (tens? hundreds? thousands?) of independent PDP-like systems, each of which can only settle into a single state at a time.

Learning and Consciousness

The two problems discussed in the previous section come together into a general statement of the problems of learning and consciousness. To be skilled at something requires that the requisite actions be performed smoothly, effortlessly, with minimum conscious thought or effort. In many ways, skilled performance consists of the execution of automated routines. The PDP structures seem like ideal candidates for mechanisms that produce these automated, subconscious behaviors. But how do these structures come about? What are the procedures for learning? This view suggests that we consider "learning" to be the programming of automated routines.

Under this view, when we learn some new task or skill, two different systems are involved. One involves conscious control and awareness of the task and the contingencies. Call this system DCC (for *deliberate conscious control*—after Norman & Shallice, in press). The other involves a PDP mechanism, a pre-established, preprogrammed set of procedures that get followed with minimum need for conscious control or even awareness.

The basic idea is this: Learning involves the setting of weights on the associative network structures of a PDP system. However, this weight setting requires some evaluative mechanism that determines when things are going well and when they are not, and perhaps other kinds of judgments that can be used to affect the weight setting and establishment of appropriate connections. This is one role for DCC. In addition, during the time while the weights are being established, before the PDP structure is properly set up, if the activity is to be performed at all, it must be through DCC. Hence, two roles for DCC are to control activities when the final "programming" has not yet been completed, and also to control the "programming." Norman and Shallice suggested that DCC only controls activities by influencing activation values. Among other things, this means that the control by PDP and DCC mechanisms are compatible, so that both may operate at the same time. Thus, partially learned activities may be under control of both modes. Conscious control of activities acts by activating or inhibiting the PDP networks, attempting to direct them into appropriate states, but without full power to cause such a result: Conscious control is simply one more source of activation and inhibition that adds and subtracts from the other influences on total activation values.

These views of learning and of direct conscious control are consistent with a number of qualitative phenomena:

- Control by DCC is slow and serial, hence not very effective for some kinds of skills. If DCC control is exerted after appropriate PDP mechanisms exist, then conscious control of activity will be inferior to subconscious control through the PDP mechanisms. Performance improves if done subconsciously, blocking conscious control. This is indeed the procedure taught by a number of athletic training methods: "Stop thinking about your actions and performance will improve."

- Introspection results from the interpretive attempts of deliberate conscious control—by the requirements of the "evaluator" or "trainer" for learning. Introspection is therefore essential for learning, but it is based upon observation of the outputs of a subconscious (PDP) system. As a result, introspections are only capable of accurate descriptions of system states. Because there is no information available about how that state was reached, introspection cannot give reasons for the resulting states. People usually do give reasons however, but these reasons are of necessity constructions and self-hypotheses: We have no conscious awareness of how PDP mechanisms settle into states and no vocabulary for describing the weights and synaptic interactions even were we to have that awareness.

- Consider the role of meaning and understanding in learning. Much has been made of the need for the learner to understand the task and activity. Yet, understanding does not seem relevant to the way that skilled performers operate. Moreover, many things we learn have arbitrary relations: the relation between the name of an object and the object, the order of the letters of the alphabet, the fingering of musical instruments, command sequences of computers. I believe the point is that PDP mechanisms can set up almost any arbitrary relationship. Hence, to the expert, once a skill has been acquired, meaningfulness of the relationships is irrelevant. However, the conscious, supervisory mechanisms require meaningful and understandable relationships: The DCC mechanisms must watch over actions and evaluate them, and for this, understandable and intelligible relationships are important. Understanding is also essential for troubleshooting, but in many ways, this is what the learning process is all about: troubleshooting the performance in order to establish the appropriate connections.

TOWARD A NEW UNDERSTANDING OF HUMAN INFORMATION PROCESSING

The work presented in this book provides the framework for new directions in our understanding of the human information processing system. In the older view, processing was done by separate components that communicated by passing messages—symbols—among themselves. The perceptual system communicated with the memory system. Memory communicated with perception and problem solving. Language systems communicated with the others. The communications consisted of symbolic messages, interpreted through a representational system, that implied a correspondence between the symbol and the referent and, therefore, an interpretive mechanism (see Rumelhart & Norman, in press). The emphasis was on symbolic representation, and hence the emphasis was on the rules of thought, problem solving, and memory. This led to the development of frame-based schema systems, with slots, fillers, default values, and inference mechanisms as well as to production-theory descriptions of human problem solving.

The new view is quite different. Under this new view, processing is done by PDP networks that configure themselves to match the arriving data with minimum conflict or discrepancy. The systems are always tuning themselves (adjusting their weights). Learning is continuous, natural, and fundamental to the operation. New conceptualizations are reflected by qualitatively different state configurations. Information is passed among the units, not by messages, but by activation values, by scalars not symbols. The interpretation of the processing is *not* in terms of the messages being sent but rather by what states are active. Thus, it is what units are active that is important, not what messages are sent. In the conventional system, learning takes place through changes in the representational structures, in the information contained in memory. In this new approach, learning takes place by changes in the system itself. Existing connections are modified, new connections are formed, old ones are weakened. In the conventional system, we distinguished between the information being processed and the processing structures. In the PDP system, they are the same: The information is reflected in the very shape, form, and operation of the processing structures.

A large number of issues are now naturally addressed that were difficult to deal with before. In turn, a large number of things that were easy to do before now are difficult. This is a combination of evolution and revolution. Many of our old ideas are still applicable. Many of the essential, critical problems are still problems. But we have a new set of tools, powerful tools that can help make major advances on a number of fronts. These new mechanisms are fundamentally subconscious ones. Learning results from the natural adjustments to states: It results naturally from the operation of the system. But in the process of examining what extra mechanisms are necessary to guide the learning and the overall behavior of the system, we appear to be led directly to consideration of the interacting roles of processing mechanisms and evaluative mechanisms, of subconscious and conscious processes. A nice fallout.

I still have concerns. I do not know how to map much of my understanding of cognition onto these new structures. I believe the type-token distinction is critical, that the cognitive system can and does manipulate variables, that there must be levels of processes, some of which oversee others. None of these problems seem insurmountable to me. Some may go away, being problems forced upon us (me) only because of old habits of thinking, old views of computation. Some may be solved by new developments. Some may get handled by an amalgamation of methods, combining the best of the more conventional mechanisms of human information processing with these newer views of information processing.

Some old questions are put to rest. To ask "why" a particular action was performed or "why" a particular interpretation was placed on an event means to ask "why" a given internal stable state was reached. But there is no simple answer to these questions. To force the questions is to risk confabulation: answers made up after the fact with no necessary connection to reality. In general, *there is no single reason why any given cognitive state occurs.* The system is multiply connected, multiply controlled. The states are a result of many different factors all impinging upon one another: The state is simply the best match to all the sources of information. This provides a rather nice resolution of many of the arguments that have taken place over the years within cognitive psychology.

ACKNOWLEDGMENTS

I thank Jay McClelland for thoughtful comments on the manuscript.

Future Directions

Parallel distributed processing has come a long way since we began our explorations. Four years ago, we had no clear idea how PDP models were going to overcome the limitations of the perceptron; how they could be related to higher level concepts such as the schema; or how they could be applied fruitfully to language processing. We think it is fair to say that the chapters in both volumes of this book document considerable progress on these and other problems. During this same period, work emerging from other research groups has also produced major advances. Yet, in spite of the progress, these explorations are not complete. The more we have learned, the more questions we have uncovered. Thus, we think of these volumes more as a progress report than as a final statement.

Even as the book is going to press, the work is continuing and expanding. Our own ideas are evolving, and many other researchers are beginning to join in these explorations, bringing new viewpoints and new issues with them. In the following few paragraphs we sketch what we see as the major emerging themes of the next several years of work—the most promising directions and the biggest challenges that lie ahead.

From where we stand, three general areas stand out as major foci of future work. These are:

- The application of parallel distributed processing to higher-level cognitive processes and sequential symbol processing.

- Further development of mechanisms of learning and a concommitant analysis of network architectures.

- The closer binding of PDP models with neuroscience.

We discuss each of these developments below.

Higher-Level Processes

We began our work on PDP models with applications that were more perceptual in nature and that focused more on those areas in which conventional sequential symbol-processing models have been least successful. We obviously believe this is only a first step. We believe that PDP models will provide useful insights into such higher-level cognitive processes as thinking and langauge processing. Chapters 14 and 19 offer our initial explorations of these areas. We do not mean to suggest that we believe these chapters provide all of the answers. Indeed, we have already begun to elaborate these ideas. More important, many other researchers have begun to appreciate the benefits of massively parallel processing in these areas, and are beginning to develop models that go beyond the first steps we have taken in these two chapters. Here we indicate a bit more of what we see as the general form these new directions will take.

Sequential symbol processing. It is becoming obvious to many others besides ourselves that parallel processing and sequential symbol processing must somehow occur in the same system. There appear to be three major schemes for integrating these two processing modalities. One of these is our approach to the problem: to develop PDP models in which sequential behavior is captured in the successive settlings of a parallel network or set of networks. Touretzky and Hinton (1985), have been working somewhat along the same lines, developing a PDP production system. They hope eventually to do variable binding and instantiation and, at the same time, preserve the benefits of parallel distributed processing in their symbol-processing machine.

Beyond this approach, there appear to be two other approaches to what might be called *massive parallelism in symbol processing*. The first of these, suggested by Norman in Chapter 26, is the idea of a hybrid system—a PDP net or set of nets, coupled with a symbol processing controller. A similar approach is taken by Schneider (1985) in his recent model of controlled and automatic processes in attention. A slight variant of this approach is to embed calls to parallel modules inside a conventional sequential-processing system.

We applaud this kind of approach since it strives to preserve the best of both worlds. It also captures the fact that the cognitive system is complex and consists of many parts—a point that no one would dispute. We do not take this road ourselves, in part because we see no reason to suppose that the mechanisms that control cognitive processing are not themselves constructed of the same underlying parallel hardware as the other aspects of the cognitive systems, and in part because we prefer to view the system not so much in terms of controlled and controlling modules, but in terms of more distributed forms of control.

The final approach is to modify other computational frameworks so that they incorporate many of the features of PDP models. This approach is exemplified by John R. Anderson's (1983) production system implementation of our interactive activation model, and Thibadeau, Just, and Carpenter's (1982) CAPS production system.

These developments are signs that other researchers, coming from other computational viewpoints, have begun to appreciate many of the benefits of interactive parallel processing. By building parallel capabilities directly into production systems, however (that is, by effectively allowing large numbers of productions to partially match at the same time), these models have some tendency of becoming top-heavy. It remains to be seen whether the direct computational benefits these models have (e.g., explicit, direct variable binding of many productions at the same time) will turn out to outweigh the extra computational overhead they introduce. Our own hope is that we can succeed in achieving models that capture the extent of human capabilities more naturally as emergent properties of parallel networks, but in the meantime, these massively parallel production systems provide another way to bring parallel distributed processing together with the sequential symbol-processing tradition in cognitive science.

Language processing. We began our work on PDP models with a strong interest in linguistic information processing. It was an attempt to deal with the problems we saw with more conventional accounts of language that contributed to our original attraction to PDP models. Some of the basic directions of our account of language are indicated in Chapters 18 and 19. We see the fleshing out of these sketches as central to our understanding of language acquisition, syntactic processing, and the interaction between syntactic and semantic processing.

Although it is not yet represented directly in our work, we believe that our approach will offer new insight into the role of metaphor and analogy in language. Chapter 19 begins to indicate that the notion of literal meaning, like the notions of grammaticality in syntax and regularity in morphology, will turn out to be a matter of degree and will best be explained in terms of coalitions of units interacting to produce the observed regularities. The notions of figurative and literal meanings will best be seen as coarse categories describing the nature of the meanings synthesized by PDP networks rather than fundamentally different meaning types arising from fundamentally different processes. Meanwhile, it is clear to everyone that language processing provides many challenges and many opportunities to test and elaborate our developing conception of how to build powerful processing machinery out of simple computational elements, while at the same time preserving the beneficial aspects of parallel distributed processing.

Learning and Architecture

Learning has always been a key issue in the development of network models. One of the most appealing features of PDP networks is that very simple, homogeneous learning procedures can be used to allow the networks to self-modify and adapt to their environments. Until very recently, these self-organizing systems have been of limited complexity. No multilayer learning algorithm has existed. Two such learning procedures have been developed in this book (Chapters 7 and 8) and others are under development elsewhere (cf. Barto & Anandan, 1985). These learning schemes are already generating a large amount of activity. With the development of the error propagation learning procedure, for example (Chapter 8), we can study the kinds of internal representations that evolve to support processing in particular problems and, in that way, come to understand better the problems under study as well as the nature of the networks required to solve them. Moreover, we can use these powerful learning procedures for genuine artificial intelligence—that is, as a tool for designing systems capable of discovering new ways of solving difficult problems.

Neuroscience

In spite of the fact that PDP models exploit brain-style processing, our own work has not yet moved far toward making direct contact with neuroscience. We believe that the opportunity is there, however, and that the future will bring more and more models specifically designed with data from neuroscience in mind. There are two natural areas of application—the application to data from people with brain pathologies and the application of PDP models to detailed modeling of neurophysiological and neuroanatomical data.

Neuropsychology. The study of cognitive deficits that arise from brain damage has focused on dissociations of function. If one patient shows one deficit, and another shows a different deficit, it is reasonable to infer that their different pathologies damaged different, separable components of the cognitive system. From this kind of data, one can try to characterize the separate components of the cognitive system and to identify the role each component plays in the functioning of the system as a whole. This approach has been successful in leading to several important insights about the macrostructure of the cognitive system. But there has been little work, until now, that focuses on the development of models that attempt to characterize the microstructure of the components in such as way as to provide an account of the way in which performance degrades with damage. Cognitive dysfunction is a matter of degree, so different patients can manifest widely different degrees of what appears to be a common deficit. Mild cases often show their deficits in subtle ways, and make relatively subtle errors—confusing similar concepts but not confusing distinct concepts. Some pathologies that produce highly diffuse damage (e.g., Alzheimer's disease) are progressive and increase gradually in severity as the disease process continues.

PDP models hold out considerable promise of providing detailed accounts of the way in which cognitive function degrades with damage. This point has been made by J. A. Anderson (1977; Anderson, Silverstein, Ritz, & Jones, 1977) and by Wood (1978), and models of aspects of neuropsychological phenomena can be found in two places in this book (Chapters 3 and 25). Much more can now be done, we think, to further the application of PDP models to neuropsychological phenomena. Equally important, neuropsychological data can, in some cases, help us make decisions about the macrostructure of the cognitive systems we construct out of PDP networks. Thus, there is hope that we will soon see real synergistic interactions between neuropsychology and parallel modeling.

Physiology and anatomy. Our primary focus has been on "cognitive" data, but the language of PDP is, by design, well suited as a medium for actual neural modeling. We expect that as these theoretical tools become more familiar and better understood, they will prove more and more useful for organizing data from neuroscience. One clear example of this lies in the paper by Gluck and Thompson (in press), in which PDP modeling techniques are applied to an analysis of conditioning in the simple nervous system of *Aplysia*, as studied by Kandel and his collaborators (Hawkins, Abrams, Carew, & Kandel, 1983; Kandel & Schwartz, 1982). Although we are a long way from connecting our more abstract networks with particular brain structures, we believe that the application of PDP models to cognitive and neural modeling enterprises will serve to both clarify the specifically neuroscience questions and lead us to a better understanding of the actual neural substructure underlying cognitive processes in the brain.

In addition to the neurophysiology of the brain, we expect that neuroanatomical facts will have an increasing impact on the way we build our PDP models. We have learned that the major determiner of what a particular system learns is largely a function of the possible connections. It is the anatomy that determines the possible connections. It seems likely that in subsystems in which the anatomy is relatively well understood, as for example in the visual system, that we can use the known anatomical structure in our models to constrain them and thereby explain why information is processed in the way that it is.

The four years we have spent working on these volumes have been exciting ones for all of us. We have learned much. It is our hope in publishing these volumes that they will communicate some of this sense of excitement, and that at least some of our readers will take up the problems and issues where we have left them, will help us correct some of our errors and misconceptions, and will help advance our understanding of the microstructure of cognition.

David E. Rumelhart
James L. McClelland

References

Abeles, M. (1982). *Studies of brain function: Vol. 6. Local cortical circuits.* New York: Springer-Verlag.

Abeles, M., de Riboupierre, F., & de Riboupierre, Y. (1983). Detection of single unit reponses which are loosely time-locked to a stimulus. *IEEE Transactions on Systems, Man, and Cybernetics, 13,* 683-691.

Allman, J., Miezen, J., & McGuinness, E. (1985). Stimulus specific responses from beyond the classical receptive field: Neurophysiological mechanisms for local-global comparisons in visual neurons. *Annual Review of Neuroscience, 8,* 407-430.

Amari, S., & Takeuchi, A. (1978). Mathematical theory on formation of category detecting nerve cells. *Biological Cybernetics, 29,* 127-136.

Andén, N. E., Dahlström, A., Fuxe, K., Larsson, K., Olson, L., & Ungerstredt, U. (1966). Ascending monoamine neurons to the telencephalon and diencephalon. *Acta Physiologica Scandinanvica, 67,* 313-326.

Andersen, P., Brooks, C. McC., Eccles, J. C., & Sears, T. A. (1964). The ventro-basal nucleus of the thalamus: Potential fields, synaptic transmission and excitability of both presynaptic and postsynaptic components. *Journal of Physiology, 174,* 348-369.

Anderson, J. A. (1973). A theory for the recognition of items from short memorized lists. *Psychological Review, 80,* 417-438.

Anderson, J. A. (1977). Neural models with cognitive implications. In D. LaBerge & S. J. Samuels (Eds.), *Basic processes in reading: Perception and comprehension* (pp. 27-90). Hillsdale, NJ: Erlbaum.

Anderson, J. A. (1983). Cognitive and psychological computation with neural models. *IEEE Transactions on Systems, Man, and Cybernetics, SMC-13,* 799-815.

Anderson, J. A., & Hinton, G. E. (1981). Models of information processing in the brain. In G. E. Hinton & J. A. Anderson (Eds.), *Parallel models of associative memory* (pp. 9-48). Hillsdale, NJ: Erlbaum.

Anderson, J. A., Silverstein, J. W., Ritz, S. A., & Jones, R. S. (1977). Distinctive features, categorical perception, and probability learning: Some applications of a neural model. *Psychological Review, 84*, 413-451.

Anderson, J. R. (1983). *The architecture of cognition.* Cambridge, MA: Harvard University Press.

Anderson, J. R., & Ross, B. H. (1980). Evidence against a semantic-episodic distinction. *Journal of Experimental Psychology: Human Learning and Memory, 6*, 441-465.

Anlezark, G. M., Crow, T. J., & Greenway, A. P. (1973). Impaired learning and decreased cortical norepinephrine after bilateral locus coeruleus lesions. *Science, 181*, 682-684.

Asanuma, H., & Rosén, I. (1972). Topographical organization of cortical efferent zones projecting to distal forelimb muscles in the monkey. *Experimental Brain Research, 14*, 243-256.

Bagley, W. C. (1900). The apperception of the spoken sentence: A study in the psychology of language. *American Journal of Psychology, 12*, 80-130.

Baker, J. F., Petersen, S. E., Newsome, W. T., & Allman, J. (1981). Visual response properties of neurons in four extrastriate visual areas of the owl monkey (Aotus trivirgatus): A quantitative comparison of the medial (M), dorsomedial (DM), dorsolateral (DL), and middle temporal (MT) areas. *Journal of Neuropsychology, 45*, 397-406.

Ballard, D. H. (1984). Parameter nets. *Artificial Intelligence, 22*, 235-267.

Ballard, D. H. (in press). Cortical connections and parallel processing: Structure and function. *Behavioral and Brain Sciences.*

Ballard, D. H., Hinton, G. E., & Sejnowski, T. J. (1983). Parallel visual computation. *Nature, 306*, 21-26.

Barlow, H. B. (1972). Single units and sensation: A neuron doctrine for perceptual psychology? *Perception, 1*, 371-394.

Barrow, H. G., & Tenenbaum, J. M. (1978). In A. R. Hanson & E. M. Riseman (Eds.), *Computer vision systems* (pp. 3-26). New York: Academic Press.

Bartlett, F. C. (1932). *Remembering.* Cambridge: Cambridge University Press.

Barto, A. G., & Anandan, P. (1985). Pattern recognizing stochastic learning automata. *IEEE Transactions on Systems, Man, and Cybernetics, 15*, 360-375.

Barto, A. G., Sutton, R. S., & Anderson, C. W. (1983). Neuronlike adaptive elements that can solve difficult learning control problems. *IEEE Transactions on Systems, Man, and Cybernetics, 13*, 835-846.

Bates, E. (1979). *Emergence of symbols.* New York: Academic Press.

Bates, E., & MacWhinney, B. (in press). Competition, variation, and language learning: What is not universal in language acquisition. In B. MacWhinney (Ed.), *Mechanisms of language acquisition.* Hillsdale, NJ: Erlbaum.

Bear, M. F., Paradiso, M. A., Schwartz, M., Nelson, S. B., Carnes, K. M., & Daniels, J. D. (1983). Two methods of catecholamine depletion in kitten visual cortex yield different effects on plasticity. *Nature, 302*, 245-247.

Becker, J. T., Walker, J. A., & Olton, D. S. (1980). Neuroanatomical bases of spatial memory. *Brain Research, 200*, 307-320.

Berko, J. (1958). The child's learning of English morphology. *Word, 14,* 150-177.

Bever, T. G. (1970). The cognitive basis for linguistic structures. In J. R. Hayes (Ed.), *Cognition and the development of language.* New York: Wiley.

Bienenstock, E. L., Cooper, L. N., & Munro, P. W. (1982). Theory for the development of neuron selectivity: Orientation specificity and binocular interaction in visual cortex. *Journal of Neuroscience, 2,* 32-48.

Blakemore, C., Garey, L. J., & Vital-Durand, F. (1978). The physiological effects of monocular deprivation and their reversal in the monkey's visual cortex. *Journal of Physiology, 283,* 223-262.

Blakemore, C., Hawken, M. J., & Mark, R. F. (1982). Brief monocular deprivation leaves subthreshold synaptic input on neurones of the cat's visual cortex. *Journal of Physiology, 327,* 489-505.

Blakemore, C., & Van Sluyters, R. C. (1974). Reversal of the physiological effects of monocular deprivation in kittens: Further evidence for a sensitive period. *Journal of Physiology, 237,* 195-216.

Blasdel, G. G., & Pettigrew, J. D. (1978). Effect of prior visual experience on cortical recovery from the effects of unilateral eyelid suture in kittens. *Journal of Physiology, 274,* 601-619.

Blasdel, G. G., & Pettigrew, J. D. (1979). Degree of interocular synchrony required for maintenance of binocularity in kitten's visual cortex. *Journal of Neurophysiology, 42,* 1692-1710.

Bobillier, P., Seguin, S., Petitjean, F., Salvert, D., Touret, M., & Jouvet, M. (1976). The raphé nuclei of the cat brain stem: A topographical atlas of their efferent projections as revealed by autoradiography. *Brain Research, 113,* 449-486.

Bobrow, D. G., & Norman, D. A. (1975). Some principles of memory schemata. In D. G. Bobrow & A. Collins (Eds.), *Representation and understanding: Studies in cognitive science* (pp. 131-149). New York: Academic Press.

Bond, Z. S., & Garnes, S. (1980). Misperceptions of fluent speech. In R. Cole (Ed.), *Perception and production of fluent speech.* Hillsdale, NJ: Erlbaum.

Brachman, R. J., & Schmolze, J. G. (1985). An overview of the KL-ONE knowledge representation system. *Cognitive Science, 9,* 171-216.

Broadbent, D. E. (1967). Word frequency effect and response bias. *Psychological Review, 74,* 1-15.

Broadbent, D. E. (1985). A question of levels: Comment on McClelland and Rumelhart. *Journal of Experimental Psychology: General, 114,* 189-192.

Brock, L. G., Coombs, J. S., & Eccles, J. C. (1952a). The nature of the monosynaptic excitatory and inhibitory processes in the spinal cord. *Proceedings of the Royal Society of London, Series B, 140,* 169-176.

Brock, L. G., Coombs, J. S., & Eccles, J. C. (1952b). The recording of potentials from motoneurons with an intracellular electrode. *Journal of Physiology, 117,* 431-460.

Brodmann, K. (1909). *Vergleichende localisationslehre der grosshirnrinde in ihren prinzipien dargestellt auf grund des zellenbaues [Principles of comparative localization in the cerebral cortex presented on the basis of cytoarchitecture].* Leipzig: Barth.

Brooks, L. R. (1978). Nonanalytic concept formation and memory for instances. In E. Rosch & B. B. Lloyd (Eds.), *Cognition and categorization.* Hillsdale, NJ: Erlbaum.

Brooks, L. R., Jacoby, L. L., & Whittlesea, B. W. A. (1985). *The influence of specific familiarity on picture identification.* Manuscript submitted for publication.

Brown, R. (1973). *A first language.* Cambridge, MA: Harvard University Press.

Bruce, C., Desimone, R., & Gross, C. G. (1981). Visual properties of neurons in a polysensory area in superior temporal sulcus of the macaque. *Journal of Neurophysiology, 46,* 369-384.

Buisseret, P., Gary-Bobo, E., & Imbert, M. (1978). Ocular motility and recovery of orientational properties of visual cortical neurones in dark-reared kittens. *Nature, 272,* 816-817.

Bybee, J. L., & Slobin, D. I. (1982). Rules and schemas in the development and use of the English past tense. *Language, 58,* 265-289.

Cajal, S. Ramón y (1892). *Die retina der wirbelthiere [The retina of vertebrates].* Wiesbaden: Bergman.

Cajal, S. Ramón y (1911). *Histologie du systeme nerveux de l'homme et des vertébrés T. 2. [Histology of the nervous system of man and of vertebrates, Vol. 2].* Paris: Maloine.

Carman, J. B., Cowan, W. M., & Powell, T. P. S. (1964). The cortical projections upon the claustrum. *Journal of Neurology, Neurosurgery, & Psychiatry, 27,* 46-51.

Changeux, J. -P., Heidmann, T., & Patte, P. (1984). Learning by selection. In P. Marler & H. S. Terrace (Eds.), *The biology of learning* (pp. 115-133). Berlin: Springer-Verlag.

Chapman, R. A. (1966). The repetitive responses of isolated axons from the crab, *Carcinus maenas. Journal of Experimental Biology, 45,* 475-488.

Charniak, E. (1983). Passing markers: A theory of contextual influence in language comprehension. *Cognitive Science, 7,* 171-190.

Cohen, M. A., & Grossberg, S. (1983). Absolute stability of global pattern formation and parallel memory storage by competitive neural networks. *IEEE Transactions on Systems, Man, and Cybernetics, 13,* 815-825.

Cohen, N. J. (1981). *Neuropsychological evidence for a distinction between procedural and declarative knowledge in human memory and amnesia.* Unpublished doctoral dissertation, University of California, San Diego.

Cohen, N. J., & Squire, L. R. (1980). Preserved learning and retention of pattern-analyzing skill in amnesia: Dissociation of knowing how and knowing that. *Science, 210,* 207-210.

Cohen, N. J., Eichenbaum, H., Deacedo, B. S., & Corkin, S. (1985). Different memory systems underlying acquisition of procedural and declarative knowledge. *Annals of the New York Academy of Sciences, 444,* 54-71.

Cole, K. S. (1968). *Membranes, ions, and impulses.* Berkeley: University of California Press.

Cole, R. A. (1973). Listening for mispronunciations: A measure of what we hear during speech. *Perception & Psychophysics, 13,* 153-156.

Cole, R. A., & Jakimik, J. (1978). Understanding speech: How words are heard. In G. Underwood (Ed.), *Strategies of information processing.* New York: Academic Press.

Cole, R. A., & Jakimik, J. (1980). A model of speech perception. In R. A. Cole (Ed.), *Perception and production of fluent speech.* Hillsdale, NJ: Erlbaum.

Cole, R. A., & Rudnicky, A. (1983). What's new in speech perception? The research and ideas of William Chandler Bagley, 1874-1946. *Psychological Review, 90,* 94-101.

Colonnier, M. (1968). Synaptic patterns on different cell types in the different laminae of the cat visual cortex: An electron microscope study. *Brain Research, 9,* 268-287.

Conel, J. LeRoy (1939). *The postnatal development of the human cerebral cortex: Vol. 1. The cortex of the newborn.* Cambridge: Harvard University Press.

Connors, B. W., Gutnick, M. J., & Prince, D. A. (1982). Electrophysiological properties of neocortical neurons in vitro. *Journal of Neurophysiology, 48,* 1302-1320.

Cooper, L. N., Lieberman, F., & Oja, E. (1979). A theory for the acquisition and loss of neuron specificity in visual cortex. *Biological Cybernetics, 33,* 9-28.

Cotton, S., & Grosjean, F. (1984). The gating paradigm: A comparison of successive and individual presentation formats. *Perception & Psychophysics, 35,* 41-48.

Cottrell, G. (1985). *A connectionist approach to word sense disambiguation* (TR-154). Rochester, NY: University of Rochester, Department of Computer Science.

Cottrell, G., & Small, S. (1983). A connectionist scheme for modeling word sense disambiguation. *Cognition and Brain Theory, 1,* 89-120.

Crain, S., & Steedman, M. (1985). On not being led up the garden-path: The use of context by the psychological parser. In D. Dowty, L. Karttunen, & A. Zwicky (Eds.), *Natural language parsing.* Cambridge, England: Cambridge University Press.

Crick, F. H. C. (1984). The function of the thalamic reticular complex: The searchlight hypothesis. *Proceedings of the National Academy of Sciences, USA, 81,* 4586-4590.

Crow, T. J. (1968). Cortical synapses and reinforcement: A hypothesis. *Nature, 219,* 736-737.

Crowder, R. G. (1976). *Principles of learning and memory.* Hillsdale, NJ: Erlbaum.

Crowder, R. G. (1978). Mechanisms of auditory backward masking in the stimulus suffix effect. *Psychological Review, 85,* 502-524.

Crowder, R. G. (1981). The role of auditory memory in speech perception and discrimination. In T. Myers, J. Laver, & J. Anderson (Eds.), *The cognitive representation of speech* (pp. 167-179). New York: North-Holland.

Cynader, M. (1983). Prolonged sensitivity to monocular deprivation in dark-reared cats: Effects of age and visual exposure. *Developmental Brain Research, 8,* 155-164.

Cynader, M., Berman, N., & Hein, A. (1976). Recovery of function in cat visual cortex prolonged deprivation. *Experimental Brain Research, 25,* 131-137.

Cynader, M., Timney, B. N., & Mitchell, D. E. (1980). Period of susceptibility of kitten visual cortex to the effects of monocular deprivation extends beyond six months of age. *Brain Research, 191,* 545-550.

Daniel, P. M., & Whitteridge, D. (1961). The representation of the visual field on the cerebral cortex in monkeys. *Journal of Physiology, 159,* 203-221.

Davis, E. (1981). *Organizing spatial knowledge* (Research Rep. No. 193). New Haven, CT: Yale University, Department of Computer Science.

Daw, N. W., & Wyatt, H. J. (1976). Kittens reared in a unidirectional environment: Evidence for a critical period. *Journal of Physiology, 257,* 155-170.

Denes, P. (1955). Effect of duration on the perception of voicing. *Journal of the Acoustical Society of America, 27,* 761-764.

Desimone, R., Albright, T. D., Gross, C. G., & Bruce, C. J. (1980). Responses of inferior temporal neurons to complex visual stimuli. *Society for Neuroscience Abstracts, 6,* 581.

Desimone, R., Schein, S. J., Moran, J., & Ungerleider, L. G. (1985). Contour, color, and shape analysis beyond the striate cortex. *Vision Research, 25,* 441-452.

Dowling, J. E., & Boycott, B. B. (1966). Organization of the primate retina: Electron microscopy. *Proceedings of the Royal Society of London, Series B, 166,* 80-111.

Dräger, U. C. (1978). Observations on monocular deprivation in mice. *Journal of Neurophysiology, 41,* 28-42.

Duda, R. O., & Hart, P. E. (1973). *Pattern classification and scene analysis.* New York: Wiley.

Durham, D., & Woolsey, T. A. (1984). Effects of neonatal whisker lesions on mouse central trigeminal pathways. *Journal of Comparative Neurology, 223,* 424-447.

Eccles, J. C., Llinás, R., & Sakaski, K. (1966). The excitatory synaptic action of climbing fibres on the Purkinje cells of the cerebellum. *Journal of Physiology, 182,* 268-296.

Edelman, G. M. (1981). Group selection as the basis for higher brain function. In F. O. Schmitt, F. G. Worden, G. Adelman, & S. G. Dennis (Eds.), *The organization of cerebral cortex.* Cambridge, MA: MIT Press.

Eimas, P. D. (1978). Developmental aspects of speech perception. In R. Held, H. W. Liebowitz, & H.-L. Teuber (Eds.), *Handbook of sensory physiology: Vol. 8. Perception* (pp. 357-374). Berlin: Springer.

Elio, R., & Anderson, J. R. (1981). The effects of category generalizations and instance similarity on schema abstraction. *Journal of Experimental Psychology: Human Learning and Memory, 7,* 397-417.

Elman, J. L. (1983). [Phonotactic rule effects]. Unpublished data.

Elman, J. L., & McClelland, J. L. (1984). The interactive activation model of speech perception. In N. Lass (Ed.), *Language and speech* (pp. 337-374). New York: Academic Presss.

Elman, J. L., & McClelland, J. L. (in press). Exploiting the lawful variability in the speech wave. In J. S. Perkell & D. H. Klatt (Eds.), *Invariance and variability of speech processes*. Hillsdale, NJ: Erlbaum.

Emson, P. C., & Hunt, S. P. (1981). Anatomical chemistry of the cerebral cortex. In F. O. Schmitt, F. G. Worden, G. Adelman, & S. G. Dennis (Eds.), *The organization of the cerebral cortex* (pp. 325-345). Cambridge, MA: MIT Press.

Erman, L. D., & Lesser, U. R. (1980). The Hearsay-II speech understanding system: A tutorial. In W. A. Lea (Ed.), *Trends in speech recognition* (pp. 361-381). Englewood Cliffs, NJ: Prentice-Hall.

Ervin, S. (1964). Imitation and structural change in children's language. In E. Lenneberg (Ed.), *New directions in the study of language*. Cambridge, MA: MIT Press.

Fairén, A., & Valverde, F. (1980). A specialized type of neuron in the visual cortex of cat: A Golgi and electron microscope study of chandelier cells. *Journal of Comparative Neurology, 194*, 761-779.

Famiglietti, E. V. (1970). Dendro-dendritic synapses in the lateral geniculate nucleus of the cat. *Brain Research, 20*, 181-191.

Fanty, M. (1985). *Context-free parsing in connectionist networks* (TR-174). Rochester, NY: University of Rochester, Department of Computer Science.

Feldman, J. A. (1981). A connectionist model of visual memory. In G. E. Hinton & J. A. Anderson (Eds.), *Parallel models of associative memory* (pp. 49-81). Hillsdale, NJ: Erlbaum.

Feldman, J. A., & Ballard, D. H. (1982). Connectionist models and their properties. *Cognitive Science, 6*, 205-254.

Feldman, M. L., & Peters, A. (1978). The forms of nonpyramidal neuron in the visual cortex of the rat. *Journal of Comparative Neurology, 179*, 761-794.

Feustel, T. C., Shiffrin, R. M., & Salasoo, A. (1983). Episodic and lexical contributions to the repetition effect in word identification. *Journal of Experimental Psychology: General, 112*, 309-346.

Fillmore, C. J. (1968). The case for case. In E. Bach & R. T. Harms (Eds.), *Universals in linguistic theory*. New York: Holt, Rinehart, & Winston.

Fodor, J. A., Bever, T. G., & Garrett, M. F. (1974). *The psychology of language*. New York: McGraw-Hill.

Foley, J. D., & Van Dam, A. (1983). *Fundamentals of interactive computer graphics*. Reading, MA: Addison-Wesley.

Ford, M., Bresnan, J., & Kaplan, R. M. (1982). A competence-based theory of syntactic closure. In J. Bresnan (Ed.), *The mental representation of grammatical relations*. Cambridge, MA: MIT Press.

Foss, D. J., & Blank, M. A. (1980). Identifying the speech codes. *Cognitive Psychology, 12*, 1-31.

Foss, D. J., & Gernsbacher, M. A. (1983). Cracking the dual code: Toward a unitary model of phoneme identification. *Journal of Verbal Learning and Verbal Behavior, 22*, 609-633.

Fowler, C. A. (1984). Segmentation of coarticulated speech in perception. *Perception & Psychophysics, 36*, 359-368.

Fox, R. A. (1982). Unpublished manuscript, Vanderbilt University.

Fox, R. A. (1984). Effect of lexical status on phonetic categorization. *Journal of Experimental Psychology: Human Perception and Performance, 10*, 526-540.

Frazier, L., & Rayner, K. (1982). Making and correcting errors during sentence comprehension: Eye movements in the analysis of structurally ambiguous sentences. *Cognitive Psychology, 14*, 178-210.

Freeman, R. D., & Bonds, A. B. (1979). Cortical plasticity in monocularly deprived immobilized kittens depends on eye movement. *Science, 206*, 1093-1095.

Freeman, R. D., Sclar, G., & Ohzawa, I. (1982). Cortical binocularity is disrupted by strabismus more slowly than by monocular deprivation. *Developmental Brain Research, 3*, 311-316.

Friedlander, M. J., Lin, C.-S., Stanford, L. R., & Sherman, S. M. (1981). Morphology of functionally identified neurons in lateral geniculate nucleus of the cat. *Journal of Neurophysiology, 46*, 80-129.

Fry, D. B., Abramson, A. S., Eimas, P. D., & Liberman, A. M. (1962). The identification and discrimination of synthetic vowels. *Language & Speech, 5*, 171-189.

Fujimura, O., & Lovins, J. B. (1982). Syllables as concatenative phonetic units. In A. Bell & J. B. Hooper (Eds.), *Syllables and segments* (pp. 107-120). Amsterdam: North-Holland.

Fujisaki, H., & Kawashima, T. (1968, August). *The influence of various factors on the identification and discrimination of synthetic vowel sounds.* Paper presented at the Sixth International Congress on Acoustics, Tokyo, Japan.

Fuxe, K., Hökfelt, T., Said, S., & Mutt, V. (1977). Vasoactive intestinal polypeptide and the nervous system: Immunohistochemical evidence for localization in central and peripheral neurons particularly intracortical neurons of the cerebral cortex. *Neuroscience Letters, 5*, 241-246.

Ganong, W. F. (1980). Phonetic categorization in auditory word perception. *Journal of Experimental Psychology: Human Perception and Performance, 6*, 110-115.

Garey, L. J., & Powell, T. P. S. (1971). An experimental study of the termination of the lateral geniculo-cortical pathway in the cat and monkey. *Proceedings of the Royal Society of London, Series B, 179*, 41-63.

Gati, I., & Tversky, A. (1982). Representations of qualitative and quantitative dimensions. *Journal of Experimental Psychology: Human Perception and Performance, 8*, 325-340.

Georgopoulos, A. P., Caminiti, R., Kalaska, J. F., & Massey, T. (1983). Spatial coding of movement: A hypothesis concerning the coding of movement direction by motor cortical populations. *Experimental Brain Research, Suppl. 1.*

Gerstein, G. L., Bloom, M. J., Espinosa, I. E., Evanczuk, S., & Turner, M. R. (1983). Design of a laboratory for multineuron studies. *IEEE Transactions on Systems, Man, and Cybernetics, 13*, 668-676.

Gilbert, C. D. (1983). Microcircuitry of the visual cortex. *Annual Review of Neuroscience, 6*, 217-247.

Gilbert, C. D., & Kelly, J. P. (1975). The projection of cells in different layers of the cat's visual cortex. *Journal of Comparative Neurology, 163*, 81-105.

Gilbert, C. D., & Wiesel, T. N. (1981). Projection bands in visual cortex. *Society for Neuroscience, Abstract, 7*, 356.

Gilbert, C. D., & Wiesel, T. N. (1983). Clustered intrinsic connections in cat visual cortex. *Journal of Neuroscience, 3*, 1116-1133.

Gluck, M. A., & Thompson, R. F. (in press). Modeling the neural substrates of associative learning and memory: A computational approach. *Psychological Review.*

Glushko, R. J. (1979). The organization and activation of orthographic knowledge in reading aloud. *Journal of Experimental Psychology: Human Perception and Performance, 5*, 674-691.

Gray, E. G. (1959). Axo-somatic and axo-dendritic synapses of the cerebral cortex: An electron microscope study. *Journal of Anatomy, 93*, 420-433.

Green, D. M., & Swets, J. A. (1966). *Signal detection theory and psychophysics.* New York: Wiley.

Grinvald, A. (1985). Real-time optical mapping of neuronal activity: From single growth cones to the intact mammalian brain. *Annual Review of Neuroscience, 8*, 263-305.

Grosjean, F. (1980). Spoken word recognition processes and the gating paradigm. *Perception & Psychophysics, 28*, 267-283.

Grosjean, F. (1985). *The recognition of a word after its acoustic offset: Evidence and implications.* Unpublished manuscript, Northeastern University.

Grosjean, F., & Gee, J. (1984). *Another view of spoken word recognition.* Unpublished manuscript, Northeastern University.

Grossberg, S. (1976). Adaptive pattern classification and universal recoding: Part 1. Parallel development and coding of neural feature detectors. *Biological Cybernetics, 23*, 121-134.

Grossberg, S. (1980). How does the brain build a cognitive code? *Psychological Review, 87*, 1-51.

Grossberg, S. (1982). *Studies of mind and brain.* Dordrecht, Holland: Reidel.

Guillery, R. W. (1969). The organization of synaptic interconnections in the laminae of the dorsal lateral geniculate nucleus of the cat. *Zeitschrift für Zellforschung und Mikroskopische Anatomie, 96*, 1-38.

Halle, M., & Stevens, K. N. (1964). Speech recognition: A model and a program for research. In J. A. Fodor & J. J. Katz (Eds.), *The structure of language: Readings in the psychology of language.* Englewood Cliffs, NJ: Prentice-Hall.

Harik, S. J., Duckrow, R. B., LaManna, J. C., Rosenthal, M., Sharma, V. K., & Banerjee, S. P. (1981). Cerebral compensation for chronic noradrenergic denervation induced by locus coerulus lesion: Recovery of receptor binding, isoproterenol-induced adenylate cyclase activity, and oxidative metabolism. *Journal of Neuroscience, 1*, 641-649.

Hawkins, R. D., Abrams, T. W., Carew, T. J., & Kandel, E. R. (1983). A cellular mechanism of classical conditioning in Aplysia: Activity dependent amplification of presynaptic facilitation. *Science, 219*, 400-404.

Henneberg, R. (1910). Messung der Oberflachenausdehnung der Grosshirnrinde [Measurement of the surface area of the cerebral cortex]. *Journal für Psychologie und Neurologie, 17*, 144-158.

Hinton, G. E. (1981a). Implementing semantic networks in parallel hardware. In G. E. Hinton & J. A. Anderson (Eds.), *Parallel models of associative memory* (pp. 161-188). Hillsdale, NJ: Erlbaum.

Hinton, G. E. (1981b). A parallel computation that assigns canonical object-based frames of reference. *Proceedings of the 7th International Joint Conference on Artificial Intelligence*, 683-685.

Hinton, G. E. (1981c). Shape representation in parallel systems. *Proceedings of the 7th International Joint Conference on Artificial Intelligence*, 1088-1096.

Hinton, G. E. (1984). Parallel computations for controlling an arm. *Journal of Motor Behavior, 16*, 171-194.

Hinton, G. E., & Anderson, J. A. (Eds.). (1981). *Parallel models of associative memory*. Hillsdale, NJ: Erlbaum.

Hinton, G. E., & Sejnowski, T. J. (1983). Optimal perceptual inference. *Proceedings of the IEEE Conference on Computer Vision Pattern Recognition*, 448-453.

Hinton, G. E., Sejnowski, T. J., & Ackley, D. H. (1984). *Boltzmann machines: Constraint satisfaction networks that learn* (Tech. Rep. CMU-CS-84-119). Pittsburgh: Carnegie-Mellon University, Department of Computer Science.

Hintzman, D. (1983). *Schema abstraction in a multiple trace memory model*. Paper presented at conference on "The Priority of the Specific," Elora, Ontario, Canada.

Hofstadter, D. (1979). *Gödel, Escher, Bach: An eternal golden braid*. New York: Basic Books.

Hogg, T., & Huberman, B. A. (1984). Understanding biological computation: Reliable learning and recognition. *Proceedings of the National Academy of Sciences, USA, 81*, 6871-6875.

Hopfield, J. J. (1982). Neural networks and physical systems with emergent collective computational abilities. *Proceedings of the National Academy of Sciences, USA, 79*, 2554-2558.

Hopfield, J. J., & Tank, D. (1985). "Neural" computation of decisions in optimization problems. *Biological Cybernetics, 52*, 141-152.

Hopkins, W. F., & Johnston, D. (1984). Frequency-dependent noradrenergic modulation of long-term potentiation in the hippocampus. *Science, 226*, 350-352.

Houser, C. R., Vaughn, J. E., Barber, R. P., & Roberts, E. (1980). GABA neurons are the major cell type of the nucleus reticularis thalami. *Brain Research, 200*, 341-354.

Hubel, D. H., & Wiesel, T. N. (1962). Receptive fields, binocular interaction, and functional architecture in the cat's visual cortex. *Journal of Physiology, 160*, 106-154.

Hubel, D. H., & Wiesel, T. N. (1965). Binocular interaction in striate cortex of kittens reared with artificial squint. *Journal of Neurophysiology, 28*, 1041-1059.

Hubel, D. H., & Wiesel, T. N. (1968). Receptive fields and functional architecture of monkey striate cortex. *Journal of Physiology, 195*, 215-243.

Hubel, D. H., & Wiesel, T. N. (1970a). Cells sensitive to binocular depth in area 18 of the macaque monkey cortex. *Nature, 225*, 41-42.

Hubel, D. H., & Wiesel, T. N. (1970b). The period of susceptibility to the physiological effects of unilateral eye closure in kittens. *Journal of Physiology*, *206*, 419-436.

Hubel, D. H., & Wiesel, T. N. (1972). Laminar and columnar distribution of geniculo-cortical fibers in the macaque monkey. *Journal of Comparative Neurology*, *146*, 421-450.

Hubel, D. H., & Wiesel, T. N. (1977). Functional architecture of macaque monkey visual cortex. *Proceedings of the Royal Society of London, Series B*, *198*, 1-59.

Hubel, D. H., Wiesel, T. N., & LeVay, S. (1977). Plasticity of ocular dominance columns in monkey striate cortex. *Philosophical Transactions of the Royal Society of London, Series B*, *278*, 377-409.

Huppert, F. A., & Piercy, M. (1978). Dissociation between learning and remembering in organic amnesia. *Nature*, *275*, 317-318.

Imig, T. J., & Brugge, J. F. (1978). Sources and terminations of callosal axons related to binaural and frequency maps in primary auditory cortex of the cat. *Journal of Comparative Neurology*, *182*, 637-660.

Iverson, L. L. (1984). Amino acids and peptides: Fast and slow chemical signals in the nervous system? *Proceedings of the Royal Society of London, Series B*, *221*, 245-260.

Jacoby, L. L. (1983a). Perceptual enhancement: Persistent effects of an experience. *Journal of Experimental Psychology: Learning, Memory, and Cognition*, *9*, 21-38.

Jacoby, L. L. (1983b). Remembering the data: Analyzing interaction processes in reading. *Journal of Verbal Learning and Verbal Behavior*, *22*, 485-508.

Jenkins, W. M., Merzenich, M. M., & Ochs, M. T. (1984). Behaviorally controlled differential use of restricted hand surfaces induces changes in the cortical representation of the hand in area 3b of adult owl monkeys. *Society for Neuroscience Abstracts*, *10*, 665.

Johnson, E., Sickels, E. R., & Sayers, F. C. (Eds.). (1970). *Anthology of children's literature* (4th ed.). Boston: Houghton-Mifflin.

Johnston, J. C., Hale, B. L., & van Santen, J. P. H. (1983). *Resolving letter position uncertainty in words* (TM 83-11221-19). Murray Hill, NJ: Bell Labs.

Jones, E. G. (1975a). Lamination and differential distribution of thalamic afferents within the sensory-motor cortex of the squirrel monkey. *Journal of Comparative Neurology*, *160*, 167-204.

Jones, E. G. (1975b). Some aspects of the organization of the thalamic reticular complex. *Journal of Comparative Neurology*, *162*, 285-308.

Jones, E. G. (1975c). Varieties and distribution of nonpyramidal cells in the somatic sensory cortex of the squirrel monkey. *Journal of Comparative Neurology*, *160*, 205-268.

Jones, E. G., & Leavitt, R. Y. (1974). Retrograde axonal transport and the demonstration of nonspecific projections to the cerebral cortex and striatum from thalamic intralaminar nuclei in rat, cat, and monkey. *Journal of Comparative Neurology*, *154*, 349-378.

Jones, E. G., & Powell, T. P. S. (1970a). An anatomical study of converging sensory pathways within the cerebral cortex of the monkey. *Brain*, *93*, 793-820.

Jones, E. G., & Powell, T. P. S. (1970b). An electron microscopic study of the laminar pattern and mode of termination of afferent fibre pathways in the somatic sensory cortex of the cat. *Philosophical Transactions of the Royal Society of London, Series B, 257,* 45-62.

Jones, E. G., & Wise, S. P. (1977). Size, laminar, and columnar distribution of efferent cells in the sensory-motor cortex of monkey. *Journal of Comparative Neurology, 175,* 391-438.

Just, M. A., & Carpenter, P. A. (1980). A theory of reading: Eye fixations to comprehension. *Psychological Review, 87,* 329-354.

Kahneman, D., & Miller, D. T. (in press). Norm theory: Comparing reality to its alternatives. *Psychological Review.*

Kandel, E. R., & Schwartz, J. H. (1981). *Principles of neural science.* New York: Elsevier/North Holland.

Kandel, E. R., & Schwartz, J. H. (1982). Molecular biology of memory: Modulation of transmitter release. *Science, 218,* 433-443.

Kant, E. (1963). *Critique of pure reason* (2nd ed., N. Kemp Smith, Trans.). London: Macmillan. (Original work published 1787)

Kaplan, R., & Bresnan, J. (1982). Lexical functional grammar: A formal system for grammatical representation. In J. Bresnan (Ed.), *The mental representation of grammatical relations.* Cambridge, MA: MIT Press.

Kasamatsu, T. (1983). Neuronal plasticity by the central norepinephrine system in the cat visual cortex. *Progress in Psychobiology & Physiological Psychology, 10,* 1-112.

Kasamatsu, T., & Heggelund, P. (1982). Single cell responses in cat visual cortex to visual stimulation during iontophoresis of noradrenaline. *Experimental Brain Research, 45,* 317-327.

Kasamatsu, T., & Pettigrew, J. (1979). Preservation of binocularity after monocular deprivation in the striate cortex of kittens treated with 6-hydroxydopamine. *Journal of Comparative Neurology, 185,* 139-161.

Kasamatsu, T., Pettigrew, J., & Ary, M. (1979). Restoration of visual cortical plasticity by local microperfusion of norepinephrine. *Journal of Comparative Neurology, 185,* 163-182.

Kawamoto, A. H. (1985). *Dynamic processes in the (re)solution of lexical ambiguity.* Unpublished doctoral dissertation, Brown University.

Kewley-Port, D. (1982). Measurement of formant transitions in naturally produced stop consonant-vowel syllables. *Journal of the Acoustical Society of America, 72,* 379-389.

Klatt, D. H. (1980). Speech perception: A model of acoustic-phonetic analysis and lexical access. In R. Cole (Ed.), *Perception and production of fluent speech* (pp. 243-288). Hillsdale, NJ: Erlbaum.

Knapp, A., & Anderson, J. A. (1984). A signal averaging model for concept formation. *Journal of Experimental Psychology: Learning, Memory, and Cognition, 10,* 616-637.

Koch, C., & Poggio, T. (1985). Biophysics of computation: Nerves, synapses, and membranes. In G. M. Edelman, W. E. Gall, & W. M. Cowan (Eds.), *New insights into synaptic function.* New York: Neuroscience Research Foundation/Wiley.

Koch, C., Poggio, T., & Torre, V. (1982). Retinal ganglion cells: A functional interpretation of dendritic morphology. *Philosophical Transactions of the Royal Society of London, Series B, 298*, 227-264.

Kohonen, T. (1977). *Associative memory: A system theoretical approach.* New York: Springer.

Kohonen, T. (1984). *Self-organization and associative memory.* Berlin: Springer-Verlag.

Konishi, M. (1978). Ethological aspects of auditory pattern recognition. In R. Held, H. W. Liebowitz, & H.-L. Teuber (Eds.), *Handbook of sensory physiology: Vol. 8. Perception* (pp. 289-309). Berlin: Springer.

Kratz, K. E., Spear, P. D., & Smith, D. C. (1976). Postcritical-period reversal of effects of monocular deprivation on striate cortex cells in the cat. *Journal of Neurophysiology, 39*, 501-511.

Krnjevíc, K., & Phillis, J. W. (1963). Iontophoretic studies of neurons in the mammalian cerebral cortex. *Journal of Physiology, 165*, 274-304.

Kubie, J., Muller, R. U., & Ranck, J. B. Jr. (1983). Manipulations of the geometry of environmental enclosures control the spatial firing patterns of rat hippocampal neurons. *Society for Neuroscience Abstracts, 9*, 647.

Kucera, H., & Francis, W. (1967). *Computational analysis of present-day American English.* Providence, RI: Brown University Press.

Kuczaj, S. A. (1977). The acquisition of regular and irregular past tense forms. *Journal of Verbal Learning and Verbal Behavior, 16*, 589-600.

Kuczaj, S. A. (1978). Children's judgments of grammatical and ungrammatical irregular past tense verbs. *Child Development, 49*, 319-326.

Kuffler, S. W. (1953). Discharge patterns and functional organization of mammalian retina. *Journal of Neurophysiology, 16*, 37-68.

Kuipers, B. (1978). Modeling spatial knowledge. *Cognitive Science, 2*, 129-153.

Kuipers, B. (1983). Modeling human knowledge of routes: Partial knowledge and individual variation. *Proceedings of the National Conference on Artificial Intelligence*, 216-219.

Kuno, M. (1964). Mechanism of facilitation and depression of the excitatory synaptic potential in spinal motoneurons. *Journal of Physiology, 175*, 100-112.

Kuperstein, M., & Eichenbaum, H. (1985). Unit activity, evoked potentials, and slow waves in the rat hippocampus and olfactory bulb recorded with a 24-channel microelectrode. *Neuroscience, 15*, 703-712.

Kurtzman, H. S. (1985). *Studies in syntactic ambiguity resolution.* Unpublished doctoral dissertation, Massachusetts Institute of Technology.

Lane, H. L. (1965). The motor theory of speech perception: A critical review. *Psychological Review, 72*, 275-309.

Lehiste, I. (1960). An acoustic-phonetic study of internal open juncture. *Phonetica, 5*, 1-54.

Lehiste, I. (1964). Juncture. In *Proceedings of the 5th International Congress of Phonetic Sciences, Munster* (pp. 172-200). New York: Karger.

Lenneberg, E. H. (1967). *Biological foundations of language.* New York: Wiley.

LeVay, S. (1973). Synaptic patterns in the visual cortex of the cat and monkey. *Journal of Comparative Neurology, 150*, 53-86.

LeVay, S., Connolly, M., Houde, J., & Van Essen, D. C. (1985). The complete pattern of ocular dominance stripes in the striate cortex and visual field of the macaque monkey. *Journal of Neuroscience, 5*, 486-501.

LeVay, S., & Sherk, H. (1981a). The visual claustrum of the cat: I. Structure and connections. *Journal of Neuroscience, 1*, 956-980.

LeVay, S., & Sherk, H. (1981b). The visual claustrum of the cat: II. The visual field map. *Journal of Neuroscience, 1*, 981-992.

LeVay, S., Wiesel, T. N., & Hubel, D. H. (1980). The development of ocular dominance columns in normal and visually deprived monkeys. *Journal of Comparative Neurology, 191*, 1-51.

Levin, J. A. (1976). *Proteus: An activation framework for cognitive process models* (Tech. Rep. No. ISI/WP-2). Marina del Rey: University of Southern California, Information Sciences Institute.

Lewis, B. (1878). On the comparative structur of the cortex cerebri. *Brain, 1*, 79-96.

Liberman, A. M. (1970). The grammars of speech and language. *Cognitive Psychology, 1*, 301-323.

Liberman, A. M., Cooper, F. S., Shankweiler, D., & Studdert-Kennedy, M. (1967). Perception of the speech code. *Psychological Review, 84*, 452-471.

Licklider, J. C. R., & Miller, G. A. (1951). The perception of speech. In S. S. Stevens (Ed.), *Handbook of Experimental Psychology*. New York: Wiley.

Llinás, R., & Nicholson, C. (1969). Electrophysiological analysis of alligator cerebellum: A study of dendritic spikes. In R. Llinás (Ed.), *Neurobiology of cerebellar evolution and development* (pp. 431-465). Chicago: American Medical Association.

Loeb, G. E. (1985, February). The functional replacement of the ear. *Scientific American*, pp. 104-111.

Loftus, E. F. (1977). Shifting human color memory. *Memory & Cognition, 6*, 696-699.

Loomis, L. H., & Sternberg, S. (1968). *Advanced calculus*. Reading, MA: Addison-Wesley.

Lorente de Nó, R. (1943). Cerebral cortex: Architecture, intracortical connections, motor projections. In J. F. Fulton (Ed.), *Physiology of the nervous system* (2nd ed., pp. 274-301). New York: Oxford University Press.

Luce, R. D. (1959). *Individual choice behavior*. New York: Wiley.

Luce, R. D. (1963). Detection and recognition. In R. D. Luce, R. R. Bush, & E. Galanter (Eds.), *Handbook of mathematical psychology* (Vol. 1). New York: Wiley.

Lund, J. S. (1973). Organization of neurons in the visual cortex, area 17, of the monkey (*Macaca mulatta*). *Journal of Comparative Neurology, 147*, 455-496.

Lund, J. S., Lund, R. D., Hendrickson, A. E., Bunt, A. H., & Fuchs, A. F. (1975). The origin of efferent pathways from the primary visual cortex, area 17, of the macaque monkey as shown by retrograde transport of horseradish peroxidase. *Journal of Comparative Neurology, 164*, 287-304.

Lund, R. D. (1972). Synaptic patterns in the superficial layer of the superior colliculus of the monkey, *Macaca mulatta*. *Experimental Brain Research, 15*, 194-211.

Lynch, J. C., Mountcastle, V. B., Talbot, W. H., & Yin, T. C. T. (1977). Parietal lobe mechanisms for directed visual attention. *Journal of Neurophysiology, 40,* 362-389.

Ma, S. -K. (1976). *Modern theory of critical phenomena.* Reading, MA: Benjamin.

Macchi, G., Bentivoglio, M., Minciacchi, D., & Molinari, M. (1981). The organization of the claustroneocortical projections in the cat studied by means of the HRP retrograde axonal transport. *Journal of Comparative Neurology, 195,* 681-695.

MacWhinney, B., Bates, E., & Kliegl, R. (1984). Cue validity and sentence interpretation in English, German, and Italian. *Journal of Verbal Learning and Verbal Behavior, 23,* 127-150.

MacWhinney, B., & Sokolov, J. L. (in press). The competition model of the acquisition of syntax. In B. MacWhinney (Ed.), *Mechanisms of language acquisition.* Hillsdale, NJ: Erlbaum.

Macy, A., Ohzawa, I., & Freeman, R. D. (1982). A quantitative study of the classification and stability of ocular dominance in the cat's visual cortex. *Experimental Brain Research, 48,* 401-408.

Mann, V. A., & Repp, B. H. (1980). Influence of vocalic context on perception of the [s]-[s] distinction. *Perception & Psychophysics, 28,* 213-228.

Marcus, M. P. (1980). *A theory of syntactic recognition for natural language.* Cambridge, MA: MIT Press.

Marin-Padilla, M. (1969). Origin of the pericellular baskets of the pyramidal cells of the human motor cortex: A Golgi study. *Brain Research, 14,* 633-646.

Marr, D. (1982). *Vision.* San Francisco: Freeman.

Marrocco, R. T., & McClurkin, J. W. (1979). Binocular interaction in the lateral geniculate nucleus of the monkey. *Brain Research, 168,* 633-637.

Marslen-Wilson, W. D. (1973). Linguistic structure and speech shadowing at very short latencies. *Nature, 244,* 522-523.

Marslen-Wilson, W. D. (1980). Speech understanding as a psychological process. In J. C. Simon (Ed.), *Spoken language generation and understanding* (pp. 39-67). New York: Reidel.

Marslen-Wilson, W. D., & Tyler, L. K. (1975). Processing structure of sentence perception. *Nature, London, 257,* 784-786.

Marslen-Wilson, W. D., & Tyler, L. K. (1980). The temporal structure of spoken language understanding. *Cognition, 8,* 1-71.

Marslen-Wilson, W. D., & Tyler, L. K. (1981). Central processes in speech understanding. *Philosophical Transactions of the Royal Society, London B, 295,* 317-332.

Marslen-Wilson, W. D., & Welsh, A. (1978). Processing interactions and lexical access during word recognition in continuous speech. *Cognitive Psychology, 10,* 29-63.

Martin, K. A. C., Somogyi, P., & Whitteridge, D. (1983). Physiological and morphological properties of identified basket cells in the cat's visual cortex. *Experimental Brain Research, 50,* 193-200.

Massaro, D. W. (1975). *Experimental psychology and information processing.* Chicago: Rand McNally.

Massaro, D. W. (1981). Sound to representation: An information-processing analysis. In T. Myers, J. Laver, & J. Anderson (Eds.), *The cognitive representation of speech* (pp. 181-193). New York: North-Holland.

Massaro, D. W., & Cohen, M. M. (1977). The contribution of voice-onset time and fundamental frequency as cues to the /zi/-/si/ distinction. *Perception & Psychophysics, 22,* 373-382.

Massaro, D. W., & Cohen, M. M. (1983). Phonological constraints in speech perception. *Perception & Psychophysics, 34,* 338-348.

Massaro, D. W., & Oden, G. C. (1980a). Evaluation and integration of acoustic features in speech perception. *Journal of the Acoustical Society of America, 67,* 996-1013.

Massaro, D. W., & Oden, G. C. (1980b). Speech perception: A framework for research and theory. In N. Lass (Ed.), *Speech and language: Advances in basic research and practice* (Vol. 3, pp. 129-165). New York: Academic Press.

Maunsell, J. H. R., Newsome, W. T., & Van Essen, D. C. (1980). The spatial organization of connections between V1 and V2 in the macaque: Patchy and nonpatchy projections. *Society for Neuroscience Abstracts, 6,* 580.

Maunsell, J. H. R., & Van Essen, D. C. (1983). The connections of the middle temporal visual area (MT) and their relation to a cortical hierarchy in the macaque monkey. *Journal of Neuroscience, 3,* 2563-2586.

McClelland, J. L. (1981). Retrieving general and specific information from stored knowledge of specifics. *Proceedings of the Third Annual Meeting of the Cognitive Science Society,* 170-172.

McClelland, J. L. (1985). Putting knowledge in its place: A scheme for programming parallel processing structures on the fly. *Cognitive Science, 9,* 113-146.

McClelland, J. L., & Elman, J. L. (1986). The TRACE model of speech perception. *Cognitive Psychology, 18,* 1-86.

McClelland, J. L., & Rumelhart, D. E. (1981). An interactive activation model of context effects in letter perception: Part 1. An account of basic findings. *Psychological Review, 88,* 375-407.

McClelland, J. L., & Rumelhart, D. E. (1985). Distributed memory and the representation of general and specific information. *Journal of Experimental Psychology: General, 114,* 159-188.

McConkie, G. W., Zola, D., Blanchard, H. E., & Wolverton, G. S. (1982). Perceiving words during reading. *Perception & Psychophysics, 32,* 271-281.

McDermott, D. (1980). *Spatial inferences with ground, metric formulas on simple objects* (Research Rep. No. 173). New Haven, CT: Yale University, Department of Computer Science.

McNaughton, B. L., Barnes, C. A., & O'Keefe, J. (1983). The contributions of position, direction, and velocity to single unit activity in the hippocampus of freely moving rats. *Experimental Brain Research, 52,* 41-49.

Medin, D. L., & Schwanenflugel, P. J. (1981). Linear separability in classification learning. *Journal of Experimental Psychology: Human Learning and Memory, 7,* 355-368.

Medin, D. L., & Schaffer, M. M. (1978). Context theory of classification learning. *Psychological Review, 85,* 207-238.

Merzenich, M. M., & Kaas, J. H. (1982). Reorganization of mammalian somatosensory cortex following peripheral nerve injury. *Trends in Neuroscience, 5,* 434-436.

Merzenich, M. M., Kaas, J. H., Wall, J. T., Sur, M., Nelson, R. J., & Felleman, D. J. (1983). Progression of change following median nerve section in the cortical representation of the hand in areas 3b and 1 in adult owl and squirrel monkeys. *Neuroscience, 10,* 639-665.

Mesulam, M.-M., & Van Hoesen, G. W. (1976). Acetylcholinesterase-rich projections from the basal forebrain of the rhesus monkey to neocortex. *Brain Research, 109,* 152-157.

Michael, C. R. (1978). Color vision mechanisms in monkey striate cortex: Dual-opponent cells with concentric receptive fields. *Journal of Neurophysiology, 41,* 572-588.

Miller, G. A. (1962). Some psychological studies of grammar. *American Psychologist, 17,* 748-762.

Miller, G., Heise, G., & Lichten, W. (1951). The intelligibility of speech as a function of the context of the test materials. *Journal of Experimental Psychology, 41,* 329-335.

Miller, J. L. (1981). Effects of speaking rate on segmental distinctions. In P. D. Eimas & J. L. Miller (Eds.), *Perspectives on the study of speech* (pp. 39-74). Hillsdale, NJ: Erlbaum.

Miller, J. L., Green, K., & Schermer, T. M. (1984). A distinction between the effects of sentential speaking rate and semantic congruity on word identification. *Perception & Psychophysics, 36,* 329-337.

Miller, J. P., Rall, W., & Rinzel, J. (1985). Synaptic amplification by active membrane in dendritic spines. *Brain Research, 325,* 325-330.

Milner, B. (1966). Amnesia following operation on the temporal lobes. In C. M. W. Whitty & O. L. Zangwill (Eds.), *Amnesia.* London: Butterworth.

Minsky, M. (1975). A framework for representing knowledge. In P. H. Winston (Ed.), *The psychology of computer vision* (pp. 211-277). New York: McGraw-Hill.

Minsky, M., & Papert, S. (1969). *Perceptrons.* Cambridge, MA: MIT Press.

Miyuwaki, K., Strange, W., Verbrugge, R., Liberman, A. M., Jenkins, J. J., & Fujimura, O. (1975). An effect of linguistic experience: The discrimination of [r] and [l] by native speakers of Japanese and English. *Perception & Psychophysics, 18,* 331-340.

Montero, V. M., & Scott, G. L. (1981). Synaptic terminals in the dorsal lateral geniculate nucleus from neurons of the thalamic reticular nucleus: A light and electron microscope autoradiographic study. *Neuroscience, 6,* 2561-2577.

Morrell, F. (1972). Integrative properties of parastriate neurons. In A. G. Karczmar & J. C. Eccles (Eds.), *Brain and human behavior.* Berlin: Springer-Verlag.

Morris, R. G. M., Garrud, P., Rawlins, J. N. P., & O'Keefe, J. (1982). Place navigation impaired in rats with hippocampal lesions. *Nature, 297,* 681-683.

Morrison, J. H., Magistretti, P. J., Benoit, R., & Bloom, F. E. (1984). The distribution and morphological characteristics of the intracortical VIP-positive cell: An immunohistochemical analysis. *Brain Research, 292,* 269-282.

Morton, J. (1969). Interaction of information in word recognition. *Psychological Review*, 76, 165-178.

Morton, J. (1979). Facilitation in word recognition: Experiments causing change in the logogen model. In P. A. Kohlers, M. E. Wrolstal, & H. Bouma (Eds.), *Processing visible language I*. New York: Plenum.

Moser, M. G. (1983). An overview of NIKL: The new implementation of KL-ONE. In C. Sidner, M. Bates, R. Bobrow, B. Goodman, A. Haas, R. Ingria, D. Israel, D. McAllester, M. Moser, J. Schmolze, & M. Vilain (Eds.), *Research in knowledge representation for natural language understanding: Annual report* (BBN Report No. 5421). Cambridge, MA: Bolt, Beranek, and Newman.

Mountcastle, V. B. (1957). Modality and topographic properties of single neurons of cat's somatic sensory cortex. *Journal of Neurophysiology*, 20, 408-434.

Mountcastle, V. B. (1978). An organizing principle for cerebral function: The unit module and the distributed system. In G. M. Edelman & V. B. Mountcastle (Eds.), *The mindful brain* (pp. 7-50). Cambridge, MA: MIT Press.

Mower, G. D., Christen, W. G., & Caplan, C. J. (1983). Very brief visual experience eliminates plasticity in the cat visual cortex. *Science*, 221, 178-180.

Mozer, M. C. (1983). Letter migration in word perception. *Journal of Experimental Psychology: Human Perception and Performance*, 9, 531-546.

Muller, R. U., Kubie, J. L., & Ranck, J. B. Jr. (1983). High resolution mapping of the spatial fields of hippocampal neurons in the freely moving rat. *Society for Neuroscience Abstracts*, 9, 646.

Munro, P. W. (1984). A model for generalization and specification by single neurons. *Biological Cybernetics*, 51, 169-179.

Murray, E. A., & Coulter, J. D. (1981). Organization of corticospinal neurons in the monkey. *Journal of Comparative Neurology*, 195, 339-365.

Nakatani, L., & Dukes, K. (1977). Locus of segmental cues for word juncture. *Journal of the Acoustical Society of America*, 62, 714-719.

Nauta, W. J. H., & Feirtag, M. (1979). The organization of the brain. *Scientific American*, 241, 88-111.

Nelson, R. J., Sur, M., Felleman, D. J., & Kaas, J. H. (1980). Representations of the body surface in postcentral parietal cortex of *Macaca fascicularis*. *Journal of Comparative Neurology*, 192, 611-644.

Norman, D. A., & Bobrow, D. G. (1975). On data-limited and resource-limited processes. *Cognitive Psychology*, 7, 44-64.

Norman, D. A., & Bobrow, D. G. (1976). On the role of active memory processes in perception and cognition. In C. N. Cofer (Ed.), *The structure of human memory*. San Francisco: Freeman.

Norman, D. A., & Bobrow, D. G. (1979). Descriptions: An intermediate stage in memory retrieval. *Cognitive Psychology*, 11, 107-123.

Norman, D. A., & Shallice, T. (in press). Attention to action: Willed and automatic control of behavior. In R. J. Davidson, G. E. Schwartz, & D. Shapiro (Eds.), *Consciousness and self regulation: Advances in research, Vol. IV*. New York: Plenum Press.

Norris, D. (1982). Autonomous processes in comprehension: A reply to Marslen-Wilson and Tyler. *Cognition*, *11*, 97-101.

Nusbaum, H. C., & Slowiaczek, L. M. (1982). *An activation model of auditory word recognition* (Research on Speech Perception, Progress Report No. 8). Bloomington: Indiana University, Department of Psychology.

O'Keefe, J. (1976). Place units in the hippocampus of the freely moving rat. *Experimental Neurobiology*, *51*, 78-109.

O'Keefe, J. (1979). A review of the hippocampal place cells. *Progress in Neurobiology*, *13*, 429-439.

O'Keefe, J., & Dostrovsky, J. (1971). The hippocampus as a spatial map: Preliminary evidence from unit activity in the freely moving rat. *Brain Research*, *34*, 171-175.

O'Keefe, J., & Nadel, L. (1979). The hippocampus as a cognitive map. *Behavioral and Brain Sciences*, *2*, 487-533.

O'Kusky, J., & Colonnier, M. (1982). A laminar analysis of the number of neurons, glia, and synapses in the visual cortex (area 17) of adult macaque monkeys. *Journal of Comparative Neurology*, *210*, 278-290.

Oden, G. C. (1978). Semantic constraints and judged preference for interpretations of ambiguous sentences. *Memory & Cognition*, *6*, 26-37.

Oden, G. C., & Massaro, D. W. (1978). Integration of featural information in speech perception. *Psychological Review*, *85*, 172-191.

Olson, C. R., & Freeman, R. D. (1975). Progressive changes in kitten striate cortex during monocular vision. *Journal of Neurophysiology*, *38*, 26-32.

Olton, D. S. (1984). Animal models of human amnesia. In L. R. Squire & N. Butters (Eds.), *The neuropsychology of memory*. New York: Guilford Press.

Olton, D. S., Branch, M., & Best, P. (1978). Spatial correlates of hippocampal unit activity. *Experimental Neurobiology*, *58*, 387-409.

Pandya, D. N., & Kuypers, H. G. J. M. (1969). Cortico-cortical connections in the rhesus monkey. *Brain Research*, *13*, 13-36.

Peck, C. K., & Blakemore, C. (1975). Modification of single neurons in the kitten's visual cortex after brief periods of monocular visual experience. *Experimental Brain Research*, *22*, 57-68.

Perez, R., Glass, L., & Shlaer, R. (1975). Development of specificity in the cat visual cortex. *Journal of Mathematical Biology*, *1*, 275-288.

Perkel, D. H., & Perkel, D. J. (1985). Dendritic spines: Role of active membrane in modulating synaptic efficacy. *Brain Research*, *325*, 331-335.

Perrett, D. I., Rolls, E. T., & Caan, W. (1982). Visual neurones responsive to faces in the monkey temporal cortex. *Experimental Brain Research*, *47*, 329-342.

Peters, A., & Kaiserman-Abramof, I. R. (1969). The small pyramidal neuron of the rat cerebral cortex: The synapses upon dendritic spines. *Zietschrift für Zellforschung und Mikrokopische Anatomie*, *100*, 487-506.

Peters, A., Palay, S. L., & Webster, H. deF. (1976). *The fine structure of the nervous system*. Philadelphia: W. B. Saunders.

Peters, A., Proskauer, C. C., & Kaiserman-Abramof, I. R. (1968). The small pyramidal neuron of the rat cerebral cortex: The axon hillock and initial segment. *Journal of Cell Biology*, *39*, 604-619.

Peters, A., Proskauer, C. C., & Ribak, C. E. (1982). Chandelier cells in rat visual cortex. *Journal of Comparative Neurology, 206,* 397-416.

Piaget, J. (1952). *The origins of intelligence in children.* New York: International University Press.

Pinker, S. (1984). *Language learnability and language development.* Cambridge, MA: Harvard University Press.

Pisoni, D. B. (1973). Auditory and phonetic memory codes in the discrimination of consonants and vowels. *Perception & Psychophysics, 13,* 253-260.

Pisoni, D. B. (1975). Auditory short-term memory and vowel perception. *Memory & Cognition, 3,* 7-18.

Pisoni, D. B., & Lazarus, J. H. (1974). Categorical and noncategorical modes of speech perception along the voicing continuum. *Journal of the Acoustical Society of America, 55,* 328-333.

Pisoni, D. B., & Tash, J. (1974). Reaction times to comparisons within and across phonetic categories. *Perception & Psychophysics, 15,* 285-290.

Poggio, T., & Torre, V. (1978). A new approach to synaptic interactions. In R. Heim & G. Palm (Eds.), *Approaches to complex systems.* Berlin: Springer-Verlag.

Porrino, L. J., Crane, A. M., & Goldman-Rakic, P. S. (1981). Direct and indirect pathways from the amygdala to the frontal lobe in rhesus monkeys. *Journal of Comparative Neurology, 198,* 121-136.

Poschel, B. P. H., & Ninteman, F. W. (1963). Norepinephrine: A possible excitatory neurohormone of the reward system. *Life Sciences, 2,* 782-788.

Posner, M. I., & Keele, S. W. (1968). On the genesis of abstract ideas. *Journal of Experimental Psychology, 77,* 353-363.

Posner, M. I., & Keele, S. W. (1970). Retention of abstract ideas. *Journal of Experimental Psychology, 83,* 304-308.

Powell, T. P. S., & Cowan, W. M. (1967). The interpretation of the degenerative changes in the intralaminar nuclei of the thalamus. *Journal of Neurology, Neurosurgery, and Psychiatry, 30,* 140-153.

Powell, T. P. S., & Hendrickson, A. E. (1981). Similarity in number of neurons through the depth of the cortex in the binocular and monocular parts of area 17 of the monkey. *Brain Research, 216,* 409-413.

Rall, W. (1970). Cable properties of dendrites and effects of synaptic location. In P. Andersen & J. Jansen (Eds.), *Excitatory synaptic mechanisms* (pp. 175-187). Oslo Universitersforlag.

Rall, W., Shepherd, G. M., Reese, T. S., & Brightman, M. W. (1966). Dendrodendritic synaptic pathway for inhibition in the olfactory bulb. *Experimental Neurology, 14,* 44-56.

Ralston, H. J. (1968). The fine structure of neurons in the dorsal horn of the cat spinal cord. *Journal of Comparative Neurology, 132,* 275-302.

Ramachandran, V. S., & Kupperman, B. (in press). Reversal of the physiological effects of monocular deprivation in adult dark-reared cats. *Developmental Brain Research.*

Rayner, K. (1975). The perception span and peripheral cues in reading. *Cognitive Psychology, 7,* 65-81.

Rayner, K., Carlson, M., & Frazier, L. (1983). The interaction of syntax and semantics during sentence processing: Eye movements in the analysis of semantically biased sentences. *Journal of Verbal Learning and Verbal Behavior, 22,* 358-374.

Rayner, K., McConkie, G. W., & Ehrlich, S. (1978). Eye movements and integrating information across fixations. *Journal of Experimental Psychology: Human Perception and Performance, 4,* 529-544.

Reddy, D. R. (1976). Speech recognition by machine: A review. *Proceedings of the IEEE, 64,* 501-531.

Reddy, D. R., Erman, L. D., Fennell, R. D., & Neely, R. B. (1973). The Hearsay speech understanding system: An example of the recognition process. *Proceedings of the International Conference on Artificial Intelligence,* 185-194.

Reich, P. A. (1969). The finiteness of natural language. *Language, 45,* 831-843.

Reilly, D. L., Cooper, L. N, & Elbaum, C. (1982). A neural model for category learning. *Biological Cybernetics, 45,* 35-41.

Reitboek, H. J. P. (1983). A 19-channel matrix drive with individually controllable fiber microelectrodes for neurophysiological applications. *IEEE Transactions on Systems, Man, and Cybernetics, 13,* 676-683.

Repp, B. H., & Liberman, A. M. (1984). Phonetic categories are flexible. *Haskins Laboratories Status Report on Speech Research, SR-77/78,* 31-53.

Rockel, A. J., Hiorns, R. W., & Powell, T. P. S. (1980). The basic uniformity in structure of the neocortex. *Brain, 103,* 221-244.

Rockland, K. S., & Lund, J. S. (1983). Intrinsic laminar lattice connections in macaque prestriate visual cortex. *Journal of Comparative Neurology, 216,* 303-318.

Rockland, K. S., & Pandya, D. N. (1979). Laminar origins and terminations of cortical connections of the occipital lobe in the rhesus monkey. *Brain Research, 179,* 3-20.

Rosch, E. (1975). Cognitive representations of semantic categories. *Journal of Experimental Psychology: General, 104,* 192-233.

Rose, J. E. (1949). The cellular structure of the auditory region of the cat. *Journal of Comparative Neurology, 91,* 409-439.

Rosenblatt, F. (1962). *Principles of neurodynamics.* New York: Spartan.

Rumelhart, D. E. (1975). Notes on a schema for stories. In D. G. Bobrow & A. Collins (Eds.), *Representation and understanding* (pp. 211-236). New York: Academic Press.

Rumelhart, D. E. (1977). Understanding and summarizing brief stories. In D. LaBerge & S. J. Samuels (Eds.), *Basic processes in reading: Perception and comprehension* (pp. 265-303). Hillsdale, NJ: Erlbaum.

Rumelhart, D. E. (1979). Some problems with the notion of literal meanings. In A. Ortony (Ed.), *Metaphor and thought.* Cambridge, England: Cambridge University Press.

Rumelhart, D. E. (1980). Schemata: The building blocks of cognition. In R. Spiro, B. Bruce, & W. Brewer (Eds.), *Theoretical issues in reading comprehension* (pp. 33-58). Hillsdale, NJ: Erlbaum.

Rumelhart, D. E., & McClelland, J. L. (1981). Interactive processing through spreading activation. In A. M. Lesgold & C. A. Perfetti (Eds.), *Interactive processes in reading*. Hillsdale, NJ: Erlbaum.

Rumelhart, D. E., & McClelland, J. L. (1982). An interactive activation model of context effects in letter perception: Part 2. The contextual enhancement effect and some tests and extensions of the model. *Psychological Review*, *89*, 60-94.

Rumelhart, D. E., & McClelland, J. L. (1985). Levels indeed! A response to Broadbent. *Journal of Experimental Psychology: General*, *114*, 193-197.

Rumelhart, D. E., & Norman, D. A. (1982). Simulating a skilled typist: A study of skilled cognitive-motor performance. *Cognitive Science*, *6*, 1-36.

Rumelhart, D. E., & Norman, D. A. (in press). Representation in memory. In R. C. Atkinson, R. J. Herrnstein, G. Lindzey, & R. D. Luce (Eds.), *Handbook of experimental psychology*. New York: Wiley.

Rumelhart, D. E., & Ortony, A. (1977). The representation of knowledge in memory. In R. C. Anderson, R. J. Spiro, & W. E. Montague (Eds.), *Schooling and the acquisition of knowledge* (pp. 99-135). Hillsdale, NJ: Erlbaum.

Russell, W. R., & Nathan, P. W. (1946). Traumatic amnesia. *Brain*, *69*, 280-300.

Sachs, M. B., Voigt, H. F., & Young, E. D. (1983). Auditory nerve representation of vowels in background noise. *Journal of Neurophysiology*, *50*, 27-45.

Salasoo, A., & Pisoni, D. B. (1985). Interaction of knowledge sources in spoken word identification. *Journal of Memory and Language*, *24*, 210-231.

Salasoo, A., Shiffrin, R. M., & Feustel, T. C. (1985). Building permanent memory codes: Codification and repetition effects in word identification. *Journal of Experimental Psychology: General*, *114*, 50-77.

Samuel, A. G. (1977). The effect of discrimination training on speech perception: Noncategorical perception. *Perception & Psychophysics*, *22*, 321-330.

Samuel, A. G., van Santen, J. P. H., & Johnston, J. C. (1982). Length effects in word perception: We is better than I but worse than you or them. *Journal of Experimental Psychology: Human Perception and Performance*, *8*, 91-105.

Sanderson, K. J., Bishop, P. O., & Darien-Smith, I. (1971). The properties of the binocular receptive fields of lateral geniculate neurons. *Experimental Brain Research*, *13*, 178-207.

Schacter, D. (1985). Priming of old and new knowledge in amnesic patients and normal subjects. *Annals of the New York Academy of Sciences*, *444*, 41-53.

Schank, R. C. (1973). Identification of conceptualizations underlying natural language. In R. C. Schank & K. M. Colby (Eds.), *Computer models of thought and language*. San Francisco: Freeman.

Schank, R. C., & Abelson, R. P. (1977). *Scripts, plans, goals, and understanding*. Hillsdale, NJ: Erlbaum.

Scheibel, M. E., & Scheibel, A. B. (1966). The organization of the nucleus reticularis thalami: A Golgi study. *Brain Research*, *1*, 43-62.

Schiller, P. H., Finlay, B. L., & Volman, S. F. (1976). Quantitative studies of single-cell properties in monkey striate cortex: III. Spatial frequency. *Journal of Neurophysiology*, *39*, 1334-1351.

Schneider, W. (1985). Toward a model of attention and the development of automatic processing. In M. I. Posner & O. S. M. Marin (Eds.), *Attention and performance XI* (pp. 474-492). Hillsdale, NJ: Erlbaum.

Scott, J. P. (1962). Critical periods in behavioral development. *Science, 138,* 949-958.

Scott, J. P. (1968). *Early experience and the organization of behavior.* Belmont, CA: Wadsworth.

Sejnowski, T. J. (1976). On global properties of neuronal interaction. *Biological Cybernetics, 22,* 85-95.

Sejnowski, T. J. (1981). Skeleton filters in the brain. In G. E. Hinton & J. A. Anderson (Eds.), *Parallel models of associative memory* (pp. 49-82). Hillsdale, NJ: Erlbaum.

Sejnowski, T. J. (in press). Computational neuroscience. *Behavioral and Brain Sciences.*

Sejnowski, T. J., & Hinton, G. E. (in press). Separating figure from ground with a Boltzmann machine. In M. A. Arbib & A. R. Hanson (Eds.), *Vision, brain, and cooperative computation.* Cambridge: MIT Press.

Selman, B. (1985). *Rule-based processing in a connectionist system for natural language understanding* (TR CSRI-168). Toronto: University of Toronto, Computer Systems Research Institute.

Selman, B., & Hirst, G. (1985). A rule-based connectionist parsing system. *Proceedings of the Seventh Annual Conference of the Cognitive Science Society,* 212-221.

Shallice, T., & McGill, J. (1978). The origins of mixed errors. In J. Reguin (Ed.), *Attention and Performance VII.* Hillsdale, NJ: Erlbaum.

Shaw, G. L., Silverman, D. J., & Pearson, J. C. (1985). Model of cortical organization embodying a basis for a theory of information processing and memory recall. *Proceedings of the National Academy of Sciences, USA, 82,* 2364-2368.

Shepard, R. N. (1984). Ecological constraints on internal representation: Resonant kinematics of perceiving, imagining, thinking, and dreaming. *Psychological Review, 91,* 417-447.

Shepard, R. N., & Podgorny, P. (1975). Cognitive processes that resemble perceptual processes. In W. K. Estes (Ed.), *Handbook of learning and cognitive processes* (pp. 189-237). Hillsdale, NJ: Erlbaum.

Shepherd, G. M. (1978, February). Microcircuits in the nervous system. *Scientific American,* p. 92.

Shepherd, G. M. (1979). *The synaptic organization of the brain.* New York: Oxford University Press.

Shepherd, G. M. (1985). *Modifiable logic operations are properties of simulated interactions between excitable dendritic spines.* Unpublished manuscript.

Shepherd, G. M., Brayton, R. K., Miller, J. P., Segev, I., Rinzel, J., & Rall, W. (1985). Signal enhancement in distal cortical dendrites by means of interactions between active dendritic spines. *Proceedings of the National Academy of Sciences, USA, 82,* 2192-2195.

Singer, W. (1979). Central-core control of visual cortex functions. In F. O. Schmitt & F. G. Worden (Eds.), *The neurosciences* (pp. 1093-1109). Cambridge, MA: MIT Press.

Singer, W. (1982). Central core control of developmental plasticity in the kitten visual cortex: 1. Diencephalic lesions. *Experimental Brain Research, 47,* 209-222.

Singer, W., & Rauschecker, J. P. (1982). Central core control of developmental plasticity in the kitten visual cortex: 2. Electrical activation of mesencephalic and diencephalic projections. *Experimental Brain Research, 47,* 223-233.

Small, S. L. (1980). *Word expert parsing: A theory of distributed word-based natural language understanding* (TR- 954). Baltimore: University of Maryland, Department of Computer Science.

Somogyi, P. (1977). A specific 'axo-axonal' interneuron in the visual cortex of the rat. *Brain Research, 136,* 345-350.

Somogyi, P., Freund, T. F., & Cowey, A. (1982). The axo-axonic interneuron in the cerebral cortex of the rat, cat, and monkey. *Neuroscience, 7,* 2577-2607.

Somogyi, P., Kisvárday, Z. F., Martin, K. A. C., & Whitteridge, D. (1983). Synaptic connections of morphologically identified and physiologically characterized large basket cells in the striate cortex of cat. *Neuroscience, 10,* 261-294.

Spencer, W. A., & Kandel, E. R. (1961). Electrophysiology of hippocampal neurons: IV. Fast prepotentials. *Journal of Neurophysiology, 24,* 272-285.

Squire, L. R. (1981). Two forms of amnesia: An analysis of forgetting. *Journal of Neuroscience, 1,* 635-640.

Squire, L. R. (1982). The neuropsychology of human memory. *Annual Review of Neuroscience, 5,* 241-273.

Squire, L. R., Cohen, N. J., & Nadel, L. (1984). The medial temporal region and memory consolidation: A new hypothesis. In H. Weingartner & E. Parker (Eds.), *Memory consolidation.* Hillsdale, NJ: Erlbaum.

Squire, L. R., Slater, P. C., & Chace, P. (1975). Retrograde amnesia: Temporal gradient in very long-term memory following electroconvulsive therapy. *Science, 187,* 77-79.

Squire, L. R., Slater, P. C., & Miller, P. (1981). Retrograde amnesia following ECT: Long term follow-up studies. *Archives of General Psychiatry, 38,* 89-95.

Stemberger, J. P. (1981). Morphological haplology. *Language, 57,* 791-817.

Stevens, K., & Blumstein, S. (1981). The search for invariant acoustic correlates of phonetic features. In P. D. Eimas & J. L. Miller (Eds.), *Perspectives on the study of speech* (pp. 1-38). Hillsdale, NJ: Erlbaum.

Studdert-Kennedy, M., Liberman, A. M., Harris, K. S., & Cooper, F. S. (1970). Motor theory of speech perception: A reply to Lane's critical review. *Psychological Review, 77,* 234-249.

Summerfield, Q., & Haggard, M. (1977). On the dissociation of spatial and temporal cues to the voicing distinction in initial stop consonants. *Journal of the Acoustical Society of America, 62,* 435-448.

Sutton, R. S., & Barto, A. G. (1981). Toward a modern theory of adaptive networks: Expectation and prediction. *Psychological Review, 88,* 135-171.

Swanson, L. W. (1981). A direct projection from Ammon's horn to prefrontal cortex in the rat. *Brain Research, 217,* 150-154.

Swinney, D. A. (1982). The structure and time-course of information interaction during speech comprehension: Lexical segmentation, access, and interpretation. In J. Mehler, E. C. T. Walker, & M. Garrett (Eds.), *Perspectives on mental representation*. Hillsdale, NJ: Erlbaum.

Szentágothai, J., & Arbib, M. A. (1974). Conceptual models of neural organization. *Neuroscience Research Program Bulletin, 12*, 307-510.

Szentágothai, J., Hámori, J., & Tömböl, T. (1966). Degeneration and electron microscope analysis of the synaptic glomeruli in the lateral geniculate body. *Experimental Brain Research, 2*, 283-301.

Thibadeau, R., Just, M. A., & Carpenter, P. A. (1982). A model of the time course and content of reading. *Cognitive Science, 6*, 157-204.

Thompson, H. (1984). Word recognition: A paradigm case in computational (psycho-)linguistics. *Proceedings of the Sixth Annual Meeting of the Cognitive Science Society*.

Tigges, M., & Tigges, J. (1979). Types of degenerating geniculocortical axon terminals and their contribution to layer IV of area 17 in the squirrel monkey (Saimiri). *Cell and Tissue Research, 196*, 471-486.

Tömböl, T. (1974). An electron microscopic study of the neurons of the visual cortex. *Journal of Neurocytology, 3*, 525-531.

Touretzky, D., & Hinton, G. E. (1985). Symbols among the neurons: Details of a connectionist inference architecture. *Proceedings of the Ninth International Joint Conference on Artificial Intelligence*.

Ts'o, D. Y., Gilbert, C. D., & Wiesel, T. N. (1985). Relationships between horizontal connections revealed by cross-correlation analysis. *Society for Neurosciences Abstracts, 11*, 17.

Tyler, L. K., & Wessels, J. (1983). Quantifying contextual contributions to word-recognition processes. *Perception & Psychophysics, 34*, 409-420.

Uchizono, K. (1965). Characteristics of excitatory and inhibitory synapses in the central nervous system of the cat. *Nature, 207*, 642-643.

Van der Loos, H., & Glaser, E. M. (1972). Autapses in neocortex cerebri: Synapses between a pyramidal cell's axon and its own dendrites. *Brain Research, 48*, 355-360.

Van Essen, D., & Maunsell, J. H. R. (1983). Hierarchical organization and functional streams in visual cortex. *Trends in Neuroscience, 6*, 370-375.

Van Essen, D. C., Maunsell, J. H. R., & Bixby, J. L. (1981). The middle temporal visual area in the macaque: Myeloarchitecture, connections, functional properties, and topographic organization. *Journal of Comparative Neurology, 199*, 293-326.

Van Essen, D. C., & Zeki, S. M. (1978). The topographic organization of rhesus monkey prestriate cortex. *Journal of Physiology, 277*, 193-226.

Videen, T. O., Daw, N. W., & Rader, R. K. (1984). The effect of norepinephrine on visual cortical neurons in kittens and adult cats. *Journal of Neuroscience, 4*, 1607-1617.

Vincent, S. R., Hökfelt, T., Skirboll, L. R., & Wu, T. Y. (1983). Hypothalamic γ-aminobutyric acid neurons project to the neocortex. *Science, 220*, 1309-1310.

Vital-Durand, F., Garey, L. J., & Blakemore, C. (1978). Monocular and binocular deprivation in the monkey: Morphological effects and reversibility. *Brain Research, 158*, 45-64.

von der Heydt, R., Peterhans, E., & Baumgartner, G. (1984). Illusory contours and cortical neuron responses. *Science, 224*, 1260-1262.

von der Malsburg, C. (1973). Self-organization of orientation selective cells in the striate cortex. *Kybernetik, 14*, 85-100.

von der Malsburg, C. (in press). Nervous structures with dynamical links. *Berichte der Bunsen-Gesellschaft fur Physikalische Chemie.*

Vygotsky, L. S. (1962). *Thought and language* (E. Hanfmann & G. Vakar, Eds. and Trans.). Cambridge, MA: MIT Press. (Original work published 1934)

Walker, A. E. (1938). *The primate thalamus.* Chicago: The University of Chicago Press.

Waltz, D. L., & Pollack, J. B. (1985). Massively parallel parsing. *Cognitive Science, 9*, 51-74.

Weller, R. E., & Kaas, J. H. (1983). Retinotopic patterns of connections of area 17 with visual areas V-II and MT in macaque monkeys. *Journal of Comparative Neurology, 220*, 253-279.

Werblin, F. S., & Dowling, J. E. (1969). Organization of the retina of the mudpuppy, *Necturus maculosus*: II. Intracellular recording. *Journal of Neurophysiology, 32*, 339-355.

Westrum, L. E. (1966). Synaptic contacts on axons in the cerebral cortex. *Nature, 210*, 1289-1290.

Whittlesea, B. W. A. (1983). *Representation and generalization of concepts: The abstractive and episodic perspectives evaluated.* Unpublished doctoral dissertation, MacMaster University.

Wickelgren, W. A. (1969). Context-sensitive coding, associative memory, and serial order in (speech) behavior. *Psychological Review, 76*, 1-15.

Wickelgren, W. A. (1979). Chunking and consolidation: A theoretical synthesis of semantic networks, configuring in conditioning, S-R versus cognitive learning, normal forgetting, the amnesic syndrome, and the hippocampal arousal system. *Psychological Review, 86*, 44-60.

Wiesel, T. N., & Gilbert, C. D. (1983). Morphological basis of visual cortical function. *Quarterly Journal of Experimental Physiology, 68*, 523-543.

Wiesel, T. N., & Hubel, D. H. (1963). Single-cell responses in striate cortex of kittens deprived of vision in one eye. *Journal of Neurophysiology, 26*, 1003-1017.

Wiesel, T. N., & Hubel, D. H. (1965). Extent of recovery from the effects of visual deprivation in kittens. *Journal of Neurophysiology, 28*, 1060-1072.

Wilson, H. R., & Cowan, J. D. (1972). Excitatory and inhibitory interactions in localized populations of model neurons. *Biophysics Journal, 12*, 1-24.

Winston, P. H. (1975). Learning structural descriptions from examples. In P. H. Winston (Ed.), *The psychology of computer vision* (pp. 157-209). New York: McGraw-Hill.

Wise, R. A. (1978). Catecholamine theories of reward: A critical review. *Brain Research, 152*, 215-247.

Wood, C. (1978). Variations on a theme by Lashley: Lesion experiments on the neural models of Anderson, Silverstein, Ritz, and Jones. *Psychological Review, 85*, 582-591.

Woods, W. (1970). Transition network grammars for natural language analysis. *Communications of the ACM, 13*, 591-606.

Woolsey, C. N., Marshall, W. H., & Bard, P. (1942). Representation of cutaneous tactile sensibility in cerebral cortex of monkey as indicated by evoked potentials. *Bulletin of Johns Hopkins Hospital, 70,* 399-441.

Woolsey, C. N., Settlage, P. H., Meyer, D. R., Sencer, W., Pinto-Hamuy, T., & Travis, A. M. (1952). Patterns of localization in precentral and 'supplementary' motor areas and their relation to concept of premotor area. *Research Publications Association for Research in Nervous and Mental Disease, 30,* 238- 264.

Woolsey, T. A. (1978). Some anatomical bases of cortical somatotopic organization. *Brain, Behavior, and Evolution, 15,* 325-371.

Zeki, S. M. (1978). The cortical projections of foveal striate cortex in the rhesus monkey. *Journal of Physiology, 277,* 227-244.

Zeki, S. M. (1983a). Colour coding in the cerebral cortex: The reaction of cells in monkey visual cortex to wavelengths and colours. *Neuroscience, 9,* 741-765.

Zeki, S. M. (1983b). Colour coding in the cerebral cortex: The responses of wavelength-selective and colour-coded cells in monkey visual cortex to changes in wavelength composition. *Neuroscience, 9,* 767-781.

Zipser, D. (1985). A computational model of hippocampal place fields. *Behavioral Neuroscience, 99,* 1006-1018.

Index

Page numbers in roman type refer to Volume 1; page numbers in italic type refer to Volume 2.